df	TWO-TAILED AREA							df
	.20	.10	.05	.02	.01	.001	.0001	
26	1.315	1.706	2.056	2.479	2.779	3.707	4.587	26
27	1.314	1.703	2.052	2.473	2.771	3.690	4.558	27
28	1.313	1.701	2.048	2.467	2.763	3.674	4.530	28
29	1.311	1.699	2.045	2.462	2.756	3.659	4.506	29
30	1.310	1.697	2.042	2.457	2.750	3.646	4.482	30
40	1.303	1.684	2.021	2.423	2.704	3.551	4.321	40
60	1.296	1.671	2.000	2.390	2.660	3.460	4.169	60
100	1.290	1.660	1.984	2.364	2.626	3.390	4.053	100
140	1.288	1.656	1.977	2.353	2.611	3.361	4.006	140
∞	1.282	1.645	1.960	2.326	2.576	3.291	3.891	∞

STATISTICS FOR THE LIFE SCIENCES

STATISTICS FOR THE LIFE SCIENCES

MYRA L. SAMUELS
Purdue University

DELLEN PUBLISHING COMPANY
San Francisco

COLLIER MACMILLAN PUBLISHERS
London

divisions of Macmillan, Inc.

On the cover: Don Kingfisher is a monoprint executed in 1988 by Texas native John Alexander. It was published by Landfall Press, Inc. Alexander's dense and explosive visionary paintings explore the landscapes of southern primitive America. His expressionist bravado investigates the wild imagery of nature with vivid energy. He is represented by the Marlborough Gallery, New York. His work is in the permanent collections of the Metropolitan Museum, New York; the Museum of Fine Art, Houston; and the Museum of Contemporary Art, Chicago.

Permissions: Dellen Publishing Company
400 Pacific Avenue
San Francisco, California 94133

Orders: Dellen Publishing Company
c/o Macmillan Publishing Company
Front and Brown Streets
Riverside, New Jersey 08075

Collier Macmillan Canada, Inc.

LIBRARY OF CONGRESS CATALOGING IN PUBLICATION DATA

Samuels, Myra L.
    Statistics for the life sciences/Myra L. Samuels.
    Bibliography
    Includes index.
    ISBN 0-02-405501-8
    1. Biometry.   2. Medical statistics.   3. Agriculture—Statistics.
    4. Life sciences—Statistics.   I. Title.
    QH323.5.S23   1989
    574'.072—dc 19   88-18710

Printing: 2 3 4 5 6 7 8 9 Year: 9 0 1 2

ISBN 0-02-405501-8

C O N T E N T S

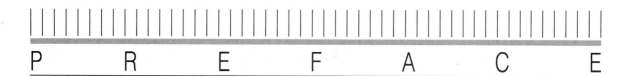

Statistics for the Life Sciences is an introductory text in statistics, specifically addressed to students specializing in the life sciences. Its primary aims are (1) to show students how statistical reasoning is relevant to biological, medical, and agricultural research; (2) to enable students confidently to carry out simple statistical analyses and to interpret the results; and (3) to raise students' consciousness concerning basic statistical issues such as randomization, confounding, and the role of independent replication.

STYLE AND APPROACH

The style of *Statistics for the Life Sciences* is informal and uses only minimal mathematical notation. There are no prerequisites except elementary algebra; anyone who can read a biology or chemistry textbook can read this text. It is suitable for use by graduate or undergraduate students in biology, agronomy, medical and health sciences, nutrition, pharmacy, animal science, physical education, forestry, and other life sciences. For several years, previous versions of this text have been used for a one-semester course for such students at Purdue University.

USE OF REAL DATA   Real examples are more interesting and often more enlightening than artificial ones. *Statistics for the Life Sciences* includes hundreds of examples and exercises that use real data, representing a wide variety of research in the life sciences. Each example has been chosen to illustrate a particular statistical issue. The exercises have been designed to reduce computational effort and focus students' attention on concepts and interpretations.

EMPHASIS ON IDEAS   This text emphasizes statistical ideas rather than computations or mathematical formulations. Probability theory is included only where it is statistically relevant; conditional probability, independence, and conditional distributions are introduced in context, rather than in a separate chapter. Throughout the discussion of descriptive and inferential statistics, interpretation is stressed. By means of salient examples, the student is shown why it is important that an analysis be appropriate for the research question to be answered, for the statistical design of the study, and for the nature of the underlying distributions. The student is warned against the common blunder of confusing statistical nonsignificance with practical insignificance, and is encouraged to use confidence intervals to assess the magnitude of an effect. The student is led to recognize the impact on real research of design concepts such as random sampling, randomization, efficiency, and the control of extraneous variation by blocking or adjustment. Numerous exercises amplify and reinforce the student's grasp of these ideas.

THE ROLE OF THE COMPUTER   The analysis of research data is usually carried out with the aid of a computer. In studying statistics, however, it is desirable for the student to gain experience working directly with data, using paper and pencil and a hand-held calculator. This experience will help the student appreciate the nature and purpose of the statistical computations. The student is thus prepared to make intelligent use of the computer—to give it appropriate instructions and properly interpret the output. Accordingly, most of the exercises in this text are intended for hand calculation. Selected exercises are labeled *Computer exercises*; these can be used in classes that have access to statistical computing facilities. (Typically, the computer exercises require calculations that would be unduly burdensome if carried out by hand.)

ORGANIZATION

This text is organized to permit coverage in one semester of the maximum number of important statistical ideas, including power, multiple inference, and the basic principles of design. By including or excluding optional sections, the instructor can also use the text for a one-quarter course or a two-quarter course. It is suitable for a terminal course or for the first course of a sequence.

The following is a brief outline of the text.

Chapter 1:   Introduction. The nature and impact of variability in biological data.

Chapter 2:   Orientation. Frequency distributions, descriptive statistics, the concept of population vs sample.

Chapters 3, 4, 5:   Theoretical preparation. Probability, binomial and normal distributions, sampling distributions.

Chapters 6 *and* 7:   Inference for mean of one sample and comparison of two independent samples. Confidence intervals, t test, Mann–Whitney test.

Chapter 8:   Design. Randomization, blocking, hazards of observational studies, defining the experimental unit.

Chapter 9:   Inference for paired samples. Confidence interval, t test, sign test.

Chapters 10 *and* 11:   Categorical data. Confidence interval for a proportion, chi-square goodness-of-fit test, conditional probability, contingency tables.

Chapter 12:   Analysis of variance. One-way layout. Optional sections treat (a) the Newman–Keuls procedure: (b) contrasts, including interaction in two-factor designs.

Chapter 13:   Regression and correlation. Descriptive and inferential aspects of simple linear regression and correlation and the relationship between them.

Chapter 14:   Multiplicity. Pitfalls of multiple inference and of data-dredging, and some remedies.

Statistical tables are provided at the back of the book. The tables of critical values are especially easy to use, because they follow mutually consistent layouts and so are used in essentially the same way.

Optional appendices at the back of the book give the interested student a deeper look into such matters as the derivation of computational formulas, how pseudo-random numbers are generated, and how the Mann–Whitney null distribution is calculated.

ACKNOWLEDGMENTS

Many colleagues and students have contributed to this book by sharing with me the fruits of their labor—their data—and by thoughtful and stimulating discussion of their work. I would like especially to thank Donald A. Holt, William F. Jacobson, Roger P. Maickel, Edward W. Mills, Cary A. Mitchell, Jay F. Nash, Jr., Larry A. Nelson, Ralph L. Nicholson, Carl H. Noller, Wyman E. Nyquist, Nicholas G. Popovich, Edward S. Simon, Terry S. Stewart, and Jane L. Wolfson.

I am grateful to Erich Klinghammer for positive reinforcement throughout this project. For critiquing portions of the manuscript, I would like to thank David S. Moore, Virgil L. Anderson, George P. McCabe, and Jane L. Wolfson.

For reviewing the entire manuscript and providing substantial constructive feedback, I am indebted to Kenneth J. Koehler of Iowa State University, Thomas A. Louis of the University of Minnesota, and Dana Quade of the University of North Carolina. I also appreciate reviews of the early chapters by Theodore B. Bailey, Jr. of Iowa State University, George M. Barnwell of the University of Texas Health Science Center at San Antonio, Beth Dawson-Saunders of Southern Illinois University, Erich L. Lehmann of the University of California at Berkeley, Terry S. Stewart of Purdue University, and William M. Vollmer of the Oregon Health Sciences University.

I gratefully acknowledge the Longman Group and Butterworth Scientific Publications for permission to use Table 5, and the Institute of Mathematical Statistics for permission to use Table 10. The remaining tables were computed using standard computational methods.

This book has benefited greatly from the careful editing and production work of Susan L. Reiland and the generous encouragement of Donald E. Dellen. Jordan A. Samuels, Jaime San Martin, Fu-Chuen Chang, and Gregory P. Donoho helped to prepare solutions to the exercises. Tai-Fang C. Lu and Regina P. Becker provided programming assistance. Finally, I am grateful to my husband Steve and my children Jordan and Ellen for their extraordinary and continuing support.

STATISTICS FOR THE LIFE SCIENCES

C H A P T E R 1

CONTENTS

INTRODUCTION

S E C T I O N 1.1

**STATISTICS AND THE
LIFE SCIENCES**

Researchers in the life sciences carry out investigations in various settings: in the clinic, in the laboratory, in the greenhouse, in the field. Frequently, the resulting data exhibit some *variability*. For instance, patients given the same drug respond somewhat differently; cell cultures prepared identically develop somewhat differently; adjacent plots of genetically identical wheat plants yield somewhat different amounts of grain. Often the degree of variability is substantial even when experimental conditions are held as constant as possible.

The challenge to the life scientist is to discern the patterns that may be more or less obscured by the variability of responses in living systems. The scientist must try to distinguish the "signal" from the "noise."

Statistics is a discipline that has evolved in response to the needs of scientists and others whose data exhibit variability. The concepts and methods of statistics enable the investigator to describe variability, to plan research so as to take variability into account, and to analyze data so as to extract the maximum information and also to quantify the reliability of that information.

S E C T I O N 1.2

**EXAMPLES AND
OVERVIEW**

In this section we give some examples to illustrate the degree of variability found in biological data, and the ways in which variability poses a challenge to the biological researcher. We will briefly mention some of the statistical issues raised by each example, and indicate where in this book the issues are addressed.

The first two examples provide a contrast between an experiment that showed no variability, and another that showed considerable variability.

EXAMPLE 1.1
VACCINE FOR ANTHRAX

Anthrax is a serious disease of sheep and cattle. In 1881 Louis Pasteur conducted a famous experiment to demonstrate the effect of his vaccine against anthrax. A group of 24 sheep were vaccinated; another group of 24 unvaccinated sheep served as controls. Then, all 48 animals were inoculated with a virulent culture of anthrax bacillus. Table 1.1 shows the results.[1] The data of Table 1.1 show no variability; all the vaccinated animals survived and all the unvaccinated animals died.

TABLE 1.1
Response of Sheep to
Anthrax

	TREATMENT	
RESPONSE	Vaccinated	Not Vaccinated
Died of Anthrax	0	24
Survived	24	0
Total	24	24
Percent survival	100%	0%

■

EXAMPLE 1.2
BACTERIA AND CANCER

To study the effect of bacteria on tumor development, researchers used a strain of mice with a naturally high incidence of liver tumors. One group of mice were maintained entirely germ-free, while another group were exposed to the intestinal bacteria *Escherichia coli*. The incidence of liver tumors is shown in Table 1.2.[2]

TABLE 1.2

Incidence of Liver
Tumors in Mice

RESPONSE	TREATMENT	
	E. coli	Germ-free
Liver Tumors	8	19
No Liver Tumors	5	30
Total	13	49
Percent with liver tumors	62%	39%

In contrast to Table 1.1, the data of Table 1.2 show variability; mice given the same treatment did not all respond the same way. Because of this variability, the results in Table 1.2 are equivocal; the data suggest that exposure to *E. coli* increases the risk of liver tumors, but the possibility remains that the observed difference in percentages (62% versus 39%) might reflect only chance variation rather than an effect of *E. coli*. If the experiment were replicated with different animals, the percentages might be substantially changed; note especially that the 62% is based on only 13 animals. ∎

In Chapter 11 we will discuss statistical techniques for evaluating data such as those in Tables 1.1 and 1.2. Of course, in some experiments variability is minimal and the message in the data stands out clearly without any special statistical analysis. It is worth noting, however, that absence of variability is itself an experimental result, which must be justified by sufficient data. For instance, because Pasteur's anthrax data (Table 1.1) show no variability at all, it is intuitively plausible to conclude that the data provide "solid" evidence for the efficacy of the vaccination. But note that this conclusion involves a judgment; consider how much *less* "solid" the evidence would be if Pasteur had included only 3 animals in each group, rather than 24. In fact, a judgment that variability is negligible can be justified by an appropriate statistical analysis. Thus, a statistical view can be helpful even in the absence of variability.

The next two examples illustrate some of the questions that a statistical approach can help to answer.

EXAMPLE 1.3

FLOODING AND ATP

In an experiment on root metabolism, a plant physiologist grew birch tree seedlings in the greenhouse. He flooded four seedlings with water for one day and kept four others as controls. He then harvested the seedlings and analyzed the roots for adenosine triphosphate (ATP). The measured amounts of ATP (nmoles per mg tissue) are given in Table 1.3 and displayed in Figure 1.1 on page 4.[3]

TABLE 1.3

ATP Concentration in
Birch Roots (nmol/mg)

FLOODED	CONTROL
1.45	1.70
1.19	2.04
1.05	1.49
1.07	1.91

FIGURE 1.1
ATP Concentration in
Birch Tree Roots

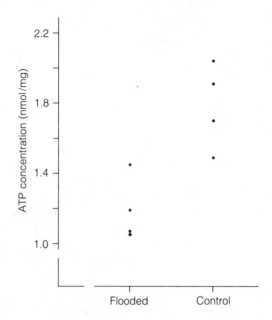

The data of Table 1.3 raise several questions: How should one summarize the ATP values in each experimental condition? How much information do the data provide about the effect of flooding? How confident can one be that the reduced ATP in the flooded group is really a response to flooding rather than just random variation? What size experiment would be required in order to firmly corroborate the apparent effect seen in these data?   ■

Chapters 2, 6, 7, and 8 address questions like those posed in Example 1.3.

EXAMPLE 1.4
MAO AND
SCHIZOPHRENIA

Monoamine oxidase (MAO) is an enzyme which is thought to play a role in the regulation of behavior. To see whether different categories of schizophrenic patients have different levels of MAO activity, researchers collected blood specimens from 42 patients and measured the MAO activity in the platelets. The results are given in Table 1.4 and displayed in Figure 1.2. (Values are expressed as nmol benzylaldehyde product per 10^8 platelets per hour.)[4]

To analyze the MAO data, one would naturally want to make comparisons among the three groups of patients, to describe the reliability of those comparisons, and to characterize the variability within the groups. To go beyond the data to a biological interpretation, one must also consider more subtle issues, such as the following: How were the patients selected? Were they chosen from a common hospital population, or were the three groups obtained at different times or places? Were precautions taken so that the person measuring the MAO was unaware of the patient's diagnosis? Did the investigators consider various ways of subdividing the patients before choosing the particular diagnostic categories used in Table 1.4? At first glance, these questions may seem irrelevant—can we not let the measurements speak for themselves? We will see, however,

TABLE 1.4
MAO Activity in
Schizophrenic Patients

DIAGNOSIS	MAO ACTIVITY				
I:	6.8	4.1	7.3	14.2	18.8
Chronic undifferentiated	9.9	7.4	11.9	5.2	7.8
schizophrenic	7.8	8.7	12.7	14.5	10.7
(18 patients)	8.4	9.7	10.6		
II:	7.8	4.4	11.4	3.1	4.3
Undifferentiated with	10.1	1.5	7.4	5.2	10.0
paranoid features	3.7	5.5	8.5	7.7	6.8
(16 patients)	3.1				
III:	6.4	10.8	1.1	2.9	4.5
Paranoid schizophrenic	5.8	9.4	6.8		
(8 patients)					

FIGURE 1.2
MAO Activity in
Schizophrenic Patients

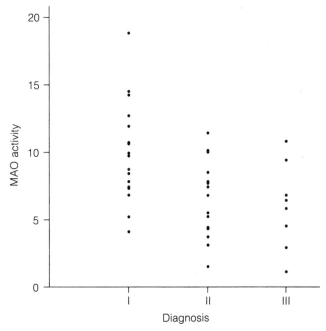

that the proper interpretation of data always requires careful consideration of how the data were obtained. ∎

Chapters 2, 3, 8, and 9 include discussions of selection of experimental subjects and of guarding against unconscious investigator bias. In Chapter 14 we will show how sifting through a data set in search of patterns can lead to serious misinterpretations, and we will give guidelines for avoiding the pitfalls in such searches.

The next example shows how the effects of variability can distort the results of an experiment, and how this distortion can be minimized by careful design of the experiment.

EXAMPLE 1.5

FOOD CHOICE BY
INSECT LARVAE

The clover root curculio, *Sitona hispidulus*, is a root-feeding pest of alfalfa. An entomologist conducted an experiment to study food choice by *Sitona* larvae. She wished to investigate whether larvae would preferentially choose alfalfa roots which were nodulated (their natural state) over roots whose nodulation had been suppressed. Larvae were released in a dish where both nodulated and non-nodulated roots were available. After 24 hours the investigator counted the larvae that had clearly made a choice between root types. The results are shown in Table 1.5.[5]

TABLE 1.5

Food Choice by *Sitona*
Larvae

CHOICE	NUMBER OF LARVAE
Chose nodulated roots	46
Chose nonnodulated roots	12
Other (no choice, died, or lost)	62
Total	120

The data in Table 1.5 appear to suggest rather strongly that *Sitona* larvae prefer nodulated roots. But our description of the experiment has obscured an important point—we have not stated how the roots were arranged. To see the relevance of the arrangement, suppose the experimenter had used only one dish, placing all the nodulated roots on one side of the dish and all the nonnodulated roots on the other side, as shown in Figure 1.3(a), and had then released 120 larvae in the center of the dish. This experimental arrangement would be seriously deficient, because the data of Table 1.5 would then permit several competing interpretations—for instance, (a) perhaps the larvae really do prefer nodulated roots; or (b) perhaps the two sides of the dish were at slightly different temperatures, and the larvae were responding to temperature rather than nodulation; or (c) perhaps one larva chose the·nodulated roots just by chance, and the other larvae followed its trail. Because of these possibilities, the experimental arrangement shown in Figure 1.3(a) can yield only weak information about larval food preference.

FIGURE 1.3

Possible Arrangements of
Food Choice Experiment.
The dark-shaded areas
contain nodulated roots
and the light-shaded
areas contain non-
nodulated roots.
(a) A poor arrangement;
(b) A good arrangement.

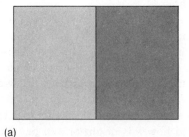

(a)

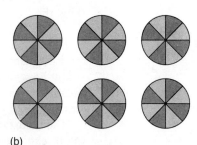

(b)

The experiment was actually arranged as in Figure 1.3(b), using six dishes with nodulated and nonnodulated roots arranged in a symmetric pattern. Twenty larvae were released into the center of each dish. This arrangement avoids the pitfalls of the arrangement in Figure 1.3(a). Because of the alternating regions of nodulated and nonnodulated roots, any fluctuation in environmental conditions (such as temperature) would tend to affect the two root types equally. By using several dishes, the experimenter has generated data that can be interpreted even if the larvae do tend to follow each other. To properly analyze the experiment, we would need to know the results in each dish; the condensed summary in Table 1.5 is not adequate. ∎

In Chapter 8 we will describe various ways of arranging experimental material in space and time so as to yield the most informative experiment. In later chapters we will discuss how to analyze the data to extract as much information as possible, and yet to resist the temptation to overinterpret patterns that may represent only random variation.

The next example describes an investigation of *two* experimental factors.

EXAMPLE 1.6
AIR POLLUTION AND PLANT GROWTH

In a study of the mutual effect of the air pollutants ozone and sulfur dioxide, Blue Lake snap beans were grown in open-top field chambers. Some chambers were fumigated repeatedly with sulfur dioxide. The air in some chambers was carbon-filtered to remove ambient ozone. There were three chambers per treatment combination, allocated at random. After one month of treatment, the total yield of bean pods was measured for each chamber. The results are given in Table 1.6 and displayed in Figure 1.4[6] on page 8. The researchers were particularly interested in determining whether the effect of sulfur dioxide is greater in the presence of ozone than in its absence.

TABLE 1.6
Total Pod Yield of Snap Beans (kg)

OZONE ABSENT		OZONE PRESENT	
SULFUR DIOXIDE		SULFUR DIOXIDE	
Absent	Present	Absent	Present
1.52	1.49	1.15	.65
1.85	1.55	1.30	.76
1.39	1.21	1.57	.69

The experiment in Example 1.6 is designed to investigate the interaction between two experimental factors—ozone and sulfur dioxide. Techniques for studying such interactions will be considered in Chapter 12.

The following example is a study of the relationship between two measured quantities.

FIGURE 1.4
Yield of Snap Beans
Under Four Different
Treatments

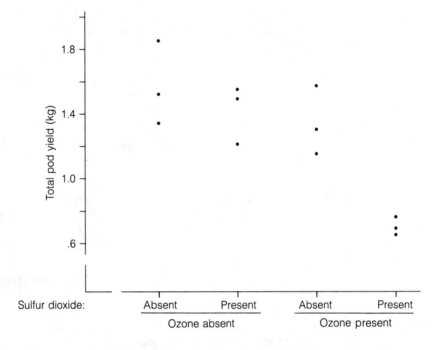

FIGURE 1.4
Yield of Snap Beans
Under Four Different
Treatments

EXAMPLE 1.7

BODY SIZE AND ENERGY EXPENDITURE

How much food does a person need? To investigate the dependence of nutritional requirements on body size, researchers used underwater weighing techniques to determine the fat-free body mass for each of seven men. They also measured the total 24-hour energy expenditure during conditions of quiet sedentary activity; this was repeated twice for each subject. The results are shown in Table 1.7 and plotted in Figure 1.5.[7]

TABLE 1.7

Fat-Free Mass and Energy Expenditure

SUBJECT	FAT-FREE MASS (kg)	24-HOUR ENERGY EXPENDITURE (Kcal)	
1	49.3	1,851	1,936
2	59.3	2,209	1,891
3	68.3	2,283	2,423
4	48.1	1,885	1,791
5	57.6	1,929	1,967
6	78.1	2,490	2,567
7	76.1	2,484	2,653

A primary goal in the analysis of these data would be to describe the relationship between fat-free mass and energy expenditure—to characterize not only the overall trend of the relationship, but also the degree of scatter or variability in the relationship. (Note also that, to analyze the data, one needs to decide how to handle the duplicate observations on each subject.)

FIGURE 1.5 Fat-Free Mass and Energy Expenditure in Seven Men. Each man is represented by a different symbol.

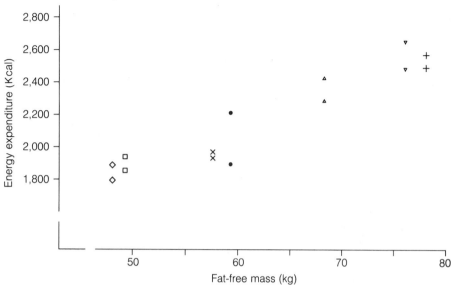

The focus of Example 1.7 is on the relationship between two variables: fat-free mass and energy expenditure. Chapter 13 deals with methods for describing such relationships, and also for quantifying the reliability of the descriptions.

A LOOK AHEAD

The preceding examples have illustrated the kind of data to be considered in this book. In fact, each of the examples will reappear as an exercise or example in an appropriate chapter.

As the examples show, research in the life sciences is usually concerned with the comparison of two or more groups of observations, or with the relationship between two or more variables. We will begin our study of statistics by focusing on a simpler situation—observations of a *single* variable for a *single* group. Many of the basic ideas of statistics will be introduced in this oversimplified context. Two-group comparisons and more complicated analyses will then be discussed in Chapter 7 and later chapters.

CHAPTER 2

CONTENTS

DESCRIPTION OF SAMPLES AND POPULATIONS

INTRODUCTION

A first step toward understanding a set of data is to describe the data in summary form. In this chapter we discuss three mutually complementary aspects of summary data description: frequency distributions, measures of center, and measures of dispersion.

SAMPLES

We first consider the description of observations taken under homogeneous conditions. Such a data set is called a **sample**. Later in this chapter we will adopt the view that the sample has been drawn from a population, and we will discuss the matter of describing a population.

The number of observations in a sample is called the **sample size** and is denoted by the letter n. The following are some examples of samples:

The birthweights of 150 babies born in a certain hospital

The sexes of 73 *Cecropia* moths caught in a trap

The flower colors of 81 plants that are progeny of a single parental cross

The number of bacterial colonies in each of six petri dishes

In conceptualizing a sample, it is helpful to be aware of the following elements:

(a) The *observed variable*. For example:
birthweight
sex
flower color
number of colonies

(b) The *observational unit*. For example:
baby
moth
plant
petri dish

(c) The *sample size*. For example:
$n = 150$
$n = 73$
$n = 81$
$n = 6$

Remark   There is some potential for confusion between the statistical meaning of the term *sample* and the sense in which this word is sometimes used in biology. If a biologist draws blood from 20 people and measures the glucose concentration in each, he might say he has 20 samples of blood. However, the statistician says he has *one* sample of 20 glucose measurements; the sample size is $n = 20$. In the interest of clarity, throughout this book we will use the term *specimen* where a biologist might prefer *sample*. So we would speak of glucose measurements on **20 specimens of blood**.

TYPES OF VARIABLES

The observation that is made on each observational unit can be a qualitative observation of a **categorical** variable or it can be a quantitative observation. Examples of categorical variables are:

Blood type of a person: A, B, AB, O

Sex of a fish: male, female

Color of a flower: red, pink, white

Shape of a seed: wrinkled, smooth

For some categorical variables, the categories can be arrayed in a meaningful rank order. Such a variable is said to be **ordinal**. Examples of ordinal categorical variables are:

Response of a patient to therapy: none, partial, complete

Tenderness of beef: tough, slightly tough, tender, very tender

Size of a wheat plant: short, intermediate, tall

Quantitative observations are expressed as numerical values. Often they are measurements on a **continuous** scale. Examples of continuous variables are:

Weight of a rat

Cholesterol concentration in a blood specimen

Optical density of a solution

A variable such as weight is continuous because, in principle, two weights can be arbitrarily close together. Some types of quantitative variables are not continuous but fall on a **discrete** scale, with spaces between the possible values. For example, the number of eggs in a bird's nest is a discrete variable because only the values 0, 1, 2, 3, ..., are possible. Other examples of discrete variables are:

Number of leaves on a plant

Number of cancerous lymph nodes detected in a patient

The distinction between continuous and discrete variables is not a rigid one. After all, physical measurements are always rounded off. We may measure the weight of a steer to the nearest kilogram, of a rat to the nearest gram, or of an insect to the nearest milligram. The scale of the actual measurements is always discrete, strictly speaking. The continuous scale can be thought of as an approximation to the actual scale of measurement.

In summary, observed variables can be of the following types:

1. Categorical variables
   a. Ordinal
   b. Not ordinal
2. Quantitative variables
   a. Discrete
   b. Continuous

We will sometimes find it useful to discuss these types separately when considering methods of data analysis.

NOTATION FOR VARIABLES AND OBSERVATIONS

In the early part of this book, we will adopt a notational convention to distinguish between a variable and an observed value of that variable. We will denote variables by uppercase letters such as Y. We will denote the observations themselves (that is, the data) by lowercase letters such as y. Thus, we distinguish, for example, between Y = birthweight (the variable) and y = 7.9 lb (the observation). This distinction will be helpful in explaining some fundamental ideas concerning variability. In later chapters the notation will become too cumbersome and we will abandon it.

S E C T I O N 2.2

FREQUENCY DISTRIBUTIONS: TECHNIQUES FOR DATA

A frequency distribution is a form of data description which provides an overview of a set of observations.

UNGROUPED FREQUENCY DISTRIBUTIONS

An ungrouped frequency distribution is simply a display of the **frequency**, or number of occurrences, of each value in the data set. The information can be presented in tabular form or more vividly as a bar graph. Here are two examples of ungrouped frequency distributions for categorical data.

EXAMPLE 2.1
COLOR OF POINSETTIAS

Poinsettias can be red, pink, or white. In one investigation of the hereditary mechanism controlling the color, 182 progeny of a certain parental cross were categorized by color.[1] The bar graph in Figure 2.1 is a visual display of the results given in Table 2.1.

FIGURE 2.1
Bar Graph of Color of 182 Poinsettias

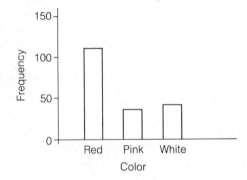

TABLE 2.1
Color of 182 Poinsettias

COLOR	FREQUENCY (NUMBER OF PLANTS)
Red	108
Pink	34
White	40
Total	182

EXAMPLE 2.2
CLUMPING OF BLOOD

The strength of reaction of a blood specimen to a certain antigen is categorized into one of six classes according to the degree of clumping of the red blood cells: Class I, complete clumping; Class II, marked clumping; ...; Class VI, no clumping. The results for specimens from 70 type-B people are given in Table 2.2 and displayed as a bar graph in Figure 2.2.[2]

TABLE 2.2
Strength of Clumping
Reaction of 70 Blood
Specimens

STRENGTH OF REACTION	FREQUENCY (NUMBER OF SPECIMENS)
I	6
II	35
III	15
IV	3
V	3
VI	8
Total	70

FIGURE 2.2
Bar Graph of Strength of
Clumping Reaction of 70
Blood Specimens

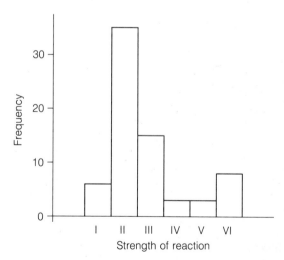

A bar graph for the distribution of a quantitative variable is called a **histogram**. The following example shows a histogram for an ungrouped frequency distribution.

EXAMPLE 2.3

LITTER SIZE OF SOWS

A group of 36 2-year-old sows of the same breed ($\frac{3}{4}$ Duroc, $\frac{1}{4}$ Yorkshire) were bred to Yorkshire boars. The number of piglets surviving to 21 days of age was recorded for each sow.[3] The results are given in Table 2.3 and displayed as a histogram in Figure 2.3.

TABLE 2.3

Number of Surviving
Piglets of 36 Sows

NUMBER OF PIGLETS	FREQUENCY (NUMBER OF SOWS)
5	1
6	0
7	2
8	3
9	3
10	9
11	8
·12	5
13	3
14	2
Total	36

FIGURE 2.3

Histogram of Number of
Surviving Piglets of 36
Sows

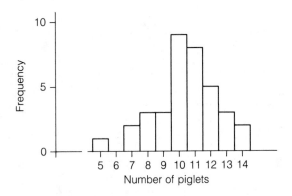

Remark Note the spaces between the bars in Figure 2.1. Whether to use such spaces for categorical and/or discrete data is a matter of taste; note that the spaces have been omitted in Figures 2.2 and 2.3. For continuous data (discussed below) the spaces are always omitted.

RELATIVE FREQUENCY

The frequency scale is often replaced by a **relative frequency** scale:

$$\text{Relative frequency} = \frac{\text{Frequency}}{n}$$

The relative frequency scale is useful if several data sets of different sizes (n's) are to be displayed together for comparison. As another option, a relative frequency can be expressed as a percentage frequency. The shape of the display is not affected by the choice of frequency scale, as the following example shows.

EXAMPLE 2.4
COLOR OF POINSETTIAS

The poinsettia color distribution of Example 2.1 is expressed as frequency, relative frequency, and percent frequency in Table 2.4 and Figure 2.4.

TABLE 2.4
Color of 182 Poinsettias

COLOR	FREQUENCY	RELATIVE FREQUENCY	PERCENT FREQUENCY
Red	108	.59	59
Pink	34	.19	19
White	40	.22	22
Total	182	1.00	100

FIGURE 2.4
Histogram of Poinsettia
Colors on Three Scales:
(a) Frequency;
(b) Relative frequency;
(c) Percent frequency.

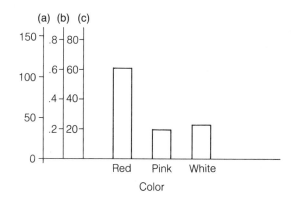

GROUPED FREQUENCY DISTRIBUTIONS

In the preceding examples, simple ungrouped frequency distributions provided concise summaries of the data. For many data sets, it is necessary to **group** the data in order to condense the information adequately. (This is usually the case with continuous variables.) The following example shows a grouped frequency distribution.

EXAMPLE 2.5
SERUM CK

Creatine phosphokinase (CK) is an enzyme related to muscle and brain function. As part of a study to determine the natural variation in CK concentration, blood was drawn from 36 male volunteers. Their serum concentrations of CK (measured in U/l) are given in Table 2.5.[4] Table 2.6 shows these data grouped into **classes**. For instance, the frequency of the class 20–39 is 1, which means that one CK value fell in this range. The grouped frequency distribution is displayed as a histogram in Figure 2.5.

TABLE 2.5

Serum CK Values for 36 Men

121	82	100	151	68	58
95	145	64	201	101	163
84	57	139	60	78	94
119	104	110	113	118	203
62	83	67	93	92	110
25	123	70	48	95	42

TABLE 2.6

Frequency Distribution of Serum CK Values for 36 Men

SERUM CK (U/l)	FREQUENCY (NUMBER OF MEN)
20–39	1
40–59	4
60–79	7
80–99	8
100–119	8
120–139	3
140–159	2
160–179	1
180–199	0
200–219	2
Total	36

FIGURE 2.5

Histogram of Serum CK Concentrations of 36 Men

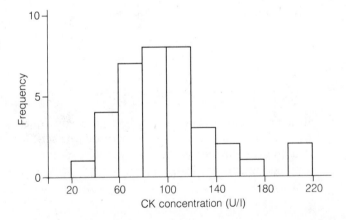

A grouped frequency distribution should display the essential features of the data. For instance, the histogram of Figure 2.5 shows that the average CK value is about 100 U/l, with the majority of the values falling between 60 and 140 U/l. In addition, the histogram shows the *shape* of the distribution. Note that the CK values are piled up around a central peak, or **mode**. On either side of this mode, the frequencies decline and ultimately form the **tails** of the distribution. These shape features are labelled in Figure 2.6. The CK distribution is not symmetric but is **skewed** to the right, which means that the right tail is more stretched out than the left.

FIGURE 2.6

Shape Features of the CK Distribution

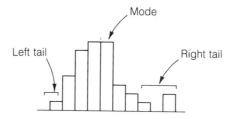

Technical note   Notice that the histogram in Figure 2.5 has been drawn with the bars touching. The boundaries of these bars are 19.5, 39.5, 59.5, ... These values, known as **class boundaries**, are not the same as the **class limits** shown in the tabulated distribution (Table 2.6). Table 2.7 shows the class limits and the class boundaries for the CK distribution; note that the class boundaries are midway between adjacent class limits. The distance between class boundaries is called the **class width**. For the CK distribution, the class width is 20 (*not* 19).

TABLE 2.7

Class Limits and Class Boundaries for the CK Distribution

CLASS LIMITS	CLASS BOUNDARIES
20–39	19.5–39.5
40–59	39.5–59.5
60–79	59.5–79.5
⋮	⋮
200–219	199.5–219.5

Ticks and labels for the horizontal axis of a histogram should be chosen to create a clear and uncluttered display. Usually this means labelling class limits rather than class boundaries. For instance, selected class limits have been labelled in Figure 2.5; technically, the ticks are offset slightly to the right of the boundaries between the bars. For some data sets, the ticks will be quite noticeably offset; as an extreme example, notice in Figure 2.3 that the ticks are centered in the bars.

HOW TO CONSTRUCT A GROUPED FREQUENCY DISTRIBUTION

There is obviously more than one way to group a given set of data. The goal is to highlight the essential features of the data while suppressing unimportant detail. The first step in constructing a grouped frequency distribution is to choose the number of classes. This choice involves a certain amount of judgment. If too many classes are used, the histogram has a jagged and ill-defined shape, as illustrated by the histogram of Figure 2.7, which uses 20 classes for the CK data of Example 2.5. On the other hand, if too few classes are used, important features of the distribution may be glossed over.

FIGURE 2.7

Histogram of CK Data, Using 20 Classes

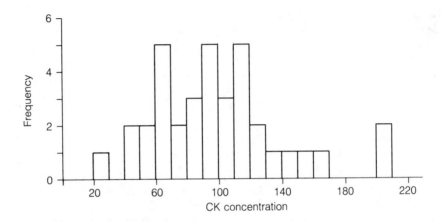

The choice of an appropriate number of classes depends partly on the sample size. For example, if we had CK data on 3,600 men rather than only 36, then a histogram based on 20 classes would probably have a fairly smooth appearance. For moderate sample sizes (say, $n \leq 50$), it is usually appropriate to use between 5 and 15 classes. For larger samples, one can use 20 classes or even more.

The second step in constructing a grouped frequency distribution is to choose a convenient class width. This is accomplished by determining the range of the data, perhaps rounding upward somewhat, and then dividing by the desired number of classes. The range is the difference between the largest and the smallest observation. For example, for the CK data, we have

Largest observation = 203
Smallest observation =   25
                Range = 178

If we want 10 classes, it is convenient to round the range upward to 200; then the class width will be

$$\text{Class width} = \frac{200}{10} = 20$$

Alternatively, we might have chosen to divide the same range into 8 classes of width 25.

The third step is to determine convenient class limits. In Example 2.5, the class limits were 20, 39, 40, 59, 60, ... Other choices could have been made.

For the same data the limits 10, 29, 30, 49, 50, . . . , would also be appropriate. Limits such as 23, 42, 43, 62, 63, . . . , would not be wrong but are awkward to work with. Notice that the class limits *must* be chosen in such a way that each observation falls into one and only one class. The choice of class limits therefore depends on the accuracy of measurement of the raw data. For instance, if the CK measurements were specified to the nearest tenth (.1 U/l), then classes such as 20.0–39.9, 40.0–59.9, . . . , would be necessary so that no observation could "fall through the cracks" between the classes.

The final step in constructing a grouped frequency distribution is to determine the frequencies. This can be done by constructing a tally, as illustrated in Figure 2.8. The quickest way to make a tally is to read through the observations in their natural order, just as they happen to occur in the raw data; as you come to each observation, make a tally mark adjacent to the appropriate class.

FIGURE 2.8
Tally for CK Data

Class	Tally
20–39	I
40–59	IIII
60–79	HIT II
80–99	HIT III
100–119	HIT III
120–139	III
140–159	II
160–179	I
180–199	
200–219	II

Computer note   Some statistical computer programs will produce frequency distributions and histograms. In some programs you can specify the class limits yourself; this often produces better results than letting the computer choose them. (Sometimes there is no substitute for human judgment.)

INTERPRETING AREAS IN A HISTOGRAM

A histogram can be looked at in two ways. The tops of the bars sketch out the shape of the distribution. But the *areas* within the bars also have a meaning. The area of each bar is proportional to the corresponding frequency. Consequently, the area of one or several bars can be interpreted as expressing the number of observations in the classes represented by the bars. For example, Figure 2.9 (page 22) shows a histogram of the CK distribution of Example 2.5. The shaded area is 42% of the total area in all the bars. Accordingly, 42% of the CK values are in the corresponding classes; that is, 15 of 36 or 42% of the values are between 60 U/l and 100 U/l.*

*Strictly speaking, between 60 U/l and 99 U/l, inclusive.

FIGURE 2.9

Histogram of CK
Distribution. The shaded
area is 42% of the total
area and represents 42%
of the observations.

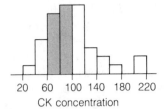

The area interpretation of histograms is a simple but important idea. In our later work with distributions we will find the idea to be indispensable.

FREQUENCY DISTRIBUTIONS WITH UNEQUAL CLASS WIDTHS

When a grouped frequency distribution is formed, classes are usually chosen to be of equal width. Occasionally classes of unequal width are used, for example, to smooth the distribution in a region where the data are sparse. If the classes are of unequal width, the method for drawing the histogram must be modified. To take an exaggerated example, suppose the last four classes of the CK grouping of Table 2.6 were coalesced into one class: 140–219. This class would have a frequency of 5. If the resulting distribution were plotted using raw frequencies, the histogram would be distorted in shape, as illustrated in Figure 2.10(a). Furthermore, the areas of the bars would no longer be proportional to the frequencies of the corresponding classes. The distortion can be removed by dividing the frequency of the coalesced class by 4, since it is 4 times as wide as the other classes. This gives the histogram of Figure 2.10(b). Notice that in this modified histogram the height of the wide bar is the *average* of the heights of the four narrow bars which it has replaced. This averaging process tends to retain the approximate shape of the original histogram; also, the proportionality between area and frequency is preserved. [Of course, the vertical axis in Figure 2.10(b) can no longer be labelled "frequency"; this will be discussed further in Section 2.9.]

FIGURE 2.10

Histograms of CK Distribution with Unequal Class
Widths. (a) Distorted;
(b) Appropriate.

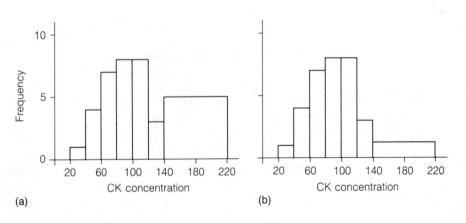

Even if you are not actually drawing a histogram, it is important to check the class widths when intepreting a tabulated distribution. If they are unequal, the frequencies do not indicate the shape of the distribution.

STEM-AND-LEAF DISPLAYS

A disadvantage of using a tally to form a grouped frequency distribution is that one loses track of the original data values. An alternative approach that does not share this disadvantage is the **stem-and-leaf display**. The construction of a stem-and-leaf display is illustrated in the following example.

EXAMPLE 2.6
WATER CONSUMPTION

As part of a pharmacological experiment, 31 lab rats were deprived of water for 23 hours and then permitted access to water for 1 hour. The amounts of water consumed (ml) during that hour are shown in Table 2.8. A stem-and-leaf display for these data appears in Figure 2.11.[5]

TABLE 2.8
One-Hour Water Consumption (ml) for 31 Rats

10.6	14.3	11.5	18.4	11.8
14.1	13.0	9.4	17.4	15.8
13.7	12.6	16.5	11.1	13.5
15.2	12.0	13.7	15.8	
15.4	14.0	14.7	17.0	
12.5	10.0	16.6	13.6	
12.9	18.2	11.4	16.6	

FIGURE 2.11
Stem-and-Leaf Display for Water Consumption Data

```
 9 | 4
10 | 6 0
11 | 5 4 1 8
12 | 5 9 6 0
13 | 7 0 7 6 5
14 | 1 3 0 7
15 | 2 4 8 8
16 | 5 6 6
17 | 4 0
18 | 2 4
```

To interpret the stem-and-leaf display of Figure 2.11, note that each data value has been split into a "stem" and a "leaf," as follows:

		STEM	LEAF
10.6	$\rightarrow$	10	6
14.1	$\rightarrow$	14	1

and so on.

In the display, each stem is accompanied by all its leaves. Notice that a stem-and-leaf display can be viewed as a histogram by turning it sideways. Unlike a histogram, however, the stem-and-leaf display retains the original data values.

To construct a stem-and-leaf display, simply read through the data values and write down each leaf next to its stem. In general, the last digit of an observation is the leaf, and the rest is the stem. It may be necessary to round the data so that this principle will produce a satisfactory display. Suppose, for instance, that the water consumption data had been measured to the nearest .01 ml, and the values were 10.63, 14.12, 13.68, . . . ; then we would need to round the data to one decimal place before constructing the stem-and-leaf display.

Note that the construction of a stem-and-leaf display does not depend on the location of the decimal point in the data For instance, if the water consumption data of Table 2.8 were expressed in dl rather than in ml, then the observations would be .106, .141, .137, . . . , but the stem-and-leaf display would be exactly the same as Figure 2.11.

In a research report, a frequency distribution would usually be presented as a table or a histogram. The stem-and-leaf display has many advantages as a working tool during the analysis of data. Stem-and-leaf displays can take more complicated forms than we have presented here. You may enjoy devising your own.

| | | | | | | | | | | | |
S E C T I O N 2.3

FREQUENCY DISTRIBUTIONS: SHAPES AND EXAMPLES

In this section we illustrate some of the diversity of frequency distributions encountered in the life sciences.

SHAPES OF FREQUENCY DISTRIBUTIONS

One interesting aspect of a set of data is the shape of its frequency distribution. The shape of a distribution can be indicated by a smooth curve that approximates the histogram, as shown in Figure 2.12.

FIGURE 2.12
Approximation of a Histogram by a Smooth Curve

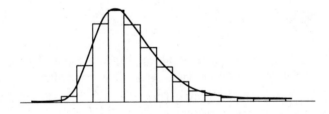

Some distributional shapes are shown in Figure 2.13. A common shape for biological data is **unimodal** (has one mode) and is somewhat skewed to the right, as in (c). Approximately bell-shaped distributions, as in (a), also occur. Sometimes

FIGURE 2.13
Shapes of Distributions

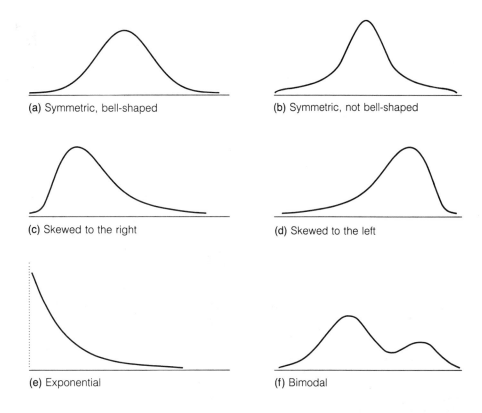

(a) Symmetric, bell-shaped   (b) Symmetric, not bell-shaped

(c) Skewed to the right   (d) Skewed to the left

(e) Exponential   (f) Bimodal

a distribution is symmetric but differs from a bell in having long tails; an exaggerated version is shown in (b). Left-skewed (d) and exponential (e) shapes are less common. **Bimodality** (two modes), as in (f), can indicate the existence of two distinct subgroups of observational units.

Notice that the shape characteristics we are emphasizing, such as number of modes and degree of symmetry, are *scale-free*; that is, they are not affected by the arbitrary choices of vertical and horizontal scale in plotting the distribution on paper. By contrast, a characteristic such as whether the distribution appears short and fat, or tall and skinny, *is* affected by how the distribution is plotted, and so is not an inherent feature of the biological variable.

The following three examples illustrate biological frequency distributions with various shapes. In the first example, the shape provides evidence that the distribution is in fact biological rather than nonbiological.

EXAMPLE 2.7
MICROFOSSILS

In 1977 paleontologists discovered microscopic fossil structures, resembling algae, in rocks $3\frac{1}{2}$ billion years old. A central question was whether these structures were biological in origin. One line of argument focused on their size distribution, which is shown in Figure 2.14. This distribution, with its unimodal and rather symmetric shape, resembles that of known microbial populations but not that of known nonbiological structures.[6]

FIGURE 2.14

Sizes of Microfossils

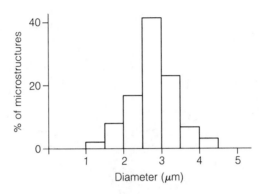

EXAMPLE 2.8

CELL FIRING TIMES

A neurobiologist observed discharges from rat muscle cells grown in culture together with nerve cells. The time intervals between 308 successive discharges were distributed as shown in Figure 2.15. Note the exponential shape of the distribution.[7]

FIGURE 2.15

Time Intervals Between Electrical Discharges in Rat Muscle Cells

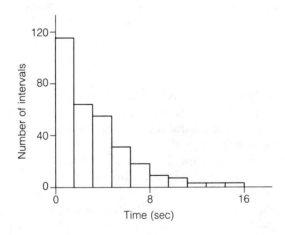

EXAMPLE 2.9

HEIGHTS OF AMERICANS

As part of a national survey, heights were measured for a large sample of American adults. The results for males and females are shown in Figure 2.16(a) and (b). These distributions are fairly symmetric and bell-shaped. Part (c) of the figure shows the height distribution for males and females combined. It may seem surprising that the combined height distribution is unimodal. In fact, it requires a substantial degree of dimorphism to produce a bimodal distribution.[8]

FIGURE 2.16
Heights of Americans

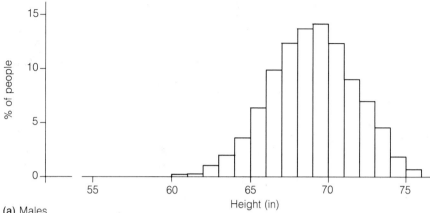

(a) Males

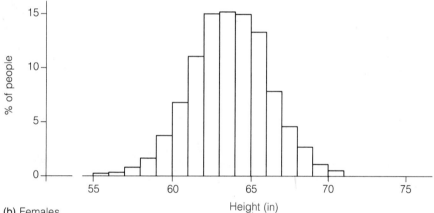

(b) Females

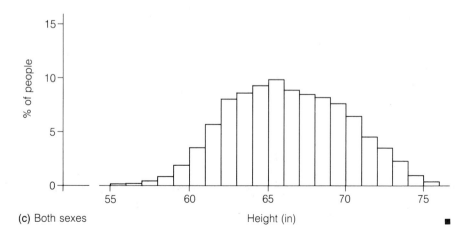

(c) Both sexes

SOURCES OF VARIATION

In interpreting biological data, it is helpful to be aware of sources of variability. The variation among observations in a data set often reflects the combined effects of several underlying factors. The following two examples illustrate such situations.

EXAMPLE 2.10

WEIGHTS OF BEANS

In a classic experiment to distinguish environmental from genetic influence, a geneticist weighed seeds of the princess bean *Phaseolus vulgaris*. Figure 2.17 shows the weight distributions of (a) 5,494 seeds from a commercial seed lot, and (b) 712 seeds from a highly inbred line that was derived from a single seed from the original lot. The variability in (a) is due to both environmental and genetic factors; in (b), because the plants are nearly genetically identical, the variation in weights is due largely to environmental influence.[9]

FIGURE 2.17

Weights of Princess Beans. (a) From an open-bred population; (b) From an inbred line.

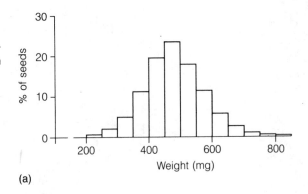

(a)

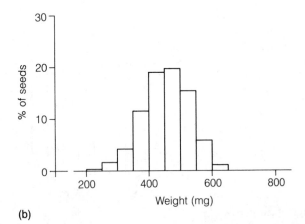

(b)

EXAMPLE 2.11

SERUM ALT

Alanine aminotransferase (ALT) is an enzyme found in most human tissues. Part (a) of Figure 2.18 shows the serum ALT concentrations for 129 adult volunteers. The following are potential sources of variability among the measurements:

1. Inter-individual
   a. Genetic
   b. Environmental
2. Intra-individual
   a. Biological: changes over time
   b. Analytical: imprecision in assay

The effect of the last source—analytical variation—can be seen in part (b) of Figure 2.18, which shows the frequency distribution of 109 assays of the *same* specimen of serum; the figure shows that the ALT assay is fairly imprecise.[10]

FIGURE 2.18

Distribution of Serum ALT Measurements (a) For 129 volunteers; (b) For 109 assays of the same specimen.

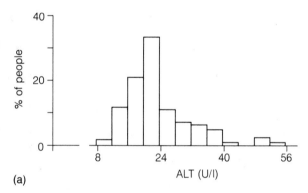

(a)

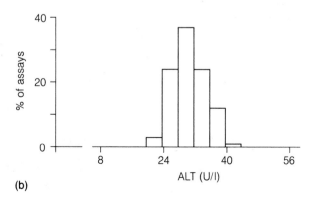

(b)

2.1 A paleontologist measured the width (in mm) of the last upper molar in 36 specimens of the extinct mammal *Acropithecus rigidus*. The results were as follows:[11]

6.1	5.7	6.0	6.5	6.0	5.7
6.1	5.8	5.9	6.1	6.2	6.0
6.3	6.2	6.1	6.2	6.0	5.7
6.2	5.8	5.7	6.3	6.2	5.7
6.2	6.1	5.9	6.5	5.4	6.7
5.9	6.1	5.9	5.9	6.1	6.1

Construct a frequency distribution and display it as a table and as a histogram.

2.2 In a study of schizophrenia, researchers measured the activity of the enzyme monoamine oxidase (MAO) in the blood platelets of 18 patients. The results (expressed as nmoles benzylaldehyde product per 10^8 platelets) were as follows:[12]

6.8	8.4	8.7	11.9	14.2	18.8
9.9	4.1	9.7	12.7	5.2	7.8
7.8	7.4	7.3	10.6	14.5	10.7

Construct a frequency distribution and display it as a table and as a histogram.

2.3 A dendritic tree is a branched structure that emanates from the body of a nerve cell. As part of a study of brain development, 36 nerve cells were taken from the brains of newborn guinea pigs. The investigators counted the number of dendritic branch-segments emanating from each nerve cell. The numbers were as follows:[13]

23	30	54	28	31	29	34	35	30
27	21	43	51	35	51	49	35	24
26	29	21	29	37	27	28	33	33
23	37	27	40	48	41	20	30	57

Construct a frequency distribution and display it as a table and as a histogram.

2.4 The total amount of protein produced by a dairy cow can be estimated from periodic testing of her milk. The following are the total annual protein production values (lb) for 28 2-year-old Holstein cows. Diet, milking procedures, and other conditions were the same for all the animals.[14]

425	481	477	434	410	397	438
545	528	496	502	529	500	465
539	408	513	496	477	445	546
471	495	445	565	499	508	426

Construct a frequency distribution and display it as a table and as a histogram.

2.5 For each of 31 healthy dogs, a veterinarian measured the glucose concentration in the anterior chamber of the right eye, and also in the blood serum. The following data are the anterior chamber glucose measurements, expressed as a percentage of the blood glucose.[15]

81	85	93	93	99	76	75	84
78	84	81	82	89	81	96	82
74	70	84	86	80	70	131	75
88	102	115	89	82	79	106	

Construct a frequency distribution and display it as a table and as a histogram.

2.6 Refer to the glucose data of Exercise 2.5. Prepare a stem-and-leaf display of the data.

2.7 In a behavioral study of the fruitfly *Drosophila melanogaster*, a biologist measured, for individual flies, the total time spent preening during a six-minute observation period. The following are the preening times (sec) for 20 flies:[16]

34	24	10	16	52
76	33	31	46	24
18	26	57	32	25
48	22	48	29	19

Construct a stem-and-leaf display for these data.

| | | | | | | | | | | | | |

SECTION 2.4

DESCRIPTIVE STATISTICS: MEASURES OF CENTER

For categorical data, the frequency distribution provides a concise and complete summary of a sample. For quantitative variables, the frequency distribution can be usefully supplemented by numerical measures of important features of the data.

A numerical measure calculated from data is called a **statistic**. "Descriptive statistics" are statistics that describe a set of data. We will discuss the major types of simple descriptive statistics. In this section we discuss measures of the center of the data.

There are several different ways to define the "center" or "typical value" of the observations in a sample. We will consider two widely used measures of center—the mean and the median—and, more briefly, a less well-known one—the trimmed mean.

THE MEAN

The most familiar measure of center is the ordinary average or **mean** (sometimes called the arithmetic mean). The mean of a sample (or the sample mean) is denoted by the symbol $\bar{y}$ (read "y-bar"). Example 2.12 illustrates the computation of the sample mean.

EXAMPLE 2.12

WEIGHT GAIN OF LAMBS

The following are the two-week weight gains (lb) of six young lambs of the same breed who had been raised on the same diet:[17]

11	13	19	2	10	1

The mean weight gain of the lambs in this sample is

$$\bar{y} = \frac{11 + 13 + 19 + 2 + 10 + 1}{6}$$

$$= \frac{56}{6}$$

$$= 9.33 \text{ lb}$$ ∎

The mean is the "point of balance" of the data. Figure 2.19 shows the lamb weight-gain data, plotted as dots, and the location of $\bar{y}$. If the data points were children on a weightless seesaw, then the seesaw would exactly balance if supported at $\bar{y}$.

FIGURE 2.19
Plot of the Lamb
Weight-Gain Data

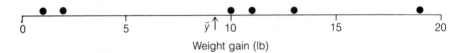

Weight gain (lb)

The general definition of the sample mean is

$$\bar{y} = \frac{\text{Sum of the } y\text{'s}}{n}$$

where the y's are the observations in the sample, and n is the sample size (that is, the number of y's). This definition can be formalized using the notation Σ (read "summation"), which means to add up.

THE SAMPLE MEAN

The **sample mean** is defined as

$$\bar{y} = \frac{\Sigma y}{n}$$

The numerator of this expression denotes the sum of all the observations in the sample.

THE MEDIAN

The sample **median** is the value that most nearly lies in the middle of the sample. To find the median, you first arrange the observations in increasing order. In the array of ordered observations, the median is the middle value (if n is odd) or midway between the two middle values (if n is even). Example 2.13 illustrates these definitions.

EXAMPLE 2.13
WEIGHT GAIN OF LAMBS

a. For the weight-gain data of Example 2.12, the ordered observations are

1 2 10 11 13 19
   ↑ ↑

The median weight gain is

$$\text{Median} = \frac{10 + 11}{2} = 10.5 \text{ lb}$$

b. Suppose the sample contained one more lamb, with the seven ranked observations as follows:

1 2 10 10 11 13 19
    ↑

For this sample, the median weight gain is

$$\text{Median} = 10 \text{ lb}$$

Notice that in this example there are two lambs whose weight gain is equal to the median. ■

A more formal way to define the median is in terms of rank position in the ordered array (counting the smallest observation as rank 1, the next as 2, and so on). The rank position of the median is equal to

$$(.5)(n + 1)$$

Thus, if $n = 7$, we calculate $(.5)(n + 1) = 4$, so that the median is the fourth largest observation; if $n = 6$, we have $(.5)(n + 1) = 3.5$, so that the median is midway between the third and fourth largest observations.

RESISTANCE A descriptive statistic is called **resistant** if the value of the statistic is relatively unaffected by changes in a small portion of the data, even if the changes are large ones. The median is a resistant statistic, but the mean is not resistant because it can be greatly shifted by changes in even one observation. Example 2.14 illustrates this behavior.

EXAMPLE 2.14
WEIGHT GAIN OF LAMBS

Recall that for the lamb weight-gain data

1 2 10 11 13 19

we found

$$\bar{y} = 9.3 \qquad \text{Median} = 10.5$$

Suppose now that the observation 19 is changed, or even omitted. How would the mean and median be affected? You can visualize the effect by imagining moving or removing the right-hand dot in Figure 2.19. Clearly the mean could change a great deal; the median would generally be less affected. For instance:

If the 19 is changed to 12, the mean becomes 8.2 and the median does not change.

If the 19 is omitted, the mean becomes 7.4 and the median becomes 10.

The above changes are not wild ones; that is, the changed samples might well have arisen from the same feeding experiment. Of course, a huge change, such as changing the 19 to 100, would shift the mean very drastically; note that it would not shift the median at all.   ∎

VISUALIZING THE MEAN AND MEDIAN

For samples large enough to construct frequency distributions, we can visualize the mean and the median in relation to the histogram. The median divides the area under the histogram roughly in half because it divides the observations roughly in half ["roughly" because some observations may be tied at the median, as in Example 2.13(b), and because the observations within each class are not uniformly distributed across the class]. The mean can be visualized as the point of balance of the histogram: If the histogram were made out of plywood, it would roughly balance if supported at the mean.

If the frequency distribution is symmetric, the mean and the median are equal and fall in the center of the distribution. If the frequency distribution is skewed, both measures are pulled toward the longer tail, but the mean is usually pulled farther than the median. The effect of skewness is illustrated by the following example.

EXAMPLE 2.15
CRICKET SINGING TIMES

Male Mormon crickets (*Anabrus simplex*) sing to attract mates. A field researcher measured the duration of 51 unsuccessful songs—that is, the time until the singing male gave up and left his perch.[18] Figure 2.20 shows the histogram of the 51 singing times. Table 2.9 gives the raw data. The median is 3.7 min and the mean is 4.3 min. The discrepancy between these measures is due largely to the long straggly tail of the distribution; the few unusually long singing times influence the mean but not the median.

FIGURE 2.20
Histogram of Cricket
Singing Times

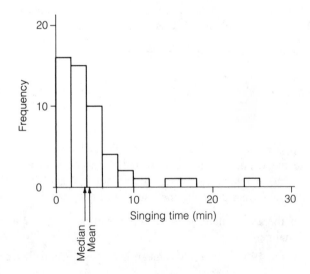

TABLE 2.9							
4.3	3.9	17.4	2.3	.8	1.5	.7	3.7
24.1	9.4	5.6	3.7	5.2	3.9	4.2	3.5
6.6	6.2	2.0	.8	2.0	3.7	4.7	
7.3	1.6	3.8	.5	.7	4.5	2.2	
4.0	6.5	1.2	4.5	1.7	1.8	1.4	
2.6	.2	.7	11.5	5.0	1.2	14.1	
4.0	2.7	1.6	3.5	2.8	.7	8.6	

TABLE 2.9
51 Cricket Singing Times (min)

MEAN VERSUS MEDIAN

Both the mean and the median are usually reasonable measures of the center of a data set. The mean is related to the sum; for example, if the mean weight gain of 100 lambs is 9 lb, then the total weight gain is 900 lb, and this total may be of primary interest since it translates more or less directly into profit for the farmer. In some situations the mean makes very little sense. Suppose, for example, that the observations are survival times of cancer patients on a certain treatment protocol, and that most patients survive less than 1 year, while a few respond well and survive for 5 or even 10 years. In this case, the mean survival time might be greater than that of most patients; the median would more nearly represent the experience of a "typical" patient. Note also that the mean survival time cannot be computed until the last patient has died; the median does not share this disadvantage. Situations in which the median can readily be computed, but the mean cannot, are not uncommon in bioassay, survival, and toxicity studies.

We have noted that the median is more resistant than the mean. If a data set contains a few observations rather distant from the main body of the data—that is, a long "straggly" tail—then the mean may be unduly influenced by these few unusual observations. Thus, the "tail" may "wag the dog." In such cases, the resistance of the median may be advantageous.

An advantage of the mean is that in some circumstances it is more **efficient** than the median. Efficiency is a technical notion in statistical theory; roughly speaking, a method is efficient if it takes full advantage of all the information in the data. Partly because of its efficiency, the mean has played a major role in classical methods in statistics.

THE TRIMMED MEAN

Recently, measures of center have been proposed that are more resistant than the mean and more efficient than the median. A simple example of such a "compromise" measure is the **trimmed mean**, which is a mean computed after "trimming" some observations from each tail. For example, a 10% trimmed mean is computed as follows: delete the smallest 10% of the observations and the largest 10% of the observations and compute the mean of the remaining 80%. Thus, if $n = 20$, the 10% trimmed mean would be the mean of the middle 16 observations.

We should emphasize that the use of the trimmed mean is entirely different from the practice of "editing" the data by throwing away "questionable" observations. When the trimmed mean is used, the trimming policy is chosen before inspecting the data, and the subsequent statistical analysis must be modified to take account of the trimming. We will return to this issue in Chapter 6.

| | | | | | | | | | | | | | |

EXERCISES 2.8–2.14

2.8 Invent a sample of size 5 for which the sample mean is 20 and not all the observations are equal.

2.9 Invent a sample of size 5 for which the sample mean is 20 and the sample median is 15.

2.10 A researcher applied the carcinogenic (cancer-causing) compound benzo(a)pyrene to the skin of five mice, and measured the concentration in the liver tissue after 48 hours. The results (nmoles/gm) were as follows:[19]

   6.3   5.9   7.0   6.9   5.9

Determine the mean and the median.

2.11 Six men with high serum cholesterol participated in a study to evaluate the effects of diet on cholesterol level. At the beginning of the study their serum cholesterol levels (mg/dl) were as follows:[20]

   366   327   274   292   274   230

Determine the mean and the median.

2.12 The weight gains of beef steers were measured over a 140-day test period. The average daily gains (lb/day) of 9 steers on the same diet were as follows:[21]

   3.89   3.51   3.97   3.31   3.21
   3.36   3.67   3.24   3.27

Determine the mean and median.

2.13 As part of a classic experiment on mutations, ten aliquots of identical size were taken from the same culture of the bacterium *E. coli*. For each aliquot, the number of bacteria resistant to a certain virus was determined. The results were as follows:[22]

   14   15   13   21   15
   14   26   16   20   13

a. Construct a frequency distribution of these data and display it as a histogram.
b. Determine the mean and the median of the data and mark their locations on the histogram.

2.14 The accompanying table gives the litter size (number of piglets surviving to 21 days) for each of 36 sows (as in Example 2.3). Determine the median litter size.

LITTER SIZE (NUMBER OF PIGLETS)	FREQUENCY (NUMBER OF SOWS)
5	1
6	0
7	2
8	3
9	3
10	9
11	8
12	5
13	3
14	2
Total	36

SECTION 2.5

MEASURES OF DISPERSION: METHODS

A measure of center is an incomplete description of a data set. At the very least, a good description should also characterize the variability or dispersion of the observations—are the observations in the sample all nearly equal, or do they differ substantially from each other? We will consider the following measures of dispersion: the range, the interquartile range, the standard deviation, and the coefficient of variation.

THE RANGE

The sample **range** is the difference between the largest and smallest observations in a sample. Here is an example.

EXAMPLE 2.16
BLOOD PRESSURE

The systolic blood pressures (mm Hg) of seven middle-aged men were as follows:[23]

151   124   132   170   146   124   113

For these data, the sample range is

170 − 113 = 57 mm Hg   ■

QUARTILES AND THE INTERQUARTILE RANGE

The sample **quartiles** are the values that divide the ranked sample into quarters, just as the median divides it in half. The sample **interquartile range** is the difference between the first and third quartiles of the sample. The following example illustrates these definitions.

EXAMPLE 2.17

BLOOD PRESSURE

For the blood pressure data of Example 2.16, the rank-ordered sample is

113   124   124   132   146   151   170

The quartiles are marked by arrows. The first quartile is 124 mm Hg, the second quartile (which is the median) is 132 mm Hg, and the third quartile is 151 mm Hg. The interquartile range is

$$151 - 124 = 27 \text{ mm Hg}$$

The quartiles can be formally defined in terms of rank position in the ordered array. The rank positions of the quartiles are

First quartile: $(.25)(n + 1)$

Second quartile: $(.50)(n + 1)$

Third quartile: $(.75)(n + 1)$

When $n = 7$, this definition shows that the quartiles occupy rank positions 2, 4, and 6, as illustrated in Example 2.17. When $(n + 1)$ is not divisible by 4, interpolation may be necessary in determining the quartiles.

When data are being analyzed by hand, a stem-and-leaf display is a useful aid in determining the quartiles, because it facilitates quick rank-ordering of the data, as illustrated in the next example.

EXAMPLE 2.18

WATER CONSUMPTION

The stem-and-leaf display of Figure 2.21 represents the water consumption data of Example 2.6. Notice that the values in the display have been rank-ordered, so it is easy to locate the quartiles. The three quartiles are 12.0 ml, 13.7 ml, and 15.8 ml. The rank-ordered stem-and-leaf display of Figure 2.21 can be constructed very quickly from the unordered display of Figure 2.11.

FIGURE 2.21

Rank-Ordered Stem-and-Leaf Display for Water Consumption of 31 Rats. The quartiles are circled.

```
 9 | 4
10 | 0  6
11 | 1  4  5  8
12 | Ⓞ  5  6  9
13 | 0  5  6  7  ⑦
14 | 0  1  3  7
15 | 2  4  8  ⑧
16 | 5  6  6
17 | 0  4
18 | 2  4
```

THE STANDARD DEVIATION

The standard deviation is the classical and most widely used measure of dispersion. To define the standard deviation, we first consider the **deviation** of each observation from the mean:

$$\text{Deviation} = y - \bar{y}$$

The standard deviation of the sample, or sample **standard deviation**, is determined by combining the deviations in a special way, as described in the accompanying box.

DEFINITIONAL FORMULA FOR s

The sample standard deviation is denoted by the letter s and is defined by the following formula:

$$s = \sqrt{\frac{\sum(y - \bar{y})^2}{n - 1}}$$

In this formula, the expression $\sum(y - \bar{y})^2$ denotes the sum of the squared deviations.

To illustrate the use of the formula, we have chosen a data set that is especially simple to handle because the mean happens to be an integer.

EXAMPLE 2.19
GROWTH OF
CHRYSANTHEMUMS

In an experiment on chrysanthemums, a botanist measured the stem elongation (mm in 7 days) of five plants grown on the same greenhouse bench. The results were as follows:[24]

76   72   65   70   82

The data are tabulated in the first column of Table 2.10. The sample mean is

$$\bar{y} = \frac{365}{5} = 73 \text{ mm}$$

The deviations $(y - \bar{y})$ are tabulated in the second column of Table 2.10; the first observation is 3 mm above the mean, the second is 1 mm below the mean, and so on. Notice that the sum of the deviations (taking into account the signs) is zero; this will be true for any data set.

TABLE 2.10

Illustration of the
Definitional Formula for
the Sample Standard
Deviation

OBSERVATION y	DEVIATION $y - \bar{y}$	SQUARED DEVIATION $(y - \bar{y})^2$
76	3	9
72	−1	1
65	−8	64
70	−3	9
82	9	81
Sum 365 = $\sum y$	0	164 = $\sum (y - \bar{y})^2$

The third column of Table 2.10 shows that the sum of the squared deviations is

$$\sum(y - \bar{y})^2 = 164$$

Since $n = 5$, the standard deviation is

$$s = \sqrt{\frac{164}{4}}$$
$$= \sqrt{41}$$
$$= 6.4 \text{ mm}$$

Note that the units of s (mm) are the same as the units of Y. This is because we have squared the deviations and then later taken the square root.   ∎

VARIANCE

The square of the standard deviation is called the **variance**. The sample variance is denoted by s^2:

$$\text{Variance} = s^2$$
$$\text{Standard deviation} = \sqrt{\text{Variance}}$$

EXAMPLE 2.20

CHRYSANTHEMUM
GROWTH

The variance of the chrysanthemum growth data is

$$s^2 = 41 \text{ mm}^2$$

Note that the units of the variance (mm²) are *not* the same as the units of Y.   ∎

An abbreviation   We will frequently abbreviate "standard deviation" as "SD"; the symbol "s" will be used in formulas.

INTERPRETATION OF THE DEFINITION OF s

The magnitude (disregarding sign) of each deviation $(y - \bar{y})$ can be interpreted as the *distance* of the corresponding observation from the sample mean $\bar{y}$. Figure 2.22 shows a plot of the chrysanthemum growth data (Example 2.19) with each distance marked.

FIGURE 2.22

Plot of Chrysanthemum
Growth Data with
Deviations Indicated as
Distances

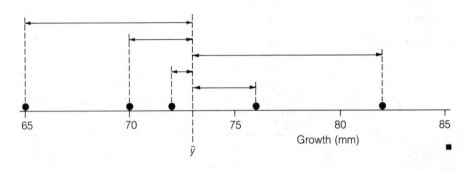

∎

From the formula for s, you can see that each deviation contributes to the SD. Thus, a sample of the same size but with less dispersion will have a smaller SD, as illustrated in the following example.

EXAMPLE 2.21
CHRYSANTHEMUM
GROWTH

If the chrysanthemum growth data of Example 2.19 are changed to

75 72 73 75 70

then the mean is the same ($\bar{y} = 73$ mm), but the SD is smaller ($s = 2.1$ mm), because the observations lie closer to the mean. The relative dispersion of the two samples can be easily seen from Figure 2.23.

FIGURE 2.23
Two Samples of
Chrysanthemum Growth
Data with the Same Mean
but Different Standard
Deviations. (a) $s = 6.3$
mm; (b) $s = 2.1$ mm.

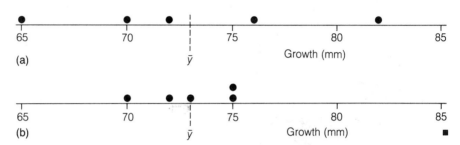

Let us look more closely at the way in which the deviations are combined to form the SD. The formula calls for dividing by $(n - 1)$. If the divisor were n instead of $(n - 1)$, then the quantity inside the square root sign would be the average (mean) of the squared deviations. Unless n is very small, the inflation due to dividing by $(n - 1)$ instead of n is not very great, so that the SD can be interpreted approximately as

$$s \approx \sqrt{\text{Sample average value of } (y - \bar{y})^2}$$

Thus, it is roughly appropriate to think of the SD as a "typical" distance of the observations from their mean.

Why $n - 1$? Since dividing by n seems more natural, you may wonder why the formula for the SD specifies dividing by $(n - 1)$. The reasons are rather technical, but we can give one intuitive justification by considering the extreme case when $n = 1$, as in the following example.

EXAMPLE 2.22
CHRYSANTHEMUM
GROWTH

Suppose the chrysanthemum growth experiment of Example 2.19 had included only one plant, so that the sample consisted of the single observation

73

For this sample, $n = 1$ and $\bar{y} = 73$. However, the SD formula breaks down (giving $\frac{0}{0}$), so the SD cannot be computed. This is reasonable, because the sample gives no information about variability in chrysanthemum growth under the experimental conditions. If we were to divide by n, we would obtain a SD of zero, suggesting that there is little or no variability; such a conclusion hardly seems justified by observation of only one plant.

THE COEFFICIENT OF VARIATION

The **coefficient of variation** is the standard deviation expressed as a percentage of the mean. Here is an example.

EXAMPLE 2.23

CHRYSANTHEMUM
GROWTH

For the chrysanthemum growth data of Example 2.19, we have $\bar{y} = 73.00$ mm, and $s = 6.40$ mm. Thus,

$$\frac{s}{\bar{y}} = \frac{6.40}{73.00} = .088 \quad \text{or} \quad 8.8\%$$

The sample coefficient of variation is 8.8%.   ■

Note that the coefficient of variation is not affected by multiplicative changes of scale. For example, if the chrysanthemum data were expressed in inches instead of mm, then both $\bar{y}$ and s would be in inches, and the coefficient of variation would be unchanged. Because of its imperviousness to scale change, the coefficient of variation is a useful measure for comparing the dispersions of two or more variables that are measured on different scales.

STATISTICAL COMPUTATIONS WITH A CALCULATOR

Of course, the most practical and painless way to perform statistical computations is to use a computer. Indeed, this is increasingly the approach taken by researchers in the life sciences. As a statistics student, however, you may often be working with a calculator rather than a computer.

Modern calculators with statistical functions can calculate the mean and standard deviation automatically. In order to use your calculator effectively, you will find it helpful to understand how this is done.

The calculator *cannot* use the definitional formula to calculate the SD, for the following reason: If you enter observations one at a time into the calculator, it cannot calculate the deviations $(y - \bar{y})$ until you have entered all the data. To use the definitional formula, the calculator would need to store all the observations simultaneously, and it lacks the memory capacity to do this. Rather than the definitional formula, the calculator uses the following formula for the SD:

COMPUTATIONAL FORMULA FOR s

$$s = \sqrt{\frac{\sum y^2 - \frac{\left(\sum y\right)^2}{n}}{n - 1}}$$

In this formula, the expression $\sum y^2$ denotes the sum of the squared observations, while the expression $(\sum y)^2$ denotes the squared sum of the observations.

The calculator has three special memory registers whose purpose is to accumulate n, Σy, and Σy^2. As you enter each observation into the calculator, it updates each of these registers to include the new observation. When you press the "mean" key, the calculator uses the contents of the first two registers to calculate the mean. When you press the "SD" key, the calculator applies the computational formula to the contents of the three registers.

To clarify the meaning of the computational formula, we illustrate it with an example.

EXAMPLE 2.24
CHRYSANTHEMUM
GROWTH

Table 2.11 shows the application of the computational formula to the chrysanthemum growth data of Example 2.19.

TABLE 2.11
Illustration of the
Computational Formula for
the Standard Deviation

y	y^2
76	5,776
72	5,184
65	4,225
70	4,900
82	6,724
Sum 365 $= \Sigma y$	26,809 $= \Sigma y^2$

The computational formula gives

$$s = \sqrt{\dfrac{26{,}809 - \dfrac{(365)^2}{5}}{4}}$$

$$= \sqrt{\dfrac{26{,}809 - 26{,}645}{4}} = \sqrt{\dfrac{164}{4}}$$

$$= \sqrt{41}$$

which is the same as the answer given by the definitional formula in Example 2.19. ∎

The computational formula is exactly equivalent to the definitional formula. (This fact is demonstrated in Appendix 2.1.) In the days of pencil-and-paper computation, the computational formula was used because it reduced the labor of computing the SD. Today its practical application is in the calculator.

SECTION 2.6

**MEASURES OF
DISPERSION:
INTERPRETATION AND
COMPARISON**

In this section we discuss and compare the range, the interquartile range, and the standard deviation.

VISUALIZING THE RANGE AND THE QUARTILES

The range and the interquartile range are easy to interpret. The range is the spread of all the observations and the interquartile range is the spread of (roughly) the middle 50% of the observations. In terms of the histogram of a data set, the range can be visualized as (roughly) the width of the histogram and the quartiles as (roughly) the values that divide the area into four equal parts. The following example illustrates these ideas.

EXAMPLE 2.25

DAILY GAIN OF CATTLE

The performance of beef cattle was evaluated by measuring their weight gain during a 140-day testing period on a standard diet. Table 2.12 gives the average daily gains (kg/day) for 39 bulls of the same breed (Charolais); the observations are listed in increasing order.[25] The values range from 1.18 kg/day to 1.92 kg/ day. The quartiles are 1.29, 1.41, and 1.58 kg/day. Figure 2.24 shows a histogram of the data, the range, the quartiles, and the interquartile range (IQR). The shaded area represents the middle 50% (approximately) of the observations.

TABLE 2.12

Average Daily Gain (kg/day) of 39 Charolais Bulls

1.18	1.24	1.29	1.37	1.41	1.51	1.58	1.72
1.20	1.26	1.33	1.37	1.41	1.53	1.59	1.76
1.23	1.27	1.34	1.38	1.44	1.55	1.64	1.83
1.23	1.29	1.36	1.40	1.48	1.57	1.64	1.92
1.23	1.29	1.36	1.41	1.50	1.58	1.65	

FIGURE 2.24

Histogram of 39 Daily Gain Measurements, Showing the Range, the Quartiles, and the Interquartile Range (IQR). The shaded area represents about 50% of the observations.

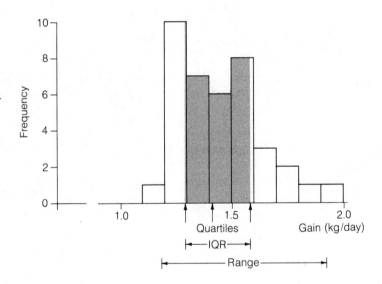

VISUALIZING THE STANDARD DEVIATION

We have seen that the SD is a combined measure of the distances of the observations from their mean. It is natural to ask how many of the observations are within ±1 SD of the mean, within ±2 SDs of the mean, and so on. There are no universal answers to these questions, but we can give some general guidelines. The following is an example.

EXAMPLE 2.26

DAILY GAIN OF CATTLE

For the daily-gain data of Example 2.25, the mean is $\bar{y} = 1.445$ kg/day and the SD is $s = .183$ kg/day. In Figure 2.25 the intervals $\bar{y} \pm s$, $\bar{y} \pm 2s$, and $\bar{y} \pm 3s$ have been marked on a histogram of the data. The interval $\bar{y} \pm s$ is

$$1.445 \pm .183 \quad \text{or} \quad 1.262 \text{ to } 1.628$$

You can verify from Table 2.12 that this interval contains 25 of the 39 observations. Thus, $\frac{25}{39}$ or 64% of the observations are within ± 1 SD of the mean; the corresponding area is shaded in Figure 2.25. The interval $\bar{y} \pm 2s$ is

$$1.445 \pm .366 \quad \text{or} \quad 1.079 \text{ to } 1.811$$

This interval contains $\frac{37}{39}$ or 95% of the observations. You may verify that the interval $\bar{y} \pm 3s$ contains all of the observations.

FIGURE 2.25

Histogram of Daily-Gain Data Showing Intervals 1, 2, and 3 Standard Deviations from the Mean. The shaded area represents about 64% of the observations.

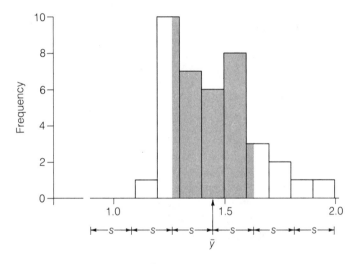

It turns out that the percentages found in Example 2.26 are fairly typical of distributions that are observed in the life sciences.

TYPICAL PERCENTAGES

For "nicely shaped" distributions—that is, unimodal distributions that are not too skewed and whose tails are not overly long or short—we usually expect to find

about 68% of the observations within ± 1 SD of the mean;

about 95% of the observations within ± 2 SDs of the mean;

> 99% of the observations within ± 3 SDs of the mean.

The typical percentages enable us to construct a rough mental image of a frequency distribution if we know just the mean and SD. (The value 68% may seem to come from nowhere. Its origin will become clear in Chapter 4.)

The typical percentages given in the box may be grossly wrong if the sample is small or if the shape of the frequency distribution is not "nice." For instance, the cricket singing-time data (Table 2.9 and Figure 2.20) has $s = 4.4$ min, and the interval $\bar{y} \pm s$ contains 90% of the observations. This is much higher than the "typical" 68% because the SD has been inflated by the long straggly tail of the distribution.

CHEBYSHEV'S RULE (OPTIONAL)

The above typical percentages are not universal; they apply only to "nicely shaped" distributions. The Russian mathematician P. L. Chebyshev discovered an interpretation of the SD which is applicable to *any* data set. Chebyshev's rule places limits on how much of the data can lie more than 2 or 3 standard deviations from the mean.

CHEBYSHEV'S RULE

For any data set,

at least 75% of the observations are within ± 2 SDs of the mean;

at least 89% of the observations are within ± 3 SDs of the mean.

Because Chebyshev's rule is universal, the percentage ranges in the rule are very broad; for instance, "at least 75%" means anywhere between 75% and 100%. Nevertheless, the rule can be used to limit the extremes of a data set, as illustrated by the following example.

EXAMPLE 2.27

THYMIDINE IN BLOOD CELLS

Cyclosporin A is a drug that acts to suppress immune response. As part of a study of the drug, researchers measured the amount of radioactively labelled thymidine incorporated by human white blood cells in the presence of cyclosporin A. The results, expressed in counts per minute per 10^6 cells, were reported as follows:[26]

Mean: 1,100

SD: 50

n: 3

Even though we do not have the original data, we know from Chebyshev's rule that all three measurements must have been between

1,000 cpm   and   1,200 cpm

If one measurement were outside this interval, then only $\frac{2}{3}$ or 67% of the observations would be within it, which would be less than the 75% required by Chebyshev's rule.   ■

COMPARISON OF MEASURES OF DISPERSION

The dispersion, or spread, of the data in a sample can be described by the standard deviation, the range, or the interquartile range. The range is simple to understand but it can be a poor descriptive measure because it depends only on the extreme tails of the distribution. The interquartile range, by contrast, describes the spread in the central "body" of the distribution. The standard deviation takes account of all the observations, and can be roughly interpreted in terms of the spread of the observations around their mean. However, the SD can be inflated by observations in the extreme tails. The interquartile range is a resistant measure, while the SD is nonresistant. Of course, the range is very highly nonresistant.

The descriptive interpretation of the SD is less straightforward than that of the range and the interquartile range. Nevertheless, the SD is the basis for most standard classical statistical methods. The SD enjoys this classic status for various technical reasons, including efficiency in certain situations.

The developments in later chapters will emphasize classical statistical methods, in which the mean and SD play a central role. Consequently, in this book we will rely primarily on the mean and SD rather than other descriptive measures.

| | | | | | | | | | | | | |

EXERCISES 2.15–2.30

[*Note:* Exercises preceded by an asterisk refer to optional sections.]

2.15 In a study of milk production in sheep (for use in making cheese), a researcher measured the three-month milk yield for each of 11 ewes. The yields (liters) were as follows:[27]

56.5	89.8	110.1	65.6	63.7	82.6
75.1	91.5	102.9	44.4	108.1	

Determine the range, the quartiles, and the interquartile range.

2.16 A botanist grew 15 pepper plants on the same greenhouse bench. After 21 days, she measured the total stem length (cm) of each plant, and obtained the following values:[28]

12.4	12.2	13.4
10.9	12.2	12.1
11.8	13.5	12.0
14.1	12.7	13.2
12.6	11.9	13.1

a. Construct a stem-and-leaf display for these data, and use it to determine the quartiles.

b. Calculate the interquartile range.

2.17 Use the definitional formula to calculate the standard deviation of each of the following fictitious samples:

a. 16, 13, 18, 13

b. 38, 30, 34, 38, 35

c. 1, −1, 5, −1

d. 4, 6, −1, 4, 2

2.18 The following are samples of fictitious data. For each sample, calculate the standard deviation two ways—using the definitional formula, and using the computational formula—and verify that the answers agree.
  a. 8, 6, 9, 4, 8
  b. 4, 7, 5, 4
  c. 9, 2, 6, 7, 6

2.19 a. Invent a sample of size 5 for which the deviations $(y - \bar{y})$ are $-3, -1, 0, 2, 2$.
  b. Use the definitional formula to compute the standard deviation of your sample.
  c. Should everyone get the same answer for part **b**? Why?

2.20 Four plots of land, each 346 square feet, were planted with the same variety ("Beau") of wheat. The plot yields (lb) were as follows:[29]

  35.1   30.6   36.9   29.8

  a. Calculate the mean and the standard deviation.
  b. Calculate the coefficient of variation.

2.21 A plant physiologist grew birch seedlings in the greenhouse and measured the ATP content of their roots. (See Example 1.3.) The results (nmol ATP/mg tissue) were as follows for four seedlings that had been handled identically.[30]

  1.45   1.19   1.05   1.07

  a. Calculate the mean and the standard deviation.
  b. Calculate the coefficient of variation.

2.22 Ten patients with high blood pressure participated in a study to evaluate the effectiveness of the drug Timolol in reducing their blood pressure. The accompanying table shows systolic blood pressure measurements taken before and after two weeks of treatment with Timolol.[31] Calculate the mean and standard deviation of the *change* in blood pressure (note that some values are negative).

PATIENT	BLOOD PRESSURE (mm Hg)		
	Before	After	Change
1	172	159	−13
2	186	157	−29
3	170	163	−7
4	205	207	2
5	174	164	−10
6	184	141	−43
7	178	182	4
8	156	171	15
9	190	177	−13
10	168	138	−30

2.23 Dopamine is a chemical that plays a role in the transmission of signals in the brain. A pharmacologist measured the amount of dopamine in the brain of each of seven rats. The dopamine levels (nmoles/g) were as follows:[32]

    6.8   5.3   6.0   5.9   6.8   7.4   6.2

a. Calculate the mean and standard deviation.
b. Determine the median and the interquartile range.
c. Calculate the coefficient of variation.
d. Replace the observation 7.4 by 10.4 and repeat parts **a** and **b**. Which of the descriptive measures display resistance and which do not?

2.24 In a study of the lizard *Sceloporus occidentalis*, biologists measured the distance (m) run in two minutes for each of 15 animals. The results (listed in increasing order) were as follows:[33]

    18.4   22.2   24.5   26.4   27.5   28.7   30.6   32.9
    32.9   34.0   34.8   37.5   42.1   45.5   45.5

a. Determine the quartiles and the interquartile range.
b. Determine the range.

2.25 Refer to the running-distance data of Exercise 2.24. The sample mean is 32.23 m and the SD is 8.07 m. What percentage of the observations are within
a. 1 SD of the mean?
b. 2 SDs of the mean?

2.26 Listed below in increasing order are the serum creatine phosphokinase (CK) levels (U/l) of 36 healthy men (these are the data of Example 2.5):

25	62	82	95	110	139
42	64	83	95	113	145
48	67	84	100	118	151
57	68	92	101	119	163
58	70	93	104	121	201
60	78	94	110	123	203

The sample mean CK level is 98.28 U/l and the SD is 40.38 U/l. What percentage of the observations are within
a. 1 SD of the mean?
b. 2 SDs of the mean?
c. 3 SDs of the mean?

2.27 Refer to the CK data of Exercise 2.26. Construct a histogram of the data (you can use the frequency distribution given in Table 2.6). Mark the horizontal axis with intervals 1, 2, and 3 SDs from the mean (as in Figure 2.25).

*2.28 Compare the results of Exercise 2.25 with the predictions of Chebyshev's rule.

*2.29 Compare the results of Exercise 2.26 with the predictions of Chebyshev's rule.

*2.30 The IQs of ten people were measured.[34] The mean was 128.4 and the standard deviation was 10.4. Is it possible that three of the people had IQs over 150? Explain. [*Hint:* Use Chebyshev's rule.]

EFFECT OF TRANSFORMATION OF VARIABLES (OPTIONAL)

To be more comfortable working with data, it is helpful to know how the features of a distribution are affected if the observed variable is transformed. **Transformation**, or re-expression, of a variable Y means replacing Y by a new variable, say Y'.

The simplest transformations are **linear** transformations, so called because a graph of Y against Y' would be a straight line. A familiar reason for linear transformation is a change in the scale of measurement, as illustrated in the following two examples.

EXAMPLE 2.28

WEIGHT

Suppose Y represents the weight of an animal in kg, and we decide to re-express the weight in lb. Then

$$Y = \text{Weight in kg}$$
$$Y' = \text{Weight in lb}$$

so

$$Y' = 2.2Y$$

This is a **multiplicative** transformation, because Y' is calculated from Y by multiplying by the constant value 2.2.   ∎

EXAMPLE 2.29

BODY TEMPERATURE

Measurements of basal body temperature (temperature on waking) were made on 47 women.[35] Typical observations Y, in °C, were:

Y: 36.23, 36.41, 36.77, 36.15, . . .

Suppose we convert these data from °C to °F, and call the new variable Y':

Y':   97.21, 97.54, 98.19, 97.07, . . .

The relation between Y and Y' is

$$Y' = 1.8Y + 32$$

The combination of **additive** ($+32$) and multiplicative ($\times 1.8$) changes indicates a linear relationship.   ∎

Another reason for linear transformation is **coding**, which means transforming the data for convenience in handling the numbers. The following is an example.

EXAMPLE 2.30

BODY TEMPERATURE

Consider the temperature data of Example 2.29. If we subtract 36 from each observation, the data become

.23   .41   .77   .15   . . .

This is additive coding, since we added a constant value (-36) to each observation. Now suppose we further transform the data to the form

23   41   77   15   . . .

This step of the coding is multiplicative, since each observation is multiplied by a constant value (100).   ∎

As the foregoing examples illustrate, a linear transformation consists of (1) multiplying all the observations by a constant, or (2) adding a constant to all the observations, or (3) both.

HOW LINEAR TRANSFORMATIONS AFFECT $\bar{y}$ AND s

The effect of a linear transformation on $\bar{y}$ is "natural"; that is, **under a linear transformation, $\bar{y}$ changes like Y**. For instance, if temperatures are converted from °C to °F, then the mean is similarly converted.

The effect of a multiplicative transformation on s is "natural"; **under a multiplicative transformation, s changes like Y**. For instance, if weights are converted from kg to lb, the SD is similarly converted. However, **an additive transformation does not affect s**. Thus, for example, we would *not* convert the SD of temperature data from °C to °F in the same way as we convert each observation; we would multiply the SD by 1.8 but we would not add 32. The fact that the SD is unchanged by additive transformation will appear less surprising if you recall (from the definitional formula) that s depends only on the deviations $(y - \bar{y})$, and these are not changed by an additive transformation. The following example illustrates this idea.

EXAMPLE 2.31   Consider a simple set of fictitious data, coded by subtracting 20 from each observation. The original and transformed observations are shown in Table 2.13.

TABLE 2.13
Effect of Additive
Transformation

ORIGINAL OBSERVATIONS y	DEVIATIONS $y - \bar{y}$	TRANSFORMED OBSERVATIONS y'	DEVIATIONS $y' - \bar{y}'$
25	−1	5	−1
26	0	6	0
28	2	8	2
25	−1	5	−1
Mean 26		6	

The SD for the original observations is

$$s = \sqrt{\frac{(-1)^2 + (0)^2 + (2)^2 + (-1)^2}{3}}$$

$$= 1.4$$

Because the deviations are unaffected by the transformation, the SD for the transformed observations is the same:

$$s' = 1.4$$

If you choose to linearly transform your data, you can easily trace and retrace the effect on the mean and standard deviation, without algebraic formalities. Simply follow the path of a typical observation, and then mimic this with the mean and SD, remembering to omit the additive step when transforming the SD. The following example illustrates this procedure.

EXAMPLE 2.32
BODY TEMPERATURE

Consider the transformed data of Example 2.30. Call these Y'.

Y: 36.23 36.41 36.77 36.15 . . .
Y': 23   41   77   15   . . .

Suppose the mean and SD of the transformed data have been computed to be

$$\bar{y}' = 49.7 \qquad s' = 17.2$$

and we want to determine the mean and SD of the original data. We trace the first transformed observation ($y' = 23$) back to its corresponding y, and then mimic with $\bar{y}$ and s. Notice that the addition step is omitted when retracing s.

Retrace y': $(23 \times .01) + 36 = 36.23 \ = y$
Retrace $\bar{y}'$: $(49.7 \times .01) + 36 = 36.497 = \bar{y}$
Retrace s': $(17.2 \times .01) \qquad\quad = .172 \ = s$

Thus, the mean and SD of the original data are

$$\bar{y} = 36.497 \qquad s = .172 \qquad\qquad\blacksquare$$

Other statistics   Under linear transformations, other measures of center (for instance, the median) change like $\bar{y}$, and other measures of dispersion (for instance, the interquartile range) change like s. The quartiles themselves change like $\bar{y}$.

HOW LINEAR TRANSFORMATIONS AFFECT THE FREQUENCY DISTRIBUTION

A linear transformation of the data does not change the essential shape of its frequency distribution; by suitably scaling the horizontal axis, you can make the transformed histogram identical to the original histogram. Example 2.33 illustrates this idea.

EXAMPLE 2.33
BODY TEMPERATURE

Figure 2.26 shows the distribution of 47 temperature measurements that have been transformed as in Example 2.32. The figure shows that the two distributions can be represented by the same histogram with different horizontal scales.

FIGURE 2.26

Distribution of 47
Temperature
Measurements, Showing
Original and Linearly
Transformed Scales

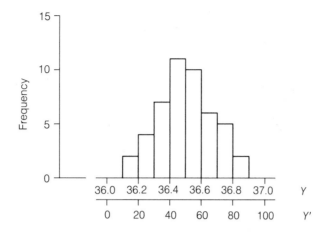

NONLINEAR TRANSFORMATIONS

Data are sometimes re-expressed in a **nonlinear** way. Examples of nonlinear transformations are

$$Y' = \log(Y)$$
$$Y' = \sqrt{Y}$$
$$Y' = Y^2$$

These transformations are termed "nonlinear" because a graph of Y' against Y would be a curve rather than a straight line. The logarithmic transformation is especially common in biology because many important relationships can be simply expressed in terms of logs. For instance, there is a phase in the growth of a bacterial colony when log(colony size) increases at a constant rate with time. Observe that whenever you plot a graph (not a histogram) on semi-log paper you are in effect making a logarithmic transformation of one of the variables.

Nonlinear transformations can affect data in complex ways. For example, the mean does not change "naturally" under a log transformation; the log of the mean is not the same as the mean of the logs. Furthermore, nonlinear transformations (unlike linear ones) *do* change the essential shape of a frequency distribution, as illustrated in the following example.

EXAMPLE 2.34

CRICKET SINGING TIMES

Figure 2.27(a) on page 54 shows the distribution of the cricket singing-time data of Table 2.9. If we transform these data by taking logs (base 10), the transformed data have the distribution shown in Figure 2.27(b). Notice that the transformation has the effect of "pulling in" the straggly upper tail and "stretching out" the clumped values on the lower end of the original distribution. [*Note:* The distribution in Figure 2.27(b) was constructed by grouping logs of the raw data into classes of equal width. This is *not* the same as simply plotting Figure 2.27(a) on semi-log paper.]

FIGURE 2.27

Distribution of Y and
of log(Y), for 51
Observations of
Y = Singing Time

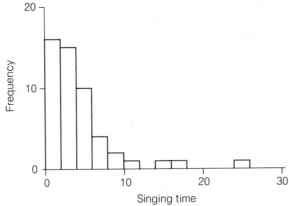

(a)

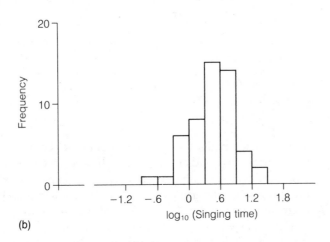

(b)

■

EXERCISES 2.31–2.33   **2.31** A biologist made a certain pH measurement in each of 24 frogs; typical values were[36]

7.43,   7.16,   7.51,   . . .

Suppose he transformed the data to

43,   16,   51,   . . .

and he calculated a mean of 37.3 and a standard deviation of 12.9 for the transformed
data. What were the mean and standard deviation of the original pH measurements?

2.32 The mean and SD of a set of 47 body temperature measurements were as follows:[37]

$\bar{y} = 36.497\ °C$ $s = .172\ °C$

If the 47 measurements were converted to °F,
a. What would be the new mean and SD?
b. What would be the new coefficient of variation?

2.33 A researcher measured the average daily gains (in kg/day) of 20 beef cattle; typical values were[38]

1.39, 1.57, 1.44, . . .

Suppose he transformed the data to

39, 57, 44, . . .

and he calculated a mean of 46.1 and a standard deviation of 17.8 for the transformed data.

a. What were the mean and standard deviation of the original measurements in kg/day?

b. Express the mean and standard deviation in lb/day. [*Hint:* 1 kg = 2.20 lb]

c. Calculate the coefficient of variation when the data are expressed (i) in kg/day; (ii) in lb/day.

| | | | | | | | | | | | | |

SECTION 2.8

SAMPLES AND POPULATIONS: STATISTICAL INFERENCE

In the preceding sections we have examined several ways of describing a set of observations. We have called the data set a "sample." Now we discuss the reason for this terminology.

The description of a data set is sometimes of interest for its own sake. Usually, however, the researcher hopes to generalize, to extend the findings beyond the limited scope of the particular group of animals, plants, or other units which were actually observed. Statistical theory provides a rational basis for this process of generalization, a basis that takes account of the variability of the data. The key idea of the statistical approach is to view the particular data in an experiment as a sample from a larger **population**; the population is the real focus of scientific and/or practical interest. The following example illustrates this idea.

EXAMPLE 2.35

BLOOD TYPES

In an early study of the ABO blood-typing system, researchers determined blood types of 3,696 people in England. The results are given in Table 2.14.[39]

TABLE 2.14

Blood Types of 3,696 People

BLOOD TYPE	FREQUENCY
A	1,634
B	327
AB	119
O	1,616
Total	3,696

These data were not collected for the purpose of learning about the blood types of those particular 3,696 people. Rather, they were collected for their scientific value as a source of information about the distribution of blood types in a larger population. For instance, one might presume that the blood type distribution of *all* English people should resemble the distribution for these 3,696 people. In particular, the observed relative frequency of Type A blood was

$$\frac{1,634}{3,696} \quad \text{or} \quad 44\% \text{ Type A}$$

One might conclude from this that approximately 44% of the people in England have Type A blood.   ■

STATISTICAL INFERENCE

The process of drawing conclusions about a population, based on observations in a sample from that population, is called **statistical inference**. For instance, in Example 2.35 the conclusion that approximately 44% of the people in England have Type A blood would be a statistical inference. The inference is shown schematically in Figure 2.28. Of course, such an inference might be entirely wrong—perhaps the 3,696 people are not at all representative of English people in general. Two possible sources of difficulty are: (1) the 3,696 people might have been selected in a way that was systematically biased for (or against) Type A people, and (2) the number of people examined might have been too small to permit generalization to a population of many millions.

FIGURE 2.28

Schematic Representation of Inference from Sample to Population Regarding Prevalence of Blood Type A

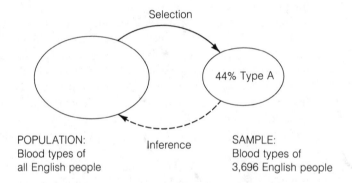

Selection

44% Type A

POPULATION:
Blood types of
all English people

Inference

SAMPLE:
Blood types of
3,696 English people

In making a statistical inference, one would prefer that the sample resemble the population closely—that the sample be *representative* of the population. However, one must ask about the likelihood of this desirable state of affairs. In other words, one must ask the important question: *How representative (of the population) is a sample likely to be?* We will see in Chapters 3 and 5 how statistical theory can help to answer this question. But the question itself becomes meaningful only if the population has been defined, a process that we now discuss in more detail.

DEFINING THE POPULATION

Ideally, the population should be defined in such a way that it is plausible to believe that a sufficiently large sample *would* be representative of the population. The first step in defining the population is to ask how the observations were obtained. Two important issues are: How were the observational units selected? What was the observed variable? The following example illustrates the reasoning involved in defining the population.

EXAMPLE 2.36
BLOOD TYPES

How were the 3,696 English people of Example 2.35 actually chosen? It appears from the original paper that this was a "sample of convenience," that is, friends of the investigators, employees, and sundry unspecified sources. There is little basis for believing that the *people* themselves would be representative of the entire English population. Nevertheless, one might argue that their *blood types* might be (more or less) representative of the population. The argument would be that the biases that entered into the selection of those particular people were probably not related to blood type (although an objection might be made on the basis of race). The argument for representativeness would be much *less* plausible if the observed variable were blood pressure rather than blood type; we know that blood pressure tends to increase with age, and the selection procedure was undoubtedly biased against certain age groups (for example, elderly people). ∎

As Example 2.36 shows, whether a sample is likely to be representative of a population depends not only on how the observational units (in this case people) were chosen, but also on what variable was observed. Generally, therefore, it is most appropriate to think of the population as consisting of observations, rather than of people or other observational units. We can conceptualize the population as an indefinitely large extension of the sample. In other words, **in order to try to define the population from which our data came, we try to describe the set of observations which we would obtain if the process generating the data were repeated indefinitely.** The following is another example.

EXAMPLE 2.37
ALCOHOL AND MOPEG

The biochemical MOPEG plays a role in brain function. Seven healthy male volunteers participated in a study to determine whether drinking alcohol might elevate the concentration of MOPEG in the cerebrospinal fluid. The MOPEG concentration was measured twice for each man—once at the start of the experiment, and again after he drank 80 gm of ethanol. The results (in pmol/ml) are given in Table 2.15.[40]

TABLE 2.15
Effect of Alcohol on MOPEG

| VOLUNTEER | MOPEG CONCENTRATION | | |
	Before	After	Change
1	46	56	10
2	47	52	5
3	41	47	6
4	45	48	3
5	37	37	0
6	48	51	3
7	58	62	4

Let us focus on the rightmost column, which shows the change in MOPEG concentration (that is, the difference between the "after" and the "before" measurement). In thinking of these values as a sample from a population, we need to specify all the details of the experimental conditions—how the cerebrospinal specimens were obtained, the exact timing of the measurements and the alcohol consumption, and so on—as well as relevant characteristics of the volunteers themselves. Thus, the definition of the population might be something like this:

Population: Change in cerebrospinal MOPEG concentration in healthy young men when measured before and after drinking 80 gm of ethanol, both measurements being made at 8:00 A.M., . . . (other relevant experimental conditions are specified here).

There is no single "correct" definition of a population for an experiment like this. A scientist reading a report of the experiment might find the above definition too narrow (for instance, perhaps it does not matter that the volunteers were male) or too broad. He might use his knowledge of alcohol and brain chemistry to formulate his own definition, and he would then use that definition as a basis for interpreting these seven observations.　　　　　　　　　　　　　　■

A DYNAMIC EXAMPLE

The population–sample concept is at the heart of statistical thinking. In the following example we dramatize this concept by looking at larger and larger samples from the same population. (Of course, in practice, one usually takes only one sample from a population rather than samples of various sizes.)

EXAMPLE 2.38
SUCROSE CONSUMPTION

An entomologist is interested in the mechanism controlling feeding behavior in the black blowfly (*Phormia regina*). One variable of interest to him is the amount of sucrose (sugar) solution a fly will drink in 30 minutes. The measurement procedure is such that a given fly can be measured only once. To study the inherent variability of the system, the researcher has measured hundreds of flies under standardized conditions. Figure 2.29 shows histograms of sucrose consumption values (mg) for samples of various numbers of individuals.[41] The means and standard deviations of the samples are as follows:

n	20	40	100	400	900
$\bar{y}$	15.5	14.7	14.3	15.0	14.9
s	6.5	5.9	5.0	5.4	5.4

Notice that, as the sample size is increased, the frequency distribution tends to stabilize and, similarly, the mean and the SD tend to stabilize.

It is natural to define a population from which the samples came, as follows:

Population: Sucrose consumption values for all *P. regina* individuals under the standardized conditions

Just as each sample has a distribution, a mean, and an SD, so also we can envision a population distribution, a population mean, and a population SD.

FIGURE 2.29 Histograms of Various Samples of Sucrose Consumption Data

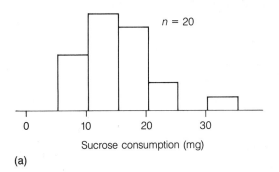

n = 20

Sucrose consumption (mg)

(a)

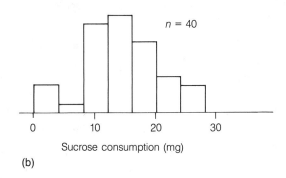

n = 40

Sucrose consumption (mg)

(b)

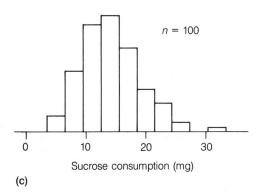

n = 100

Sucrose consumption (mg)

(c)

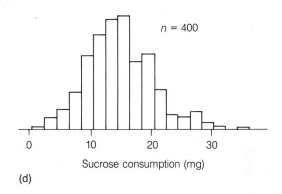

n = 400

Sucrose consumption (mg)

(d)

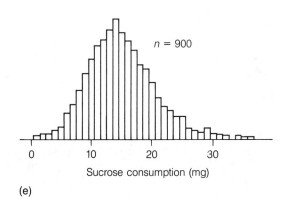

n = 900

Sucrose consumption (mg)

(e)

Remark   As noted in Example 2.38, the SD tends to stabilize as the sample size is increased. To see intuitively why this should happen, recall from Section 2.5 that

$$s \approx \sqrt{\text{Sample average value of } (y - \bar{y})^2}$$

The right-hand side of this expression depends only on the *composition* of the sample, not on its size; thus, **samples of different sizes but with similar compositions (relative frequency distributions) will have similar SDs.** Increasingly larger samples from the same population will tend to have compositions increasingly similar to the population, and so also to have means and SDs increasingly similar to the mean and SD of the population.

SECTION 2.9

DESCRIBING A POPULATION

Because observations are made only on a sample, characteristics of biological populations are almost never known exactly. Typically, our knowledge of a population characteristic comes from a sample. In statistical language, we say that the sample characteristic is an **estimate** of the corresponding population characteristic. Thus, estimation is a type of statistical inference.

In order to discuss inference from a sample to a population, we will need a language for describing the population. This language parallels the language that describes the sample.

PROPORTIONS

For a categorical variable, we can describe a population by simply stating the proportion, or relative frequency, of the population in each category. The following is a simple example.

EXAMPLE 2.39
OAT PLANTS

In a certain population of oat plants, resistance to crown rust disease is distributed as shown in Table 2.16.[42]

TABLE 2.16
Disease Resistance
in Oats

RESISTANCE	PROPORTION OF PLANTS
Resistant	.47
Intermediate	.43
Susceptible	.10
Total	1.00

Remark   The population described in Example 2.39 is realistic, but it is not a specific real population; the exact proportions for any real population are not known. For similar reasons, we will use fictitious but realistic populations in several other examples, here and in Chapters 3, 4, and 5.

For categorical data, the sample proportion of a category is an estimate of the corresponding population proportion. Because these two proportions are not necessarily the same, it is essential to have a notation that distinguishes between them. **We denote the population proportion of a category by π (pi) and the sample proportion by $\hat{\pi}$ (read "pi-hat"):**

$\hat{\pi}$ = Sample proportion

π = Population proportion

The symbol " ^ " can be interpreted as "estimate of." Thus,

$\hat{\pi}$ is an estimate of π.

We illustrate this notation with an example.

EXAMPLE 2.40

LUNG CANCER

Eleven patients suffering from adenocarcinoma (a type of lung cancer) were treated with the chemotherapeutic agent Mitomycin. Three of the patients showed a positive response (defined as shrinkage of the tumor by at least 50%).[43] Suppose we define the population for this study as "responses of all adenocarcinoma patients." Then we can represent the sample and population proportions of the category "positive response" as follows:

π = Proportion of positive responders among all adenocarcinoma patients

$\hat{\pi}$ = Proportion of positive responders among the 11 patients in the study

$\hat{\pi} = \frac{3}{11} = .27$

Note that π is unknown, and $\hat{\pi}$, which is known, is an estimate of π. ∎

We should emphasize that an "estimate," as we are using the term, may or may not be a *good* estimate. For instance, the estimate $\hat{\pi}$ in Example 2.40 is undoubtedly a very poor estimate of π, if for no other reason than that it is based on so few patients. Of course, the question of whether an estimate is good or poor is an important one, and we will show in later chapters how this question can be answered.

OTHER DESCRIPTIVE MEASURES

If the observed variable is quantitative, one can consider descriptive measures other than proportions—the mean, the quartiles, the SD, and so on. Each of these quantities can be computed for a sample of data, and each is an estimate of its corresponding population analog. For instance, the sample median is an estimate of the population median. In later chapters, we will focus especially on the mean and the SD, and so we will need a special notation for the population mean and SD. **The population mean is denoted by μ (mu), and the population SD is denoted by σ (sigma).** We may define these as follows for a quantitative variable Y:

μ = Population average value of Y

$\sigma = \sqrt{\text{Population average value of } (Y - \mu)^2}$

The following example illustrates this notation.

EXAMPLE 2.41

TOBACCO LEAVES

An agronomist counted the number of leaves on each of 150 tobacco plants of the same strain (Havana). The results are shown in Table 2.17.[44]

TABLE 2.17

Number of Leaves on Tobacco Plants

NUMBER OF LEAVES	FREQUENCY (NUMBER OF PLANTS)
17	3
18	22
19	44
20	42
21	22
22	10
23	6
24	1
Total	150

The sample mean is

$\bar{y} = 19.78 =$ Mean number of leaves on the 150 plants

The population mean is

$\mu =$ Mean number of leaves on Havana tobacco plants grown under these conditions

We do not know μ, but we can regard $\bar{y} = 19.78$ as an estimate of μ. The sample SD is

$s = 1.38 =$ SD of number of leaves on the 150 plants

The population SD is

$\sigma =$ SD of number of leaves on Havana tobacco plants grown under these conditions

We do not know σ, but we can regard $s = 1.38$ as an estimate of σ.*   ■

FREQUENCY DISTRIBUTION: RELATIVE FREQUENCY HISTOGRAM

In addition to numerical measures such as the mean and SD, a set of quantitative data can be described by a frequency distribution. The frequency distribution of the sample gives us information about the frequency distribution of the population.

One way to describe a population frequency distribution is with a relative frequency histogram. In this case the inference from the sample to the population is straightforward, because each sample relative frequency (proportion) is an estimate of the corresponding population relative frequency. Thus, the entire

*You may wonder why we use $\bar{y}$ and s instead of $\hat{\mu}$ and $\hat{\sigma}$. The answer is: tradition.

sample histogram is an estimate of the population histogram. The following is an example.

EXAMPLE 2.42
TOBACCO LEAVES

Figure 2.30 shows a relative frequency histogram for the tobacco leaf data of Example 2.41. The histogram is an estimate of the corresponding population relative frequency histogram.

FIGURE 2.30
Relative Frequency Histogram for Number of Leaves on 150 Tobacco Plants

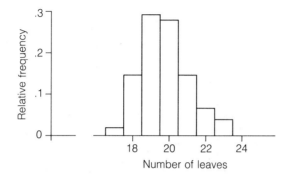

FREQUENCY DISTRIBUTION: DENSITY CURVE

It is often desirable, especially when the observed variable is continuous, to describe a population frequency distribution by a smooth curve. We may visualize the curve as an idealization of a histogram with very narrow classes. The following example illustrates this idea.

EXAMPLE 2.43
BLOOD GLUCOSE

A glucose tolerance test can be useful in diagnosing diabetes. The blood level of glucose is measured one hour after the subject has drunk 50 mg of glucose dissolved in water. Figure 2.31 (page 64) shows the distribution of responses to this test for a certain population of women.[45] The distribution is represented by histograms with class widths equal to (a) 10 and (b) 5, and by (c) a smooth curve. ■

A smooth curve representing a frequency distribution is called a **density curve**. The vertical coordinates of a density curve are plotted on a scale called a **density scale**. When the density scale is used, relative frequencies are represented as areas under the curve. Formally, the relation is as follows:

INTERPRETATION OF DENSITY
For any two numbers a and b,
Area under density curve between a and b $=$ Proportion of Y values between a and b
This relation is indicated in Figure 2.32 for an arbitrary distribution.

FIGURE 2.31
Different Representations
of the Distribution of
Blood Glucose Levels in
a Population of Women

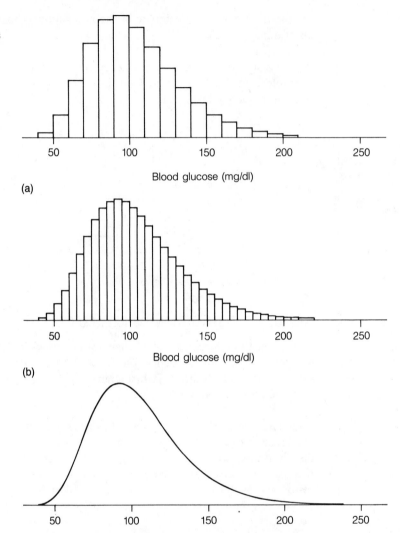

(a)

(b)

(c)

FIGURE 2.32
Interpretation of Area
Under a Density Curve

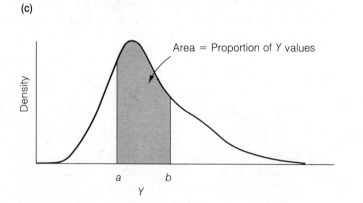

Because of the way the density curve is interpreted, the area under the entire curve must be equal to 1, as shown in Figure 2.33.

FIGURE 2.33
The Area Under an Entire Density Curve Must Be 1.

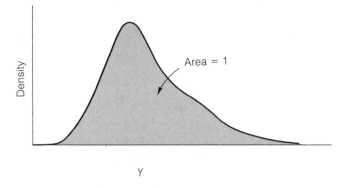

The interpretation of density curves in terms of areas is illustrated concretely in the following example.

EXAMPLE 2.44
BLOOD GLUCOSE

Figure 2.34 shows the density curve for the blood glucose distribution of Example 2.43, with the vertical scale explicitly shown. The shaded area is equal to .42, which indicates that about 42% of the glucose levels are between 100 mg/dl and 150 mg/dl. The area under the density curve to the left of 100 mg/dl is equal to .50; this indicates that the population median glucose level is 100 mg/dl. The area under the entire curve is 1.

FIGURE 2.34
Interpretation of an Area Under the Blood Glucose Density Curve

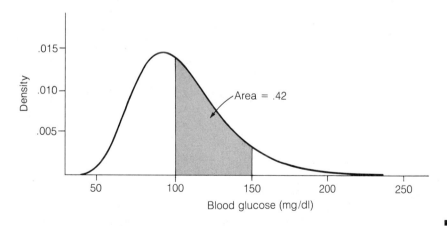

THE CONTINUUM PARADOX The area interpretation of a density curve has a paradoxical element. If we ask for the relative frequency of a single specific Y value, the answer is zero. For example, suppose we want to determine from Figure 2.34 the relative frequency of blood glucose levels *equal* to 150. The area interpretation gives an answer of zero. This seems to be nonsense—how can every value of Y have a relative frequency of zero? Let us look more closely at the question. If blood glucose is measured to the nearest mg/dl, then we are really asking for the relative frequency of glucose levels between 149.5 and 150.5 mg/dl, and the corresponding area is not zero. On the other hand, if we are thinking of blood glucose as an *idealized* continuous variable, then the relative frequency of any particular value (such as 150) *is* zero. This is admittedly a paradoxical situation. It is similar to the paradoxical fact that an idealized straight line can be 1 centimeter long, and yet each of the idealized points of which the line is composed has length equal to zero. In practice, the continuum paradox does not cause any trouble; we simply do not discuss the relative frequency of a single Y value (just as we do not discuss the length of a single point).

USES OF DENSITY CURVES   We have introduced density curves as a convenient description of a population distribution. We will see in later chapters that density curves also play a very important role in statistical methods of data analysis.

DETERMINING AREAS UNDER DENSITY CURVES   In reading a plotted density curve, you can estimate areas roughly by eye; you estimate how large a given area is, *relative* to the area under the entire curve. For example, the shaded area in Figure 2.34 is 42% of the entire area under the curve. (Because only relative area is needed, density curves are usually plotted without explicitly labelling the density scale.) In applying statistical methods, one needs more precise information about areas under density curves; this information is available in tables (and from computer programs). We will first encounter such tables in Chapter 4.

DENSITY SCALE FOR HISTOGRAMS   Histograms can also be plotted with density (rather than frequency or relative frequency) on the vertical scale.* When the density scale is used, equal areas represent equal relative frequencies. For a single histogram with equal class widths, conversion to the density scale will not affect the appearance of the histogram. But for a histogram with unequal class widths the density scale should be used to avoid distortion of the shape of the distribution. (In fact, the adjustment for unequal class widths shown in Figure 2.10 is essentially a conversion to the density scale.) When several histograms of the same variable Y, with different class widths, are plotted together for comparison, the density scale should be used to avoid distorted comparisons; the areas under all the histograms will then be the same. For instance, the histograms and the curve in Figure 2.31 are plotted on the density scale; the scale is explicitly shown in Figure 2.34. The density scale was also used in plotting Figure 2.29.

POPULATION SIZE

In discussing methods for describing a population, we have not mentioned the size of the population. In fact, most populations arising in biological research are of large but indefinite size. The statistical methods to be considered in this

---

*The relation between the scales is as follows:

$$\text{Density} = \frac{\text{Relative frequency}}{\text{Class width}}$$

book are all designed for situations where the population size is very large compared to the sample size, and where quantities of interest are expressed as averages over the populations, or proportions of the population. In this case, the population size plays no explicit role; as long as it is large, its actual magnitude is irrelevant.

EXERCISES 2.34–2.35

2.34 A sample of 100 length measurements has mean 106 mm and standard deviation 14 mm. Typical observations are

103,   87,   115,   ...

a. Find the value of $\Sigma(y - \bar{y})^2$ for this sample.
b. Suppose we duplicate each value in the above sample, thus creating a new sample of 200 measurements with typical observations

103,   103,   87,   87,   115,   115,   ...

(i) Find the mean of the new sample.
(ii) Find the standard deviation of the new sample. [*Hint:* First find the value of $\Sigma(y - \bar{y})^2$ for the new sample.] Has duplication of the sample increased the SD substantially, decreased it substantially, or left it approximately the same?

2.35 The diameter of a tree trunk is an important variable in forestry. The density curve shown here represents the distribution of diameters (measured 4.5 feet above the ground) in a population of 30-year-old Douglas fir trees. Areas under the curve are shown in the figure.[46] What percentage of the trees have diameters
a. between 4 inches and 10 inches?
b. less than 4 inches?
c. more than 8 inches?

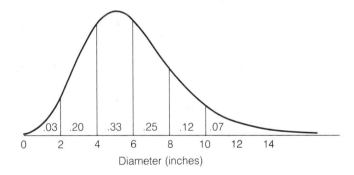

Diameter (inches)

SECTION 2.10

PERSPECTIVE

In this chapter we have considered various ways of describing a set of data. We have also introduced the notion of regarding a data set as a sample from a suitably defined population, and regarding features of the sample as estimates of corresponding features of the population.

PARAMETERS AND STATISTICS

Some features of a distribution—for instance, the mean—can be represented by a single number, while some—for instance, the shape—cannot. We have noted that a numerical measure that describes a sample is called a statistic. Correspondingly, a numerical measure that describes a population is called a **parameter**. For the most important numerical measures, we have defined notations to distinguish between the statistic and the parameter. These notations are summarized in Table 2.18 for convenient reference.

TABLE 2.18

Notation for Some Important Statistics and Parameters

MEASURE	SAMPLE VALUE (STATISTIC)	POPULATION VALUE (PARAMETER)
Proportion	$\hat{\pi}$	π
Mean	$\bar{y}$	μ
Standard deviation	s	σ

A LOOK AHEAD

It is natural to view a sample characteristic (for instance, $\bar{y}$) as an estimate of the corresponding population characteristic (for instance, μ). But in taking such a view one must guard against unjustified optimism. Of course, if the sample were perfectly representative of the population, then the estimate would be perfectly accurate. But this raises the central question: How representative (of the population) is a sample likely to be? Intuition suggests that, if the observational units are appropriately selected, then the sample should be more or less representative of the population. Intuition also suggests that larger samples should tend to be more representative than smaller samples. These intuitions are basically correct, but they are too vague to provide practical guidance for research in the life sciences. Practical questions which need to be answered are:

1. How can an investigator judge whether a sample can be viewed as "more or less" representative of a population?
2. How can an investigator quantify "more or less" in a specific case?

In Chapter 3 we will describe a theoretical model—the random sampling model—that provides a framework for the judgment in question (1), and in Chapter 6 we will see how this model can provide a concrete answer to question (2). Specifically, in Chapter 6 we will see how to analyze a set of data so as to quantify how closely the sample mean ($\bar{y}$) estimates the population mean (μ). But before returning to data analysis in Chapter 6, we will need to lay some groundwork in Chapters 3, 4, and 5; the developments in these chapters are an essential prelude to understanding the techniques of statistical inference.

2.36 To calibrate a standard curve for assaying protein concentrations, a plant pathologist used a spectrophotometer to measure the absorbance of light (wavelength 500 nm) by a protein solution. The results of 27 replicate assays of a standard solution containing 60 μg protein per ml water were as follows:[47]

.111	.115	.115	.110	.099
.121	.107	.107	.100	.110
.106	.116	.098	.116	.108
.098	.120	.123	.124	.122
.116	.130	.114	.100	.123
.119	.107			

Construct a frequency distribution and display it as a table and as a histogram.

2.37 Refer to the absorbance data of Exercise 2.36.
   a. Prepare a stem-and-leaf display of the data.
   b. Use the stem-and-leaf display of part **a** to determine the median and the interquartile range.

2.38 Twenty patients with severe epilepsy were observed for eight weeks. The following are the numbers of major seizures suffered by each patient during the observation period:[48]

5	0	9	6	0	0	5	0	6	1
5	0	0	0	0	7	0	0	4	7

   a. Determine the median number of seizures.
   b. Determine the mean number of seizures.
   c. Construct a histogram of the data (use the ungrouped frequency distribution). Mark the positions of the mean and the median on the histogram.
   d. What feature of the frequency distribution suggests that neither the mean nor the median is a meaningful summary of the experience of these patients?

2.39 Use the definitional formula to calculate the standard deviation of each of the following fictitious samples:
   a. 11, 8, 4, 10, 7
   b. 23, 29, 24, 21, 23
   c. 6, 0, −3, 2, 5

2.40 To study the spatial distribution of Japanese beetle larvae in the soil, researchers divided a 12 × 12-foot section of a cornfield into 144 one-foot squares. They counted the number of larvae Y in each square, with the results shown in the table at the top of page 70.[49]

   a. The mean and standard deviation of Y are $\bar{y} = 2.23$ and $s = 1.47$. What percentage of the observations are within
      (i) 1 standard deviation of the mean?
      (ii) 2 standard deviations of the mean?
   b. Determine the total number of larvae in all 144 squares. How is this number related to $\bar{y}$?
   c. Determine the median value of Y.

NUMBER OF LARVAE	FREQUENCY (NUMBER OF SQUARES)
0	13
1	34
2	50
3	18
4	16
5	10
6	2
7	1
Total	144

2.41 One measure of physical fitness is maximal oxygen uptake, which is the maximum rate at which a person can consume oxygen. A treadmill test was used to determine the maximal oxygen uptake of nine college women before and after participation in a ten-week program of vigorous exercise. The accompanying table shows the before and after measurements and the change; all values are in ml O_2 per min per kg body weight.[50]

	MAXIMAL OXYGEN UPTAKE		
PARTICIPANT	Before	After	Change
1	48.6	38.8	−9.8
2	38.0	40.7	2.7
3	31.2	32.0	.8
4	45.5	45.4	−.1
5	41.7	43.2	1.5
6	41.8	45.3	3.5
7	37.9	38.9	1.0
8	39.2	43.5	4.3
9	47.2	45.0	−2.2

The following computations are to be done on the *change* in maximal oxygen uptake (the right-hand column).

a. Calculate the mean and the standard deviation.

b. Determine the median.

c. Eliminate participant #1 from the data and repeat parts **a** and **b**. Which of the descriptive measures display resistance and which do not?

2.42 A veterinary anatomist investigated the spatial arrangement of the nerve cells in the intestine of a pony. He removed a block of tissue from the intestinal wall, cut the block into many equal sections, and counted the number of nerve cells in each of 23 randomly selected sections. The counts were as follows:[51]

```
35   19   33   34   17   26   16   40
28   30   23   12   27   33   22   31
28   28   35   23   23   19   29
```

Construct a frequency distribution and display it as a table and as a histogram.

2.43 Refer to the nerve-cell data of Exercise 2.42.

   a. Prepare a stem-and-leaf display of the data.

   b. Use the stem-and-leaf display of part **a** to determine the median, the quartiles, and the interquartile range.

2.44 Trypanosomes are parasites that cause disease in humans and animals. In an early study of trypanosome morphology, researchers measured the lengths of 500 individual trypanosomes taken from the blood of a rat. The results are summarized in the accompanying frequency distribution.[52]

LENGTH (μm)	FREQUENCY (NO. OF INDIVIDUALS)	LENGTH (μm)	FREQUENCY (NO. OF INDIVIDUALS)
15	1	27	36
16	3	28	41
17	21	29	48
18	27	30	28
19	23	31	43
20	15	32	27
21	10	33	23
22	15	34	10
23	19	35	4
24	21	36	5
25	34	37	1
26	44	38	1

   a. Construct a histogram of the data. Use the given frequency distribution without further grouping.

   b. What feature of the histogram suggests the interpretation that the 500 individuals are a mixture of two distinct types?

2.45 A geneticist counted the number of bristles on a certain region of the abdomen of the fruitfly *Drosophila melanogaster*. The results for 119 individuals were as shown in the table.[53]

NUMBER OF BRISTLES	NUMBER OF FLIES	NUMBER OF BRISTLES	NUMBER OF FLIES
29	1	38	18
30	0	39	13
31	1	40	10
32	2	41	15
33	2	42	10
34	6	43	2
35	9	44	2
36	11	45	3
37	12	46	2

   a. Find the median number of bristles.

   b. Find the first and third quartiles of the sample.

c. The sample mean is 38.45 and the standard deviation is 3.20. What percentage of the observations fall within 1 standard deviation of the mean?

2.46 The carbon monoxide in cigarettes is thought to be hazardous to the fetus of a pregnant woman who smokes. In a study of this theory, blood was drawn from pregnant women before and after smoking a cigarette. Measurements were made of the percent of blood hemoglobin bound to carbon monoxide as carboxyhemoglobin (COHb). The results for ten women are shown in the table.[54]

	BLOOD COHb (%)		
SUBJECT	Before	After	Increase
1	1.2	7.6	6.4
2	1.4	4.0	2.6
3	1.5	5.0	3.5
4	2.4	6.3	3.9
5	3.6	5.8	2.2
6	.5	6.0	5.5
7	2.0	6.4	4.4
8	1.5	5.0	3.5
9	1.0	4.2	3.2
10	1.7	5.2	3.5

a. Calculate the mean and standard deviation of the *increase* in COHb.
b. Calculate the mean COHb before and the mean after. Is the mean increase equal to the increase in means?
c. Construct a stem-and-leaf display of the increase in COHb. Use the display to determine the median increase.
d. Repeat part c for the before measurements and for the after measurements. Is the median increase equal to the increase in medians?

2.47 (*Computer problem*)   A medical researcher in India obtained blood specimens from 31 young children, all of whom were infected with malaria. The following data, listed in increasing order, are the numbers of malarial parasites found in 1 ml of blood from each child.[55]

100	140	140	271	400	435	455	770
826	1,400	1,540	1,640	1,920	2,280	2,340	3,672
4,914	6,160	6,560	6,741	7,609	8,547	9,560	10,516
14,960	16,855	18,600	22,995	29,800	83,200	134,232	

a. Construct a frequency distribution of the data, using a class width of 10,000; display the distribution as a histogram.
b. Transform the data by taking the logarithm (base 10) of each observation. Construct a frequency distribution of the transformed data and display it as a histogram. How does the log transformation affect the shape of the frequency distribution?
c. Determine the mean of the original data and the mean of the log-transformed data. Is the mean of the logs equal to the log of the mean?
d. Determine the median of the original data and the median of the log-transformed data. Is the median of the logs equal to the log of the median?

C H A P T E R 3

CONTENTS

RANDOM SAMPLING, PROBABILITY, AND THE BINOMIAL DISTRIBUTION

SECTION 3.1

PROBABILITY AND THE LIFE SCIENCES

Probability, or chance, plays an important role in scientific thinking about living systems. Some biological processes are affected directly by chance. A familiar example is the segregation of chromosomes in the formation of gametes; another example is the occurrence of mutations.

Even when the biological process itself does not involve chance, the results of an experiment are always somewhat affected by chance—chance fluctuations in environmental conditions, chance variation in the genetic makeup of experimental animals, and so on. Often, chance also enters directly through the design of an experiment; for instance, varieties of wheat may be randomly allocated to plots in a field. (Random allocation will be discussed in Chapter 8.)

The conclusions of a statistical data analysis are often stated in terms of probability. Probability enters statistical analysis, not only because chance influences the results of an experiment, but also because of theoretical frameworks, or *models*, which are used as a basis for statistical inference. In this chapter we will describe the most fundamental of these theoretical models, the random sampling model. In addition, we will introduce the language of probability and develop some simple tools for manipulating probabilities.

SECTION 3.2

RANDOM SAMPLING

The first step in developing a basis for statistical inference is to define what is meant by **random sampling**.

DEFINITION OF RANDOM SAMPLING

Informally, the process of random sampling can be visualized in terms of labelled tickets, such as those used in a lottery or raffle. Suppose that each member of the population is represented by one ticket, and that the tickets are placed in a large box and thoroughly mixed. Then n tickets are drawn from the box by a blindfolded assistant, with new mixing after each ticket is removed; these n tickets constitute the sample. (Equivalently, we may visualize that n assistants reach in the box simultaneously, each assistant drawing one ticket.)

More abstractly, we may define random sampling as follows:

DEFINITION

A sample of n items is said to be chosen by **random sampling** from a population if (a) every member of the population has the same chance of being included in the sample; and (b) the members of the sample are chosen independently of each other. [Requirement (b) means that the chance of a given population member being chosen does not depend on which other members are chosen.]

Random sampling can be defined in other, equivalent, ways. We may envision the sample members being chosen one at a time from the population; we say that the sampling is random if, at each stage of the drawing, every remaining member of the population is equally likely to be the next one chosen. Another view is to consider the totality of possible samples of size n; if all possible samples are equally likely to be obtained, then the sampling process is random.

The type of random sampling we have defined is technically called **simple random sampling**. There are other kinds of sampling which are random in a sense, but are not simple. For example, consider sampling from a human population as follows: First choose some families at random, and then include in the sample all members of those families. With this kind of sampling, which is called *cluster sampling*, all members of the population have the same chance of being in the sample, but the various members of the sample are not chosen independently of each other.

A sample chosen by random sampling is often called a *random sample*. But note that it is actually the *process* of sampling rather than the sample itself that is defined as random. Randomness is not a property of the particular sample that happens to be chosen.

CHOOSING A RANDOM SAMPLE

The technique of actually choosing a random sample from a concrete population has two types of application in biological studies: (1) choosing a sample of units for study from a larger population which is available; and (2) random allocation of units to treatment groups (as will be explained in Chapter 8). In addition, some of the exercises ("sampling exercises") in this book require random sampling; by giving you some experience with drawing random samples and looking at the results, these exercises help you to feel more comfortable with statistical reasoning.

The technique of random sampling is easy to learn. First, you need a source of random digits. A calculator or computer can supply random digits. Alternatively, you can use a table of random digits, such as Table 1 at the end of this book.

HOW TO USE THE TABLE OF RANDOM DIGITS For ease of reading, the rows and columns of Table 1 are numbered, and the digits in the table are grouped into 5×5 blocks. To use Table 1, begin reading at a random place in the table.* If you need single-digit numbers, just read down the table; if you need two-digit numbers, read two columns across, down the table; and so on. When you get to the bottom, go back to the top, move over an appropriate number of columns so that you will not use the same column twice, and continue reading.

---

*There are various ways to choose a random starting place. One simple method is to close your eyes and drop a paper clip onto Table 1; start reading at the digit closest to the outer end of the paper clip wire.

HOW TO READ RANDOM DIGITS FROM YOUR CALCULATOR   Many calculators generate random numbers expressed as decimal numbers between 0 and 1. To convert these to random digits, simply ignore the decimal and just read the individual digits in each random number. If you need single-digit numbers, read only the first digit; if you need two-digit numbers, read the first two digits; and so on.

Remark   In calling the digits in Table 1 or your calculator *random digits*, we are using the term *random* loosely. Strictly speaking, random digits are digits produced by a random *process*—for example, tossing a ten-sided die. While ten-sided dice do exist (for instance, they are used in the game "Dungeons and Dragons"), it is not practical to use them to generate random digits. The digits in Table 1 and in your calculator are actually *pseudo-random digits*; they are generated by a deterministic (although possibly very complex) process that is designed to produce sequences of digits which mimic randomly generated sequences. For those readers who are curious about this, a simple example of a procedure for generating pseudo-random digits is given in Appendix 3.1.

HOW TO CHOOSE A RANDOM SAMPLE   The following is a simple procedure for choosing a random sample of n items from a finite population of items.

a. Label the members of the population with identification numbers. All identification numbers must have the same number of digits; for instance, if the population contains 75 items, the identification numbers could be 01, 02, . . . , 75.
b. Read numbers from Table 1 or your calculator. Reject any numbers that do not correspond to any population member. Continue until n numbers have been acquired. (Ignore any repeated occurrence of the same number.)
c. The population members with the chosen identification numbers constitute the sample.

The following example illustrates this procedure.

EXAMPLE 3.1

Suppose we are to choose a random sample of size 6 from a population of 75 members. Label the population members 01, 02, . . . , 75. Suppose we start reading Table 1 at row 04, column 12; we obtain the numbers shaded in Table 3.1, which is a reproduction of part of Table 1.

We ignore the numbers greater than 75, and we ignore the second occurrence of 23. Thus, the population members with the following identification numbers will constitute the sample:

23   38   59   21   08   09   ■

Remark   In many situations, the sampling procedure described above is not the most efficient. For example, suppose the population contains 150 items. Then the procedure would be cumbersome because most three-digit identification numbers would be rejected. In this situation it would be more efficient to let more

TABLE 3.1
Reproduction of Part
of Table 1

	01	06	11	16	21
01	06048	96063	22049	86532	75170
02	25636	73908	85512	78073	19089
03	61378	45410	43511	54364	97334
04	15919	71559	12310	00727	54473
05	47328	20405	88019	82276	33679
06	72548	80667	53893	64400	81955
07	87154	04130	55985	44508	37515
08	68379	96636	32154	94718	22845
09	89391	54041	70806	36012	30833
10	15816	60231	28365	61924	66934
11	29618	55219	18394	11625	27673
12	30723	42988	30002	95364	45473
13	54028	04975	92323	53836	76128
14	40376	02036	48087	05216	26684
15	64439	37357	90935	57330	79738

than one identification number correspond to each population member (for instance, let 001, 201, 401, 601, and 801 correspond to the first member of the population, 002, 202, 402, 602, and 802 to the second member, and so on).

THE RANDOM SAMPLING MODEL

We saw in Chapter 2 that, in order to generalize beyond a particular set of data, an investigator may view the data as a sample from a population. But how can we provide a rationale for inference from a limited sample to a very much larger population? The approach of statistical theory is to refer to an idealized model of the sample–population relationship. In this model, which is called the **random sampling model**, the sample is chosen from the population by random sampling. The model is represented schematically in Figure 3.1.

FIGURE 3.1
The Random Sampling
Model

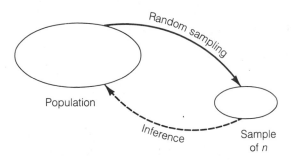

In Chapter 2 we posed a central question of statistical inference: How representative (of the population) is a sample likely to be? The random sampling model is useful because it provides a basis for answering this question. The model can be used to determine how much an inference might be influenced by chance, or "luck of the draw." More explicitly, a randomly chosen sample usually will not exactly resemble the population from which it was drawn. The discrepancy between the sample and the population is called **sampling error**. We will see in later chapters how statistical theory derived from the random sampling model enables us to set limits on the likely amount of sampling error in an experiment. The quantification of sampling error is a major contribution which statistical theory has made to scientific thinking.

How does the random sampling model relate to reality? In some studies in the life sciences, the observational units are literally chosen by random sampling. In much biological research, however, the observations in a data set are not chosen by an actual random sampling procedure. Before applying the random sampling model to a real study, it is necessary to ask the question: Can the data in this study reasonably be viewed *as if* they were obtained by random sampling from some population? The first step in answering this question is to define the population. As discussed in detail in Section 2.8, in defining the population one tries to identify those factors which are relevant to the observed variable Y. The next step is to scrutinize the procedure by which the observational units were selected and to ask: Could the *observations* have been chosen at random?

The most clear-cut kind of nonrandomness is **bias**, that is, a systematic tendency for some values of Y to be selected more readily than others. The following two examples illustrate biased selection.

EXAMPLE 3.2
LENGTHS OF FISH

A biologist plans to study the distribution of body length in a certain population of fish in the Chesapeake Bay. The sample will be collected using a fishing net. Smaller fish can more easily slip through the holes in the net. Thus, smaller fish are less likely to be caught than larger ones, so that the sampling procedure is biased.   ■

EXAMPLE 3.3
SIZES OF NERVE CELLS

A neuroanatomist plans to measure the sizes of individual nerve cells in cat brain tissue. In examining a tissue specimen, the investigator must decide which of the hundreds of cells in the specimen should be selected for measurement. Some of the nerve cells are incomplete because the microtome cut through them when the tissue was sectioned. If the size measurement can be made only on complete cells, a bias arises because the smaller cells had a greater chance of being missed by the microtome blade.   ■

When the sampling procedure is biased, the sample mean is a poor estimate of the population mean, because it is systematically distorted. For instance, in Example 3.2 smaller fish will tend to be underrepresented in the sample, so that the sample mean length will be an overestimate of the population mean length.

The following example illustrates a kind of nonrandomness that is different from bias.

EXAMPLE 3.4
SUCROSE IN BEET
ROOTS

An agronomist plans to sample beet roots from a field in order to measure their sucrose content. Suppose he were to take all his specimens from a randomly selected small area of the field. This sampling procedure would not be biased, but would tend to produce *too homogeneous* a sample, because environmental variation across the field would not be reflected in the sample. ∎

Example 3.4 illustrates an important principle that is sometimes overlooked in the analysis of data: In order to check applicability of the random sampling model, one needs to ask not only whether the sampling procedure might be biased, but also whether the sampling procedure will adequately reflect the variability inherent in the population. Faulty information about variability can distort scientific conclusions just as seriously as bias can.

We now consider some examples where the random sampling model might reasonably be applied.

EXAMPLE 3.5
NITRITE METABOLISM

To study the conversion of nitrite to nitrate in the blood, researchers injected four New Zealand White rabbits with a solution of radioactively labelled nitrite molecules. Ten minutes after injection, they measured for each rabbit the percentage of the nitrite that had been converted to nitrate.[1] Although the four animals were not literally chosen at random from a specified population, nevertheless it might be reasonable to view the measurements of nitrite metabolism as a random sample from similar measurements made on all New Zealand White rabbits. (This formulation assumes that age and sex are irrelevant to nitrite metabolism.) ∎

EXAMPLE 3.6
FUNGUS RESISTANCE
IN CORN

A certain variety of corn is resistant to fungus disease. To study the inheritance of this resistance, an agronomist crossed the resistant variety with a nonresistant variety and measured the degree of resistance in the progeny plants. The actual progeny in the experiment can be regarded as a random sample from a conceptual population of all *potential* progeny of that particular cross. ∎

When the purpose of a study is to *compare* two or more experimental conditions, a very narrow definition of the population may be satisfactory, as illustrated in the next example.

EXAMPLE 3.7
TREATMENT OF
ULCERATIVE COLITIS

A medical team conducted a study of two therapies, A and B, for treatment of ulcerative colitis. All the patients in the study were referral patients in a clinic in a large city. Each patient was observed for satisfactory "response" to therapy. In applying the random sampling model, the researchers might want to make an inference to the population of all ulcerative colitis patients in urban referral clinics. First consider inference about the actual probabilities of response; such an inference would be valid if the probability of response to each therapy is the same at all urban referral clinics. However, this assumption might be somewhat questionable, and the investigators might believe that the population should be defined very narrowly—for instance, as "the type of ulcerative colitis patients who are referred to *this* clinic." Even such a narrow population can be of interest in a comparative study. For instance, if treatment A is better than treatment B for the narrow population, it might be reasonable to infer that A would be better than B for a broader population (even if the actual response probabilities might be different in the broader population). In fact, it might even be argued that the broad population should include all ulcerative colitis patients, not merely those in urban referral clinics.   ∎

It often happens in research that, for practical reasons, the population actually studied is narrower than the population which is of real interest. In order to apply the kind of rationale illustrated in Example 3.7, one must argue that the results in the narrowly defined population (or, at least, some aspects of those results) can be meaningfully extrapolated to the population of interest. This extrapolation is not a *statistical* inference; it must be defended on biological, not statistical, grounds.

EXERCISES 3.1–3.2

3.1 (*Sampling exercise*)   Refer to the collection of 100 ellipses shown in the accompanying figure, which can be thought of as representing a natural population of the mythical organism *C. ellipticus*. The ellipses have been given identification numbers 00, 01, . . . , 99 for convenience in sampling. Certain individuals of *C. ellipticus* are mutants and have two tail bristles.

a. Use random digits (from Table 1 or your calculator) to choose a random sample of size 5 from the population and note the number of mutants in the sample.

b. Repeat part **a** nine more times, for a total of ten samples. (Some of the ten samples may overlap.)

To facilitate pooling of results from the entire class, report your results in the following format:

NUMBER OF MUTANTS	NONMUTANTS	FREQUENCY (NO. OF SAMPLES)
0	5	
1	4	
2	3	
3	2	
4	1	
5	0	
		Total: 10

3.2 (*Sampling exercise*)   Proceed as in Exercise 3.1, but obtain a total of 20 samples (each of size 5) rather than 10.

| | | | | | | | | | | |

SECTION 3.3

INTRODUCTION TO PROBABILITY

In this section we introduce the language of probability and its interpretation.

BASIC CONCEPTS

A **probability** is a numerical quantity that expresses the likelihood of an **event**. The probability of an event E is written as

$$\Pr\{E\}$$

The probability $\Pr\{E\}$ is always a number between 0 and 1, inclusive.

We can speak meaningfully about a probability $\Pr\{E\}$ only in the context of a **chance operation**—that is, an operation whose outcome is determined at least partially by chance. The chance operation must be defined in such a way that *each time the chance operation is performed, the event E either occurs or does not occur*. The following two examples illustrate these ideas.

EXAMPLE 3.8

COIN TOSSING

Consider the familiar chance operation of tossing a coin, and define the event

E: Heads

Each time the coin is tossed, either it falls heads or it does not. If the coin is equally likely to fall heads or tails, then

$$\Pr\{E\} = \tfrac{1}{2} = .5$$

Such an ideal coin is called a "fair" coin. If the coin is not fair (perhaps because it is slightly bent) then $\Pr\{E\}$ will be some value other than .5, for instance:

$$\Pr\{E\} = .6$$

∎

EXAMPLE 3.9

COIN TOSSING

Consider the event

E: 10 heads in a row

The chance operation "toss a coin" is not adequate for this event, because we cannot tell from one toss whether E has occurred. A chance operation that would be adequate is

Chance operation: Toss a coin 10 times.

Another chance operation that would be adequate is

Chance operation: Toss a coin 10,000 times,

with the understanding that E occurs if there is a run of 10 heads anywhere in the 10,000 tosses. Intuition suggests that E would be more likely with the second definition of the chance operation (10,000 tosses) than with the first (10 tosses). This intuition is correct, and serves to underscore the importance of the chance operation in interpreting a probability.

∎

The language of probability can be used to describe the results of random sampling from a population. The simplest application of this idea is a sample of size $n = 1$—that is, choosing one member at random from a population. The following is an illustration.

EXAMPLE 3.10
SAMPLING FRUITFLIES

A large population of the fruitfly *Drosophila melanogaster* is maintained in a lab. In the population, 30% of the individuals are black because of a mutation, while 70% of the individuals have the normal grey body color. Suppose one fly is chosen at random from the population. Then the probability that a black fly is chosen is .3. More formally, define

E: Sampled fly is black.

Then

$Pr\{E\} = .3$ ∎

The preceding example illustrates the basic relationship between probability and random sampling: *The probability that a randomly chosen individual has a certain characteristic is equal to the proportion of population members with the characteristic.*

FREQUENCY INTERPRETATION OF PROBABILITY

The **frequency interpretation** of probability provides a link between probability and the real world by relating the probability of an event to a measurable quantity, namely, the long-run relative frequency of occurrence of the event.*

According to the frequency interpretation, the probability of an event E is meaningful only in relation to a chance operation which can in principle be repeated indefinitely often. Each time the chance operation is repeated, the event E either occurs or does not occur. *The probability $Pr\{E\}$ is interpreted as the relative frequency of occurrence of E in an indefinitely long series of repetitions of the chance operation.*

Specifically, suppose that the chance operation is repeated a large number of times, and that for each repetition the occurrence or nonoccurrence of E is noted. Then we may write

$$Pr\{E\} \leftrightarrow \frac{\# \text{ of times } E \text{ occurs}}{\# \text{ of times chance operation is repeated}}$$

The arrow in the above expression indicates "approximate equality in the long run"; that is, if the chance operation is repeated many times, the two sides of the expression will be approximately equal. Here is a simple example.

*Some statisticians prefer a different view, namely that the probability of an event is a subjective quantity expressing a person's "degree of belief" that the event will happen. Statistical methods based on this "subjectivist" interpretation are rather different from those presented in this book.

EXAMPLE 3.11
COIN TOSSING

Consider again the chance operation of tossing a coin, and the event

  E: Heads

If the coin is fair, then

$$\Pr\{E\} = .5 \leftrightarrow \frac{\#\text{ of heads}}{\#\text{ of tosses}}$$

The arrow in the above expression indicates that, in a long series of tosses of a fair coin, we expect to get heads about 50% of the time. ■

The following two examples illustrate the relative frequency interpretation for more complex events.

EXAMPLE 3.12
COIN TOSSING

Suppose that a fair coin is tossed twice. For reasons that will become clear in Section 3.4, the probability of getting heads both times is .25. This probability has the following relative frequency interpretation.

  Chance operation: Toss a coin twice

  E: Both tosses are heads

$$\Pr\{E\} = .25 \leftrightarrow \frac{\#\text{ of times both tosses are heads}}{\#\text{ of pairs of tosses}}$$

    ■

EXAMPLE 3.13
SAMPLING FRUITFLIES

In the *Drosophila* population of Example 3.10, 30% of the flies are black and 70% are grey. Suppose that two flies are randomly chosen from the population. We will see in Section 3.4 that the probability that both flies are the same color is .58. This probability can be interpreted as follows:

  Chance operation: Choose a random sample of size $n = 2$

  E: Both flies in the sample are the same color

$$\Pr\{E\} = .58 \leftrightarrow \frac{\#\text{ of times both flies are same color}}{\#\text{ of times a sample of }n = 2\text{ is chosen}}$$

We can relate this interpretation to a concrete sampling experiment. Suppose that the *Drosophila* population is in a large container, and that we have some mechanism for choosing a fly at random from the container. We choose one fly at random, and then another; these two constitute the first sample of $n = 2$. After recording their colors, we put the two flies back into the container, and we are ready to repeat the sampling operation once again. Such a sampling experiment would be tedious to carry out physically, but it can be readily simulated using a computer. Table 3.2 shows a partial record of the results of choosing 10,000 random samples of size $n = 2$ from a simulated *Drosophila* population. After

TABLE 3.2

Partial Results of
Simulated Sampling from
a *Drosophila* Population

SAMPLE NUMBER	COLOR 1ST FLY	COLOR 2ND FLY	DID *E* OCCUR?	RELATIVE FREQUENCY OF *E* (CUMULATIVE)
1	G	B	No	.000
2	B	B	Yes	.500
3	B	G	No	.333
4	G	G	Yes	.500
5	G	G	Yes	.600
6	G	B	No	.500
7	B	G	No	.429
8	G	B	No	.375
9	G	G	Yes	.444
10	B	B	Yes	.500
⋮	⋮	⋮	⋮	⋮
20	B	B	Yes	.550
⋮	⋮	⋮	⋮	⋮
100	G	G	Yes	.670
⋮	⋮	⋮	⋮	⋮
1,000	G	B	No	.592
⋮	⋮	⋮	⋮	⋮
10,000	B	G	No	.582

each repetition of the chance operation (that is, after each sample of $n = 2$), the cumulative relative frequency of occurrence of the event *E* was updated, as shown in the rightmost column of the table. Figure 3.2 (page 86) shows the cumulative relative frequency plotted against the number of samples. Notice that, as the number of samples becomes large, the relative frequency of occurrence of *E* approaches .58 (which is $\Pr\{E\}$). In other words, the percentage of color-homogeneous samples among all the samples approaches 58% as the number of samples increases. It should be emphasized, however, that the *absolute* number of color-homogeneous samples generally does *not* tend to get closer to 58% of the total number. For instance, if we compare the results shown in Table 3.2 for the first 100 and the first 1,000 samples, we find the following:

	COLOR-HOMOGENEOUS		DEVIATION FROM 58% OF TOTAL		
First 100 samples:	67	or 67%	+9	or	+9%
First 1,000 samples:	592	or 59.2%	+12	or	+1.2%

FIGURE 3.2
Results of Sampling from
Fruitfly Population. Note
that the axes are scaled
differently in **(a)** and **(b)**.

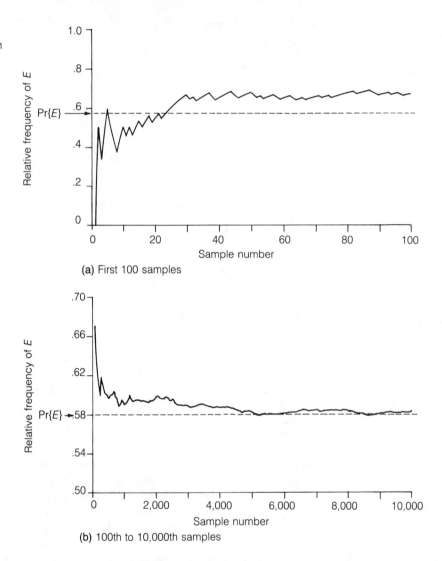

FIGURE 3.2
Results of Sampling from
Fruitfly Population. Note
that the axes are scaled
differently in **(a)** and **(b)**.

(a) First 100 samples

(b) 100th to 10,000th samples

Note that the deviation from 58% is larger in absolute terms, but smaller in
relative terms, for 1,000 samples than for 100 samples. The excess of 9 color-
homogeneous samples among the first 100 samples is not *cancelled* by a corresponding deficiency in later samples, but rather is swamped, or overwhelmed,
by a larger denominator.   ■

RANDOM VARIABLES

A **random variable** is simply a variable whose value depends on the outcome of
a chance operation. The following examples illustrate this idea.

EXAMPLE 3.14
DICE

Consider the chance operation of tossing a die. Let the random variable Y represent the number of spots showing. The possible values of Y are $Y = 1$, 2, 3, 4, 5, and 6. We do not know the value of Y until we have tossed the die. If we know how the die is weighted, then we can specify the probability that Y has a particular value, say $\Pr\{Y = 4\}$, or a particular set of values, say $\Pr\{2 \leq Y \leq 4\}$.* For instance, if the die is perfectly balanced so that each of the six faces is equally likely, then

$$\Pr\{Y = 4\} = \tfrac{1}{6} = .17$$

and

$$\Pr\{2 \leq Y \leq 4\} = \tfrac{3}{6} = .5 \qquad \blacksquare$$

EXAMPLE 3.15
FAMILY SIZE

Suppose a family is chosen at random from a certain population, and let the random variable Y denote the number of children in the chosen family. The possible values of Y are 0, 1, 2, 3, ... The probability that Y has a particular value is equal to the percentage of families with that many children. For instance, if 25% of the families have 2 children, then

$$\Pr\{Y = 2\} = .25 \qquad \blacksquare$$

EXAMPLE 3.16
HEIGHTS OF MEN

Let the random variable Y denote the height of a man chosen at random from a certain population. If we know the distribution of heights in the population, then we can specify the probability that Y has a certain value or falls within a certain interval of values. For instance, if 52% of the men are between 65 and 70 inches tall (inclusive), then

$$\Pr\{65 \leq Y \leq 70\} = .52 \qquad \blacksquare$$

The random variables in the preceding examples are fairly simple. In later chapters we will see that random variables can provide a convenient shorthand for describing even highly complex events.

| | | | | | | | | | | | | |

EXERCISES 3.3–3.8

3.3 (*Sampling exercise*) Consider a string of five randomly generated digits; that is, each digit is equally likely to be any of 0, 1, 2, . . . , 9, regardless of the other digits. Let E be the event that all five digits are different. It can be shown that $\Pr\{E\} = .30$. Use Table 1 (or your calculator) to generate 20 strings of 5 random digits each. Keep a record of your results, and tabulate the cumulative relative frequency of occurrence of E (as in Table 3.2). To facilitate pooling the results from the entire class, also report the total number of occurrences of E.

3.4 (*Sampling exercise*) Proceed as in Exercise 3.3, but generate 50 strings of 5 random digits each. Calculate the cumulative relative frequency of E only after every tenth string.

*The notation $2 \leq Y \leq 4$ means that Y is between 2 and 4, inclusive.

3.5 In a certain population of the freshwater sculpin, *Cottus rotheus*, the distribution of the number of tail vertebrae is as shown in the table.[2]

NO. OF VERTEBRAE	PERCENT OF FISH
20	3
21	51
22	40
23	6
Total	100

Let Y represent the number of tail vertebrae in a fish randomly chosen from the population. Find
a. $\Pr\{Y = 21\}$ **b.** $\Pr\{Y \le 22\}$

3.6 In a certain population of the European starling, there are 5,000 nests with young. The distribution of brood size (number of young in a nest) is given in the accompanying table.[3]

BROOD SIZE	FREQUENCY (NO. OF BROODS)
1	90
2	230
3	610
4	1,400
5	1,760
6	750
7	130
8	26
9	3
10	1
Total	5,000

Suppose one of the 5,000 broods is to be chosen at random, and let Y be the size of the chosen brood. Find
a. $\Pr\{Y = 3\}$ **b.** $\Pr\{Y \ge 7\}$ **c.** $\Pr\{4 \le Y \le 6\}$

3.7 In the starling population of Exercise 3.6, there are 22,435 young in all the broods taken together. Suppose one of the *young* is to be chosen at random, and let Y' be the size of the chosen individual's brood.
a. Find $\Pr\{Y' = 3\}$.
b. Find $\Pr\{Y' \ge 7\}$.
c. Explain why choosing a young at random and then observing its brood is not equivalent to choosing a brood at random. Your explanation should show why the answer to part **b** is greater than the answer to part **b** of Exercise 3.6.

3.8 In a certain population of the parasite *Trypanosoma*, the lengths of individuals are distributed as indicated by the density curve shown here. Areas under the curve are shown in the figure.[4]

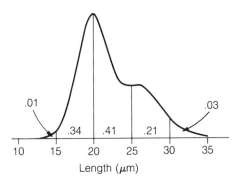

Let Y represent the length of an individual trypanosome chosen at random from the population. Find
a. $\Pr\{20 < Y < 30\}$ **b.** $\Pr\{Y > 20\}$ **c.** $\Pr\{Y < 20\}$

To add some depth to the notion of probability, we now consider simple manipulations of probabilities. A basic tool for such manipulations is the **binomial distribution**. The binomial distribution is a **probability distribution**—that is, a set of probabilities for the various possible events associated with a certain chance operation. The chance operation is defined in terms of a set of assumptions called the independent-trials model.

THE INDEPENDENT-TRIALS MODEL

The **independent-trials model** relates to a sequence of chance "trials." Each trial is assumed to have two possible outcomes, which are arbitrarily labelled "success" and "failure." The probability of success on each individual trial is denoted by the Greek letter π (pi), and is assumed to be constant from one trial to the next. In addition, the trials are assumed to be **independent**, which means that the chance of success or failure on each trial is independent of what happens on the other trials. The total number of trials is denoted by n. These assumptions are summarized in the following definition of the model.

INDEPENDENT-TRIALS MODEL

A series of n trials is conducted. Each trial results in success or failure. The probability of success is equal to the same quantity, π, for each trial, regardless of the outcomes of the other trials.

The following examples illustrate situations that can be described by the independent-trials model.

EXAMPLE 3.17
ALBINISM

If two carriers of the gene for albinism marry, each of their children has probability $\frac{1}{4}$ of being albino. The chance that the second child is albino is the same $(\frac{1}{4})$ whether or not the first child is albino; similarly, the outcome for the third child is independent of the first two, and so on. Using the labels "success" for albino and "failure" for non-albino, the independent-trials model applies with $\pi = \frac{1}{4}$ and $n = $ the number of children in the family. ∎

EXAMPLE 3.18
MUTANTS

Suppose that 39% of the individuals in a large population have a certain mutant trait and that a random sample of individuals is chosen from the population. As each individual is chosen for the sample, the probability is .39 that the chosen individual will be mutant. This probability is the same as each choice is made, regardless of the results of the other choices, because the percentage of mutants in the large population remains equal to .39 even when a few individuals have been removed. Using the labels "success" for mutant and "failure" for nonmutant, the independent-trials model applies with $\pi = .39$ and $n = $ the sample size. ∎

AN EXAMPLE OF THE BINOMIAL DISTRIBUTION

The binomial distribution specifies the probabilities of various numbers of successes and failures when the basic chance operation consists of n independent trials. Before giving the general formula for the binomial distribution, we consider a simple example.

EXAMPLE 3.19
ALBINISM

Suppose two carriers of the gene for albinism marry (see Example 3.17) and have two children. Then the probability that both of their children are albino is

$$\text{Pr\{both children are albino\}} = \left(\frac{1}{4}\right)\left(\frac{1}{4}\right) = \frac{1}{16}$$

The reason for this probability can be seen by considering the relative frequency interpretation of probability. Of a great many such families with two children, $\frac{1}{4}$ would have the first child albino; furthermore, $\frac{1}{4}$ *of these* would have the second child albino; thus, $\frac{1}{4}$ of $\frac{1}{4}$, or $\frac{1}{16}$ of all the couples would have both albino children. A similar kind of reasoning shows that the probability that both children are not albino is

$$\text{Pr\{both children are not albino\}} = \left(\frac{3}{4}\right)\left(\frac{3}{4}\right) = \frac{9}{16}$$

A new twist enters if we consider the probability that one child is albino and the other is not. There are two possible ways this can happen:

$$\text{Pr\{first child is albino, second is not\}} = \left(\frac{1}{4}\right)\left(\frac{3}{4}\right) = \frac{3}{16}$$

$$\text{Pr\{second child is albino, first is not\}} = \left(\frac{3}{4}\right)\left(\frac{1}{4}\right) = \frac{3}{16}$$

To see how to combine these possibilities, we again consider the relative frequency interpretation of probability. Of a great many such families with two children, the fraction of families with one albino and one non-albino child would be the total of the two possibilities, or

$$\frac{3}{16} + \frac{3}{16} = \frac{6}{16}$$

Thus, the corresponding probability is

$$\text{Pr\{one child is albino, the other is not\}} = \frac{6}{16}$$

These probabilities are collected in Table 3.3.

TABLE 3.3
Probability Distribution
for Number of Albino
Children

NUMBER OF		
ALBINO	NON-ALBINO	PROBABILITY
0	2	$\frac{9}{16}$
1	1	$\frac{6}{16}$
2	0	$\frac{1}{16}$
		1

The probability distribution in Table 3.3 is called the binomial distribution with $\pi = \frac{1}{4}$ and $n = 2$. Note that the probabilities add to 1. This makes sense because all possibilities have been accounted for: we expect $\frac{9}{16}$ of the families to have no albino children, $\frac{6}{16}$ to have one albino child, and $\frac{1}{16}$ to have two albino children; there are no other possible compositions for a two-child family. ∎

THE BINOMIAL DISTRIBUTION FORMULA

A general formula is available which can be used to calculate probabilities associated with the binomial distribution for any values of n and π. This formula can be proved using logic similar to that in Example 3.19. (The formula is discussed further in Appendix 3.2.) The formula is given in the accompanying box.

THE BINOMIAL DISTRIBUTION FORMULA

In the independent-trials model, the probability that the n trials result in j successes and $n - j$ failures is given by the following formula:

$$\text{Pr\{}j \text{ successes and } n - j \text{ failures\}} = C_j \pi^j (1 - \pi)^{n-j}$$

The quantity C_j appearing in the formula is called a **binomial coefficient**. Each binomial coefficient is an integer depending on j (and also on n). Values of binomial coefficients are given in Table 2 at the end of this book. For example, for $n = 5$ the binomial coefficients are as follows:

$$j: \quad 0 \quad 1 \quad 2 \quad 3 \quad 4 \quad 5$$
$$C_j: \quad 1 \quad 5 \quad 10 \quad 10 \quad 5 \quad 1$$

Thus, for $n = 5$ the binomial probabilities are as indicated in Table 3.4. Notice the pattern in Table 3.4: The powers of π ascend (0, 1, 2, 3, 4, 5) and the powers of $(1 - \pi)$ descend (5, 4, 3, 2, 1, 0). (In using the binomial distribution formula, remember that $x^0 = 1$ for any x.)

TABLE 3.4
Binomial Probabilities
for $n = 5$

NUMBER OF		
SUCCESSES j	FAILURES $n - j$	PROBABILITY
0	5	$1\pi^0 (1 - \pi)^5$
1	4	$5\pi^1 (1 - \pi)^4$
2	3	$10\pi^2 (1 - \pi)^3$
3	2	$10\pi^3 (1 - \pi)^2$
4	1	$5\pi^4 (1 - \pi)^1$
5	0	$1\pi^5 (1 - \pi)^0$

The following example shows a specific application of the binomial distribution with $n = 5$.

EXAMPLE 3.20

MUTANTS

Suppose we draw a random sample of five individuals from a large population in which 39% of the individuals are mutants (as in Example 3.18). The probabilities of the various possible samples are then given by the binomial distribution formula with $n = 5$ and $\pi = .39$; the results are displayed in Table 3.5. For instance, the probability of a sample containing 3 mutants and 2 nonmutants is

$$10(.39)^3(.61)^2 = .22$$

This means that about 22% of random samples of size 5 will contain three mutants and two nonmutants.

TABLE 3.5
Binomial Distribution with
$n = 5$ and $\pi = .39$

NUMBER OF		
MUTANTS	NONMUTANTS	PROBABILITY
0	5	.08
1	4	.27
2	3	.35
3	2	.22
4	1	.07
5	0	.01
		1.00

■

Notice that the probabilities in Table 3.3 add to 1. In fact, the probabilities in a probability distribution must always add to 1, because they account for 100% of the possibilities.

The binomial distribution of Table 3.5 is pictured graphically in Figure 3.3. Such a graphical display of a probability distribution is called a **probability histogram**.

FIGURE 3.3

Binomial Distribution with $n = 5$ and $\pi = .39$

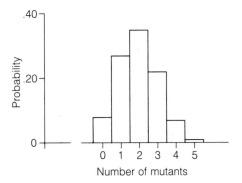

Remark In applying the independent-trials model and the binomial distribution, we assign the labels "success" and "failure" arbitrarily. For instance, in Example 3.18, we could say "success" = "mutant" and $\pi = .39$; or, alternatively, we could say "success" = "nonmutant" and $\pi = .61$. Either assignment of labels is all right; it is only necessary to be consistent.

Notes on Table 2 The following features in Table 2 are worth noting:

a. The first and last entries in each row are equal to 1. This will be true for any row; that is, $C_0 = 1$ and $C_n = 1$ for any value of n.

b. Each row of the table is symmetric; that is, C_j and C_{n-j} are equal.

c. The bottom rows of the table are left incomplete to save space, but you can easily complete them using the symmetry of the C_j's; if you need to know C_j you can look up C_{n-j} in Table 2. For instance, consider $n = 18$; if you want to know C_{15} you just look up C_3; both C_3 and C_{15} are equal to 816.

Computational note The binomial distribution formula can be computationally very inconvenient for values of n that are not small. Approximate methods are available for such cases. (One of these will be discussed in the optional Section 5.5.)

COMBINATION OF PROBABILITIES

If an event can happen in more than one way, the relative frequency interpretation of probability can be a guide to appropriate combination of the probabilities of subevents. The following example illustrates this idea.

EXAMPLE 3.21

SAMPLING FRUITFLIES

In a large *Drosophila* population, 30% of the flies are black (B) and 70% are grey (G). Suppose two flies are randomly chosen from the population (as in Example 3.13). The binomial distribution with $n = 2$ and $\pi = .3$ gives probabilities for the possible outcomes as shown in Table 3.6.

TABLE 3.6

SAMPLE COMPOSITION	PROBABILITY
Both G	.49
One B, one G	.42
Both B	.09
	1.00

Let E be the event that both flies are the same color. Then E can happen in two ways: both flies are grey or both are black. To find the probability of E, consider what would happen if we repeated the sampling procedure many times: 49% of the samples would have both flies grey, and 9% would have both flies black. Consequently, the percentage of samples with both flies the same color would be 49% + 9% = 58%. Thus, we have shown that the probability of E is

$$\Pr\{E\} = .58$$

as we claimed in Example 3.13.   ∎

Whenever an event E can happen in two or more mutually exclusive ways, a rationale such as that of Example 3.21 can be used to find $\Pr\{E\}$.

APPLICABILITY OF THE BINOMIAL DISTRIBUTION

A number of statistical procedures are based on the binomial distribution. We will study some of these procedures in later chapters. Of course, the binomial distribution is applicable only in experiments where the independent-trials model provides an adequate description of the real biological situation. We briefly discuss some aspects of this assumption.

APPLICATION TO SAMPLING   The most important application of the independent-trials model is to describe random sampling from a population when the observed variable is **dichotomous**—that is, a categorical variable with two categories (for instance, black and grey in Example 3.21). This application is valid if the sample size is a negligible fraction of the population size, so that the population composition is not altered appreciably by the removal of the individuals in the sample. However, if the sample is *not* a negligibly small part of the population, then the population composition may be altered by the sampling process, so that the "trials" involved in composing the sample are not independent, and the probabilities given by the binomial distribution are not correct. In most biological studies, the population is so large that this kind of difficulty does not arise.

CONTAGION   In some applications the phenomenon of contagion can invalidate the assumptions of the independent-trials model. The following is an example.

EXAMPLE 3.22
CHICKENPOX

Consider the occurrence of chickenpox in children. Each child in a family can be categorized according to whether he had chickenpox during a certain year. One can say that each child constitutes a "trial" and that "success" is having chickenpox during the year, but the trials are not independent because the chance of a particular child catching chickenpox depends on whether his sibling caught chickenpox. As a specific example, consider a family with five children, and suppose that the chance of an individual child catching chickenpox during the year is equal to .10. The binomial distribution gives the chance of all five children getting chickenpox as

$$\Pr\{5 \text{ children get chickenpox}\} = (.10)^5 = .00001$$

However, this answer is not correct; because of contagion, the correct probability would be much larger. ■

EXERCISES 3.9–3.15

3.9 The seeds of the garden pea (*Pisum sativum*) are either yellow or green. A certain cross between pea plants produces progeny in the ratio 3 yellow:1 green.[5] If four randomly chosen progeny of such a cross are examined, what is the probability that
a. three are yellow and one is green?
b. all four are yellow?
c. all four are the same color?

3.10 A certain drug treatment cures 90% of cases of hookworm in children.[6] Suppose that 20 children suffering from hookworm are to be treated, and that the children can be regarded as a random sample from the population. Find the probability that
a. all 20 will be cured.
b. all but one will be cured.
c. exactly 90% will be cured.

3.11 The shell of the land snail *Limocolaria martensiana* has two possible color forms: streaked and pallid. In a certain population of these snails, 60% of the individuals have streaked shells.[7] Suppose that a random sample of 10 snails is to be chosen from this population. Find the probability that the percentage of streaked-shelled snails in the *sample* will be
a. 50%
b. 60%
c. 70%

3.12 The sex ratio of newborn human infants is about 105 males:100 females.[8] If four infants are chosen at random, what is the probability that
a. two are male and two are female?
b. all four are male?
c. all four are the same sex?

3.13 Neuroblastoma is a rare, serious but treatable disease. A urine test, the VMA test, has been developed which gives a positive diagnosis in about 70% of cases of neuroblastoma.[9] It has been proposed that this test be used for large-scale screening of children. Assume that 300,000 children are to be tested, of whom 8 have the disease.

Find the probability that

a. all 8 cases will be detected.

b. only one case will be missed.

c. two or more cases will be missed. [*Hint:*   Use parts **a** and **b** to answer part **c**.]

3.14 If two carriers of the gene for albinism marry, each of their children has probability $\frac{1}{4}$ of being albino (see Example 3.17). If such a couple has six children, what is the probability that

a. none will be albino?

b. at least one will be albino? [*Hint:*   Use part **a** to answer part **b**; note that "at least one" means "one or more."]

3.15 Childhood lead poisoning is a public health concern in the United States. In a certain population, one child in eight has a high blood lead level (defined as 30 μg/dl or more).[10] In a randomly chosen group of 16 children from the population, what is the probability that

a. none has high blood lead?

b. one has high blood lead?

c. two have high blood lead?

d. three or more have high blood lead? [*Hint:*   Use parts **a–c** to answer part **d**.]

| | | | | | | | | | | | |

S E C T I O N 3.5

FITTING A BINOMIAL DISTRIBUTION TO DATA (OPTIONAL)

Occasionally it is possible to obtain data which permit a direct check of the applicability of the binomial distribution. One such case is described in the next example.

EXAMPLE 3.23

SEXES OF CHILDREN

In a classic study of the human sex ratio, families were categorized according to the sexes of the children. The data were collected in Germany in the 19th century, when large families were common. Table 3.7 shows the results for 6,115 families with 12 children.[11]

TABLE 3.7

Sex Ratios in 6,115 Families with 12 Children

NUMBER OF BOYS	GIRLS	OBSERVED FREQUENCY (NUMBER OF FAMILIES)
0	12	3
1	11	24
2	10	104
3	9	286
4	8	670
5	7	1,033
6	6	1,343
7	5	1,112
8	4	829
9	3	478
10	2	181
11	1	45
12	0	7
		6,115

It is interesting to consider whether the observed variation among families can be explained by the independent-trials model. We will explore this question by fitting a binomial distribution to the data.

The first step in fitting the binomial distribution is to determine a value for $\pi = \Pr\{\text{boy}\}$. One possibility would be to assume that $\pi = .50$. However, since it is known that the human sex ratio at birth is not exactly $1:1$ (in fact, it favors boys slightly), we will not make this assumption. Rather, we will "fit" π to the data; that is, we will determine a value for π which fits the data best. We observe that the total number of children in all the families is

$$(12)(6,115) = 73,380 \text{ children}$$

Among these children, the number of boys is

$$(3)(0) + (24)(1) + \cdots + (12)(7) = 38,100 \text{ boys}$$

Therefore, the value of π which fits the data best is

$$\pi = \frac{38,100}{73,380} = .519215$$

The next step is to compute probabilities from the binomial distribution formula with $n = 12$ and $\pi = .519215$. For instance, the probability of 3 boys and 9 girls is computed as

$$C_3(\pi)^3(1 - \pi)^9 = 220(.519215)^3(.480785)^9$$
$$= .0422691$$

For comparison with the observed data, we convert each probability to a theoretical or "expected" frequency by multiplying by 6,115 (the total number of families). For instance, the expected number of families with 3 boys and 9 girls is

$$(6,115)(.0422691) = 258.5$$

The expected and observed frequencies are displayed together in Table 3.8 (page 98). Table 3.8 shows reasonable agreement between the observed frequencies and the predictions of the binomial distribution. But a closer look reveals that the discrepancies, although not large, follow a definite pattern. The data contain more unisexual, or preponderantly unisexual, sibships than expected. In fact, the observed frequencies are higher than the expected frequencies for nine types of families in which one sex or the other predominates, while the observed frequencies are lower than the expected frequencies for four types of more "balanced" families. This pattern is clearly revealed by the last column of Table 3.8, which shows the sign of the difference between the observed frequency and the expected frequency. Thus, the observed distribution of sex ratios has heavier "tails" and a lighter "middle" than the best-fitting binomial distribution.

The systematic pattern of deviations from the binomial distribution suggests that the observed variation among families cannot be entirely explained by the independent-trials model. What factors might account for the discrepancy? This

TABLE 3.8
Sex-Ratio Data and Binomial Expected Frequencies

NUMBER OF BOYS	GIRLS	OBSERVED FREQUENCY	EXPECTED FREQUENCY	SIGN OF (OBS. − EXP.)
0	12	3	.9	+
1	11	24	12.1	+
2	10	104	71.8	+
3	9	286	258.5	+
4	8	670	628.1	+
5	7	1,033	1,085.2	−
6	6	1,343	1,367.3	−
7	5	1,112	1,265.6	−
8	4	829	854.3	−
9	3	478	410.0	+
10	2	181	132.8	+
11	1	45	26.1	+
12	0	7	2.3	+
		6,115	6,115.0	

intriguing question has stimulated several researchers to undertake more detailed analysis of these data. We briefly discuss some of the issues.

One explanation for the excess of predominantly unisexual families is that the probability of producing a boy may vary among families. If π varies from one family to another, then sex will appear to "run" in families in the sense that the number of predominantly unisexual families will be inflated. In order to clearly visualize this effect, consider the fictitious data set shown in Table 3.9.

TABLE 3.9
Fictitious Sex-Ratio Data and Binomial Expected Frequencies

NUMBER OF BOYS	GIRLS	OBSERVED FREQUENCY	EXPECTED FREQUENCY	SIGN OF (OBS. − EXP.)
0	12	2,940	.9	+
1	11	0	12.1	−
2	10	0	71.8	−
3	9	0	258.5	−
4	8	0	628.1	−
5	7	0	1,085.2	−
6	6	0	1,367.3	−
7	5	0	1,265.6	−
8	4	0	854.3	−
9	3	0	410.0	−
10	2	0	132.8	−
11	1	0	26.1	−
12	0	3,175	2.3	+
		6,115	6,115.0	

In the fictitious data set, there are $(3,175)(12) = 38,100$ males among 73,380 children, just as there are in the real data set. Consequently, the best-fitting π is the same ($\pi = .519215$) and the expected binomial frequencies are the same as in Table 3.8. The fictitious data set contains *only* unisexual sibships, and so is an extreme example of sex "running" in families. The real data set exhibits the same phenomenon more weakly. One explanation of the fictitious data set would be that some families can have only boys ($\pi = 1$) and other families can have only girls ($\pi = 0$). In a parallel way, one explanation of the real data set would be that π varies slightly among families. Variation in π is biologically plausible, even though the mechanism causing the variation has not yet been discovered.

An alternative explanation for the inflated number of sexually homogeneous families would be that the sexes of the children in a family are literally dependent on one another, in the sense that the determination of an individual child's sex is somehow influenced by the sexes of the previous children. This explanation is implausible on biological grounds because it is difficult to imagine how the biological system could "remember" the sexes of previous offspring.

In addition to between-family variation of π, other factors have been suggested as explanations for various features of the family sex-ratio data. These include twinning, variation of π within families as well as among families, and family-planning practices (the decision to stop having children might be influenced by the sexes of the children already born). It is also relevant to ask how much of the variability of the data might be attributed to chance effects (sampling error). Thorough exploration of these issues would require much further analysis of these data, as well as similar data on families of other sizes. ∎

Example 3.23 shows that poorness of fit to the independent-trials model can be biologically interesting. We should emphasize, however, that most statistical applications of the binomial distribution proceed from the assumption that the independent-trials model *is* applicable. In a typical application, the data are regarded as resulting from a *single* set of n trials. Data such as the family sex-ratio data, which refer to *many* sets of $n = 12$ trials, are not often encountered.

EXERCISES 3.16–3.18

3.16 The accompanying data on families with 6 children are taken from the same study as the families with 12 children in Example 3.23. Fit a binomial distribution to the data. (Round the expected frequencies to one decimal place.) Compare with the results in Example 3.23; what features do the two data sets share?

NUMBER OF BOYS	NUMBER OF GIRLS	NUMBER OF FAMILIES
0	6	1,096
1	5	6,233
2	4	15,700
3	3	22,221
4	2	17,332
5	1	7,908
6	0	1,579
		72,069

3.17 An important method for studying mutation-causing substances involves killing female mice 17 days after mating and examining their uteri for living and dead embryos. The classical method of analysis of such data assumes that the survival or death of each embryo constitutes an independent binomial trial. The accompanying table, which is extracted from a larger study, gives data for 310 females, all of whose uteri contained 9 embryos; all of the animals were treated alike (as controls).[12]

NUMBER OF EMBRYOS		NUMBER OF FEMALE MICE
Dead	Living	
0	9	136
1	8	103
2	7	50
3	6	13
4	5	6
5	4	1
6	3	1
7	2	0
8	1	0
9	0	0
		310

a. Fit a binomial distribution to the observed data. (Round the expected frequencies to one decimal place.)

b. Interpret the relationship between the observed and expected frequencies. Do the data cast suspicion on the classical assumption?

3.18 Students in a large botany class conducted an experiment on the germination of seeds of the Saguaro cactus. As part of the experiment, each student planted five seeds in a small cup, kept the cup near a window, and checked every day for germination (sprouting). The class results on the seventh day after planting were as displayed in the table.[13]

NUMBER OF SEEDS		NUMBER OF STUDENTS
Germinated	Not Germinated	
0	5	17
1	4	53
2	3	94
3	2	79
4	1	33
5	0	4
		280

a. Fit a binomial distribution to the data. (Round the expected frequencies to one decimal place.)

b. Two students, Fran and Bob, were talking before class. All of Fran's seeds had germinated by the seventh day, whereas none of Bob's had. Bob wondered whether he had done something wrong. With the perspective gained from seeing all 280 students' results, what would you say to Bob? [*Hint:* Can the variation among the students be explained by the hypothesis that some of the seeds were good and some were poor, with each student receiving a randomly chosen five seeds?]

c. Invent a fictitious set of data for 280 students, with the same overall percentage germination as the observed data given in the table, but with all the students getting either Fran's results (perfect) or Bob's results (nothing). How would your answer to Bob differ if the actual data had looked like this fictitious data set?

| | | | | | | | | | | | | |

SUPPLEMENTARY EXERCISES 3.19–3.27

3.19 In a certain population, 83% of the people have Rh-positive blood type.[14] Suppose two randomly chosen people from the population are to marry. Find the probability that

a. both partners are Rh-positive.

b. the partners are of opposite blood types.

3.20 In preparation for an ecological study of centipedes, the floor of a beech woods is divided into a large number of one-foot squares.[15] At a certain moment, the distribution of centipedes in the squares is as shown in the table.

NUMBER OF CENTIPEDES	PERCENT FREQUENCY (% OF SQUARES)
0	45
1	36
2	14
3	4
4	1
	100

Suppose that a square is chosen at random, and let Y be the number of centipedes in the chosen square. Find

a. $\Pr\{Y = 1\}$ **b.** $\Pr\{Y \geq 2\}$

3.21 Refer to the distribution of centipedes given in Exercise 3.20. Suppose five squares are chosen at random. Find the probability that three of the squares contain centipedes and two do not.

3.22 Wavy hair in mice is a recessive genetic trait. If mice with wavy hair are mated with straight-haired (heterozygous) mice, each offspring has probability $\frac{1}{2}$ of having wavy hair.[16] Consider a large number of such matings, each producing a litter of five offspring. What percentage of the litters will consist of

a. two wavy-haired and three straight-haired offspring?

b. all the same type (either all wavy- or all straight-haired) offspring?

3.23 A certain drug causes kidney damage in 1% of patients. Suppose the drug is to be tested on 50 patients. Find the probability that

a. none of the patients will experience kidney damage.

b. one or more of the patients will experience kidney damage. [*Hint:* Use part **a** to answer part **b**.]

3.24 Refer to Exercise 3.23. Suppose now that the drug is to be tested on n patients, and let E represent the event that kidney damage occurs in one or more of the patients. The probability $\Pr\{E\}$ is useful in establishing criteria for drug safety.
a. Find $\Pr\{E\}$ for $n = 100$.
b. How large must n be in order for $\Pr\{E\}$ to exceed .95? [*Hint:* First set up an equation for n, and then solve it either by trial and error or by using logarithms.]

3.25 To study people's ability to deceive lie detectors, researchers sometimes use the "guilty knowledge" technique.[17] Certain subjects memorize six common words; other subjects memorize no words. Each subject is then tested on a polygraph machine (lie detector), as follows. The experimenter reads, in random order, 24 words: the six "critical" words (the memorized list) and, for each critical word, three "control" words with similar or related meanings. If the subject has memorized the six words, he tries to conceal that fact. The subject is scored a "failure" on a critical word if his electrodermal response is higher on the critical word than on any of the three control words. Thus, on each of the six critical words, even an innocent subject would have a 25% chance of failing. Suppose a subject is labelled "guilty" if he fails on four or more of the six critical words. If an innocent subject is tested, what is the probability that he will be labelled "guilty"?

3.26 The density curve shown here represents the distribution of systolic blood pressures in a population of middle-aged men.[18] Areas under the curve are shown in the figure. Suppose a man is selected at random from the population, and let Y be his blood pressure. Find
a. $\Pr\{120 < Y < 160\}$
b. $\Pr\{Y < 120\}$
c. $\Pr\{Y > 140\}$

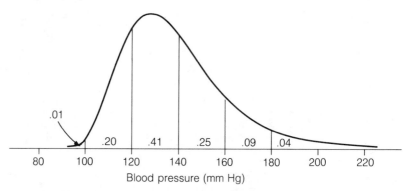

Blood pressure (mm Hg)

3.27 Refer to the blood pressure distribution of Exercise 3.26. Suppose four men are selected at random from the population. Find the probability that
a. all four have blood pressures higher than 140 mm Hg.
b. three have blood pressures higher than 140, and one has blood pressure 140 or less.

C H A P T E R 4

CONTENTS

THE NORMAL DISTRIBUTION

S E C T I O N 4.1

INTRODUCTION

In Chapter 2 we introduced the idea of regarding a set of data as a sample from a population. For a quantitative variable Y, we saw in Section 2.9 that the population distribution of Y can be described by its mean μ and its standard deviation σ, and also by a density curve, which represents relative frequencies as areas under the curve. In this chapter we study the most important type of density curve: the **normal curve**. The normal curve is a symmetric "bell-shaped" curve, whose exact form we will describe below. A distribution represented by a normal curve is called a **normal distribution**.

The normal distribution plays two roles in statistical applications. Its more straightforward use is as a convenient approximation to the distribution of an observed variable Y. The second role of the normal distribution is more theoretical and will be explored in Chapter 5.

AN EXAMPLE

Here is an example of a natural population distribution that can be approximated by a normal distribution.

EXAMPLE 4.1
SERUM CHOLESTEROL

The relationship between the concentration of cholesterol in the blood and the occurrence of heart disease has been the subject of much speculation and research. As part of a government health survey, researchers measured serum cholesterol levels for a large sample of Americans. The distribution for 14-year-old boys can be fairly well approximated by a normal curve with mean $\mu = 170$ mg/dl and standard deviation $\sigma = 30$ mg/dl. Figure 4.1 shows a histogram based on a sample of 615 boys, with the normal curve superimposed.[1]

FIGURE 4.1
Distribution of Serum
Cholesterol in 14-Year-Old
Boys

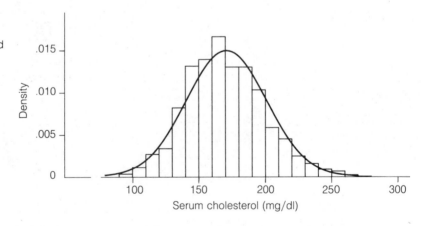

To indicate how the mean μ and standard deviation σ relate to the normal curve, Figure 4.2 shows the normal curve for the serum cholesterol distribution of Example 4.1, with tick marks at 1, 2, and 3 standard deviations from the mean.

FIGURE 4.2

Normal Distribution of
Serum Cholesterol, with
μ = 170 mg/dl and
σ = 30 mg/dl

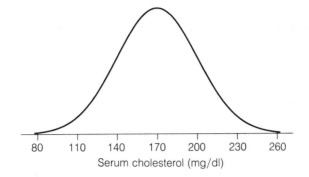

Serum cholesterol (mg/dl)

The normal curve can be used to describe the distribution of an observed variable Y in two ways: (1) as a smooth approximation to a histogram based on a sample of Y values; and (2) as an idealized representation of the population distribution of Y. The normal curve in Figure 4.1 could be interpreted either way. For simplicity, in the remainder of this chapter we will consider the normal curve as representing a population distribution.

FURTHER EXAMPLES

We now give three more examples of normal curves that approximately describe real populations. In each figure, the horizontal axis is scaled with tick marks centered at the mean, and 1 standard deviation apart.

EXAMPLE 4.2

EGGSHELL THICKNESS

In the commercial production of eggs, breakage is a major problem. Consequently, the thickness of the eggshell is an important variable. In one study, the shell thicknesses of the eggs produced by a large flock of White Leghorn hens were observed to follow approximately a normal distribution with mean μ = .38 mm, and standard deviation σ = .03 mm. This distribution is pictured in Figure 4.3.[2]

FIGURE 4.3

Normal Distribution of
Eggshell Thickness,
with μ = .38 mm and
σ = .03 mm

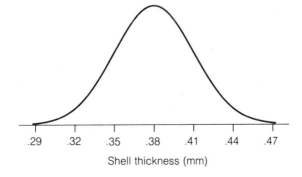

Shell thickness (mm)

EXAMPLE 4.3

INTERSPIKE TIMES IN
NERVE CELLS

In certain nerve cells, spontaneous electrical discharges are observed which are so rhythmically repetitive that they are called "clock-spikes." The timing of these spikes, even though remarkably regular, does exhibit variation. In one study, the interspike-time intervals (in milliseconds) for a single housefly (*Musca domestica*) were observed to follow approximately a normal distribution with mean $\mu = 15.6$ ms and standard deviation $\sigma = .4$ ms; this distribution is shown in Figure 4.4.[3]

FIGURE 4.4

Normal Distribution of Interspike-Time Intervals, with $\mu = 15.6$ ms and $\sigma = .4$ ms

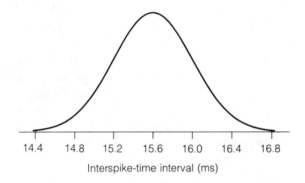

14.4   14.8   15.2   15.6   16.0   16.4   16.8

Interspike-time interval (ms)

The preceding examples have illustrated very different kinds of populations. In Example 4.3, the entire population consists of measurements on only one fly. Still another type of population is a *measurement error* population, consisting of repeated measurements of exactly the same quantity. The deviation of an individual measurement from the "correct" value is called **measurement error**. Measurement error is not the result of a mistake, but rather is due to lack of perfect precision in the measuring process or measuring instrument. Measurement error distributions are often approximately normal; in this case the mean of the distribution of repeated measurements of the same quantity is the true value of the quantity (assuming that the measuring instrument is correctly calibrated), and the standard deviation of the distribution indicates the precision of the instrument. One measurement error distribution was described in Example 2.11. The following is another example.

EXAMPLE 4.4

MEASUREMENT ERROR

When a certain electronic instrument is used for counting particles such as white blood cells, the measurement error distribution is approximately normal. For white blood cells, the standard deviation of repeated counts based on the same blood specimen is about 1.4% of the true count. Thus, if the true count of a certain blood specimen were 7,000 cells/mm³, then the standard deviation would be about 100 cells/mm³, and the distribution of repeated counts on that specimen would resemble Figure 4.5.[4]

FIGURE 4.5
Normal Distribution of
Repeated White Blood
Cell Counts of a Blood
Specimen Whose True
Value Is $\mu = 7{,}000$
cells/mm³. The standard
deviation is $\sigma = 100$
cells/mm³.

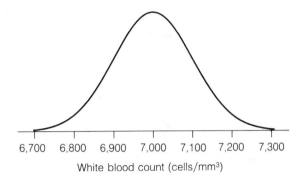

6,700   6,800   6,900   7,000   7,100   7,200   7,300

White blood count (cells/mm³)

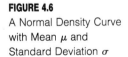

SECTION 4.2

THE NORMAL CURVES

As the examples in Section 4.1 show, there are many normal curves; each particular normal curve is characterized by its mean and standard deviation. All of the normal curves can be described by a single formula. Even though we will not make any direct use of the formula in this book, we present it here, both as a matter of interest and also to emphasize that a normal curve is not just any symmetric curve but rather a *specific* kind of symmetric curve.

If a variable Y follows a normal distribution with mean μ and standard deviation σ, then the density curve of the distribution of Y is given by the following formula:

$$\text{Density} = \frac{C}{\sigma} \cdot e^{-\frac{1}{2}\left(\frac{y-\mu}{\sigma}\right)^2}$$

This formula expresses the density as a function of the position y along the Y-axis. The quantities C and e that appear in the formula are constants; their values are approximately $C \approx .40$ and $e \approx 2.7$.

Figure 4.6 shows a graph of a normal curve. The shape of the curve is like a symmetric bell, centered at $y = \mu$. The direction of curvature is downward (like an inverted bowl) in the central portion of the curve, and upward in the tail portions. The points where the curvature changes direction are $y = \mu - \sigma$ and $y = \mu + \sigma$; notice that the curve is almost linear near these points. In principle the curve extends to $+\infty$ and $-\infty$, never actually reaching the Y-axis; however,

FIGURE 4.6
A Normal Density Curve
with Mean μ and
Standard Deviation σ

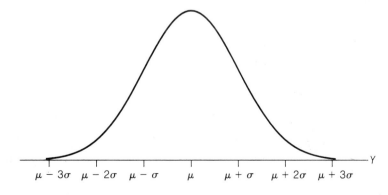

$\mu - 3\sigma$ $\mu - 2\sigma$ $\mu - \sigma$ μ $\mu + \sigma$ $\mu + 2\sigma$ $\mu + 3\sigma$

the height of the curve is very small for y values more than 3 standard deviations from the mean. The area under the curve is exactly equal to 1. [*Note:* It may seem paradoxical that a curve can enclose a finite area, even though it never descends to touch the Y-axis. This apparent paradox is clarified in Appendix 4.1.]

All normal curves have the same essential shape, in the sense that they can be made to look identical by suitable choice of the vertical and horizontal scales for each. (For instance, notice that the curves in Figures 4.2–4.5 look identical.) But normal curves with different values of μ and σ will not look identical if they are all plotted to the same scale, as illustrated by Figure 4.7. The location of the normal curve along the Y-axis is governed by μ, since the curve is centered at $y = \mu$; the width of the curve is governed by σ. The height of the curve is also determined by σ: since the area under each curve must be equal to 1, a curve with a smaller value of σ must be taller. This reflects the fact that the values of Y are more highly concentrated near the mean when the standard deviation is smaller.

FIGURE 4.7

Three Normal Curves with Different Means and Standard Deviations

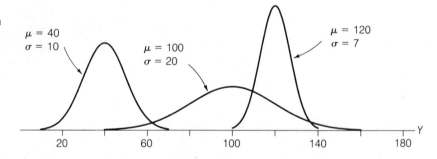

As explained in Section 2.9, a density curve can be quantitatively interpreted in terms of areas under the curve. While areas can be roughly estimated by eye, for some purposes it is desirable to have fairly precise information about areas.

S E C T I O N 4.3

AREAS UNDER A NORMAL CURVE

THE STANDARDIZED SCALE

The areas under a normal curve have been computed mathematically and tabulated for practical use. The use of this tabulated information is much simplified by the fact that all normal curves can be made equivalent with respect to areas under them by suitable rescaling of the horizontal axis. The rescaled variable is denoted by Z; the relationship between the two scales is shown in Figure 4.8.

As Figure 4.8 indicates, the Z scale measures standard deviations from the mean: $z = 1.0$ corresponds to 1.0 standard deviation above the mean, $z = -2.5$ corresponds to 2.5 standard deviations below the mean, etc. The Z scale is referred to as a **standardized scale**.

FIGURE 4.8
A Normal Curve, Showing
the Relationship Between
the Natural Scale (Y) and
the Standardized Scale
(Z)

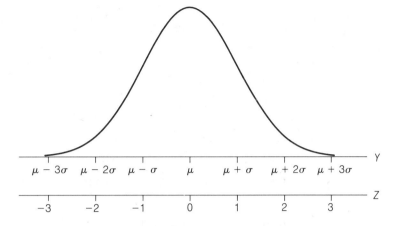

The correspondence between the Z scale and the Y scale can be expressed by
the formula given in the box.

STANDARDIZATION FORMULA

$$z = \frac{y - \mu}{\sigma}$$

Table 3 at the end of this book gives areas under a normal curve, with distances
along the horizontal axis measured in the Z scale. Each area tabled in Table 3 is
the area under a normal curve between $z = 0$ and a specified value of z. For
example, for $z = 1.00$ the tabled area is .3413; this area is shaded in Figure 4.9.

FIGURE 4.9
Illustration of the Use
of Table 3

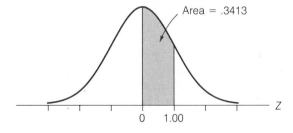

Area = .3413

The usefulness of the Z scale can be illustrated by considering the area under
the normal curve between $z = -1$ and $z = +1$. Since the normal curve is
symmetric, it follows from the result in Figure 4.9 that the area between $z = -1$ and $z = +1$ is $2(.3413) = .6826$; this area is indicated in Figure 4.10 (page
110). Thus, for any normal distribution, about 68% of the observations are within
± 1 standard deviation of the mean; for example, about 68% of the serum cholesterol values in the idealized distribution of Figure 4.2 are between 140 mg/dl
and 200 mg/dl.

FIGURE 4.10
Area Under a Normal
Curve Between $z = -1$
and $z = 1$

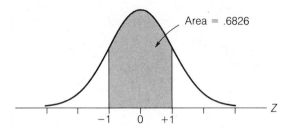

Pursuing this idea further, if we look up $z = 2.00$ and $z = 3.00$ in Table 3, we find areas of .4772 and .4987. Doubling these, we can summarize as follows:

If the variable Y follows a normal distribution, then

about 68% of the y's are within ± 1 SD of the mean;

about 95% of the y's are within ± 2 SDs of the mean;

about 99.7% of the y's are within ± 3 SDs of the mean.

These statements provide a very definite interpretation of the standard deviation in cases where a distribution is approximately normal. (In fact, the statements are often approximately true for fairly nonnormal distributions; that is why, in Section 2.6, these percentages—68%, 95%, and >99%—were described as "typical" for "nicely shaped" distributions.)

DETERMINING AREAS FOR A NORMAL CURVE

By taking advantage of the standardized scale, we can use Table 3 to answer detailed questions about any normal population, when the population mean and standard deviation are specified. The following example illustrates the use of Table 3. Of course, the population described in the example is an idealized one, since no actual population follows a normal distribution exactly.

EXAMPLE 4.5
LENGTHS OF FISH

In a certain population of the herring *Pomolobus aestivalis*, the lengths of the individual fish follow a normal distribution. The mean length of the fish is 54.0 mm, and the standard deviation is 4.5 mm.[5] We will use Table 3 to answer various questions about the population.

a. What percentage of the fish are between 54 and 60 mm long?

Figure 4.11 shows the population density curve, with the desired area indicated by shading. In order to use Table 3, we convert the limits of the area from the Y scale to the Z scale, as follows:

For $y = 54$, the value of z is

$$z = \frac{y - \mu}{\sigma} = 0$$

FIGURE 4.11
Area Under the Normal
Curve in Example 4.5(a)

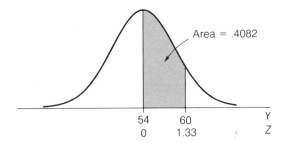

54   60   Y
0   1.33   Z

Area = .4082

For $y = 60$, the value of z is

$$z = \frac{y - \mu}{\sigma} = \frac{60 - 54}{4.5} = 1.33$$

Looking up $z = 1.33$ in Table 3, we find that the area is .4082; thus, 40.82% of the fish are between 54 and 60 mm long.

b. What percentage of the fish are more than 48 mm long?

As indicated in Figure 4.12, the desired area is the sum of two parts. The area of the shaded region to the left of $y = 54$ is equal to .4082, because this region is the symmetric (mirror) image of the shaded region in Figure 4.11. The area of the region to the right of $y = 54$ is exactly equal to .5, because the area under the entire curve is equal to 1. The total shaded area is then equal to the sum

.5000 + .4082 = .9082

Thus, 90.82% of the fish are more than 48 mm long.

FIGURE 4.12
Area Under the Normal
Curve in Example 4.5(b)

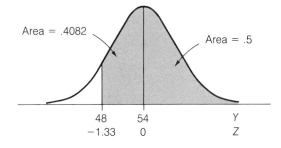

Area = .4082

Area = .5

48   54   Y
-1.33 0   Z

c. What percentage of the fish are between 50 and 60 mm long?

Figure 4.13 (page 112) shows the desired area as the sum of two parts. The area of the right-hand region is .4082 (as in part **a** above). To find the left-hand area, we note that the value of z corresponding to $y = 50$ is

$$z = \frac{y - \mu}{\sigma} = \frac{50 - 54}{4.5} = -.89$$

FIGURE 4.13

Area Under the Normal
Curve in Example 4.5(c)

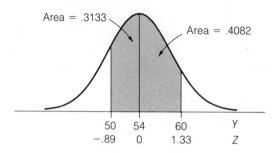

Again taking advantage of the symmetry of the normal curve, we can find this area by looking in Table 3 under $z = +.89$; the area is .3133. Consequently, the total shaded area is equal to the sum

$.4082 + .3133 = .7215$

Thus, 72.15% of the fish are between 50 and 60 mm long.

d. What percentage of the fish are between 58 and 60 mm long?

Figure 4.14 indicates the desired area. This area can be expressed as a *difference* (rather than as a sum) of two areas found from Table 3. The area from $y = 0$ to $y = 58$ is .3133, and the area from $y = 0$ to $y = 60$ is .4082; consequently, the desired area is computed as

$.4082 - .3133 = .0949$

Thus, 9.49% of the fish are between 58 and 60 mm long.

FIGURE 4.14

Area Under the Normal
Curve in Example 4.5(d)

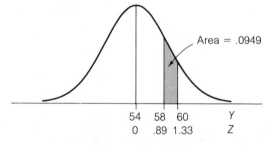

Each of the percentages found in Example 4.5 can also be interpreted in terms of probability. Let the random variable Y represent the length of a fish randomly chosen from the population. Then the results in Example 4.5 imply that

$\Pr\{54 < Y < 60\} = .4082$

$\Pr\{Y > 48\} = .9082$

and so on. Thus, the normal distribution can be interpreted as a continuous probability distribution.

Note that because the idealized normal distribution is perfectly continuous, probabilities such as

$\Pr\{Y > 48\}$ and $\Pr\{Y \geq 48\}$

are equal (see Section 2.9). If, however, the length were measured only to the nearest mm, then the measured variable would actually be discrete, and the above two probabilities would differ somewhat from each other. In cases where this discrepancy is important, the computation can be refined to take into account the discontinuity of the measured distribution (see the optional Section 4.4).

INVERSE READING OF TABLE 3

In determining facts about a normal distribution, it is sometimes necessary to read Table 3 in an "inverse" way—that is, to find the value of z corresponding to a given area rather than the other way around. This situation arises, for example, when we want to determine a *percentile* of a normal distribution. The **percentiles** of a distribution divide the distribution into 100 equal parts, just as the quartiles divide it into four equal parts [from the Latin roots *centum* (hundred) and *quartus* (fourth)]. We illustrate by determining a percentile of the fish length distribution.

EXAMPLE 4.6
LENGTHS OF FISH

Suppose we want to find the 70th percentile of the fish length distribution of Example 4.5. Let us denote the 70th percentile by y^*. By definition, y^* is the value such that 70% of the fish lengths are less than y^* and 30% are greater, as illustrated in Figure 4.15.

FIGURE 4.15
Determining the 70th
Percentile of a Normal
Distribution, Example 4.6

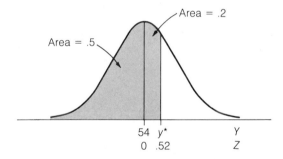

In order to find y^*, we first determine the 70th percentile in the Z scale; as Figure 4.15 illustrates, the area between $z = 0$ and the desired value of z is equal to .20. Looking in Table 3 for an area of .2000, we find that the closest value is an area of .1985, corresponding to $z = .52$. The next step is to convert this answer to the Y scale. From the standardization formula we obtain the equation

$$.52 = \frac{y^* - 54}{4.5}$$

which can be solved to give $y^* = (.52)(4.5) + 54 = 56.3$. The 70th percentile of the fish length distribution is 56.3 mm.   ∎

PROBLEM-SOLVING TIP   In solving problems that require the use of Table 3, a sketch of the distribution (as in Figures 4.11–4.15) is a very handy aid to straight thinking.

4.1 Suppose a certain population of observations is normally distributed. What percentage of the observations in the population
  a. are within ± 1.5 standard deviations of the mean?
  b. are more than 2.5 standard deviations above the mean?
  c. are more than 3.5 standard deviations away from (above or below) the mean?

4.2 a. The 90th percentile of a normal distribution is how many standard deviations above the mean?
  b. The 10th percentile of a normal distribution is how many standard deviations below the mean?

4.3 The brain weights of a certain population of adult Swedish males follow approximately a normal distribution with mean 1,400 gm and standard deviation 100 gm.[6] What percentage of the brain weights are
  a. 1,500 gm or less?   **b.** between 1,325 and 1,500 gm?
  c. 1,325 gm or more?   **d.** 1,475 gm or more?
  e. between 1,475 and 1,600 gm? **f.** between 1,200 and 1,325 gm?

4.4 Let Y represent a brain weight randomly chosen from the population of Exercise 4.3. Find
  a. $\Pr\{Y \geq 1{,}325\}$ **b.** $\Pr\{1{,}475 \leq Y \leq 1{,}600\}$

4.5 In an agricultural experiment, a large uniform field was planted with a single variety of wheat. The field was divided into many plots (each plot being 7×100 ft) and the yield (lb) of grain was measured for each plot. These plot yields followed approximately a normal distribution with mean 88 lb and standard deviation 7 lb.[7] What percentage of the plot yields were
  a. 80 lb or more?   **b.** 90 lb or more?
  c. 75 lb or less?   **d.** between 75 and 90 lb?
  e. between 90 and 100 lb? **f.** between 75 and 80 lb?

4.6 Refer to Exercise 4.5. Let Y represent the yield of a plot chosen at random from the field. Find
  a. $\Pr\{Y > 90\}$ **b.** $\Pr\{75 < Y < 90\}$

4.7 For the wheat-yield distribution of Exercise 4.5, find
  a. the 65th percentile   **b.** the 35th percentile

4.8 The serum cholesterol levels of a certain population of boys follow a normal distribution with mean 170 mg/dl and standard deviation 30 mg/dl (as in Example 4.1). What percentage of the boys have serum cholesterol values
  a. 180 or more?   **b.** 150 or less?   **c.** 210 or less?
  d. 115 or more?   **e.** between 180 and 210?   **f.** between 115 and 150?
  g. between 150 and 180?

4.9 Refer to Exercise 4.8. Suppose a boy is chosen at random from the population, and let Y be his serum cholesterol value. Find
  a. $\Pr\{Y \geq 180\}$ **b.** $\Pr\{180 < Y < 210\}$

4.10 For the serum cholesterol distribution of Exercise 4.8, find
  a. the 80th percentile   **b.** the 20th percentile

4.11 When red blood cells are counted using a certain electronic counter, the standard deviation of repeated counts of the same blood specimen is about .8% of the true value, and the distribution of repeated counts is approximately normal.[8]
   a. If the true value of the red blood count for a certain specimen is 5,000,000 cells/mm^3, what is the probability that the counter would give a reading between 4,900,000 and 5,100,000?
   b. If the true value of the red blood count for a certain specimen is μ, what is the probability that the counter would give a reading between $.98\mu$ and 1.02μ?
   c. A hospital lab performs counts of many specimens every day. For what percentage of these specimens does the reported blood count differ from the correct value by 2% or more?

S E C T I O N   4.4

THE CONTINUITY CORRECTION (OPTIONAL)

Although a normal curve theoretically represents a continuous distribution, it is common practice to use a normal curve to approximately describe the distribution of a discrete variable. Often the discreteness of the variable can be ignored without introducing any serious error; indeed, in the computations of Section 4.3 we did not distinguish between discrete and continuous variables. For greater accuracy, however, one can take account of discreteness by applying a correction, known as the **continuity correction**, when calculating areas under the normal curve. The following example illustrates the use of the continuity correction.

EXAMPLE 4.7

LITTER SIZE

Table 4.1 shows the distribution of litter size (defined as the number of live young in the first litter) for a population of female mice; the population mean is 7.8 and the standard deviation is 2.3.[9] Figure 4.16 (page 116) shows a normal curve with $\mu = 7.8$ and $\sigma = 2.3$, superimposed on the litter size distribution; the normal curve fits the distribution quite well.

TABLE 4.1

LITTER SIZE	PERCENT FREQUENCY	LITTER SIZE	PERCENT FREQUENCY
1	.4	9	15.0
2	.7	10	11.5
3	1.8	11	6.6
4	4.8	12	3.1
5	8.0	13	1.0
6	13.0	14	.5
7	15.5	15	.3
8	17.8		

Let us compare an actual population relative frequency with the relative frequency predicted by the normal curve. From Table 4.1, the percentage of litters with sizes between 5 and 10, inclusive, is

$$8.0 + 13.0 + 15.5 + 17.8 + 15.0 + 11.5 = 80.8\%$$

FIGURE 4.16
Litter Size Distribution and
Approximating Normal
Curve

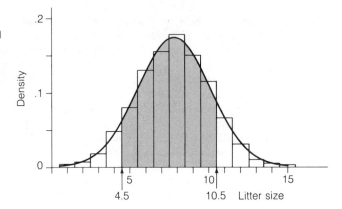

In other words, if Y represents the size of a randomly selected litter, then

$$\Pr\{5 \le Y \le 10\} = .808$$

What is the corresponding area under the normal curve? To adjust for the discreteness of Y, we should calculate the area under the curve between $y = 4.5$ and $y = 10.5$, which is shaded in Figure 4.16; the use of 4.5 instead of 5, and 10.5 instead of 10, represents the continuity correction. From Table 3, the area is found to be .8026. Thus, the value found from the normal approximation with the continuity correction (.8026) is quite close to the actual value (.808).

To see why the continuity correction is very natural, note from Figure 4.16 that the actual value of the probability (.808) is equal to the sum of the areas of the corresponding six histogram bars, which extend from $y = 4.5$ to $y = 10.5$. If we were to omit the continuity correction, we would calculate the area under the normal curve from $y = 5$ to $y = 10$; it is clear from Figure 4.16 that this answer would be too small (in fact, it is .7203).

The continuity correction is especially important if we want to consider the probability of a *single* Y value. For example, the normal approximation to $\Pr\{Y = 10\}$ is the area from $y = 9.5$ to $y = 10.5$, which is equal to .1086 or 10.9%; this is reasonably close to the actual value of 11.5% (from Table 4.1). The normal approximation *without* the continuity correction would be the area from $y = 10$ to $y = 10$, which is zero—not a sensible answer.   ∎

Whenever a discrete distribution is approximated by a normal curve, the continuity correction can be used to obtain more accurate values for predicted relative frequencies. This applies not only to variables that are inherently discrete (such as litter size) but also to variables that are actually continuous but are measured on a discrete scale because of rounding. For instance, blood pressure is a continuous variable, but it is usually measured to the nearest mm Hg, so that the actual measurements fall on a discrete scale. If the "spaces" between the possible values of a variable are small compared to the standard deviation of the distribution, then the continuity correction has little effect. For instance, if the standard deviation of a distribution of blood pressures is 20 mm Hg, then (because 1 is small compared to 20) for most purposes the continuity correction for this distribution could be ignored.

| | | | | | | | | | | |

EXERCISES 4.12–4.15

4.12 In genetic studies of the fruitfly *Drosophila melanogaster*, one variable of interest is the total number of bristles on the ventral surface of the fourth and fifth abdominal segments. For a certain *Drosophila* population,[10] the bristle count follows approximately a normal distribution with mean 38.5 and standard deviation 2.9. Find (using the continuity correction)
  a. the percentage of flies with 40 or more bristles.
  b. the percentage of flies with exactly 40 bristles.
  c. the percentage of flies whose bristle count is between 35 and 40, inclusive.

4.13 Refer to the fruitfly population of Exercise 4.12. Let Y be the bristle count of a fly chosen at random from the population.
  a. Use the continuity correction to calculate $\Pr\{35 \le Y \le 40\}$.
  b. Calculate $\Pr\{35 \le Y \le 40\}$ without the continuity correction and compare with the result of part **a**.

4.14 The litter sizes of a certain population of female mice follow approximately a normal distribution with mean 7.8 and standard deviation 2.3 (as in Example 4.7). Let Y represent the size of a randomly chosen litter. Use the continuity correction to find approximate values for each of the following probabilities:
  a. $\Pr\{Y \le 6\}$ **b.** $\Pr\{Y = 6\}$ **c.** $\Pr\{8 \le Y \le 11\}$

4.15 In a certain population of healthy people, the mean total protein concentration in the blood serum is 6.85 g/dl, the standard deviation is .42 g/dl, and the distribution is approximately normal.[11] Let Y be the total protein value of a randomly selected person, as given by an instrument that reports the value to the nearest .1 g/dl. Use the continuity correction to calculate
  a. $\Pr\{Y = 6.5\}$ **b.** $\Pr\{6.5 \le Y \le 8.0\}$

| | | | | | | | | | | |

SECTION 4.5

PERSPECTIVE

The normal distribution is also called the **Gaussian distribution**, after the German mathematician K. F. Gauss. The term *normal*, with its connotations of "typical" or "usual," can be seriously misleading. Consider, for instance, a medical context, where the primary meaning of "normal" is "not abnormal." Thus, confusingly, the phrase "the normal population of serum cholesterol levels" may refer to cholesterol levels in ideally "healthy" people, or it may refer to a Gaussian distribution such as the one in Example 4.1. In fact, for many variables the distribution in the normal (nondiseased) population is decidedly *not* normal (Gaussian).

The examples of this chapter have illustrated one use of the normal distribution—as an approximation to naturally occurring biological distributions. If a natural distribution is well approximated by a normal distribution, then the mean and standard deviation provide a complete description of the distribution: The mean is the center of the distribution, about 68% of the values are within 1 standard deviation of the mean, about 95% are within 2 standard deviations of the mean, and so on.

As noted in Section 2.6, the 68% and 95% benchmarks can be roughly applicable even to distributions that are rather skewed. (But if the distribution is skewed then the 68% is not symmetrically divided on both sides of the mean, and similarly for the 95%.) However, the benchmarks do not apply to a distribution (even a symmetric one) for which one or both tails are long and thin [see Figures 2.13(b) and 2.20].

We will see in later chapters that many classical statistical methods are specifically designed for, and function best with, data that have been sampled from normal populations. We will further see that in many practical situations these methods work very well also for samples from nonnormal populations.

The normal distribution is of central importance in spite of the fact that many, perhaps most, naturally occurring biological distributions could be described better by a skewed curve than by a normal curve. The primary use of the normal distribution is not to describe natural distributions, but to describe certain theoretical distributions, called sampling distributions, that are used in the statistical analysis of data. We will see in Chapter 5 that many sampling distributions are approximately normal; it is this property that makes the normal distribution unique.

| | | | | | | | | | | | |

SUPPLEMENTARY EXERCISES 4.16–4.29

4.16 The activity of a certain enzyme is measured by counting emissions from a radioactively labelled molecule. For a given tissue specimen, the counts in consecutive ten-second time periods may be regarded (approximately) as repeated independent observations from a normal distribution.[12] Suppose the mean ten-second count for a certain tissue specimen is 1,200 and the standard deviation is 35. Let Y denote the count in a randomly chosen ten-second time period. Find

a. $\Pr\{Y \geq 1,250\}$ **b.** $\Pr\{Y \leq 1,175\}$
c. $\Pr\{1,150 \leq Y \leq 1,250\}$ **d.** $\Pr\{1,150 \leq Y \leq 1,175\}$

4.17 The shell thicknesses of the eggs produced by a large flock of hens follow approximately a normal distribution with mean equal to .38 mm and standard deviation equal to .03 mm (as in Example 4.2). Find the 95th percentile of the thickness distribution.

4.18 Refer to the eggshell thickness distribution of Exercise 4.17. Suppose an egg is defined as thin-shelled if its shell is .32 mm thick or less.
a. What percentage of the eggs are thin-shelled?
b. Suppose a large number of eggs from the flock are randomly packed into boxes of 12 eggs each. What percentage of the boxes will contain at least one thin-shelled egg? [*Hint:* First find the percentage of boxes that will contain no thin-shelled egg.]

4.19 The heights of a certain population of corn plants follow a normal distribution with mean 145 cm and standard deviation 22 cm.[13] What percentage of the plant heights are
a. 100 cm or more?   **b.** 120 cm or less?
c. between 120 and 150 cm?   **d.** between 100 and 120 cm?
e. between 150 and 180 cm?   **f.** 180 cm or more?
g. 150 cm or less?

4.20 Refer to the corn plant population of Exercise 4.19. Find the 90th percentile of the height distribution.

4.21 Suppose four plants are to be chosen at random from the corn plant population of Exercise 4.19. Find the probability that none of the four plants will be more than 150 cm tall.

4.22 Suppose a certain population of observations is normally distributed. Find the value of z^* such that 95.00% of the observations in the population are between $-z^*$ and $+z^*$, on the Z scale.

4.23 In the nerve-cell activity of a certain individual fly, the time intervals between "spike" discharges follow approximately a normal distribution with mean 15.6 ms and standard deviation .4 ms (as in Example 4.3). Let Y denote a randomly selected interspike interval. Find
a. $\Pr\{Y > 15\}$ **b.** $\Pr\{Y > 16.5\}$
c. $\Pr\{15 < Y < 16.5\}$ **d.** $\Pr\{15 < Y < 15.5\}$

4.24 For the distribution of interspike-time intervals described in Exercise 4.23, find the quartiles and the interquartile range.

4.25 Among American women aged 18–24 years, 10% are less than 61.2 inches tall, 80% are between 61.2 and 67.4 inches tall, and 10% are more than 67.4 inches tall.[14] Assuming that the height distribution can be adequately approximated by a normal curve, find the mean and standard deviation of the distribution.

4.26 The intelligence quotient (IQ) score, as measured by the Stanford–Binet IQ test, is normally distributed in a certain population of children. The mean IQ score is 100 points, and the standard deviation is 16 points.[15] What percentage of children in the population have IQ scores
a. 140 or more?   **b.** 80 or less?
c. between 80 and 120?   **d.** between 80 and 140?
e. between 120 and 140?

4.27 Refer to the IQ distribution of Exercise 4.26. Let Y be the IQ score of a child chosen at random from the population. Find $\Pr\{80 \le Y \le 140\}$.

4.28 Refer to the IQ distribution of Exercise 4.26. Suppose five children are to be chosen at random from the population. Find the probability that one of them will have an IQ score of 80 or less and four will have scores higher than 80.

4.29 A certain assay for serum alanine aminotransferase (ALT) is rather imprecise. The results of repeated assays of a single specimen follow a normal distribution with mean equal to the true ALT concentration for that specimen and standard deviation equal to 4 U/l (see Example 2.11). Suppose that a certain hospital lab measures many specimens every day, performing one assay for each specimen, and that specimens with ALT readings of 40 U/l or more are flagged as "unusually high." If a patient's true ALT concentration is 35 U/l, what is the probability that his specimen will be flagged as "unusually high"?

CONTENTS

SAMPLING DISTRIBUTIONS

BASIC IDEAS

An important goal of data analysis is to distinguish between features of the data that reflect real biological facts and features that may reflect only chance effects. As explained in Sections 2.8 and 3.2, the random sampling model provides a framework for making this distinction. The underlying reality is visualized as a population, the data are viewed as a random sample from the population, and chance effects are regarded as sampling error—that is, discrepancy between the sample and the population.

In this chapter we develop the theoretical background which will enable us to place specific limits on the degree of sampling error to be expected in an experiment. (Although in later chapters we will distinguish between an experimental study and an observational study, for the present we will call any scientific investigation an *experiment*.) As in Chapters 2 and 3, we continue to confine the discussion to the simple context of an experiment with only one experimental group (one sample).

SAMPLING VARIABILITY

The variability among random samples from the same population is called **sampling variability**. A probability distribution that characterizes some aspect of sampling variability is termed a **sampling distribution**. While a random sample may or may not resemble the population from which it came, a sampling distribution can specify how close the resemblance is *likely* to be.

In this chapter we will discuss several aspects of sampling variability and study two important sampling distributions. We will assume throughout that the sample size is a negligibly small fraction of the population size. This assumption simplifies the theory because it guarantees that the process of drawing the sample does not change the population composition.

THE META-EXPERIMENT

According to the random sampling model, we regard the data in an experiment as a random sample from a population. But to visualize sampling variability, we must broaden our frame of reference to include, not merely a single sample, but all the possible samples that might be drawn from the population. This wider frame of reference we will call the **meta-experiment**. A meta-experiment consists of indefinitely many repetitions, or replications, of the same experiment.* Thus, if the experiment consists of drawing a random sample of size n from some population, the corresponding meta-experiment involves drawing *repeated* random samples of size n from the same population. The process of repeated drawing is carried on indefinitely, with the members of each sample being replaced before the next sample is drawn. The experiment and the meta-experiment are schematically represented in Figure 5.1.

*The term *meta-experiment* is not a standard term, but was coined by the author. It is unrelated to the term *meta-analysis*, which denotes a particular type of statistical analysis.

FIGURE 5.1

Schematic Representation
of Experiment and Meta-
Experiment

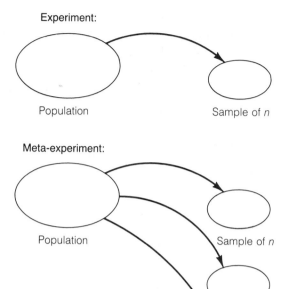

Experiment:

Population   Sample of n

Meta-experiment:

Population   Sample of n

Sample of n

Sample of n

⋮

etc.

The following two examples illustrate the notion of a meta-experiment.

EXAMPLE 5.1
BLOOD PRESSURE

An experiment consists of measuring the change in blood pressure in $n = 10$ rats after administering a certain drug. The corresponding meta-experiment would consist of repeatedly choosing groups of $n = 10$ rats from the same population, and making blood pressure measurements under the same conditions. ∎

EXAMPLE 5.2
BACTERIAL GROWTH

An experiment consists of observing bacterial growth in $n = 5$ petri dishes which have been treated identically. The corresponding meta-experiment would consist of repeatedly preparing groups of 5 petri dishes and observing them in the same way. ∎

Note that a meta-experiment is a theoretical construct rather than an operation that is actually performed by an experimenter. If the actual experiment itself consists in choosing several, say ten, random samples, then the corresponding meta-experiment would consist of repeatedly choosing sets of ten random samples from the same population.

The meta-experiment concept provides a link between sampling variability and probability. Recall from Chapter 3 that the probability of an event can be interpreted as the long-run relative frequency of occurrence of the event. Choosing a random sample is a chance operation; the meta-experiment consists of many repetitions of this chance operation, and so *probabilities concerning a random sample can be interpreted as relative frequencies in a meta-experiment.* Thus, the meta-experiment is a device for explicitly visualizing a sampling distribution: The sampling distribution describes the variability among the many random samples in a meta-experiment.

| | | | | | | | | | | | | |

SECTION 5.2

DICHOTOMOUS OBSERVATIONS

Consider an experiment in which the observed variable is dichotomous. As in Chapters 2 and 3, we will let π represent the population proportion of one category, while $\hat{\pi}$ represents the corresponding sample proportion. Then the question of how closely the sample resembles the population becomes "How close to π is $\hat{\pi}$ likely to be?" We will see how to answer this question through the **sampling distribution of $\hat{\pi}$**—that is, the collection of probabilities of all the various possible values of $\hat{\pi}$.

THE SAMPLING DISTRIBUTION OF $\hat{\pi}$

For random sampling from a large dichotomous population, we saw in Chapter 3 how to use the binomial distribution to calculate the probabilities of all the various possible sample compositions. These probabilities in turn determine the sampling distribution of $\hat{\pi}$. The following is an example.

EXAMPLE 5.3

SUPERIOR VISION

In a certain human population, 30% of the individuals have "superior" distance vision, in the sense of scoring 20/15 or better on a standardized vision test without glasses.[1] If we were to examine a random sample of 20 people from the population, how many people with superior vision might we expect to find in the sample? This question can be answered in the language of probability, by using the binomial distribution with $n = 20$ and $\pi = .30$. For instance, let us calculate the probability that 5 members of the sample would have superior vision and 15 would not:

$$\text{Pr}\{5 \text{ superior, } 15 \text{ not superior}\} = C_5(.3)^5(.7)^{15}$$
$$= 15{,}504(.3)^5(.7)^{15}$$
$$= .179$$

If we let $\hat{\pi}$ represent the sample proportion of individuals with superior vision, then a sample that contains 5 individuals with superior vision has $\hat{\pi} = \frac{5}{20} = .25$. Thus, we have found that

$$\text{Pr}\{\hat{\pi} = .25\} = .179$$

This probability can be interpreted in terms of a meta-experiment. The experiment consists of observing the distance vision of 20 randomly chosen people. The meta-experiment, as pictured in Figure 5.2, consists of repeatedly choosing 20 people and observing their distance vision. Each sample of 20 has its own value of $\hat{\pi}$. About 17.9% of the $\hat{\pi}$'s will be equal to .25.

FIGURE 5.2

Meta-Experiment for Example 5.3

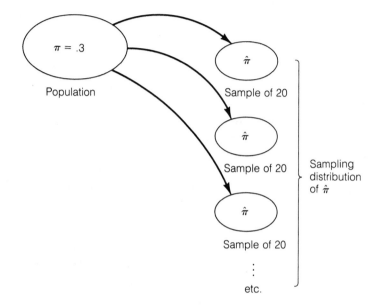

The binomial distribution can be used in a similar way to determine the entire sampling distribution of $\hat{\pi}$. The distribution is displayed in Table 5.1 and as a probability histogram in Figure 5.3 (page 126).

TABLE 5.1

Sampling Distribution of $\hat{\pi}$
When $n = 20$ and $\pi = .3$

$\hat{\pi}$	PROBABILITY	$\hat{\pi}$	PROBABILITY
.00	.001	.55	.012
.05	.007	.60	.004
.10	.028	.65	.001
.15	.072	.70	.000
.20	.130	.75	.000
.25	.179	.80	.000
.30	.192	.85	.000
.35	.164	.90	.000
.40	.114	.95	.000
.45	.065	1.00	.000
.50	.031		

FIGURE 5.3
Sampling Distribution of $\hat{\pi}$
When $n = 20$ and $\pi = .30$

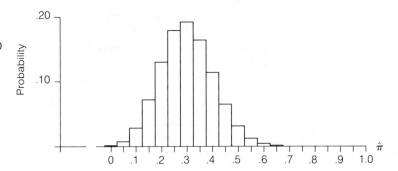

RELATIONSHIP TO STATISTICAL INFERENCE

In making a statistical inference from a sample to the population, we might use $\hat{\pi}$ as our estimate of π. The sampling distribution of $\hat{\pi}$ can be used to predict how much sampling error to expect in this estimate. For example, suppose we want to know whether the sampling error will be less than 5 percentage points, in other words, whether $\hat{\pi}$ will be within $\pm.05$ of π. We cannot predict for certain whether this event will occur, but we can assign it a probability, as illustrated in the following example.

EXAMPLE 5.4

SUPERIOR VISION

In the vision example, we see from Table 5.1 that

$$Pr\{.25 \le \hat{\pi} \le .35\} = .179 + .192 + .164$$
$$= .535 \approx .53$$

Thus, there is a 53% chance that, for a sample of size 20, $\hat{\pi}$ will be within $\pm.05$ of π.   ∎

It may have occurred to you that the preceding discussion seems to have a fatal flaw. In order to specify the sampling distribution of $\hat{\pi}$, we need to know π. But if we know π, we do not need to take a sample in order to get information about π! How, then, can the sampling distribution of $\hat{\pi}$ be a basis for statistical inference in real experiments? It turns out that there is an escape from this apparently circular reasoning. We will see in later chapters that it is not necessary to know π in order to estimate the amount of sampling error of $\hat{\pi}$.

DEPENDENCE ON SAMPLE SIZE

It is interesting to consider how the sampling distribution of $\hat{\pi}$ depends on n. You may think intuitively that a larger n should provide a more informative sample, and indeed this is true. If n is larger, then $\hat{\pi}$ is more likely to be close to π.* The following example illustrates this effect.

*This statement should not be interpreted too literally. As a function of n, the probability that $\hat{\pi}$ is close to π has an overall increasing trend, but it can fluctuate somewhat.

EXAMPLE 5.5

SUPERIOR VISION

Figure 5.4 shows the sampling distribution of $\hat{\pi}$, for three different values of n, for the vision population of Example 5.3. (Each sampling distribution is determined by a binomial distribution with $\pi = .3$. The figures are scaled so that their areas are equal.) You can see from the figure that as n increases the sampling

FIGURE 5.4 Sampling Distributions of $\hat{\pi}$ for $\pi = .30$ and Various Values of n

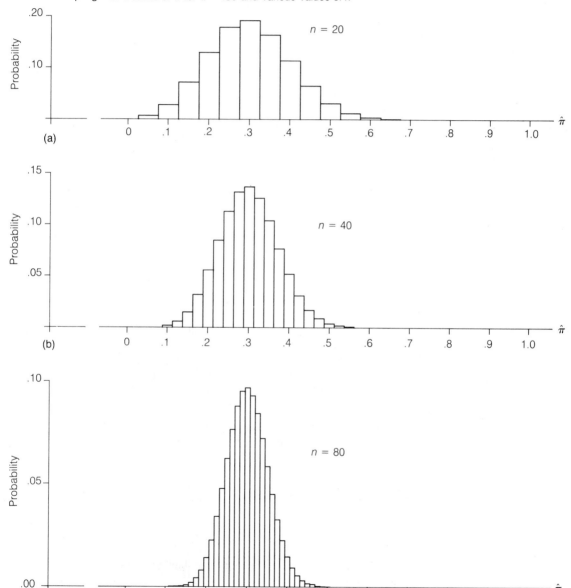

distribution becomes more compressed around the value $\hat{\pi} = .3$; thus, the probability that $\hat{\pi}$ is close to π tends to increase as n increases. For example, consider the probability that $\hat{\pi}$ is within ± 5 percentage points of π. We saw in Example 5.4 that for $n = 20$ this probability is equal to .53; Table 5.2 shows how the probability depends on n.

TABLE 5.2

n	$\Pr\{.25 \le \hat{\pi} \le .35\}$
20	.53
40	.61
80	.73
400	.97

A larger sample tends to improve the probability that $\hat{\pi}$ is *close* to π; however, the probability that $\hat{\pi}$ is *exactly equal* to π is very small for large n. In fact,

$$\Pr\{\hat{\pi} = .3\} = .097 \quad \text{for } n = 80$$

The value $\Pr\{.25 \le \hat{\pi} \le .35\} = .73$ is the sum of many small probabilities, the largest of which is .097; you can see this effect clearly in Figure 5.4(c).   ■

EXERCISES 5.1–5.7

5.1 Suppose we are to draw a random sample of five individuals from a large population in which 39% of the individuals are mutants (as in Example 3.20). Let $\hat{\pi}$ represent the proportion of mutants in the sample.
   a. Use the results in Table 3.5 to determine the probability that $\hat{\pi}$ will be equal to (i) 0; (ii) .2; (iii) .4; (iv) .6; (v) .8; (vi) 1.0.
   b. Display the sampling distribution of $\hat{\pi}$ as a histogram.

5.2 A new treatment for acquired immune deficiency syndrome (AIDS) is to be tested in a small clinical trial on 15 patients. The proportion $\hat{\pi}$ who respond to the treatment will be used as an estimate of the proportion π of (potential) responders in the entire population of AIDS patients. If in fact $\pi = .2$, and if the 15 patients can be regarded as a random sample from the population, find the probability that
   a. $\hat{\pi} = .2$ **b.** $\hat{\pi} = 0$

5.3 In a certain forest, 25% of the white pine trees are infected with blister rust. Suppose a random sample of 4 white pine trees is to be chosen, and let $\hat{\pi}$ be the sample proportion of infected trees.
   a. Compute the probability that $\hat{\pi}$ will be equal to (i) 0; (ii) .25; (iii) .50; (iv) .75; (v) 1.0.
   b. Display the sampling distribution of $\hat{\pi}$ as a histogram.

5.4 Refer to Exercise 5.3.
   a. Determine the sampling distribution of $\hat{\pi}$ for samples of size $n = 8$ white pine trees from the same forest.
   b. Construct histograms of the sampling distributions of $\hat{\pi}$ for $n = 4$ and for $n = 8$, using the same horizontal scale for both, but doubling the vertical scale for the distribution with $n = 8$ (so that the two histograms have the same area). Compare the two distributions visually. How do they differ?

5.5 The shell of the land snail *Limocolaria martensiana* has two possible color forms: streaked and pallid. In a certain population of these snails, 60% of the individuals have streaked shells (as in Exercise 3.11). Suppose a random sample of ten snails is to be chosen from the population, and let $\hat{\pi}$ be the sample proportion of streaked snails. Find
a. $\Pr\{\hat{\pi} = .5\}$ **b.** $\Pr\{\hat{\pi} = .6\}$ **c.** $\Pr\{\hat{\pi} = .7\}$
d. $\Pr\{.5 \le \hat{\pi} \le .7\}$ **e.** $\Pr\{\hat{\pi}$ is within $\pm.10$ of $\pi\}$

5.6 In a certain human population, 30% of the people have "superior" vision (as in Example 5.3). Suppose a random sample of five people is to be chosen, and their vision examined. Let $\hat{\pi}$ represent the sample proportion of people with superior vision.
a. Compute the sampling distribution of $\hat{\pi}$.
b. Construct a histogram of the distribution found in part **a**, and compare it visually with Figure 5.3. How do the two distributions differ?

5.7 Consider random sampling from a dichotomous population, and let E be the event that $\hat{\pi}$ is within $\pm.05$ of π. In Example 5.4, we found that $\Pr\{E\} = .53$ for $n = 20$ and $\pi = .3$. Calculate $\Pr\{E\}$ for $n = 20$ and $\pi = .4$. (Perhaps surprisingly, the two probabilities are roughly equal.)

SECTION 5.3

QUANTITATIVE OBSERVATIONS

If the observed variable is quantitative, then the question of similarity between sample and population is more complex than for dichotomous data. For a quantitative variable, the sample and the population can be described in various ways—by the frequency distribution, the mean, the median, the standard deviation, and so on—and the question must be answered separately for each of these descriptive measures. In this section we will focus primarily on the mean.

THE SAMPLING DISTRIBUTION OF $\bar{Y}$

The sample mean $\bar{y}$ can be used, not only as a description of the data in the sample, but also as an estimate of the population mean μ. It is natural to ask, "How close to μ is $\bar{y}$?" We cannot answer this question for the mean $\bar{y}$ of a particular sample, but we can answer it if we think in terms of the random sampling model and regard the sample mean as a random variable $\bar{Y}$. The question then becomes: "How close to μ is $\bar{Y}$ *likely* to be?" and the answer is provided by the **sampling distribution of** $\bar{Y}$—that is, a probability distribution which describes sampling variability in $\bar{Y}$.

To visualize the sampling distribution of $\bar{Y}$, imagine the meta-experiment as follows: Random samples of size n are repeatedly drawn from a fixed population with mean μ and standard deviation σ; each sample has its own mean $\bar{y}$. The variation of the $\bar{y}$'s among the samples is specified by the sampling distribution of $\bar{Y}$. This relationship is indicated schematically in Figure 5.5 (page 130).

We now state as a theorem the basic facts about the sampling distribution of $\bar{Y}$. The theorem can be proved using the methods of mathematical statistics; we will simply state it and then discuss its meaning. The theorem describes the sampling distribution of $\bar{Y}$ in terms of its mean (denoted by $\mu_{\bar{Y}}$), its standard deviation (denoted by $\sigma_{\bar{Y}}$), and its shape.

FIGURE 5.5

Schematic Representation
of the Sampling
Distribution of $\bar{Y}$

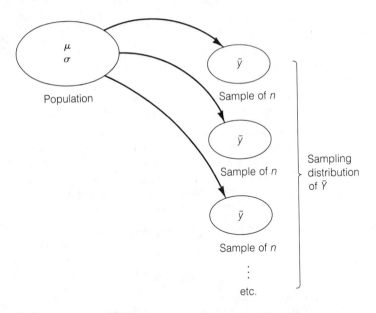

THEOREM 5.1: THE SAMPLING DISTRIBUTION OF $\bar{Y}$

1. *Mean*   The mean of the sampling distribution of $\bar{Y}$ is equal to the population mean. In symbols,

$$\mu_{\bar{Y}} = \mu$$

2. *Standard deviation*   The standard deviation of the sampling distribution of $\bar{Y}$ is equal to the population standard deviation divided by the square root of the sample size. In symbols,

$$\sigma_{\bar{Y}} = \frac{\sigma}{\sqrt{n}}$$

3. *Shape*

   a. If the population distribution of Y is normal, then the sampling distribution of $\bar{Y}$ is normal.

   b. *Central Limit Theorem*   If n is large, then the sampling distribution of $\bar{Y}$ is approximately normal, even if the population distribution of Y is not normal.

Parts 1 and 2 of Theorem 5.1 specify the relationship between the mean and standard deviation of the population being sampled, and the mean and standard deviation of the sampling distribution of $\bar{Y}$. Part 3a of the theorem states that, if the observed variable Y follows a normal distribution in the population being sampled, then the sampling distribution of $\bar{Y}$ is also a normal distribution. These relationships are indicated in Figure 5.6.

FIGURE 5.6
(a) The population distribution of a normally distributed variable Y.
(b) The sampling distribution of $\bar{Y}$ in sampling from the population of part (a).

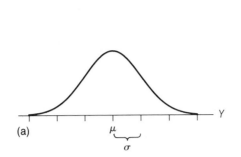

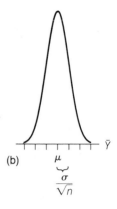

The following example illustrates the meaning of parts 1, 2, and 3a of Theorem 5.1.

EXAMPLE 5.6
WEIGHTS OF SEEDS

A large population of seeds of the princess bean *Phaseolus vulgaris* is to be sampled. The weights of the seeds in the population follow a normal distribution with mean $\mu = 500$ mg, and standard deviation $\sigma = 120$ mg.[2] Suppose now that a random sample of four seeds is to be weighed, and let $\bar{Y}$ represent the mean weight of the four seeds. Then, according to Theorem 5.1, the sampling distribution of $\bar{Y}$ will be a normal distribution with mean and standard deviation as follows:

$$\mu_{\bar{Y}} = \mu = 500 \text{ mg}$$

and

$$\sigma_{\bar{Y}} = \frac{\sigma}{\sqrt{n}} = \frac{120}{\sqrt{4}} = 60 \text{ mg}$$

The sampling distribution of $\bar{Y}$ is shown in Figure 5.7; the ticks are 1 standard deviation apart.

FIGURE 5.7
Sampling Distribution of $\bar{Y}$ for Example 5.6

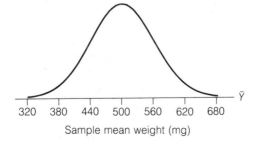

The sampling distribution of $\bar{Y}$ expresses the relative likelihood of the various possible values of $\bar{Y}$. For example, suppose we want to know the probability that the mean weight of the four seeds will be greater than 550 mg. This probability is shown as the shaded area in Figure 5.8. Notice that the value of $\bar{y} = 550$ must be converted to the Z scale using the standard deviation $\sigma_{\bar{Y}} = 60$, *not* $\sigma = 120$:

$$z = \frac{\bar{y} - \mu_{\bar{Y}}}{\sigma_{\bar{Y}}} = \frac{550 - 500}{60} = .83$$

FIGURE 5.8
Calculation of $\Pr\{\bar{Y} \geq 550\}$ for Example 5.6

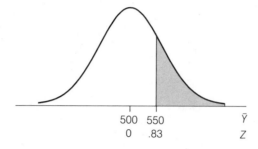

	500	550		$\bar{Y}$
	0	.83		Z

From Table 3, $z = .83$ corresponds to an area of .2967, so

$$\Pr\{\bar{Y} > 550\} = .5000 - .2967$$
$$= .2033 \approx .20$$

This probability can be interpreted in terms of a meta-experiment as follows: If we were to choose many random samples of four seeds each from the population, then about 20% of the samples would have a mean weight exceeding 550 mg.

∎

Part 3b of Theorem 5.1 is known as the **Central Limit Theorem**. The Central Limit Theorem states that, no matter what distribution Y may have in the population, if the sample size is large enough then the sampling distribution of $\bar{Y}$ will be approximately a normal distribution.

The Central Limit Theorem is of fundamental importance because it can be applied when (as often happens in practice) the form of the population distribution is not known. It is because of the Central Limit Theorem (and other similar theorems) that the normal distribution plays such a central role in statistics.

It is natural to ask how "large" a sample size is required by the Central Limit Theorem: how large must n be in order that the sampling distribution of $\bar{Y}$ be well approximated by a normal curve? The answer is that the required n depends on the shape of the population distribution. If the shape is normal, any n will do. If the shape is moderately nonnormal, a moderate n is adequate. If the shape is highly nonnormal, then a rather large n will be required. (Some specific examples of this phenomenon are given in the optional Section 5.4.)

Remark We stated in Section 5.1 that the theory of this chapter is valid if the sample size is small compared to the population size. But the Central Limit Theorem is a statement about large samples. This may seem like a contradiction: how can a large sample be a small sample? In practice, there is no contradiction. In a typical biological application, the population size might be 10^6; a sample of size $n = 100$ would be a small fraction of the population, but would nevertheless be large enough for the Central Limit Theorem to be applicable (in most situations).

DEPENDENCE ON SAMPLE SIZE

Consider the possibility of choosing random samples of various sizes from the same population. The sampling distribution of $\bar{Y}$ will depend on the sample size n in two ways. First, its standard deviation is

$$\sigma_{\bar{Y}} = \frac{\sigma}{\sqrt{n}}$$

and so is inversely proportional to $\sqrt{n}$. Second, if the population distribution is not normal, then the *form* of the sampling distribution of $\bar{Y}$ depends on n, being more nearly normal for larger n. However, if the population distribution *is* normal, then the sampling distribution of $\bar{Y}$ is always normal, and only the standard deviation depends on n.

The more important of the two effects of sample size is the first: Larger n gives a smaller value of $\sigma_{\bar{Y}}$, and consequently smaller expected sampling error if $\bar{y}$ is used as an estimate of μ. The following example illustrates this effect for sampling from a normal population.

EXAMPLE 5.7
WEIGHTS OF SEEDS

Figure 5.9 (page 134) shows the sampling distribution of $\bar{Y}$ for samples of various sizes from the princess bean population of Example 5.6. Notice that for larger n the sampling distribution is more concentrated around the population mean $\mu = 500$ mg. As a consequence, the probability that $\bar{Y}$ is close to μ is larger for larger n. For instance, consider the probability that $\bar{Y}$ is within ± 50 mg of μ, that is, $\Pr\{450 \leq \bar{Y} \leq 550\}$. Table 5.3 shows how this probability depends on n.

TABLE 5.3

n	$\Pr\{450 \leq \bar{Y} \leq 550\}$
4	.59
9	.79
16	.91
64	.999

FIGURE 5.9
Sampling Distribution of
$\bar{Y}$ for Various Sample
Sizes n

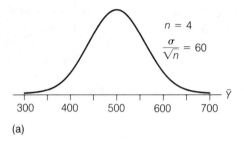

$n = 4$

$\dfrac{\sigma}{\sqrt{n}} = 60$

$\bar{Y}$

300   400   500   600   700

(a)

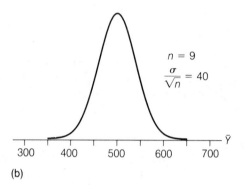

$n = 9$

$\dfrac{\sigma}{\sqrt{n}} = 40$

$\bar{Y}$

300   400   500   600   700

(b)

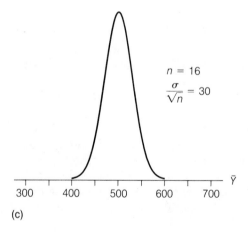

$n = 16$

$\dfrac{\sigma}{\sqrt{n}} = 30$

$\bar{Y}$

300   400   500   600   700

(c)

 Example 5.7 illustrates how the closeness of $\bar{Y}$ to μ depends on sample size. The mean of a larger sample is not *necessarily* closer to μ than the mean of a smaller sample, but it has a *greater probability* of being close. It is in this sense that a larger sample provides more information about the population mean than a smaller sample.

POPULATIONS, SAMPLES, AND SAMPLING DISTRIBUTIONS

In thinking about Theorem 5.1, it is important to distinguish clearly among three different distributions related to a quantitative variable Y: (1) the distribution of Y in the population; (2) the distribution of Y in a sample of data, and (3) the sampling distribution of $\bar{Y}$. The means and standard deviations of these distributions are summarized in Table 5.4.

TABLE 5.4

DISTRIBUTION	MEAN	STANDARD DEVIATION
Y in population	μ	σ
Y in sample	$\bar{y}$	s
$\bar{Y}$ (in meta-experiment)	μ	$\dfrac{\sigma}{\sqrt{n}}$

The following example illustrates the distinction among the three distributions.

EXAMPLE 5.8
WEIGHTS OF SEEDS

For the princess bean population of Example 5.6, the population mean and standard deviation are $\mu = 500$ mg and $\sigma = 120$ mg; the population distribution of $Y = $ weight is represented in Figure 5.10(a). Suppose we weigh a random sample of $n = 25$ seeds from the population and obtain the data in Table 5.5.

FIGURE 5.10
Three Distributions
Related to $Y = $ Seed
Weight of Princess Beans.
(a) Population distribution
of Y; (b) Distribution of 25
observations of Y;
(c) Sampling distribution
of $\bar{Y}$ for $n = 25$

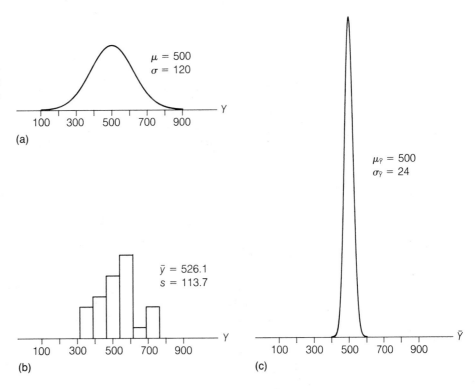

TABLE 5.5
Weights of 25 Princess
Bean Seeds

WEIGHT (mg)						
343	755	431	480	516	469	694
659	441	562	597	502	612	549
348	469	545	728	416	536	581
433	583	570	334			

For the data in Table 5.5, the sample mean is $\bar{y} = 526.1$ mg and the sample standard deviation is $s = 113.7$ mg. Figure 5.10(b) shows a histogram of the data; this histogram represents the distribution of Y in the sample. The sampling distribution of $\bar{Y}$ is a theoretical distribution which relates, not to the particular sample shown in the histogram, but rather to the meta-experiment of repeated samples of size $n = 25$; the mean and standard deviation of the sampling distribution are

$$\mu_{\bar{Y}} = 500 \text{ mg} \quad \text{and} \quad \sigma_{\bar{Y}} = \frac{120}{\sqrt{25}} = 24 \text{ mg}$$

The sampling distribution is represented in Figure 5.10(c). Notice that the distributions in Figures 5.10(a) and (b) are more or less similar; in fact, the distribution in (b) is an estimate (based on the data in Table 5.5) of the distribution in (a). By contrast, the distribution in (c) is much narrower, because it represents a distribution of means rather than of individual observations. ∎

OTHER ASPECTS OF SAMPLING VARIABILITY

The preceding discussion has focused on sampling variability in the sample mean, $\bar{Y}$. Two other important aspects of sampling variability are (1) sampling variability in the sample standard deviation, s; (2) sampling variability in the *shape* of the sample, as represented by the sample histogram. Rather than discuss these aspects formally, we illustrate them with the following example.

EXAMPLE 5.9
WEIGHTS OF SEEDS

In Figure 5.10(b) we displayed a random sample of 25 observations from the princess bean population of Example 5.6; now we display in Figure 5.11 eight additional random samples from the same population. (All nine samples were actually simulated using a computer.) Notice that, even though the samples were drawn from a normal population [pictured in Figure 5.10(a)], there is very substantial variation in the forms of the histograms. Notice also that there is considerable variation in the sample standard deviations. Of course, if the sample size were larger (say, $n = 100$ rather than $n = 25$), there would be less sampling variation; the histograms would tend to resemble a normal curve more closely, and the standard deviations would tend to be closer to the population value ($\sigma = 120$).

FIGURE 5.11
Eight Random Samples,
Each of Size $n = 25$, from
a Normal Population with
$\mu = 500$ and $\sigma = 120$

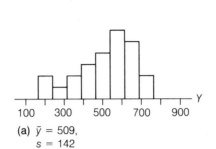

(a) $\bar{y} = 509$,
    $s = 142$

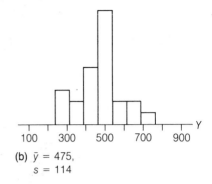

(b) $\bar{y} = 475$,
    $s = 114$

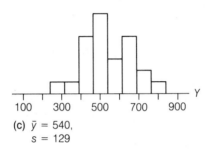

(c) $\bar{y} = 540$,
    $s = 129$

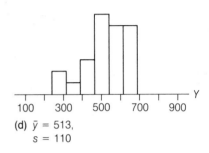

(d) $\bar{y} = 513$,
    $s = 110$

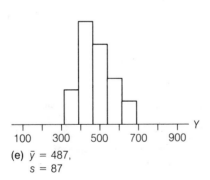

(e) $\bar{y} = 487$,
    $s = 87$

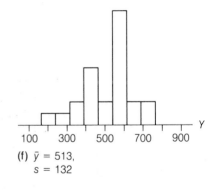

(f) $\bar{y} = 513$,
    $s = 132$

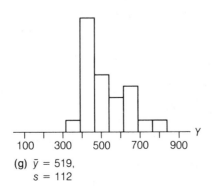

(g) $\bar{y} = 519$,
    $s = 112$

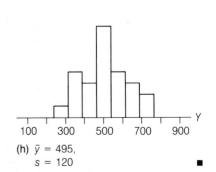

(h) $\bar{y} = 495$,
    $s = 120$ ∎

5.8 (*Sampling exercise*)   Refer to Exercise 3.1. The collection of 100 ellipses shown there can be thought of as representing a natural population of the organism *C. ellipticus*. Use random digits (from Table 1 or your calculator) to choose a random sample of five ellipses. With a metric ruler, measure the length of each ellipse in your sample. Measure only the body, excluding any tail bristles; measurements to the nearest millimeter will be adequate. Compute the mean and standard deviation of the five lengths. To facilitate the pooling of results from the entire class, express the mean and standard deviation in millimeters, keeping two decimal places.

5.9 (*Sampling exercise*)   Proceed as in Exercise 5.8, but use "judgment" sampling rather than random sampling—that is, choose a sample of 5 ellipses which in your judgment should be reasonably representative of the population. (In order to best simulate the analogous judgment in a real-life setting, you should make your choice intuitively, without any detailed preliminary study of the population.)

5.10 (*Sampling exercise*)   Proceed as in Exercise 5.8, but choose a random sample of 20 ellipses.

5.11 Refer to Exercise 5.8. The following scheme is proposed for choosing a sample of 5 ellipses from the population of 100 ellipses. (i) Choose a point at random in the ellipse "habitat" (that is, the figure); this could be done crudely by dropping a pencil point on the page, or much better by overlaying the page with graph paper and using random digits. (ii) If the chosen point is inside an ellipse, include that ellipse in the sample, otherwise start again at step (i). (iii) Continue until five ellipses have been selected. Explain why this scheme is *not* equivalent to random sampling. In what direction is the scheme biased—that is, would it tend to produce a $\bar{y}$ that is too large, or a $\bar{y}$ that is too small?

5.12 The serum cholesterol levels of a certain population of boys follow a normal distribution with mean 170 mg/dl and standard deviation 30 mg/dl (as in Example 4.1).
   a. What percentage of the boys have serum cholesterol values between 160 and 180 mg/dl?
   b. Suppose we were to choose at random from the population a large number of groups of 9 boys each. In what percentage of the groups would the group mean cholesterol value be between 160 and 180 mg/dl?
   c. If $\bar{Y}$ represents the mean cholesterol value of a random sample of 9 boys from the population, what is $\Pr\{160 \leq \bar{Y} \leq 180\}$?

5.13 An important indicator of lung function is forced expiratory volume (FEV), which is the volume of air that a person can expire in one second. Dr. Jones plans to measure FEV in a random sample of n young women from a certain population, and to use the sample mean $\bar{y}$ as an estimate of the population mean. Let E be the event that Jones' sample mean will be within ± 100 ml of the population mean. Assume that the population distribution is normal with mean 3,000 ml and standard deviation 400 ml.[3] Find $\Pr\{E\}$ if
   a. $n = 15$ **b.** $n = 60$

5.14 Refer to Exercise 5.13. Assume that the population distribution of FEV is normal with standard deviation 400 ml.
   a. Find $\Pr\{E\}$ if $n = 15$ and the population mean is 2,800 ml.
   b. Find $\Pr\{E\}$ if $n = 15$ and the population mean is 2,600 ml.
   c. How does $\Pr\{E\}$ depend on the population mean?

5.15 The heights of a certain population of corn plants follow a normal distribution with mean 145 cm and standard deviation 22 cm (as in Exercise 4.19).
a. What percentage of the plants are between 135 and 155 cm tall?
b. Suppose we were to choose at random from the population a large number of groups of 16 plants each. In what percentage of the groups would the group mean height be between 135 and 155 cm?
c. If $\bar{Y}$ represents the mean height of a random sample of 16 plants from the population, what is $\Pr\{135 \leq \bar{Y} \leq 155\}$?

5.16 The basal diameter of a sea anemone is an indicator of its age. The density curve shown here represents the distribution of diameters in a certain large population of anemones; the population mean diameter is 4.2 cm, and the standard deviation is 1.4 cm.[4] Let $\bar{Y}$ represent the mean diameter of 25 anemones randomly chosen from the population.

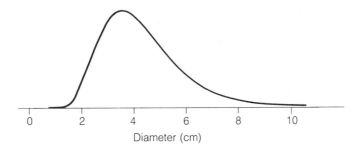

Diameter (cm)

a. Find the approximate value of $\Pr\{4 \leq \bar{Y} \leq 5\}$.
b. Why is your answer to part **a** approximately correct even though the population distribution of diameters is clearly not normal? Would the same approach be equally valid for a sample of size 2 rather than 25? Why or why not?

5.17 In a certain population of fish, the lengths of the individual fish follow approximately a normal distribution with mean 54.0 mm and standard deviation 4.5 mm. We saw in Example 4.5 that in this situation 72.15% of the fish are between 50 and 60 mm long. Suppose a random sample of four fish is chosen from the population. Find the probability that
a. all four fish are between 50 and 60 mm long.
b. the mean length of the four fish is between 50 and 60 mm.

5.18 In Exercise 5.17, the answer to part **b** was larger than the answer to part **a**. Argue that this must necessarily be true, no matter what the population mean and standard deviation might be. [*Hint:* Can it happen that the event in part **a** occurs but the event in part **b** does not?]

5.19 Professor Smith conducted a class exercise with a population of ellipses similar to the one used in Exercises 5.8 and 5.10. Professor Smith's ellipse population had a mean length of 30 mm and a standard deviation of 9 mm. Each of Smith's students took a random sample of size n and calculated the sample mean. Smith found that about 68% of the students had sample means between 28.5 and 31.5 mm. What was n? (Assume that n is large enough that the Central Limit Theorem is applicable.)

5.20 A certain assay for serum alanine aminotransferase (ALT) is rather imprecise. The results of repeated assays of a single specimen follow a normal distribution with mean equal to the ALT concentration for that specimen and standard deviation equal to 4 U/l (as in Exercise 4.29). Suppose a hospital lab measures many specimens every day, and specimens with reported ALT values of 40 or more are flagged as "unusually high." If a patient's true ALT concentration is 35 U/l, find the probability that his specimen will be flagged as "unusually high"

a. if the reported value is the result of a single assay.

b. if the reported value is the mean of three independent assays of the same specimen.

5.21 A medical researcher measured systolic blood pressure in 100 middle-aged men.[5] The results are displayed in the accompanying histogram; note that the distribution is rather skewed. According to the Central Limit Theorem, would we expect the distribution to be less skewed (and more bell-shaped) if it were based on $n = 400$ rather than $n = 100$ men? Explain.

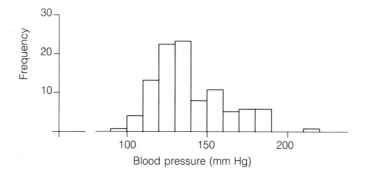

SECTION 5.4

ILLUSTRATION OF THE CENTRAL LIMIT THEOREM (OPTIONAL)

The importance of the normal distribution in statistics is due largely to the Central Limit Theorem and related theorems. In this section we take a closer look at the Central Limit Theorem.

According to the Central Limit Theorem, the sampling distribution of $\bar{Y}$ is approximately normal if n is large. If we consider larger and larger samples from a fixed nonnormal population, then the sampling distribution of $\bar{Y}$ will be more nearly normal for larger n. The following examples show the Central Limit Theorem at work for two nonnormal distributions: a moderately skewed distribution (Example 5.10) and a highly skewed distribution (Example 5.11).

EXAMPLE 5.10

EYE FACETS

The number of facets in the eye of the fruitfly *Drosophila melanogaster* is of interest in genetic studies. The distribution of this variable in a certain *Drosophila* population can be approximated by the density function shown in Figure 5.12. The distribution is moderately skewed; the population mean and standard deviation are $\mu = 64$ and $\sigma = 22$.[6]

Figure 5.13 shows the sampling distribution of $\bar{Y}$ for samples of various sizes from the eye-facet population. In order to clearly show the *shape* of these distributions, we have plotted them to different scales; the horizontal scale is stretched more for larger n. Notice that the distributions are somewhat skewed to the right, but the skewness is diminished for larger n; for $n = 32$ the distribution looks very nearly normal.

FIGURE 5.12
Distribution of Eye Facet
Number in a *Drosophila*
Population

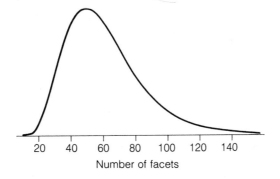

FIGURE 5.13
Sampling Distributions of
Ȳ for Samples from the
Drosophila Eye-Facet
Population

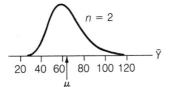

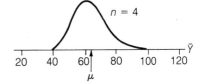

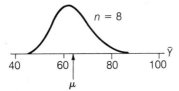

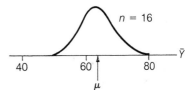

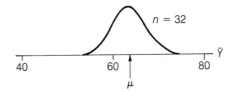

EXAMPLE 5.11

REACTION TIME

A psychologist measured the time required for a person to reach up from a fixed position and operate a pushbutton with his forefinger. The distribution of time scores (in milliseconds) for a single person is represented by the density shown in Figure 5.14 (page 142). About 10% of the time, the subject fumbled, or missed the button on the first thrust; the resulting delayed times appear as the second peak of the distribution.[7] The first peak is centered at 115 ms and the second at 450 ms; because of the two peaks, the overall distribution is violently skewed. The population mean and standard deviation are $\mu = 148$ ms and $\sigma = 105$ ms.

FIGURE 5.14

Distribution of Time
Scores in a Button-
Pushing Task

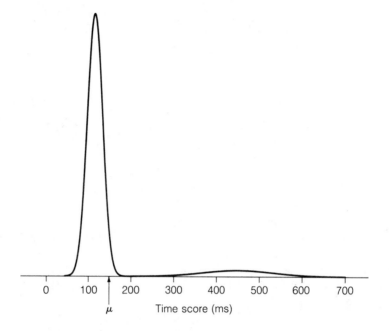

Figure 5.15 shows the sampling distribution of $\bar{Y}$ for samples of various sizes from the time-score distribution. To show the shape clearly, the Y scale has been stretched more for larger n. Notice that, for small n, the distribution has several modes. As n increases, these modes are reduced to bumps and finally disappear, and the distribution becomes increasingly symmetric.   ■

Examples 5.10 and 5.11 illustrate the fact, mentioned in Section 5.3, that the meaning of the requirement "n is large" in the Central Limit Theorem depends on the shape of the population distribution. Approximate normality of the sampling distribution of $\bar{Y}$ will be achieved for a moderate n if the population distribution is only moderately nonnormal (as in Example 5.10), while a highly nonnormal population (as in Example 5.11) will require a larger n.

Note, however, that Example 5.11 indicates the remarkable strength of the Central Limit Theorem. The skewness of the time-score distribution is so extreme that one might be reluctant to consider the mean as a summary measure. Even in this "worst case," you can see the effect of the Central Limit Theorem in the relative smoothness and symmetry of the sampling distribution for $n = 64$.

The Central Limit Theorem may seem rather like magic. To demystify it somewhat, we look at the time-score sampling distributions in more detail in the following example.

EXAMPLE 5.12

REACTION TIME

Consider the sampling distributions of $\bar{Y}$ displayed in Figure 5.15. Consider first the distribution for $n = 4$, which is the distribution of the mean of four button-pressing times. The high peak at the left of the distribution represents cases in

FIGURE 5.15

Sampling Distributions of
Ȳ for Samples from the
Time-Score Population

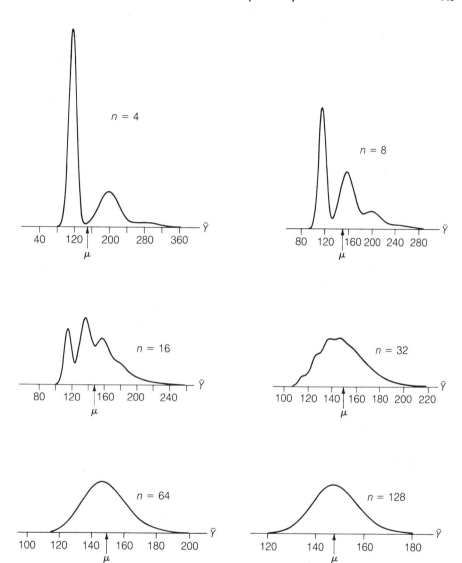

which the subject did not fumble any of the four thrusts, so that all four times
were about 115 ms; such an outcome would occur about 66% of the time [from
the binomial distribution, because $(.9)^4 = .66$]. The next lower peak represents
cases in which three thrusts took about 115 ms each, while one was fumbled
and took about 450 ms. (Notice that the average of three 115's and one 450 is
about 200, which is the center of the second peak.) Similarly, the third peak
represents cases in which the subject fumbled two of the four thrusts. The peaks
representing three and four fumbles are too low to be visible in the plot. Now
consider the plot for $n = 8$. The first peak represents eight good thrusts (no
fumbles), the second represents seven good thrusts and one fumble, the third

represents six good thrusts and two fumbles, and so on. The fourth and later peaks are blended together. For $n = 16$ the first peak is lower than the second because the occurrence of 16 good thrusts is less likely than 15 good thrusts and one fumble (as you can verify from the binomial distribution). For larger n, the first peaks are lower still and the later peaks are higher. For $n = 32$ the most likely outcome is three fumbles (about 10%) and 29 good thrusts; this outcome gives a mean time of about

$$\frac{(3)(450) + (29)(115)}{32} = 146 \text{ ms}$$

which is the location of the central peak. For similar reasons, the distribution for larger n is centered at about 148 ms, which is the population mean.   ∎

EXERCISES 5.22–5.23

5.22 Refer to Example 5.12. In the sampling distribution of $\bar{Y}$ for $n = 4$ (Figure 5.15), approximately what is the area under
a. the first peak?   **b.** the second peak?
[*Hint:*   Use the binomial distribution.]

5.23 Refer to Example 5.12. Consider the sampling distribution of $\bar{Y}$ for $n = 2$ (which is not shown in Figure 5.15).
a. Make a rough sketch of the sampling distribution. How many peaks does it have? Show the location (on the $\bar{Y}$-axis) of each peak.
b. Find the approximate area under each peak. [*Hint:*   Use the binomial distribution.]

S E C T I O N 5.5

THE NORMAL APPROXIMATION TO THE BINOMIAL DISTRIBUTION (OPTIONAL)

In Section 5.2 we saw that, for random sampling from a large dichotomous population, the sampling distribution of $\hat{\pi}$ is governed by the binomial distribution. Probabilities for the binomial distribution can be calculated from the formula

$$C_j \pi^j (1 - \pi)^{n-j}$$

However, this formula can be burdensome if n is not small. Fortunately, there is a convenient approximation available. In this section we show how the binomial distribution can be approximated by a normal distribution, if n is large.

THE NORMAL APPROXIMATION

The normal approximation to the binomial distribution can be expressed in two equivalent ways: in terms of the binomial distribution itself, or in terms of the sampling distribution of $\hat{\pi}$. We state both forms in the following theorem. In this theorem, n represents the sample size (or, more generally, the number of independent trials) and π represents the population proportion (or, more generally, the probability of success in each independent trial).

THEOREM 5.2: NORMAL APPROXIMATION TO BINOMIAL DISTRIBUTION

a. If n is large, then the binomial distribution can be approximated by a normal distribution with

$$\text{Mean} = n\pi$$

and

$$\text{Standard deviation} = \sqrt{n\pi(1 - \pi)}$$

b. If n is large, then the sampling distribution of $\hat{\pi}$ can be approximated by a normal distribution with

$$\text{Mean} = \pi$$

and

$$\text{Standard deviation} = \sqrt{\frac{\pi(1 - \pi)}{n}}$$

Remark: It is true, but not obvious, that the normal approximation to the binomial distribution is an application of the Central Limit Theorem (Section 5.3). The relationship is explained more fully in Appendix 5.1.

The following two examples illustrate the use of Theorem 5.2.

EXAMPLE 5.13

We consider a binomial distribution with $n = 20$ and $\pi = .3$. Figure 5.16(a) (page 146) shows this binomial distribution; superimposed is a normal curve with

$$\text{Mean} = n\pi = (20)(.3) = 6$$

and

$$\text{SD} = \sqrt{n\pi(1 - \pi)} = \sqrt{(20)(.3)(.7)} = 2.049$$

Note that the curve fits the distribution fairly well. Figure 5.16(b) shows the sampling distribution of $\hat{\pi}$ for $n = 20$ and $\pi = .3$ (the same distribution was shown in Figure 5.3); superimposed is a normal curve with

$$\text{Mean} = \pi = .3$$

and

$$\text{SD} = \sqrt{\frac{\pi(1 - \pi)}{n}} = \sqrt{\frac{(.3)(.7)}{20}} = .1025$$

Note that Figure 5.16(b) is just a relabelled version of Figure 5.16(a).

FIGURE 5.16

The Normal Approximation to the Binomial Distribution with $n = 20$ and $\pi = .3$

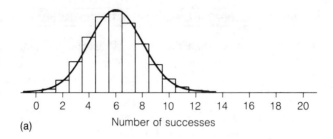

(a)

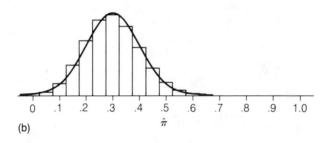

(b)

To illustrate the use of the normal approximation, let us consider the event that 20 independent trials result in 5 successes and 15 failures. In Example 5.3 we found that the exact probability of this event is .179; this probability can be visualized as the area of the bar above the "5" in Figure 5.17. The normal approximation to the probability is the corresponding area under the normal curve, which is shaded in Figure 5.17. The boundaries of the shaded area are 4.5 and 5.5, which correspond in the Z scale to

$$z = \frac{4.5 - 6}{2.049} = -.73$$

and

$$z = \frac{5.5 - 6}{2.049} = -.24$$

From Table 3, we find that the area is $.2673 - .0948 = .1725$, which is fairly close to the exact value of .179.

FIGURE 5.17

Normal Approximation to the Probability of 5 Successes

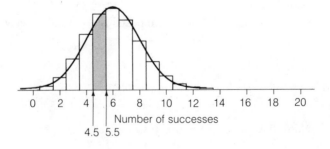

EXAMPLE 5.14

To illustrate part **b** of Theorem 5.2, we again assume that $n = 20$ and $\pi = .3$. In Example 5.4 we found that

$$\Pr\{.25 \le \hat{\pi} \le .35\} = .535$$

The normal approximation to this probability is the shaded area in Figure 5.18. The boundaries of the area are $\hat{\pi} = .225$ and $\hat{\pi} = .375$, which correspond on the Z scale to

$$z = \frac{.225 - .3}{.1025} = -.73$$

and

$$z = \frac{.375 - .3}{.1025} = .73$$

The resulting approximation (from Table 3) is then

$$\Pr\{.25 \le \hat{\pi} \le .35\} \approx 2(.2673) = .5346$$

which agrees very well with the exact value.

FIGURE 5.18

Normal Approximation to
$\Pr\{.25 \le \hat{\pi} \le .35\}$

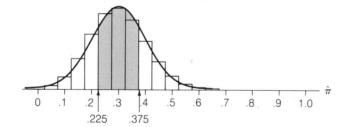

∎

THE CONTINUITY CORRECTION

Notice that the calculation in Example 5.14 used the boundaries $\hat{\pi} = .225$ and .375 rather than $\hat{\pi} = .25$ and .35; this is an example of a **continuity correction.*** The reason for the continuity correction can be seen from Figure 5.18. The exact probability is the area of the three rectangles corresponding to $\hat{\pi} = .25, .30,$ and .35; the boundaries of this region are .225 and .375. Without the continuity correction, we would calculate the area between $\hat{\pi} = .25$ and $\hat{\pi} = .35$, which is equal to .3758; this value is too small because it omits half of the $\hat{\pi} = .25$ rectangle and half of the $\hat{\pi} = .35$ rectangle.

In general, when using a continuity correction, the first step is to calculate the half-width of a histogram bar; the desired area is then extended by this amount

*The continuity correction was also discussed in the optional Section 4.4.

in each direction. For instance, in Example 5.14 the half-width of a histogram bar is equal to

$$\left(\frac{1}{2}\right)\left(\frac{1}{20}\right) = .025$$

and the boundaries of the shaded region in Figure 5.18 can be calculated as

$$.250 - .025 = .225 \quad \text{and} \quad .350 + .025 = .375$$

If the area to be calculated includes *many* bars of the probability histogram, then the continuity correction can be omitted without causing much error. If the area includes only a few bars, then the continuity correction greatly improves the accuracy of the approximation; as an extreme example, if the area includes only *one* bar (as in Figure 5.17), then omitting the continuity correction would give a probability of zero, which is not at all a useful approximation. (In Example 5.13, we applied the continuity correction by using the boundaries 4.5 and 5.5.)

HOW LARGE MUST n BE?

Theorem 5.2 states that the binomial distribution can be approximated by a normal distribution if n is "large." It is helpful to know how large n must be in order for the approximation to be adequate. The required n depends on the value of π. If $\pi = .5$, then the binomial distribution is symmetric and the normal approximation is quite good even for n as small as 10. However, if $\pi = .1$, the binomial distribution for $n = 10$ is quite skewed, and is poorly fitted by a normal curve; for larger n the skewness is diminished and the normal approximation is better. A simple rule of thumb is the following:

> The normal approximation to the binomial distribution is fairly good if both $n\pi$ and $n(1 - \pi)$ are at least equal to 5.

For example, if $n = 20$ and $\pi = .3$, as in Example 5.13, then $n\pi = 6$ and $n(1 - \pi) = 14$; since $6 \geq 5$ and $14 \geq 5$, the rule of thumb indicates that the normal approximation is fairly good.

EXERCISES 5.24–5.31

5.24 A fair coin is to be tossed 20 times. Find the probability that ten of the tosses will fall heads and ten will fall tails,
   a. using the binomial distribution formula.
   b. using the normal approximation with the continuity correction.

5.25 An epidemiologist is planning a study on the prevalence of oral contraceptive use in a certain population.[8] She plans to choose a random sample of n women and to use the sample proportion of oral contraceptive users ($\hat{\pi}$) as an estimate of the population proportion (π). Suppose that in fact $\pi = .12$. Use the normal approximation (with the continuity correction) to determine the probability that $\hat{\pi}$ will be within $\pm.03$ of π if

a. $n = 100$ **b.** $n = 200$

5.26 In a study of how people make probability judgments, college students (with no background in probability or statistics) were asked the following question[9]:

> A certain town is served by two hospitals. In the larger hospital about 45 babies are born each day, and in the smaller hospital about 15 babies are born each day. As you know, about 50% of all babies are boys. The exact percentage of baby boys, however, varies from day to day. Sometimes it may be higher than 50%, sometimes lower.
>
> For a period of 1 year, each hospital recorded the days on which more than 60% of the babies born were boys. Which hospital do you think recorded more such days?
> * The larger hospital
> * The smaller hospital
> * About the same (i.e., within 5% of each other)

a. Imagine that you are a participant in the study. Which answer would you choose, based on intuition alone?

b. Determine the correct answer by using the normal approximation (without the continuity correction) to calculate the appropriate probabilities. (For simplicity, assume that "more than 60%" means "60% or more.")

5.27 Consider random sampling from a dichotomous population with $\pi = .3$, and let E be the event that $\hat{\pi}$ is within $\pm.05$ of π. Use the normal approximation (without the continuity correction) to calculate $\Pr\{E\}$ for a sample of size $n = 400$. Your answer should agree with the value given in Table 5.2.

5.28 Refer to Exercise 5.27. Calculate $\Pr\{E\}$ for $n = 40$ (rather than 400)
a. with the continuity correction.   **b.** without the continuity correction.
Your answer to part **a** should agree with the value given in Table 5.2.

5.29 A certain cross between sweet-pea plants will produce progeny that are either purple-flowered or white flowered;[10] the probability of a purple-flowered plant is $\pi = \frac{9}{16}$. Suppose n progeny are to be examined, and let $\hat{\pi}$ be the sample proportion of purple-flowered plants. It might happen, by chance, that $\hat{\pi}$ would be closer to $\frac{1}{2}$ than to $\frac{9}{16}$. Find the probability that this misleading event would occur if
a. $n = 64$ **b.** $n = 320$
(Use the normal approximation without the continuity correction.)

5.30 A fair coin is to be tossed 10 times. Find the probability that between 30% and 40% (inclusive) of the tosses will fall heads,
a. using the binomial distribution formula.
b. using the normal approximation with the continuity correction.

5.31 In a certain population of mussels (*Mytilus edulis*), 80% of the individuals are infected with an intestinal parasite.[11] A marine biologist plans to examine 100 randomly chosen mussels from the population. Find the probability that 85% or more of the sampled mussels will be infected, using the normal approximation
a. without the continuity correction.   **b.** with the continuity correction.

In this chapter we have presented two important sampling distributions—the sampling distribution of $\hat{\pi}$ and the sampling distribution of $\bar{Y}$. Of course, there are many other important sampling distributions; examples are the sampling distribution of the sample standard deviation and the sampling distribution of the sample median.

The ethereal concept of a sampling distribution is linked to the solid reality of data through the random sampling model. Let us take another look at this model in the light of Chapter 5. As we have seen, a *random* sample is not necessarily a *representative* sample.* But using sampling distributions, one can specify the degree of representativeness to be expected in a random sample. For instance, it is intuitively plausible that a larger sample is likely to be more representative than a smaller sample from the same population. In Sections 5.2 and 5.3 we saw how a sampling distribution can make this vague intuition precise by specifying the probability that a specified degree of representativeness will be achieved by a random sample. Thus, sampling distributions provide what has been called "certainty about uncertainty."[12]

In Chapter 6 we will see for the first time how the theory of sampling distributions can be put to practical use in the analysis of data. We will find that, although the calculations of Chapter 5 seem to require the knowledge of unknowable quantities (such as μ and σ), nevertheless when analyzing data one can estimate the probable magnitude of sampling error using only information contained in the sample itself.

In addition to their application to data analysis, sampling distributions provide a basis for comparing the relative merits of different methods of analysis. For example, consider sampling from a normal population with mean μ. Of course, the sample mean $\bar{Y}$ is an estimator of μ. But since a normal distribution is symmetric, μ is also the population median, so the sample *median* is *also* an estimator of μ. How, then, can we decide which estimator is better? This question can be answered in terms of sampling distributions, as follows: Statisticians have determined that, if the population is normal, the sample median is inferior in the sense that its sampling distribution, while centered at μ, has a standard deviation larger than $\sigma/\sqrt{n}$. Consequently, the sample median is less efficient (as an estimator of μ) than the sample mean; for a given sample size n, the sample median provides less information about μ than does the sample mean. (If the population is not normal, however, the sample median can be much more efficient than the mean.)

---

*It is true, however, that sometimes the investigator can force the sample to be representative with respect to some variable (not the one under study) whose population distribution is known. For example, suppose we are sampling from a human population in order to study Y = blood pressure; since blood pressure is age-related, we might want to construct the sample so that it matches the population in age distribution. This kind of sampling is not *simple* random sampling, and the methods of analysis given in this book cannot be applied without suitable modification.

[*Note:* Exercises preceded by an asterisk refer to optional sections.]

5.32 In an agricultural experiment, a large field of wheat was divided into many plots (each plot being 7×100 ft) and the yield of grain was measured for each plot. These plot yields followed approximately a normal distribution with mean 88 lb and standard deviation 7 lb (as in Exercise 4.5). Let $\bar{Y}$ represent the mean yield of five plots chosen at random from the field. Find $\Pr\{\bar{Y} > 90\}$.

5.33 In a certain population, 83% of the people have Rh-positive blood type.[13] Suppose a random sample of $n = 10$ people is to be chosen from the population, and let $\hat{\pi}$ represent the proportion of Rh-positive people in the sample. Find
a. $\Pr\{\hat{\pi} = .8\}$ **b.** $\Pr\{\hat{\pi} = .9\}$

5.34 The heights of men in a certain population follow a normal distribution with mean 69.7 inches and standard deviation 2.8 inches.[14]
a. If a man is chosen at random from the population, find the probability that he will be more than 72 inches tall.
b. If two men are chosen at random from the population, find the probability that (i) both of them will be more than 72 inches tall; (ii) their mean height will be more than 72 inches.

5.35 Suppose a botanist grows many individually potted eggplants, all treated identically and arranged in groups of four pots on the greenhouse bench. After 30 days of growth, she measures the total leaf area Y of each plant. Assume that the population distribution of Y is approximately normal with mean = 800 cm^2 and SD = 90 cm^2.[15]
a. What percentage of the plants in the population will have leaf area between 750 cm^2 and 850 cm^2?
b. Suppose each group of four plants can be regarded as a random sample from the population. What percentage of the groups will have a group mean leaf area between 750 cm^2 and 850 cm^2?

5.36 Refer to Exercise 5.35. In a real greenhouse, what factors might tend to invalidate the assumption that each group of plants can be regarded as a random sample from the same population?

5.37 In a population of flatworms (*Planaria*) living in a certain pond, one in five individuals is adult and four are juvenile.[16] An ecologist plans to count the adults in a random sample of 20 flatworms from the pond; she will then use $\hat{\pi}$, the proportion of adults in the sample, as her estimate of π, the proportion of adults in the pond population. Find
a. $\Pr\{\hat{\pi} = \pi\}$ **b.** $\Pr\{\pi - .05 \le \hat{\pi} \le \pi + .05\}$

*5.38 Refer to Exercise 5.37. Use the normal approximation (with the continuity correction) to calculate the probabilities.

5.39 The activity of a certain enzyme is measured by counting emissions from a radio-actively labelled molecule. For a given tissue specimen, the counts in consecutive ten-second time periods may be regarded (approximately) as repeated independent observations from a normal distribution (as in Exercise 4.16). Suppose the mean ten-second count for a certain tissue specimen is 1,200 and the standard deviation is 35. For that specimen, let Y represent a ten-second count and let $\bar{Y}$ represent the mean of six ten-second counts. Find $\Pr\{1{,}175 \leq Y \leq 1{,}225\}$ and $\Pr\{1{,}175 \leq \bar{Y} \leq 1{,}225\}$, and compare the two. Does the comparison indicate that counting for one minute and dividing by six would tend to give a more precise result than merely counting for a single ten-second time period? How?

5.40 In a certain lab population of mice, the weights at 20 days of age follow approximately a normal distribution with mean weight = 8.3 gm and standard deviation = 1.7 gm.[17] Suppose many litters of 10 mice each are to be weighed. If each litter can be regarded as a random sample from the population, what percentage of the litters will have a *total* weight of 90 gm or more? [*Hint:* How is the total weight of a litter related to the mean weight of its members?]

5.41 Refer to Exercise 5.40. In reality, what factors would tend to invalidate the assumption that each litter can be regarded as a random sample from the same population?

5.42 A certain drug causes drowsiness in 20% of patients. Suppose the drug is to be given to five randomly chosen patients, and let $\hat{\pi}$ be the proportion who experience drowsiness.
a. Compute the sampling distribution of $\hat{\pi}$.
b. Display the distribution of part **a** as a histogram.

5.43 The skull breadths of a certain population of rodents follow a normal distribution with a standard deviation of 10 mm. Let $\bar{Y}$ be the mean skull breadth of a random sample of 64 individuals from this population, and let μ be the population mean skull breadth.
a. Suppose $\mu = 50$ mm. Find $\Pr\{\bar{Y}$ is within ± 2 mm of $\mu\}$.
b. Suppose $\mu = 100$ mm. Find $\Pr\{\bar{Y}$ is within ± 2 mm of $\mu\}$.
c. Suppose μ is unknown. Can you find $\Pr\{\bar{Y}$ is within ± 2 mm of $\mu\}$? If so, do it. If not, explain why not.

C H A P T E R 6

CONTENTS

ESTIMATION OF A POPULATION MEAN

S E C T I O N 6.1

**STATISTICAL
ESTIMATION**

In this chapter we undertake our first adventure into statistical inference. Recall that statistical inference is based on the random sampling model: We view our data as a random sample from some population, and we use the information in the sample to infer facts about the population. **Statistical estimation** is a form of statistical inference in which we use the data to (1) determine an estimate of some feature of the population; and (2) assess the precision of the estimate. Let us consider an example.

EXAMPLE 6.1

SOYBEAN GROWTH

As part of a study on plant growth, a plant physiologist grew 13 individually potted soybean seedlings of the type called "Wells II." She raised the plants in a greenhouse under identical environmental conditions (light, temperature, soil, and so on). She measured the total stem length (cm) for each plant after 16 days of growth. The data are given in Table 6.1.[1]

TABLE 6.1

Stem Length of Soybean Plants

STEM LENGTH (cm)				
20.2	22.9	23.3	20.0	19.4
22.0	22.1	22.0	21.9	21.5
19.7	21.5	20.9		

For these data, the mean and standard deviation are

$$\bar{y} = 21.3385 \approx 21.34 \text{ cm}$$

and

$$s = 1.2190 \approx 1.22 \text{ cm}$$

Suppose we regard the 13 observations as a random sample from a population; the population could be described by (among other things) its mean μ and its standard deviation σ. We might define μ and σ verbally as follows:

μ = Mean stem length of Wells II soybean plants grown under the specified conditions

σ = SD of stem lengths of Wells II soybean plants grown under the specified conditions

It is natural to estimate μ by the sample mean and σ by the sample standard deviation. Thus, from the data on the 13 plants,

21.34 is an estimate of μ;

1.22 is an estimate of σ.

We know that these estimates are subject to sampling error. Note that we are not speaking merely of measurement error; no matter how accurately each individual plant was measured, the sample information is imperfect in that only 13 plants were measured, rather than the entire infinite population of plants. ∎

In general, for a sample of observations on a quantitative variable Y, the sample mean and SD are estimates of the population mean and SD:

$\bar{y}$ is an estimate of μ;

s is an estimate of σ.

The notation for these means and SDs is summarized schematically in Figure 6.1.

FIGURE 6.1
Notation for Means and SDs of Sample and Population

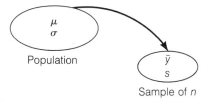

The main focus of this chapter is on the estimation of μ. We will see how to assess the reliability or precision of this estimate, and how to plan a study large enough to attain a desired precision.

SECTION 6.2

CONFIDENCE INTERVAL FOR μ

It is intuitively reasonable that the sample mean $\bar{y}$ should be an estimate of μ. It is not so obvious how to determine the reliability of the estimate. In this section we will see how the sample itself can be used to assess its own reliability.

THE STANDARD ERROR

As an estimate of μ, the sample mean $\bar{y}$ is imprecise to the extent that it is affected by sampling error. In Section 5.3 we saw that the magnitude of the sampling error—that is, the amount of discrepancy between $\bar{y}$ and μ—is described (in a probability sense) by the sampling distribution of $\bar{Y}$. The standard deviation of the sampling distribution of $\bar{Y}$ is

$$\sigma_{\bar{Y}} = \frac{\sigma}{\sqrt{n}}$$

Since s is an estimate of σ, a natural estimate of $\sigma/\sqrt{n}$ would be $s/\sqrt{n}$; this quantity is called the **standard error of the mean**. We will denote it as $SE_{\bar{y}}$, or sometimes simply SE.*

DEFINITION

The **standard error of the mean** is defined as

$$SE_{\bar{y}} = \frac{s}{\sqrt{n}}$$

---

*Some statisticians prefer to reserve the term "standard error" for $\sigma/\sqrt{n}$, and to call $s/\sqrt{n}$ the "estimated standard error."

The following example illustrates the definition.

EXAMPLE 6.2

SOYBEAN GROWTH
For the soybean growth data of Example 6.1, we have $n = 13$, $\bar{y} = 21.3385$ cm, and $s = 1.2190$ cm. The standard error of the mean is

$$SE_{\bar{y}} = \frac{s}{\sqrt{n}}$$

$$= \frac{1.2190}{\sqrt{13}} = .3381 \text{ cm}$$

■

As we have seen, the SE is an estimate of $\sigma_{\bar{Y}}$. On a more practical level, the SE can be interpreted in terms of the expected sampling error: Roughly speaking, the difference between $\bar{y}$ and μ is rarely more than a few standard errors. (We will make this statement precise in the following paragraphs.) Thus, the standard error is a measure of the reliability or precision of $\bar{y}$ as an estimate of μ; the smaller the SE, the more precise the estimate. Notice how the SE incorporates the two factors that affect reliability: (1) the inherent variability of the observations (expressed through s), and (2) the sample size (n).

CONFIDENCE INTERVAL FOR μ: BASIC IDEA

Under certain circumstances, the SE can be given a definite quantitative interpretation by using it to construct a **confidence interval** for the population mean. A confidence interval for μ is an interval whose upper and lower limits are computed from the data. The interval always contains $\bar{y}$; we hope that the interval contains μ. The first step in constructing a confidence interval is to choose a value called the **confidence coefficient**, which measures our "confidence" that the confidence interval contains μ. (The exact meaning of the confidence coefficient will be explained below.) We will introduce confidence intervals using a confidence coefficient of 95%.

Suppose that the observed variable (Y) follows a normal distribution in the population. Recall from Chapter 5 that in this case the sampling distribution of $\bar{Y}$ is also a normal distribution with mean μ and standard deviation $\sigma/\sqrt{n}$. Recall from Chapter 4 that about 95% of the values in a normal distribution lie within ± 2 standard deviations of the mean. Now consider the following statement:

$$\bar{y} \text{ lies within } \pm 2\sigma/\sqrt{n} \text{ of } \mu. \tag{6.1}$$

According to the above discussion, the statement (6.1) will be true for about 95% of samples of size n (in the meta-experiment). But if statement (6.1) is true, then the distance between $\bar{y}$ and μ does not exceed $2\sigma/\sqrt{n}$. It follows that the interval

$$\bar{y} \pm 2\sigma/\sqrt{n} \tag{6.2}$$

will contain μ for about 95% of all samples.

The interval (6.2) cannot be used for data analysis because it contains a quantity—namely, σ—that cannot be determined from the data. If we replace σ by its estimate—namely, s—then we can calculate an interval from the data, but what happens to the 95% interpretation? Fortunately, it turns out that there is an escape from this dilemma. The escape was discovered by a British scientist named W. S. Gosset, who was employed by the Guinness Brewery; he published his findings in 1908 under the pseudonym "Student," and the method has borne his name ever since.[2] "Student" discovered that, if we replace σ in the interval (6.2) by the sample SD, s, then the 95% interpretation can be preserved if the multiplier of $\sigma/\sqrt{n}$ (that is, 2) is replaced by a suitable quantity; the new quantity is denoted $t_{.05}$ and is related to a distribution known as "Student's t distribution."

STUDENT'S t DISTRIBUTION

The **Student's t distributions** are theoretical continuous distributions that are used for many purposes in statistics, including the construction of confidence intervals. The exact shape of a Student's t distribution depends on a quantity called "degrees of freedom," abbreviated "df." Figure 6.2 shows the density curves of two Student's t distributions with df = 3 and df = 10, and also a normal curve. (Note that the scaling on the t-axis corresponds to the Z-axis for the normal distribution; when dealing with Student's t distributions, it is not necessary to convert to a standardized scale.) A t curve is symmetric and bell-shaped like the normal curve, but has a larger standard deviation. As the df increase, the t curves approach the normal curve; thus, the normal curve can be regarded as a t curve with infinite df (df = ∞).

FIGURE 6.2
Two Student's t Curves
and a Normal Curve
(df = ∞)

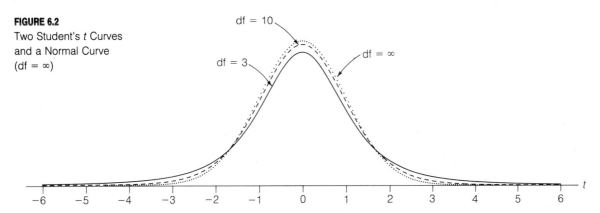

The quantity $t_{.05}$ is called the "5% critical value" of the Student's t distribution, and is defined to be the value such that the interval between $-t_{.05}$ and $+t_{.05}$ contains 95% of the area under the curve, as shown in Figure 6.3.* The total

*In some statistics textbooks, you may find other notations, such as $t_{.025}$ or $t_{.975}$, rather than $t_{.05}$.

shaded area in Figure 6.3 is equal to .05; note that the shaded area consists of two "pieces" of area .025 each.

FIGURE 6.3
Definition of the Critical Value $t_{.05}$

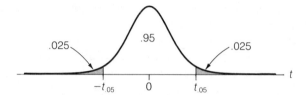

Critical values of Student's t distribution are tabulated in Table 4. The values of $t_{.05}$ are shown in the column headed ".05." If you glance down this column, you will see that the values of $t_{.05}$ decrease as the df increase; for df $= \infty$ (that is, for the normal distribution) the value is $t_{.05} = 1.960$. You can confirm from Table 3 that the interval ± 1.96 (on the Z scale) contains 95.00% of the area under a normal curve.

Other columns of Table 4 show other critical values, which are defined analogously; for instance, the interval $\pm t_{.10}$ contains 90% of the area under a Student's t curve.

CONFIDENCE INTERVAL FOR μ: METHOD

We describe Student's method for constructing a confidence interval for μ. First, suppose we have chosen a confidence coefficient equal to 95%. To construct a 95% confidence interval for μ, we compute the upper and lower limits of the interval as

$$\bar{y} \pm t_{.05}\, SE_{\bar{y}}$$

that is,

$$\bar{y} \pm t_{.05}\, \frac{s}{\sqrt{n}}$$

where the critical value $t_{.05}$ is determined from Student's t distribution with

$$df = n - 1$$

The following example illustrates the construction of a confidence interval.

EXAMPLE 6.3
SOYBEAN GROWTH

For the soybean stem length data of Example 6.1, we have $n = 13$, $\bar{y} = 21.3385$ cm, and $s = 1.2190$ cm. The value of df is

$$df = n - 1 = 13 - 1 = 12$$

so that from Table 4 we find

$$t_{.05} = 2.179$$

The 95% confidence interval for μ is

$$21.3385 \pm (2.179)\,\frac{1.2190}{\sqrt{13}}$$

$$21.3385 \pm (2.179)(.3381)$$

$$21.3385 \pm .7367$$

or, approximately,

$$21.34 \pm .74$$

The confidence interval may be left in this form. Alternatively, the endpoints of the interval may be explicitly calculated as

$$21.34 - .74 = 20.60 \quad \text{and} \quad 21.34 + .74 = 22.08$$

and the interval may be written compactly as

$$(20.6, 22.1)$$

or in a more complete form as the following "confidence statement":

$$20.6 \text{ cm} < \mu < 22.1 \text{ cm}$$

The confidence statement asserts that the population mean stem length of Wells II soybean plants, grown under the specified conditions, is between 20.6 cm and 22.1 cm. The interpretation of the "95% confidence" will be discussed after the next example.   ∎

Confidence coefficients other than 95% are used analogously. For instance, a 90% confidence interval for μ is constructed using $t_{.10}$ instead of $t_{.05}$ as follows:

$$\bar{y} \pm t_{.10}\,\frac{s}{\sqrt{n}}$$

The following is an example.

EXAMPLE 6.4
SOYBEAN GROWTH

From Table 4, we find that $t_{.10} = 1.782$ with df $= 12$. Thus, the 90% confidence interval for μ from the soybean growth data is

$$21.3385 \pm (1.782)\,\frac{1.2190}{\sqrt{13}}$$

$$21.3385 \pm .6025$$

or

$$20.7 < \mu < 21.9$$ ∎

As you see, the choice of a confidence coefficient is somewhat arbitrary. For the soybean growth data, the 95% confidence interval is

$$21.34 \pm .74$$

and the 90% confidence interval is

$$21.34 \pm .60$$

Thus, the 90% confidence interval is narrower than the 95% confidence interval; however, this does not reflect a real difference in precision, for we have more confidence that the wider interval actually contains μ. The narrower interval may look better, but this is merely a "cosmetic" improvement. In other words, it is the standard error itself which expresses the uncertainty in $\bar{y}$; the confidence interval is a formal device for interpreting the standard error.

Remark The quantity $(n - 1)$ is referred to as "degrees of freedom" because the deviations $(y - \bar{y})$ must sum to zero, and so only $(n - 1)$ of them are "free" to vary. A sample of size n provides only $(n - 1)$ independent pieces of information about variability, that is, about σ. This is particularly clear if we consider the case $n = 1$; a sample of size 1 provides some information about μ, but no information about σ, and so no information about sampling error. It makes sense, then, that when $n = 1$ we cannot use Student's t method to calculate a confidence interval: the sample standard deviation does not exist (see Example 2.21) and there is no critical value with df = 0. A sample of size 1 is sometimes called an "anecdote"; for instance, an individual medical case history is an anecdote. Of course, a case history can contribute greatly to medical knowledge, but it does not (in itself) provide a basis for judging how closely the individual case resembles the population at large.

INTERPRETATION OF A CONFIDENCE INTERVAL

In what sense can we be "confident" of a confidence interval? To answer this question, let us assume for the moment that we are dealing with a random sample from a normal population. (In Section 6.5 we will see that the assumption of normality can be relaxed.)

Consider, for instance, a 95% confidence interval. One way to interpret the confidence coefficient (95%) is to refer to the meta-experiment of repeated samples from the same population. If a 95% confidence interval for μ is constructed for each sample, then 95% of the confidence intervals will contain μ. Of course, the observed data in an experiment comprise only *one* of the possible samples; the individual researcher hopes "confidently" that his sample is one of the lucky 95%, but he will never know.

The following example provides a more concrete visualization of the meta-experiment interpretation of a confidence coefficient.

EXAMPLE 6.5

EGGSHELL THICKNESS

In a certain large population of chicken eggs (described in Example 4.2), the distribution of eggshell thickness is normal with mean $\mu = .38$ mm and standard deviation $\sigma = .03$ mm. Figure 6.4 shows some typical samples from this population; plotted on the right are the associated 90% confidence intervals. The sample sizes are $n = 5$ and $n = 20$. Notice that the second confidence interval

FIGURE 6.4 Confidence Intervals for Mean Eggshell Thickness

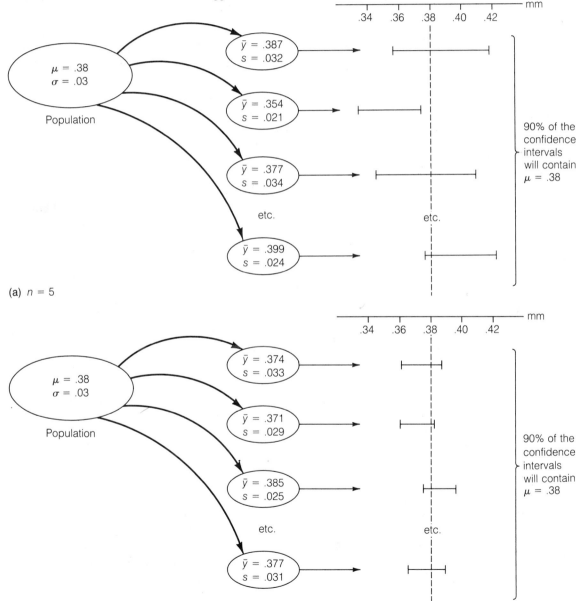

with $n = 5$ does not contain μ. In the totality of potential confidence intervals, the percentage that would contain μ is 90% for either sample size; of course, the larger samples tend to produce narrower confidence intervals.   ∎

A confidence coefficient can be interpreted as a probability, but caution is required. If we consider 95% confidence intervals, for instance, then the following statement is correct:

$$\Pr\{\text{the confidence interval contains } \mu\} = .95$$

However, one should realize that it is "the confidence interval" which is the random item in this statement, and it is not correct to replace this item with its value from the data. Thus, for instance, we found in Example 6.3 that the 95% confidence interval for the mean soybean growth is

$$20.6 \text{ cm} < \mu < 22.1 \text{ cm} \tag{6.3}$$

Nevertheless, it is *not* correct to conclude that

$$\Pr\{20.6 \text{ cm} < \mu < 22.1 \text{ cm}\} = .95$$

because this statement has no chance element. The following analogy may help to clarify this point. Suppose we let Y represent the number of spots showing when a balanced die is tossed; then

$$\Pr\{Y = 2\} = \frac{1}{6}$$

On the other hand, if we now toss the die and observe 5 spots, it is obviously *not* correct to substitute this "data" in the probability statement to conclude that

$$\Pr\{5 = 2\} = \frac{1}{6}$$

As the above discussion indicates, the confidence coefficient (for instance, 95%) is a property of the *method* rather than of a particular interval. An individual statement—such as (6.3) above—is either true or false; but in the long run, if the researcher constructs 95% confidence intervals in various experiments, each time producing a statement such as (6.3), then 95% of the statements will be true.

RELATIONSHIP TO SAMPLING DISTRIBUTION OF $\bar{Y}$

At this point it may be helpful to look back and see how a confidence interval for μ is related to the sampling distribution of $\bar{Y}$. Recall from Section 5.3 that the mean of the sampling distribution is μ and its standard deviation is $\sigma/\sqrt{n}$. Figure 6.5 shows a particular sample mean ($\bar{y}$) and its associated 95% confidence interval for μ, superimposed on the sampling distribution of $\bar{Y}$. Notice that the particular confidence interval *does* contain μ; this will happen for 95% of samples.

FIGURE 6.5

Relationship Between a
Particular Confidence
Interval for μ and the
Sampling Distribution of $\bar{Y}$

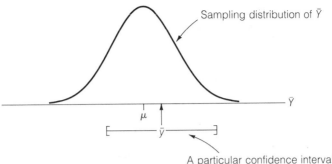

Sampling distribution of $\bar{Y}$

$\bar{Y}$

μ

$\bar{y}$

A particular confidence interval

EXERCISES 6.1–6.14

6.1 A pharmacologist measured the concentration of dopamine in the brains of several rats. The mean concentration was 1,269 ng/gm and the standard deviation was 145 ng/gm.[3] What was the standard error of the mean if
a. 8 rats were measured?   **b.** 30 rats were measured?

6.2 An agronomist measured the heights of n corn plants.[4] The mean height was 220 cm and the standard deviation was 15 cm. Calculate the standard error of the mean if
a. $n = 25$ **b.** $n = 100$

6.3 In evaluating a forage crop, it is important to measure the concentration of various constituents in the plant tissue. In a study of the reliability of such measurements, a batch of alfalfa was dried, ground, and passed through a fine screen. Five small (.3 gm) aliquots of the alfalfa were then analyzed for their content of insoluble ash.[5] The results (gm/kg) were as follows:

    10.0   8.9   9.1   11.7   7.9

For these data, calculate the mean, the standard deviation, and the standard error of the mean.

6.4 (*Sampling exercise*)   Refer to Exercise 5.8. Use your sample of five ellipse lengths to construct an 80% confidence interval for μ, using the formula $\bar{y} \pm (1.533)s/\sqrt{n}$. To facilitate the pooling of results from the entire class, compute the two endpoints explicitly.

6.5 (*Sampling exercise*)   Refer to Exercise 5.10. Use your sample of 20 ellipse lengths to construct an 80% confidence interval for μ, using the formula $\bar{y} \pm (1.328)s/\sqrt{n}$. To facilitate the pooling of results from the entire class, compute the two endpoints explicitly.

6.6 As part of a study of the development of the thymus gland, researchers weighed the glands of five chick embryos after 14 days of incubation. The thymus weights (mg) were as follows:[6]

    29.6   21.5   28.0   34.6   44.9

For these data, calculate the mean, the standard deviation, and the standard error of the mean.

6.7 Refer to Exercise 6.6. Construct
  a. a 90% confidence interval for the population mean.
  b. a 95% confidence interval for the population mean.

6.8 Six healthy three-year-old female Suffolk sheep were injected with the antibiotic Gentamicin, at a dosage of 10 mg/kg body weight. Their blood serum concentrations (μg/ml) of Gentamicin 1.5 hours after injection were as follows:[7]

  33   26   34   31   23   25

  a. Construct a 95% confidence interval for the population mean.
  b. Define in words the population mean which you estimated in part **a**. (See Example 6.1.)
  c. The interval constructed in part **a** nearly contains all of the observations; will this typically be true for a 95% confidence interval? Explain.

6.9 Human beta-endorphin (HBE) is a hormone secreted by the pituitary gland under conditions of stress. A researcher conducted a study to investigate whether a program of regular exercise might affect the resting (unstressed) concentration of HBE in the blood. He measured blood HBE levels, in January and again in May, in ten participants in a physical fitness program. The results were as shown in the table.[8] Construct a 95% confidence interval for the population mean difference in HBE levels between January and May. [*Hint:* You need to use only the values in the right-hand column.]

PARTICIPANT	HBE LEVEL (pg/ml)		
	January	May	Difference
1	42	22	20
2	47	29	18
3	37	9	28
4	9	9	0
5	33	26	7
6	70	36	34
7	54	38	16
8	27	32	−5
9	41	33	8
10	18	14	4
Mean	37.8	24.8	13.0
SD	17.6	10.9	12.4

6.10 Invertase is an enzyme that may aid in spore germination of the fungus *Colletotrichum graminicola*. A botanist incubated specimens of the fungal tissue in petri dishes and then assayed the tissue for invertase activity. The specific activity values for nine petri dishes incubated at 90% relative humidity for 24 hours are summarized as follows:[9]

  Mean = 5,111 units   SD = 818 units

Construct a 95% confidence interval for the mean invertase activity under these experimental conditions.

6.11 As part of a study of the treatment of anemia in cattle, researchers measured the concentration of selenium in the blood of 36 cows who had been given a dietary supplement of selenium (2 mg/day) for one year. The cows were all the same breed (Santa Gertrudis) and had borne their first calf during the year. The mean selenium concentration was 6.21 μg/dl and the standard deviation was 1.84 μg/dl.[10] Construct a 95% confidence interval for the population mean.

6.12 In a study of larval development in the tufted apple budmoth (*Platynota idaeusalis*), an entomologist measured the head widths of 50 larvae. All 50 larvae had been reared under identical conditions and had moulted six times. The mean head width was 1.20 mm and the standard deviation was .14 mm. Construct a 90% confidence interval for the population mean.[11]

6.13 In Table 4 we find that $t_{.05} = 1.960$ when df $= \infty$. Show how this value can be verified using Table 3.

6.14 Use Table 3 to find the value of $t_{.005}$ when df $= \infty$. (Do not attempt to interpolate in Table 4.)

SECTION 6.3

INTERPRETING AND REPORTING SUMMARY STATISTICS

Whether you are reading a scientific report, or writing one, or simply trying to understand the results of your own experiment, you will often encounter summary statistics such as means, standard deviations, and standard errors. In this section we consider some issues of interpretation and presentation of these summary statistics.

STANDARD ERROR VERSUS STANDARD DEVIATION

The terms "standard error" and "standard deviation" are sometimes confused. It is extremely important to distinguish between standard error (SE) and standard deviation (s, or SD). These two quantities describe entirely different aspects of the data. The SD describes the dispersion of the data, while the SE describes the uncertainty (due to sampling error) in the mean of the data. Let us consider a concrete example.

EXAMPLE 6.6
LAMB BIRTHWEIGHTS

A geneticist weighed 28 female lambs at birth. The lambs were all born in April, were all the same breed (Rambouillet), and were all single births (no twins). The diet and other environmental conditions were the same for all the parents. The birthweights are shown in Table 6.2.[12]

TABLE 6.2
Birthweights of 28 Rambouillet Lambs

BIRTHWEIGHT (kg)						
4.3	5.2	6.2	6.7	5.3	4.9	4.7
5.5	5.3	4.0	4.9	5.2	4.9	5.3
5.4	5.5	3.6	5.8	5.6	5.0	5.2
5.8	6.1	4.9	4.5	4.8	5.4	4.7

For these data, the mean is $\bar{y} = 5.17$ kg, the standard deviation is $s = .65$ kg, and the standard error is SE $= .12$ kg. The SD describes the variability from one lamb to the next, while the SE indicates the uncertainty associated with the mean

(5.17 kg), viewed as an estimate of the population mean birthweight. This distinction is emphasized in Figure 6.6, which shows a histogram of the lamb birthweight data; the SD is indicated as a deviation from $\bar{y}$, while the SE is indicated as uncertainty associated with $\bar{y}$ itself.

FIGURE 6.6

Birthweights of 28 Lambs

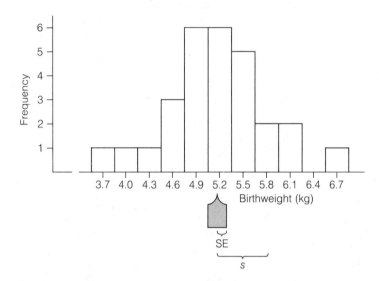

Another way to highlight the contrast between the SE and the SD is to consider samples of various sizes. As the sample size increases, the sample mean and SD tend to approach more closely the population mean and SD; indeed, the distribution of the data tends to approach the population distribution. The standard error, by contrast, tends to decrease as n increases; when n is very large the sample mean is a very precise estimate of the population mean, and so the SE is very small. The following example illustrates this effect.

EXAMPLE 6.7

LAMB BIRTHWEIGHTS

Suppose we regard the birthweight data of Example 6.6 as a sample of size $n =$ 28 from a population, and consider what would happen if we were to choose larger samples from the same population—that is, if we were to measure the birthweights of additional female Rambouillet lambs born under the specified conditions. Figure 6.7 shows the kind of results we might expect; the values given are fictitious but realistic. For very large n, $\bar{y}$ and s would be very close to μ and σ, where

   μ = Mean birthweight of female Rambouillet lambs born under the conditions described

and

   σ = Standard deviation of birthweights of female Rambouillet lambs born under the conditions described

FIGURE 6.7 Samples of Various Sizes from the Lamb Birthweight Population

	Sample size			
	$n = 28$	$n = 280$	$n = 2,800$	$n \to \infty$
$\bar{y}$	5.17	5.09	5.15	μ
s	.65	.71	.67	σ
SE	.12	.042	.013	0

Sample distribution

ROUNDING SUMMARY STATISTICS

Scientific reports often include values of means, standard deviations, and standard errors. In preparing a report, it is necessary to decide how far to round off the values. Guidelines for rounding are based on the precision of the summary statistics as estimates of their population counterparts. The box contains a simple rule of thumb for rounding that will suffice for most practical purposes. This rule is conservative, and in many cases one might report one less digit than the rule suggests.

RULE FOR ROUNDING

For reporting the mean, standard deviation, and standard error of the mean, the following procedure is recommended:

1. Round the SE to two significant digits.
2. Round $\bar{y}$ and s to match the SE with respect to the decimal position of the last significant digit.

Part (b) of the Rule for Rounding means that if, for instance, the SE is rounded to three decimal places, then $\bar{y}$ and s are also rounded to three decimal places. The following example illustrates the Rule for Rounding.*

EXAMPLE 6.8
LAMB BIRTHWEIGHTS

For the lamb birthweight data of Example 6.6, the summary statistics are

$$\bar{y} = 5.1678571... \text{ kg} \qquad s = .6543606... \text{ kg} \qquad SE = .1236625... \text{ kg}$$

How far should these be rounded in writing a report? It would clearly be silly to report $\bar{y}$ as 5.1679 kg when the SE value of .12 kg indicates that $\bar{y}$, viewed as

*The concept of significant digits is reviewed in Appendix 6.1.

an estimate of μ, is uncertain even in the first decimal place. The Rule for Rounding suggests that the SE should be rounded to two significant digits, as follows:

SE = .12 kg

Next, $\bar{y}$ and s are rounded to the same number of decimal places as the SE:

$\bar{y}$ = 5.17 kg

s =   .65 kg   ∎

Note that $\bar{y}$ and s match the SE in decimal places, but *not* necessarily in significant digits; for instance, in Example 6.8, $\bar{y}$ has three significant digits while the SE has only two.

The Rule for Rounding does not depend in a direct way on the precision of the original measurements. This may seem strange—how can we possibly know $\bar{y}$ more exactly than we know each value of y? The answer is that the errors of measurement will tend to cancel each other. In practice, experimenters sometimes make measurements that are overprecise rather than insufficiently precise. For instance, suppose the 28 lamb birthweights had been measured to the nearest gram; if the summary statistics were also expressed in grams, they would be reported as

SE =   120 gm

$\bar{y}$ = 5,170 gm

s =   650 gm

with the understanding that the final "0" in each of these values is not a significant digit. Because of the size of the standard error, the time and energy spent measuring each birthweight to the nearest gm would be wasted. Unfortunately, it is not uncommon for experimenters to err in this way—that is, to make measurements with a degree of precision that is unwarranted by the size of the standard error:

Notice that the appropriate degree of rounding does depend on the sample size. If the sample size for the lamb birthweights were $n = 2,800$ instead of $n = 28$, but $\bar{y}$ and s were the same, then the summary values would be rounded as:

SE =   .012 kg

$\bar{y}$ = 5.168 kg

s =   .654 kg

GRAPHICAL AND TABULAR PRESENTATION

The clarity and impact of a scientific report can be greatly enhanced by well-designed displays of the data. Data can be displayed graphically or in a table. We briefly discuss some of the options.

Let us first consider graphical presentation of data. Here is an example.

EXAMPLE 6.9

MAO AND
SCHIZOPHRENIA

The enzyme monoamine oxidase (MAO) is of interest in the study of human behavior. Figures 6.8 and 6.9 display measurements of MAO activity in the blood platelets in five groups of people: Groups I, II, and III are three diagnostic categories of schizophrenic patients (see Example 1.4), and groups IV and V are healthy male and female controls.[13] The MAO activity values are expressed as

FIGURE 6.8

MAO Data Displayed
as $\bar{y} \pm$ SE

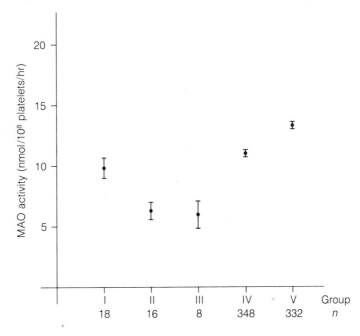

FIGURE 6.9

MAO Data Displayed
as $\bar{y} \pm$ SD

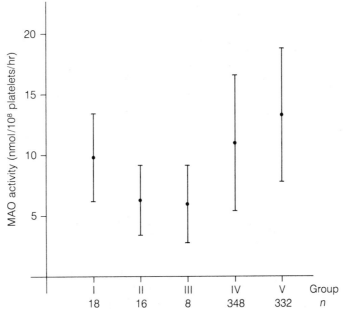

nmol benzylaldehyde product per 10^8 platelets per hour. In both Figures 6.8 and 6.9, the dots represent the group means; the vertical lines represent ±SE in Figure 6.8 and ±SD in Figure 6.9.

Figures 6.8 and 6.9 convey very different information. Figure 6.8 conveys (a) the mean MAO value in each group, and (b) the reliability of each group mean, viewed as an estimate of its respective population mean. Figure 6.9 conveys (a) the mean MAO value in each group, and (b) the variability of MAO within each group. For instance, group V shows greater variability of MAO than group I (Figure 6.9) but has a much smaller standard error (Figure 6.8) because it is a much larger group.

Figure 6.8 invites the viewer to compare the means, and gives some indication of the reliability of the comparisons. (But a full discussion of comparison of two or more means must wait until Chapter 7 and later chapters.) Figure 6.9 invites the viewer to compare the means and also to compare the standard deviations. Furthermore, Figure 6.9 gives the viewer some information about the extent of overlap of the MAO values in the various groups. For instance, consider groups IV and V; whereas they appear quite "separate" in Figure 6.8, we can easily see from Figure 6.9 that there is considerable overlap of individual MAO values in the two groups.

A third and more complete way to present data graphically is as shown in Figure 6.10, in which the data of groups I, II, and III are displayed. Each dot represents an individual observation; the data are not plotted exactly but have been grouped into narrow classes to give a neater display. Of course, this method cannot be used if the group size is too large; groups IV and V cannot be displayed in this way. Also, the method may be confusing when groups of different sizes are to be compared.

FIGURE 6.10
MAO Data Displayed as Individual Points

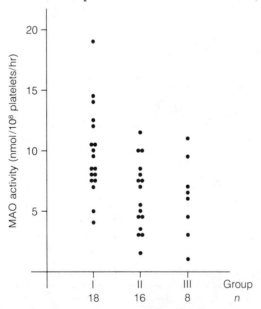

A fourth form of data display, which is coming into increasing use, is the **boxplot**. Figure 6.11 shows boxplots of the MAO data. The horizontal line through each box represents the median, the ends of the box represent the quartiles, and the endpoints of the vertical lines represent the minimum and maximum value in each group. The boxplot conveys information about skewness and extremes of the data which is missing in Figure 6.9.

FIGURE 6.11

MAO Data Displayed as Boxplots

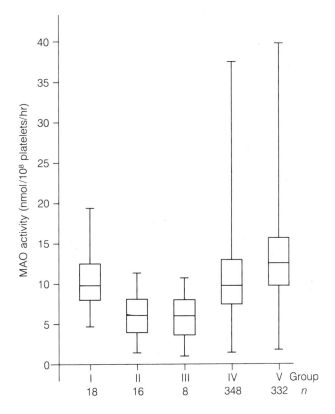

A fifth form of graphical display is shown in Figure 6.12 (page 172); here the entire relative frequency distribution of each group is shown. In this display, the Y-axis is horizontal rather than vertical as it is in Figures 6.8–6.11. ■

Notice that the five data displays in Figures 6.8–6.12 fall into two types: Figure 6.8, by showing standard errors, indicates the sampling error of the data which it displays; none of the other displays does this. On the other hand, each of the other displays conveys information about the entire distribution of each sample, rather than focusing entirely on the sample mean.

The preceding paragraph should shed some light on a question that is often asked: Should a display show the SD or the SE? There is no universal answer to this question. The choice depends on which kind of information is to be conveyed.

FIGURE 6.12
MAO Data Displayed
as Histograms

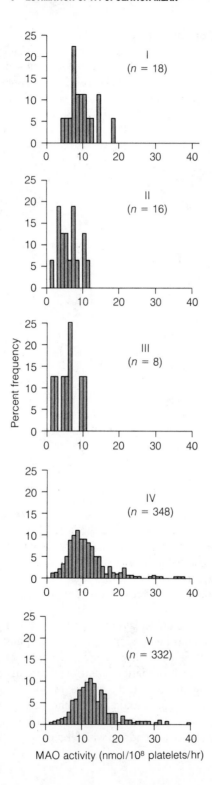

MAO activity (nmol/10^8 platelets/hr)

The SE format should be used when the emphasis is to be placed on the mean response, and variability is not of interest in itself but is perhaps merely an annoyance. This is the case in much experimental work. For instance, in graphical display of a dose–response curve, the SE format is often used because the SE tells the viewer how much credence to place in the shape of the curve drawn through the means; to show the variability of individual responses around the curve would be distracting and perhaps irrelevant. On the other hand, when the distribution of values around the mean is also of interest (as it would be for the MAO data), then a measure of dispersion such as the SD should be reported in addition to the mean. Of course, in many situations it is desirable to report *both* the SD *and* the SE. This can easily be done graphically, for instance by combining the information in Figures 6.8 and 6.9 into one display, or by incorporating standard error information into Figure 6.10 or Figure 6.12.

In some scientific reports, data are summarized in tables rather than graphically. Table 6.3 shows a tabular summary for the MAO data of Example 6.9.

TABLE 6.3
MAO Activity in Five
Groups of People

Group	n	MAO ACTIVITY (nmol/10^8 platelets/hr)		
		Mean	SE	SD
I	18	9.81	.85	3.62
II	16	6.28	.72	2.88
III	8	5.97	1.13	3.19
IV	348	11.04	.30	5.59
V	332	13.29	.30	5.50

Note that Table 6.3 shows both the SD and the SE. An author might choose to present only one of them, if he is designing a complex table. It is, however, important to include at least one measure of variability, even if variability is small.

In presenting summary statistics, whether in a table or in a graph, it is of course essential to indicate clearly whether a standard error or a standard deviation is being displayed. By the same token, when reading a scientific report you should always try to be aware of which format—SE or SD—you are looking at. In addition, the sample size (n) should always be reported when a mean is reported.

EXERCISES 6.15–6.21

6.15 Round each set of summary statistics according to the Rule for Rounding.
  a. $\bar{y} = 167.413$, $s = 27.6190$, SE = 4.36695
  b. $\bar{y} = 3{,}062.317$, $s = 286.691$, SE = 26.1712
  c. $\bar{y} = .0376152$, $s = .00624137$, SE = .000693486
  d. $\bar{y} = 759{,}241$, $s = 34{,}716.3$, SE = 11,572.1

6.16 Round each set of summary statistics according to the Rule for Rounding.
  a. $\bar{y} = 15.1982$, $s = 4.67294$, SE = .208980
  b. $\bar{y} = .047684$, $s = .0064937$, SE = .0014520
  c. $\bar{y} = 5{,}609.4320$, $s = 747.9881$, SE = 334.5104
  d. $\bar{y} = 5{,}609.4320$, $s = 747.9881$, SE = 86.3702

6.17 A zoologist measured tail length in 86 individuals, all in the 1-year age group, of the deermouse *Peromyscus*. The mean length was 60.43 mm and the standard deviation was 3.06 mm. The table presents a frequency distribution of the data.[14]

TAIL LENGTH (mm)	NUMBER OF MICE
52–53	1
54–55	3
56–57	11
58–59	18
60–61	21
62–63	20
64–65	9
66–67	2
68–69	1
Total	86

a. Calculate the standard error of the mean.
b. Construct a histogram of the data and indicate the intervals $\bar{y} \pm$ SD and $\bar{y} \pm$ SE on your histogram (see Figure 6.6).

6.18 Refer to the mouse data of Exercise 6.17. Suppose the zoologist were to measure 500 additional animals from the same population. Based on the data in Exercise 6.17, what would you predict would be the standard deviation of the 500 new measurements?

6.19 In a report of a pharmacological study, the experimental animals were described as follows:[15] "Rats weighing 150 ± 10 gm were injected..." with a certain chemical, and then certain measurements were made on the rats. If the author intends to convey the degree of homogeneity of the group of experimental animals, then should the 10 gm be the SD or the SE? Explain.

6.20 Data are often summarized in this format: $\bar{y} \pm$ SE. Suppose this interval is interpreted as a confidence interval. If the sample size is large, what would be the confidence coefficient of such an interval? That is, what is the chance that an interval computed as

$$\bar{y} \pm (1.00)\text{SE}$$

will actually contain the population mean? [*Hint:* Recall that the confidence coefficient of the interval $\bar{y} \pm (1.96)$SE is 95%.]

6.21 (*Continuation of Exercise 6.20*)
a. If the sample size is small but the population distribution is normal, is the confidence coefficient of the interval $\bar{y} \pm$ SE larger or smaller than the answer to Exercise 6.20? Explain.
b. How is the answer to Exercise 6.20 affected if the population distribution of Y is not approximately normal?

**PLANNING A STUDY
TO ESTIMATE μ**

In planning an experiment, it is wise to consider in advance whether the estimates generated from the data will be sufficiently precise. It can be painful indeed to discover after a long and expensive study that the standard errors are so large that the primary questions addressed by the study cannot be answered.

The precision with which a population mean can be estimated is determined by two factors: (1) the population variability of the observed variable Y, and (2) the sample size.

In some situations the variability of Y cannot, and perhaps should not, be reduced. For example, a wildlife ecologist may wish to conduct a field study of a natural population of fish; the heterogeneity of the population is not controllable, and in fact is a proper subject of investigation. As another example, in a medical investigation, in addition to knowing the average response to a treatment, it may also be important to know how much the response varies from one patient to another, and so it may not be appropriate to use an overly homogeneous group of patients.

On the other hand, it is often appropriate, especially in comparative studies, to reduce the variability of Y by holding *extraneous* conditions as constant as possible. For example, physiological measurements may be taken at a fixed time of day; tissue may be held at a controlled temperature; all animals used in an experiment may be the same age.

Suppose, then, that plans have been made to reduce the variability of Y as much as possible, or desirable. What sample size will be sufficient to achieve a desired degree of precision in estimation of the population mean? If we use the standard error as an indicator of precision, then this question can be approached in a straightforward manner. Recall that the SE is defined as

$$\text{SE}_{\bar{y}} = \frac{s}{\sqrt{n}}$$

In order to decide on a value of n, one must (1) specify what value of the SE is considered desirable to achieve, and (2) have available a preliminary guess of the SD, either from a pilot study or other previous experience, or from the scientific literature. The required sample size is then determined from the following equation.

$$\text{Desired SE} = \frac{\text{Guessed SD}}{\sqrt{n}}$$

The following example illustrates the use of this equation.

EXAMPLE 6.10
SOYBEAN GROWTH

The soybean stem-length data of Example 6.1 yielded the following summary statistics:

$$\bar{y} = 21.34 \text{ cm}$$
$$s = 1.22 \text{ cm}$$
$$\text{SE} = .34 \text{ cm}$$

Suppose the researcher is now planning a new study of soybean growth and has decided that it would be desirable that the SE be no more than .25 cm. As a preliminary guess of the SD, she will use the value from the old study, namely 1.22 cm. Thus, the desired n must satisfy the following relation:

$$.25 = \frac{1.22}{\sqrt{n}}$$

This equation is easily solved to give $n = 23.8$; the new experiment should include 23 or 24 plants.   ■

You may wonder how a researcher would arrive at a value such as .25 cm for the desired SE. Such a value is determined by considering how much error one is willing to tolerate in the estimate of μ. In comparative experiments, the primary consideration is usually the size of anticipated treatment effects. For instance, if one is planning to compare two experimental groups, the anticipated SE for each experimental group should be substantially smaller than (preferably less than one-fourth of) the anticipated difference between the two group means.* Thus, the soybean researcher of Example 6.10 might arrive at the value .25 cm if she were planning to compare two environmental conditions which she expected to produce stem lengths differing (on the average) by about 1 cm. She would then plan to grow 23 or 24 plants in each of the two environmental conditions.

To see how the required n depends on the specified precision, suppose the soybean researcher specified the desired SE to be .125 cm rather than .250 cm. Then the relation would be

$$.125 = \frac{1.22}{\sqrt{n}}$$

which yields $n = 95.3$, so that she would plan to include 95 or 96 plants in each group. Thus, to double the precision (by cutting the SE in half) requires not twice as many, but four times as many observations. This phenomenon of "diminishing returns" is due to the square root in the SE formula.

EXERCISES 6.22–6.24

6.22 An experiment is being planned to compare the effects of several diets on the weight gain of beef cattle, measured over a 140-day test period.[16] In order to have enough precision to compare the diets, it is desired that the standard error of the mean for each diet should not exceed 5 kg.
  a. If the population standard deviation of weight gain is guessed to be about 20 kg on any of the diets, how many cattle should be put on each diet in order to achieve a sufficiently small standard error?
  b. If the guess of the standard deviation is doubled, to 40 kg, does the required number of cattle double? Explain.

*This is a rough guideline for obtaining adequate sensitivity to discriminate between treatments. Such sensitivity, technically called *power*, is discussed in Chapter 7.

6.23 A medical researcher proposes to estimate the mean serum cholesterol level of a certain population of middle-aged men, based on a random sample of 1% of the population. He asks a statistician for advice. The ensuing discussion reveals that the researcher wants the standard error of the mean to be 3 mg/dl or less, and also that the standard deviation of serum cholesterol in the population is probably about 40 mg/dl.[17] Will the researcher's proposed sampling scheme (to examine 1% of the population) be adequate, if the population size is
a. 1,000? **b.** 10,000? **c.** 100,000?
(The answers may run counter to your intuition.)

6.24 Suppose you are planning an experiment to test the effects of various drugs on drinking behavior in the rat. The observed variable will be Y = 1-hour water consumption after 23-hour deprivation. Preliminary data (as in Example 2.6) indicate that the standard deviation of Y under control conditions is approximately 2.5 ml. Using this as a guess of σ, determine how many rats you should have in a treatment group, if you want the standard error of the group mean to be no more than
a. 1 ml **b.** .5 ml

SECTION 6.5

CONDITIONS FOR VALIDITY OF ESTIMATION METHODS

For any sample of quantitative data, one can use the methods of this chapter to compute the mean, its standard error, and various confidence intervals. However, the *interpretations* that we have given for these descriptions of the data are valid only under certain conditions.

CONDITIONS FOR VALIDITY OF THE SE FORMULA

First, the very notion of regarding the sample mean as an estimate of a population mean requires that the data be viewed "as if" they had been generated by random sampling from some population. To the extent that this is not possible, any inference beyond the actual data is questionable. The following example illustrates the difficulty.

EXAMPLE 6.11

MARIJUANA AND INTELLIGENCE

Ten people who used marijuana heavily were found to be quite intelligent; their mean IQ was 128.4, whereas the mean IQ for the general population is known to be 100. The ten people belonged to a religious group that uses marijuana for ritual purposes; since their decision to join the group might very well be related to their intelligence, it is not clear that the ten can be regarded (with respect to IQ) as a random sample from any particular population, and therefore there is no apparent basis for thinking of the sample mean (128.4) as an estimate of the mean IQ of a particular population (such as, for instance, all heavy marijuana users). An inference about the *effect* of marijuana on IQ would be even more implausible, especially because data were not available on the IQs of the ten people *before* they began marijuana use.[18] ∎

Second, the use of the standard error formula SE $= s/\sqrt{n}$ requires two further conditions:

1. The population size must be large compared to the sample size. This requirement is rarely a problem in the life sciences; the sample can be as much as 5% of the population without seriously invalidating the SE formula.
2. The observations must be independent of each other. This requirement means that the n observations actually give n independent pieces of information about the population.

Data often fail to meet the independence requirement if the experiment has a **hierarchical structure**, in which observational units are "nested" within sampling units, as illustrated by the following example.

EXAMPLE 6.12
CANINE ANATOMY

The coccygeus muscle is a bilateral muscle in the pelvic region of the dog. As part of an anatomical study, the left side and the right side of the coccygeus muscle were weighed for each of 21 female dogs. There were thus $2 \times 21 = 42$ observations, but only 21 units chosen from the population of interest (female dogs). Because of the symmetry of the coccygeus, the information contained in the right and left sides is largely redundant, so that the data contain not 42, but only 21, independent pieces of information about the coccygeus muscle of female dogs. It would therefore be incorrect to apply the SE formula as if the data comprised a sample of size $n = 42$. The hierarchical nature of the data set is indicated in Figure 6.13.[19]

FIGURE 6.13
Hierarchical Data
Structure of Example 6.12

Dog: · · ·

Hierarchical data structures are rather common in the life sciences. For instance, observations may be made on 90 nerve cells which come from only three different cats; on 80 kernels of corn which come from only four ears; on 60 young mice who come from only ten litters. A particularly clear example of nonindependent observations is replicated measurements on the same individual; for instance, if a physician makes triplicate blood pressure measurements on each of ten patients, he clearly does not have 30 independent observations. In some situations the correct treatment of hierarchical data is obvious; for instance, the triplicate blood pressure measurements should be averaged to give a single value for each patient. In other situations, however, lack of independence can be more subtle. For instance, suppose 60 young mice from ten litters are included in an experiment to compare two diets. Then the choice of a correct analysis depends on the *design* of the experiment—on such aspects as whether the diets are fed to the young mice themselves or to the mothers, and how the animals are allocated to the two diets. The subject of design of experiments will be discussed in detail in Chapter 8.

CONDITIONS FOR VALIDITY OF CONFIDENCE INTERVAL

A confidence interval for μ provides a definite quantitative interpretation for $SE_{\bar{y}}$. The validity of Student's t method for constructing confidence intervals depends on the form of the population distribution of the observed variable Y. If Y follows a normal distribution in the population, then Student's t method is exactly **valid**—that is to say, the probability that the confidence interval will contain μ is actually equal to the confidence coefficient (for example, 95%). By the same token, this interpretation is approximately valid if the population distribution is approximately normal. But many distributions encountered in the life sciences are not even approximately normal. Fortunately, a theorem of mathematical statistics, similar to the Central Limit Theorem, guarantees that the Student's t confidence interval is approximately valid even if the sampled population is not normal, *if* the sample size is large. This fact can often be used to justify the use of the confidence interval even in situations where the population distribution cannot be assumed to be approximately normal.

From a practical point of view, the important question is: How large must the sample be in order for the confidence interval to be approximately valid? Not surprisingly, the answer to this question depends on the *degree* of nonnormality of the population distribution: If the population is only moderately nonnormal, then n need not be very large. The following example illustrates this dependence.

EXAMPLE 6.13

VALIDITY OF CONFIDENCE INTERVALS

Table 6.4 shows the actual probability that a Student's t confidence interval will contain μ, for samples from three different populations.[20] The forms of the population distributions are shown in Figure 6.14. Population 1 is a normal population, population 2 is moderately skewed, and population 3 is a violently skewed, "L-shaped" distribution. (Populations 2 and 3 were discussed in optional Section 5.4.)

FIGURE 6.14

The Three Populations of Example 6.13

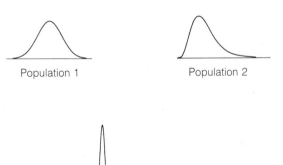

Population 1   Population 2

Population 3

TABLE 6.4

Actual Probability That Confidence Intervals Will Contain the Population Mean

(a) 95% Confidence Interval

	SAMPLE SIZE						
	2	4	8	16	32	64	Very Large
Population 1	.95	.95	.95	.95	.95	.95	.95
Population 2	.94	.93	.94	.94	.95	.95	.95
Population 3	.87	.53	.57	.80	.88	.92	.95

(b) 99% Confidence Interval

	SAMPLE SIZE						
	2	4	8	16	32	64	Very Large
Population 1	.99	.99	.99	.99	.99	.99	.99
Population 2	.99	.98	.98	.98	.99	.99	.99
Population 3	.97	.82	.60	.81	.93	.96	.99

For population 1, Table 6.4 shows that the confidence interval method is exactly valid for all sample sizes, even $n = 2$. For population 2 the method is approximately valid even for fairly small samples. For population 3 the approximation is very poor for small samples and is only fair for samples as large as $n = 64$. In a sense, population 3 is a "worst case"; it could be argued that the mean is not a meaningful measure for population 3, because of its bizarre shape. ■

SUMMARY OF CONDITIONS

In summary, Student's t method of constructing a confidence interval for μ is appropriate if the conditions stated in the box are assumed to hold.

---

1. **Conditions on the design of the study**
   a. It must be reasonable to regard the data as a random sample from a large population.
   b. The observations in the sample must be independent of each other.
2. **Conditions on the form of the population distribution**
   a. If n is small, the population distribution must be approximately normal.
   b. If n is large, the population distribution need not be approximately normal.

The required "largeness" in condition 2(b) depends (as shown in Example 6.13) on the degree of nonnormality of the population. In many practical situations, moderate sample sizes (say, $n = 20$ to 30) are large enough.

VERIFICATION OF CONDITIONS

In practice, the above "conditions" are usually "assumptions" rather than known facts. However, it is always important to check whether the assumptions are reasonable in a given case.

To determine whether the random sampling model is applicable to a particular study, the design of the study should be scrutinized, with particular attention to possible biases in the choice of experimental material and to possible noninde-pendence of the observations due to hierarchical data structures.

As to whether the population distribution is approximately normal, information on this point may be available from previous experience with similar data. If the only source of information is the data at hand, then normality can be roughly checked by making a histogram (or stem-and-leaf display) of the data. Unfor-tunately, for small or moderate sample size, this check is fairly crude; for instance, if you look back at Figure 5.11, you will see that even samples of size 25 from a normal population do not appear particularly normal. Of course, if the sample is large, then the sample histogram gives us good information about the popu-lation shape; however, if n is large the requirement of normality is less important anyway.

In any case, a crude check is better than none, and every data analysis should begin with inspection of a histogram of the data, with special attention to any observations that lie very far from the center of the distribution.

EXERCISES 6.25–6.28

6.25 SGOT is an enzyme that shows elevated activity when the heart muscle is damaged. In a study of 31 patients who underwent heart surgery, serum levels of SGOT were measured 18 hours after surgery.[21] The mean was 49.3 U/l and the standard deviation was 68.3 U/l. If we regard the 31 observations as a sample from a population, what feature of the data casts doubt on the assumption that the population distribution is normal?

6.26 A dendritic tree is a branched structure that emanates from the body of a nerve cell. In a study of brain development, researchers examined brain tissue from seven adult guinea pigs. The investigators randomly selected nerve cells from a certain region of the brain and counted the number of dendritic branch segments emanating from each selected cell. A total of 36 cells were selected, and the resulting counts were as follows:[22]

38	42	25	35	35	33	48	53	17
24	26	26	47	28	24	35	38	26
38	29	49	26	41	26	35	38	44
25	45	28	31	46	32	39	59	53

The mean of these counts is 35.67 and the standard deviation is 9.99.

Suppose we want to construct a 95% confidence interval for the population mean. We could calculate the standard error as

$$SE_{\bar{y}} = \frac{9.99}{\sqrt{36}} = 1.67$$

and obtain the confidence interval as

$$35.67 \pm (2.042)(1.67)$$

or

$$32.3 < \mu < 39.1$$

a. On what grounds might the above analysis be criticized? [*Hint:* Are the observations independent?]

b. Using the classes 15–19, 20–24, and so on, construct a frequency distribution of the data. Does the shape of the distribution support the criticism you made in part **a**? If so, explain how.

6.27 In an experiment to study the regulation of insulin secretion, blood samples were obtained from seven dogs before and after electrical stimulation of the vagus nerve. The following values show, for each animal, the increase (after minus before) in the immunoreactive insulin concentration (μU/ml) in pancreatic venous plasma.[23]

| 30 | 100 | 60 | 30 | 130 | 1,060 | 30 |

For these data, Student's *t* method yields the following 95% confidence interval for the population mean:

$$-145 < \mu < 556$$

What feature of the data suggests that Student's *t* method may not be appropriate in this case?

6.28 In a study of parasite-host relationships, 242 larvae of the moth *Ephestia* were exposed to parasitization by the Ichneumon fly. The following table shows the number of Ichneumon eggs found in each *Ephestia* larva.[24]

NUMBER OF EGGS (Y)	NUMBER OF LARVAE
0	21
1	77
2	52
3	41
4	23
5	13
6	9
7	1
8	2
9	0
10	2
11	0
12	0
13	0
14	0
15	1
Total	242

For these data, $\bar{y} = 2.368$ and $s = 1.950$. Student's t method yields the following 95% confidence interval for μ, the population mean number of eggs per larva:

$$2.12 < \mu < 2.61$$

a. Does it appear reasonable to assume that the population distribution of Y is approximately normal? Explain.

b. In view of your answer to part **a**, on what grounds can you defend the application of Student's t method to these data?

SECTION 6.6

PERSPECTIVE AND SUMMARY

In this section we place Chapter 6 in perspective by relating it to other chapters and also to other methods for analyzing a single sample of data. We also present a condensed summary of the methods of Chapter 6.

SAMPLING DISTRIBUTIONS AND DATA ANALYSIS

The theory of the sampling distribution of $\bar{Y}$ (Section 5.3) seemed to require knowledge of quantities—μ and σ—that in practice are unknown. In Chapter 6, however, we have seen how to make an inference about μ, including an assessment of the precision of that inference, using only information provided by the sample. Thus, the theory of sampling distributions has led to a practical method of analyzing data.

In later chapters we will study more complex methods of data analysis. Each method is derived from an appropriate sampling distribution; in most cases, however, we will not study the sampling distribution in detail.

CHOICE OF CONFIDENCE COEFFICIENT

In illustrating the confidence interval methods, we have usually chosen a confidence coefficient equal to 95%. However, it should be remembered that the confidence coefficient is *arbitrary*. There is nothing wrong with an 80% confidence interval, for instance.

OTHER METHODS FOR ESTIMATING A MEASURE OF CENTER

The sample mean and its associated confidence interval for μ are the "classical" methods for estimating the center of a population, and they are the methods of choice if the population distribution is approximately normal. However, if the distribution of the data is highly skewed or contains long straggly tails, then the classical methods may not be the best.

Sometimes skewness can be removed by a nonlinear transformation of the data. For instance, if the distribution of Y is skewed to the right, it often happens that the distribution of (log Y) will be much less skewed (we saw this effect in Example 2.33).

Another approach to use with skewed or long-tailed data is to supplement the sample mean by a more resistant measure of center, such as the sample median or trimmed mean. In such a situation, it is also useful to supplement the standard deviation with a more resistant measure of dispersion such as the interquartile range.*

A special difficulty arises if the data contain one or more **outliers**—that is, observations which lie so far from the main body of the data that one might suspect that they do not belong to the distribution at all. If outliers are present, it is generally not advisable to simply discard them, unless they are believed to be the result of gross errors or blunders; however, it may be desirable to minimize their effect by use of resistant analysis. Occasionally, the exceptional cases (that is, the outliers) are interesting and informative in themselves, and therefore merit special attention in the analysis.

TRIMMING VERSUS PRUNING   It is important to realize that the use of the trimmed mean is *not* the same as the regrettable practice of "pruning" the data by simply discarding observations that do not look "nice," and then using the remaining observations just as if the pruning had not occurred. For instance, suppose $n = 20$. Then the sample 5% trimmed mean is the mean of the middle 18 observations, but the associated SE should *not* be calculated as $s/\sqrt{18}$; rather, the SE formula must be modified to take account of the trimming.[25]

CHARACTERISTICS OTHER THAN MEASURES OF CENTER

This chapter has primarily discussed estimation of the center of a population. In some situations, one may wish to estimate other parameters of a population. For example, in evaluating a measurement technique, interest may focus on the repeatability of the technique, as indicated by the standard deviation of repeated determinations. As another example, in defining the limits of health, a medical researcher might want to estimate the 95th percentile of serum cholesterol levels in a certain population. Just as the precision of the mean can be indicated by a standard error or a confidence interval, statistical techniques are also available to specify the precision of estimation of parameters such as the population standard deviation or 95th percentile.

SUMMARY OF ESTIMATION METHODS

For convenient reference, we summarize in the box the confidence interval methods presented in this chapter.

---

*Methods using resistant measures are sometimes called **robust methods**, which means that their validity and efficiency are relatively insensitive to the form of the population distribution.

Standard error of the mean

$$SE_{\bar{y}} = \frac{s}{\sqrt{n}}$$

Confidence interval for μ

95% confidence interval: $\bar{y} \pm t_{.05}\, SE_{\bar{y}}$

Critical value $t_{.05}$ from Student's t distribution with df $= n - 1$.

Intervals with other confidence coefficients (such as 90%, 99%, etc.) are constructed analogously (using $t_{.10}$, $t_{.01}$, etc.).

Confidence interval formula is valid if (1) data can be regarded as a random sample from a large population, (2) observations are independent, and (3) population is normal. If n is large then condition (3) is less important.

SUPPLEMENTARY EXERCISES 6.29–6.38

6.29 To study the conversion of nitrite to nitrate in the blood, researchers injected four rabbits with a solution of radioactively labelled nitrite molecules. Ten minutes after injection, they measured for each rabbit the percentage of the nitrite that had been converted to nitrate. The results were as follows:[26]

51.1   55.4   48.0   49.5

a. For these data, calculate the mean, the standard deviation, and the standard error of the mean.
b. Construct a 95% confidence interval for the population mean percentage.

6.30 The diameter of the stem of a wheat plant is an important trait because of its relationship to breakage of the stem, which interferes with harvesting the crop. An agronomist measured stem diameter in eight plants of the Tetrastichon cultivar of soft red winter wheat. All observations were made three weeks after flowering of the plant. The stem diameters (mm) were as follows:[27]

2.3   2.6   2.4   2.2   2.3   2.5   1.9   2.0

a. For these data, calculate the mean, the standard deviation, and the standard error of the mean.
b. Construct a 95% confidence interval for the population mean.
c. Define in words the population mean that you estimated in part b (see Example 6.1).

6.31 Refer to Exercise 6.30. Suppose that the data on the eight plants are regarded as a pilot study, and that the agronomist now wishes to design a new study for which he wants the standard error of the mean to be only .03 mm. Approximately how many plants should he plan to measure in the new study?

6.32 Over a period of about nine months, 1,353 women reported the timing of each of their menstrual cycles. For the first cycle reported by each woman, the mean cycle time was 28.86 days, and the standard deviation of the 1,353 times was 4.24 days.[28]

a. Construct a 99% confidence interval for the population mean cycle time.
b. Because environmental rhythms can influence biological rhythms, one might hypothesize that the population mean menstrual cycle time is 29.5 days, the length of the lunar month. Is the confidence interval of part a consistent with this hypothesis?

6.33 Refer to the menstrual cycle data of Exercise 6.32.

a. Over the entire time period of the study, the women reported a total of 12,247 cycles. When all of these cycles are included, the mean cycle time is 28.22 days. Explain why one would expect that this mean would be smaller than the value 28.86 given in Exercise 6.32. [*Hint:* If each woman reported for a fixed time period, which women contributed more cycles to the total of 12,247 observations?]
b. Instead of using only the first reported cycle as in Exercise 6.32, one could use the first four cycles for each woman, thus obtaining $1{,}353 \times 4 = 5{,}412$ observations. One could then calculate the mean and standard deviation of the 5,412 observations and divide the SD by $\sqrt{5{,}412}$ to obtain the SE; this would yield a much smaller value than the SE found in Exercise 6.32. Why would this approach not be valid?

6.34 For the 28 lamb birthweights of Example 6.6, the mean is 5.1679 kg, the SD is .6544 kg, and the SE is .1237 kg. Construct
a. a 95% confidence interval for the population mean.
b. a 99% confidence interval for the population mean.

6.35 Refer to Exercise 6.34.
a. What assumptions are required for the validity of the confidence interval?
b. Which of the assumptions of part a can be checked (roughly) from the histogram of Figure 6.6?
c. Twin births were excluded from the lamb birthweight data. Which of the assumptions of part a might have been questionable if twins had been included along with single births?

6.36 As part of a study of natural variation in blood chemistry, serum potassium concentrations were measured in 84 healthy women. The mean concentration was 4.36 mEq/l, and the standard deviation was .42 mEq/l. The table presents a frequency distribution of the data.[29]

SERUM POTASSIUM (mEq/l)	NUMBER OF WOMEN
3.1–3.3	1
3.4–3.6	2
3.7–3.9	7
4.0–4.2	22
4.3–4.5	28
4.6–4.8	16
4.9–5.1	4
5.2–5.4	3
5.5–5.7	1
Total	84

a. Calculate the standard error of the mean.
b. Construct a histogram of the data and indicate the intervals $\bar{y} \pm SD$ and $\bar{y} \pm SE$ on the histogram (see Figure 6.6).
c. Construct a 95% confidence interval for the population mean.
d. In medical diagnosis, physicians often use "reference limits" for judging blood chemistry values; these are the limits within which we would expect to find 95% of healthy people. Would the confidence interval of part **c** be a reasonable choice of "reference limits" for serum potassium in women? Why or why not?

6.37 Refer to Exercise 6.36. Suppose a similar study is to be conducted next year, to include serum potassium measurements on 200 healthy women. Based on the data in Exercise 6.36, what would you predict would be
a. the SD of the new measurements?
b. the SE of the new measurements?

6.38 An agronomist selected six wheat plants at random from a plot, and then, for each plant, selected 12 seeds from the main portion of the wheat head; by weighing, drying, and reweighing, she determined the percent moisture in each batch of seeds. The results were as follows:[30]

62.7   63.6   60.9   63.0   62.7   63.7

a. Calculate the mean, the standard deviation, and the standard error of the mean.
b. Construct a 90% confidence interval for the population mean.

C H A P T E R 7

CONTENTS

COMPARISON OF TWO INDEPENDENT SAMPLES

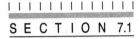

SECTION 7.1

INTRODUCTION

In Chapter 6 we considered the analysis of a single sample of quantitative data, specifically the estimation of the population mean. In practice, however, much scientific research involves the comparison of two or more samples from different populations. In the present chapter we introduce methods for comparing *two* samples.

EXAMPLES

Two-sample comparisons can arise in a variety of ways. Here are two examples.

EXAMPLE 7.1

HEMATOCRIT IN MALES AND FEMALES

Hematocrit level is a measure of the concentration of red cells in blood. Figure 7.1 shows the relative frequency distributions of hematocrit values for two samples of 17-year-old American youths—489 males and 469 females.[1] The sample means and standard deviations are given in Table 7.1.

FIGURE 7.1

Hematocrit Values in 17-Year-Old Youths. **(a)** 489 males; **(b)** 469 females.

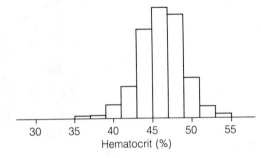

(a) Males

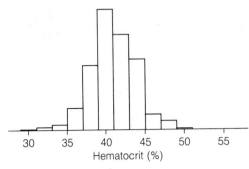

(b) Females

TABLE 7.1

Hematocrit (percent)

	MALES	FEMALES
Mean	45.8	40.6
SD	2.8	2.9

The following features can be seen from Figure 7.1 and Table 7.1. First, the males tend to have higher levels than the females. (Nevertheless, the two distributions do overlap quite a lot, so that many females have higher levels than many males.) Second, in spite of the substantial difference in means, the two distributions have very nearly the same standard deviation and are quite similar in shape. Thus, the main difference between the two distributions is a *shift* along the Y-axis.   ■

EXAMPLE 7.2

PARGYLINE AND SUCROSE CONSUMPTION

A study was conducted to determine the effect of the psychoactive drug Pargyline on feeding behavior in the black blowfly *Phormia regina*. The response variable was the amount of sucrose (sugar) solution a fly would drink in 30 minutes. The experimenters used two separate groups of flies: a group injected with Pargyline (905 flies) and a control group injected with saline (900 flies). Comparing the responses of the two groups provides an indirect assessment of the effect of Pargyline. (One might propose that a more *direct* way to determine the effect of the drug would be to measure each fly twice—on one occasion after injecting Pargyline and on another occasion after injecting saline. However, this direct method is not practical because the measurement procedure disturbs the fly so much that each fly can be measured only once.) Figure 7.2 shows the data for the two groups, and Table 7.2 shows the means and standard deviations.[2]

FIGURE 7.2

Sucrose Consumption of Flies. **(a)** 900 control flies; **(b)** 905 Pargyline-treated flies.

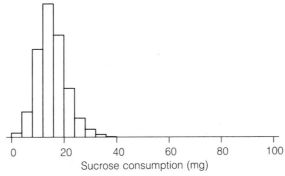

(a) Control

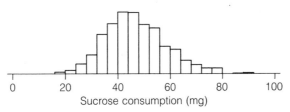

(b) Pargyline

TABLE 7.2

Sucrose Consumption (mg)

	CONTROL	PARGYLINE
Mean	14.9	46.5
SD	5.4	11.7

It is clear from Figure 7.2 that the two distributions differ in two distinct ways: first, the Pargyline distribution is shifted to the right and second, the Pargyline distribution is more dispersed. This impression is confirmed by Table 7.2, which shows that the Pargyline distribution has both a larger mean and a larger standard deviation than the control distribution.   ∎

Examples 7.1 and 7.2 both involve two-sample comparisons. But notice that the two studies differ in a fundamental way. In Example 7.1 the samples come from populations that occur naturally; the investigator is merely an observer:

Population 1:   Hematocrit values of 17-year-old U.S. males

Population 2:   Hematocrit values of 17-year old U.S. females

By contrast, the two populations in Example 7.2 do not actually exist, but rather are defined in terms of specific experimental conditions; in a sense, the populations are created by experimental intervention:

Population 1:   Sucrose consumptions of blowflies when injected with saline

Population 2:   Sucrose consumptions of blowflies when injected with Pargyline

These two types of two-sample comparisons—the observational and the experimental—are both widely used in research. The formal methods of analysis are often the same for the two types, but the interpretation of the results may be somewhat different. For instance, in Example 7.2 it might be reasonable to say that Pargyline *causes* the increase in sucrose consumption, while no such notion applies in Example 7.1. (We will discuss these two study designs further in Chapter 8.)

When the observed variable is quantitative, the comparison of two samples can include—as we saw in Examples 7.1 and 7.2—several aspects, notably (1) comparison of means, (2) comparison of standard deviations, and (3) comparison of shapes. In this chapter, and indeed throughout this book, the primary emphasis will be on comparison of means and on other comparisons related to shift.

NOTATION

Figure 7.3 presents our notation for comparison of two samples. The notation is exactly parallel to that of Chapter 6, but now a subscript (1 or 2) is used to differentiate between the two samples. The two "populations" can be naturally

FIGURE 7.3

Notation for Comparison
of Two Samples

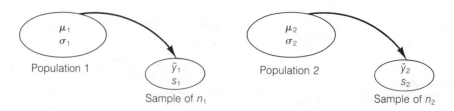

occurring populations (as in Example 7.1) or they can be conceptual populations defined by certain experimental conditions (as in Example 7.2). In either case, the data in each sample are viewed as a random sample from the corresponding population.

A LOOK AHEAD

In this chapter we will discuss two different but complementary approaches to the comparison of two means:

1. The confidence interval approach
2. The hypothesis testing approach

The confidence interval approach (Section 7.3) is a natural extension of the technique of Chapter 6. The hypothesis testing approach involves new concepts. We will first (Sections 7.4–7.9) introduce the basic ideas of hypothesis testing in the context of comparing two means. We will then (Section 7.10) discuss these ideas in more generality, and (Section 7.11) consider another hypothesis testing procedure for comparing two samples.

   We begin by describing, in the next section, some simple computations that are used both for confidence intervals and for hypothesis testing.

SECTION 7.2

STANDARD ERROR OF $(\bar{y}_1 - \bar{y}_2)$

In this section we introduce a fundamental quantity for comparing two samples: the standard error of the difference between two sample means.

BASIC IDEAS

We saw in Chapter 6 that the precision of a sample mean $\bar{y}$ can be expressed by its standard error, which is equal to

$$SE_{\bar{y}} = \frac{s}{\sqrt{n}}$$

To compare two sample means, we may consider the *difference* between them:

$$\bar{y}_1 - \bar{y}_2$$

which is an estimate of the quantity $(\mu_1 - \mu_2)$. To characterize the sampling error of estimation, we need to be concerned with the standard error of the difference $(\bar{y}_1 - \bar{y}_2)$. We illustrate with an example.

EXAMPLE 7.3

LETTUCE GROWTH

Lettuce is a crop that may be particularly suitable for use in controlled life-support systems aboard space stations. In one study of lettuce growth, plants of two different varieties were grown for 16 days in a controlled environment. Table 7.3 (page 194) shows the total dry weight of the leaves for nine plants of the variety "Salad Bowl" and six plants of the variety "Bibb."[3]

TABLE 7.3
Leaf Dry Weight (grams)
of Lettuce

	SALAD BOWL	BIBB
	3.06	1.31
	2.78	1.17
	2.87	1.72
	3.52	1.20
	3.81	1.55
	3.60	1.53
	3.30	
	2.77	
	3.62	
$\bar{y}$	3.259	1.413
s	.400	.220

The difference between the variety means is

$$\bar{y}_1 - \bar{y}_2 = 3.259 - 1.413 = 1.846 \text{ gm}$$

We know that $\bar{y}_1$ and $\bar{y}_2$ are subject to sampling error, and consequently the difference (1.846) is subject to sampling error. The question is: How much? ∎

We will present two different methods of computing a standard error for $(\bar{y}_1 - \bar{y}_2)$. The methods differ in whether the two sample standard deviations are kept separate or are "pooled." (The two methods can give different answers.)

THE UNPOOLED METHOD

The **unpooled method** for computing the standard error of $(\bar{y}_1 - \bar{y}_2)$ is easy to explain. The standard error is determined from the following formula:

$$SE_{(\bar{y}_1 - \bar{y}_2)} = \sqrt{\frac{s_1^2}{n_1} + \frac{s_2^2}{n_2}}$$

The following alternative form of the formula shows how the SE of the difference is related to the individual SEs:

$$SE_{(\bar{y}_1 - \bar{y}_2)} = \sqrt{SE_1^2 + SE_2^2}$$

where

$$SE_1 = SE_{\bar{y}_1} = \frac{s_1}{\sqrt{n_1}}$$

$$SE_2 = SE_{\bar{y}_2} = \frac{s_2}{\sqrt{n_2}}$$

We illustrate the formulas in the following example.

EXAMPLE 7.4

LETTUCE GROWTH

For the lettuce growth data, preliminary computations yield the results in Table 7.4.

TABLE 7.4

	SALAD BOWL	BIBB
s^2	.15994	.04835
n	9	6
SE	.133	.090

The unpooled SE of $(\bar{y}_1 - \bar{y}_2)$ is

$$SE_{(\bar{y}_1 - \bar{y}_2)} = \sqrt{\frac{.15994}{9} + \frac{.04835}{6}} = .161$$

Note that

$$.161 = \sqrt{(.133)^2 + (.090)^2} \qquad \blacksquare$$

Notice in Example 7.4 that the SE of the difference is greater than either of the individual SEs but less than their sum. This will always be true for the unpooled method (but not necessarily for the pooled method to be presented next).

THE POOLED METHOD

A second method of determining a standard error for $(\bar{y}_1 - \bar{y}_2)$ is based on **pooling** (that is, combining) the sample variances s_1^2 and s_2^2 into a single variance which is denoted s_c^2 ("c" for "combined"). Recall that each sample variance (s^2), which is the square of the standard deviation, is defined as

$$s^2 = \frac{\sum (y - \bar{y})^2}{n - 1}$$

The numerator of this expression is referred to as a "sum of squares" and abbreviated "SS"*; specifically,

$$SS_1 = \sum (y - \bar{y}_1)^2 \quad \text{in sample 1}$$

and

$$SS_2 = \sum (y - \bar{y}_2)^2 \quad \text{in sample 2}$$

The pooled variance is defined as

$$s_c^2 = \frac{SS_1 + SS_2}{n_1 + n_2 - 2}$$

*To avoid confusion, note that "SS" is different from "s^2" and is also different from Σy^2.

and the pooled method for computing the SE is

$$SE_{(\bar{y}_1 - \bar{y}_2)} = \sqrt{s_c^2 \left(\frac{1}{n_1} + \frac{1}{n_2}\right)}$$

We illustrate with an example.

EXAMPLE 7.5
LETTUCE GROWTH

For the lettuce growth data, the sums of squares are given in Table 7.5.

TABLE 7.5

	SALAD BOWL	BIBB
SS = $\Sigma (y - \bar{y})^2$	1.27949	.24173
n	9	6

The pooled variance is

$$s_c^2 = \frac{1.27949 + .24173}{9 + 6 - 2} = .11702$$

and the pooled SE is

$$SE_{(\bar{y}_1 - \bar{y}_2)} = \sqrt{.11702 \left(\frac{1}{9} + \frac{1}{6}\right)} = .180$$

Recall from Example 7.4 that the unpooled SE for the same data was SE = .161. ∎

To clarify the notion of pooling, let us take a closer look at the pooled variance. Each SS is related to the sample variance by the equation

$$SS = (n - 1)s^2$$

It follows that the pooled variance can be written as

$$s_c^2 = \frac{(n_1 - 1)s_1^2 + (n_2 - 1)s_2^2}{n_1 + n_2 - 2}$$

This formula shows how the pooled variance s_c^2 depends on the individual variances: s_c^2 is a weighted average of s_1^2 and s_2^2, with more weight given to the variance that is based on more observations. For example, if $n_1 = 9$ and $n_2 = 6$, then

$$s_c^2 = \left(\frac{8}{13}\right)s_1^2 + \left(\frac{5}{13}\right)s_2^2$$

If n_1 and n_2 are equal, then s_c^2 is the ordinary average (that is, the mean) of s_1^2 and s_2^2:

$$s_c^2 = \frac{s_1^2 + s_2^2}{2}$$

The pooled standard deviation s_c will always be between s_1 and s_2. For instance, for the lettuce growth data, the values were $s_1 = .40$, $s_2 = .22$, and $s_c = \sqrt{.11702} = .34$, which is between .40 and .22.

UNPOOLED METHOD VERSUS POOLED METHOD

If the sample sizes are equal $(n_1 = n_2)$ or if the sample standard deviations are equal $(s_1 = s_2)$, then the unpooled and the pooled method will give the same answer for $SE_{(\bar{y}_1 - \bar{y}_2)}$. The two answers will not differ substantially unless both the sample sizes and the sample SDs are quite discrepant.

To show the analogy between the two SE formulas, we can write them as follows:

$$
SE_{(\bar{y}_1 - \bar{y}_2)} = \begin{cases} \sqrt{\dfrac{s_1^2}{n_1} + \dfrac{s_2^2}{n_2}} & \text{(unpooled)} \\[3ex] \sqrt{\dfrac{s_c^2}{n_1} + \dfrac{s_c^2}{n_2}} & \text{(pooled)} \end{cases}
$$

In the pooled method, the separate variances—s_1^2 and s_2^2—are replaced by the single variance s_c^2, which is calculated from both samples.

Both the unpooled and the pooled SE have the same purpose—to estimate the standard deviation of the sampling distribution of $(\bar{Y}_1 - \bar{Y}_2)$. In fact, it can be shown that the standard deviation is

$$
\sigma_{(\bar{Y}_1 - \bar{Y}_2)} = \sqrt{\dfrac{\sigma_1^2}{n_1} + \dfrac{\sigma_2^2}{n_2}}
$$

Note the resemblance between this formula and the formulas for $SE_{(\bar{y}_1 - \bar{y}_2)}$.

In analyzing data when the sample sizes are unequal $(n_1 \neq n_2)$, one needs to decide whether to use the pooled or unpooled method for calculating $SE_{(\bar{y}_1 - \bar{y}_2)}$. The choice depends on whether one is willing to assume that the population SDs (σ_1 and σ_2) are equal. It can be shown that, if $\sigma_1 = \sigma_2$, then the pooled method should be used, because in this case s_c is the best estimate of the population SD. If $\sigma_1 \neq \sigma_2$, then the unpooled method should be used, because in this case s_c is not an estimate of either σ_1 or σ_2, so that pooling would accomplish nothing.

In Section 7.9 we will discuss practical guidelines for choosing between the pooled and the unpooled methods. First, we show in Sections 7.4–7.8 how the SEs can be used for statistical inference.

EXERCISES 7.1–7.12

7.1 For the following data, compute the standard error of $(\bar{y}_1 - \bar{y}_2)$ using
  a. the unpooled method.
  b. the pooled method.

SAMPLE 1	SAMPLE 2
8	6
9	8
9	4
11	
8	

7.2 Proceed as in Exercise 7.1 for the following data:

SAMPLE 1	SAMPLE 2
3	4
6	2
5	8
2	4
	7
	5

7.3 For the following data, compute the standard error of $(\bar{y}_1 - \bar{y}_2)$ using
a. the unpooled method.
b. the pooled method.
(The answers should agree.)

SAMPLE 1	SAMPLE 2
3	7
5	6
9	8
3	7

7.4 Data from two samples gave the following results:

	SAMPLE 1	SAMPLE 2
$\bar{y}$	40	50
$\Sigma\,(y - \bar{y})^2$	80	200
n	6	12

Compute the standard error of $(\bar{y}_1 - \bar{y}_2)$, using
a. the unpooled method.
b. the pooled method.

7.5 Proceed as in Exercise 7.4 for the following data:

	SAMPLE 1	SAMPLE 2
$\bar{y}$	75	55
$\Sigma\,(y - \bar{y})^2$	600	300
n	15	10

7.6 Data from two samples gave the following results:

	SAMPLE 1	SAMPLE 2
$\bar{y}$	130	100
$\Sigma\,(y - \bar{y})^2$	3,500	2,000
n	20	20

Compute the standard error of $(\bar{y}_1 - \bar{y}_2)$, using
a. the unpooled method.
b. the pooled method.
(The answers should agree.)

7.7 Proceed as in Exercise 7.6 for the following data:

	SAMPLE 1	SAMPLE 2
$\bar{y}$	90	80
$\Sigma (y - \bar{y})^2$	6,000	1,500
n	16	16

7.8 Data from two samples gave the following results:

	SAMPLE 1	SAMPLE 2
$\bar{y}$	150	160
s	20	25
n	10	5

Compute the standard error of $(\bar{y}_1 - \bar{y}_2)$, using
a. the unpooled method.
b. the pooled method.

7.9 Proceed as in Exercise 7.8 for the following data:

	SAMPLE 1	SAMPLE 2
$\bar{y}$	60	50
s	7	4
n	16	8

7.10 Data from two samples gave the following results:

	SAMPLE 1	SAMPLE 2
Mean	153	150
SE	3	4

Compute the standard error of $(\bar{y}_1 - \bar{y}_2)$, using the unpooled method.

7.11 Proceed as in Exercise 7.10 for the following data:

	SAMPLE 1	SAMPLE 2
Mean	500	450
SE	10	10

7.12 Data from two samples gave the following results:

	SAMPLE 1	SAMPLE 2
$\bar{y}$	435	400
SE	10	7
n	25	16

Compute the standard error of $(\bar{y}_1 - \bar{y}_2)$, using
a. the unpooled method.
b. the pooled method.

**CONFIDENCE
INTERVAL FOR
$(\mu_1 - \mu_2)$**

One way to compare two sample means is to construct a confidence interval for the difference in the population means—that is, a confidence interval for the quantity $(\mu_1 - \mu_2)$.

Recall from Chapter 6 that a 95% confidence interval for the mean μ of a single population is constructed as

$$\bar{y} \pm t_{.05}\, SE_{\bar{y}}$$

Analogously, a 95% confidence interval for $(\mu_1 - \mu_2)$ is constructed as

$$(\bar{y}_1 - \bar{y}_2) \pm t_{.05}\, SE_{(\bar{y}_1 - \bar{y}_2)}$$

The critical value $t_{.05}$ is determined from Student's t distribution using

$$df = n_1 + n_2 - 2$$

Notice that this df is the sum of the individual values of df:

$$n_1 + n_2 - 2 = (n_1 - 1) + (n_2 - 1)$$

Intervals with other confidence coefficients are constructed analogously; for example, for a 90% confidence interval one would use $t_{.10}$ instead of $t_{.05}$.

In using the confidence interval formula, one can calculate the standard error, $SE_{(\bar{y}_1 - \bar{y}_2)}$, by either the pooled method or the unpooled method. We will discuss the choice between the methods in Section 7.9. Roughly speaking, the pooled SE is used if s_1 and s_2 do not differ much, while the unpooled SE is used if s_1 and s_2 are very different in magnitude. Recall, however, that if $n_1 = n_2$ then the choice need not be made because the pooled SE and the unpooled SE will be equal.

The following example illustrates the construction of a confidence interval for $(\mu_1 - \mu_2)$.

EXAMPLE 7.6

TOLUENE AND
THE BRAIN

Abuse of substances containing toluene (for example, glue) can produce various neurological symptoms. In an investigation of the mechanism of these toxic effects, researchers measured the concentrations of various chemicals in the brains of rats who had been exposed to a toluene-laden atmosphere, and also in unexposed control rats. The concentrations of the brain chemical norepinephrine (NE) in the medulla region of the brain, for six toluene-exposed rats and five control rats, are given in Table 7.6.[4]

For the data in Table 7.6, the pooled method gives

$$SE_{(\bar{y}_1 - \bar{y}_2)} = 40.997$$

We determine the critical value with

$$df = 6 + 5 - 2 = 9$$

Table 4 yields $t_{.05} = 2.262$, so the confidence interval formula gives

$$(540.8 - 444.2) \pm (2.262)(40.997)$$

or

$$96.6 \pm 92.8$$

TABLE 7.6	TOLUENE (GROUP 1)	CONTROL (GROUP 2)
NE Concentration (ng/gm)	543	535
	523	385
	431	502
	635	412
	564	387
	549	
$\bar{y}$	540.8	444.2
s	66.1	69.6
n	6	5

and the 95% confidence interval for ($\mu_1 - \mu_2$) is

$$4 \text{ ng/gm} < \mu_1 - \mu_2 < 189 \text{ ng/gm}$$

According to the confidence interval, the population mean μ_1 is larger than μ_2 by an amount that might be as small as 4 ng/gm or as large as 189 ng/gm. ■

EXERCISES 7.13–7.18

7.13 Substances to be tested for cancer-causing potential are often painted on the skin of mice. The question arose whether mice might get an additional dose of the substance by licking or biting their cagemates. To answer this question, the compound benzo(a)pyrene was applied to the backs of ten mice: five were individually housed and five were group-housed in a single cage. After 48 hours, the concentration of the compound in the stomach tissue of each mouse was determined. The results (nmol/gm) were as follows:[5]

SINGLY HOUSED	GROUP-HOUSED
3.3	3.9
2.4	4.1
2.5	4.8
3.3	3.9
2.4	3.4

Construct a 95% confidence interval for the mean *excess* dose obtained by the group-housed mice.

7.14 Ferulic acid is a compound that may play a role in disease resistance in corn. A botanist measured the concentration of soluble ferulic acid in corn seedlings grown in the dark or in a light/dark photoperiod. The results (nmol acid per gm tissue) were as shown in the table.[6]

	DARK	PHOTOPERIOD
$\bar{y}$	92	115
s	13	13
n	4	4

Construct

a. a 95% confidence interval   **b.** A 90% confidence interval

for the difference in ferulic acid concentration under the two lighting conditions.

7.15 A study was conducted to determine whether relaxation training, aided by biofeedback and meditation, could help in reducing high blood pressure. Subjects were randomly allocated to a biofeedback group or a control group. The biofeedback group received training for eight weeks. The table reports the reduction in systolic blood pressure (mm Hg) after eight weeks.[7] Use the unpooled method to construct a 95% confidence interval for the difference in mean response.

TREATMENT	NO. OF PATIENTS	REDUCTION IN BP (mm Hg) Mean	SE
Biofeedback	99	13.8	1.34
Control	93	4.0	1.30

7.16 Prothrombin time is a measure of the clotting ability of blood. For ten rats treated with an antibiotic and ten control rats, the prothrombin times were reported as follows:[8]

TREATMENT	NUMBER OF RATS	PROTHROMBIN TIME (sec) Mean	SD
Antibiotic	10	25	10
Control	10	23	8

Construct a 95% confidence interval for the difference in population means.

7.17 The accompanying table summarizes the sucrose consumption (mg in 30 minutes) of black blowflies injected with Pargyline or saline (control). (These are the same data shown in Table 7.2.)

	CONTROL	PARGYLINE
$\bar{y}$	14.9	46.5
s	5.4	11.7
n	900	905

Construct

a. a 95% confidence interval
b. a 99% confidence interval

for the difference in population means. Use the unpooled SE.

7.18 In a field study of mating behavior in the Mormon cricket (*Anabrus simplex*), a biologist noted that some females mated successfully while others were rejected by the males before coupling was complete. The question arose whether some aspect of body size might play a role in mating success. The accompanying table summarizes measurements of head width (mm) in the two groups of females.[9] Use the pooled *t* method to construct a 95% confidence interval for the difference in population means.

	SUCCESSFUL	UNSUCCESSFUL	
$\bar{y}$	8.498	8.440	
s	.283	.262	$s_c = .274$
n	22	17	

SECTION 7.4

HYPOTHESIS TESTING: THE *t* TEST

We have seen that two means can be compared by using a confidence interval for the difference $(\mu_1 - \mu_2)$. In the following sections we will explore another approach to the comparison of means: the procedure known as *testing a hypothesis*. The general idea is to formulate as a hypothesis the statement that μ_1 and μ_2 do not differ, and then to see whether the data are consistent or inconsistent with that hypothesis.

THE NULL AND ALTERNATIVE HYPOTHESES

The hypothesis that μ_1 and μ_2 are equal is called a **null hypothesis**, and is abbreviated H_0. It can be written as

$$H_0: \quad \mu_1 = \mu_2$$

Its antithesis is the **alternative hypothesis**

$$H_A: \quad \mu_1 \neq \mu_2$$

which asserts that μ_1 and μ_2 are *not* equal. A researcher would usually express these hypotheses more informally, as in the following example.

EXAMPLE 7.7

TOLUENE AND THE BRAIN

For the brain NE data of Example 7.6, the observed mean NE in the toluene group ($\bar{y}_1 = 540.8$ ng/gm) was substantially higher than the mean in the control group ($\bar{y}_2 = 444.2$ ng/gm). One might ask whether this observed difference indicates a real biological phenomenon—the effect of toluene—or whether the truth might be that toluene has no effect and that the observed difference between $\bar{y}_1$ and $\bar{y}_2$ reflects only chance variation. Corresponding hypotheses, informally stated, would be

H_0^*: Toluene has no effect on NE concentration in rat medulla

H_A^*: Toluene has some effect on NE concentration in rat medulla   ■

We denote the informal statements by different symbols (H_0^* and H_A^* rather than H_0 and H_A) because they make different assertions. In Example 7.7 the informal alternative hypothesis makes a very strong claim—not only that there is a difference, but that the difference is *caused* by toluene.*

*Of course, our statements of H_0^* and H_A^* are abbreviated. Complete statements would include all relevant conditions of the experiment—adult male rats, toluene 1,000 ppm atmosphere for 8 hours, and so on. Our use of abbreviated statements should not cause any confusion.

A statistical **test of hypothesis** is a procedure for assessing the compatibility of the data with H_0. The data are considered compatible with H_0 if any discrepancies from H_0 could be readily attributed to chance (that is, to sampling error). Data judged to be *incompatible* with H_0 are taken as evidence in favor of H_A.

THE t STATISTIC

We consider the problem of testing the null hypothesis

$$H_0: \quad \mu_1 = \mu_2$$

against the alternative hypothesis

$$H_A: \quad \mu_1 \neq \mu_2$$

A classical procedure for this problem is the t test. To carry out the t test, the first step is to compute the t statistic, which is defined as

$$t_s = \frac{\bar{y}_1 - \bar{y}_2}{SE_{(\bar{y}_1 - \bar{y}_2)}}$$

The subscript "s" on t_s serves as a reminder that this value is computed from the data ("s" for "sample"). To calculate t_s, one must decide whether to compute the SE by the pooled method or the unpooled method (unless $n_1 = n_2$, in which case the two methods give the same answer). The criteria for this choice will be discussed in Section 7.9.

The quantity t_s is the **test statistic** for the t test; that is, t_s provides the data summary which is the basis for the test procedure. Notice the structure of t_s: It is a measure of the difference between the sample means ($\bar{y}$'s), expressed in relation to the SE of the difference. We illustrate with an example.

EXAMPLE 7.8

TOLUENE AND THE BRAIN

For the brain NE data of Example 7.6, the value of t_s (using the pooled SE) is

$$t_s = \frac{540.8 - 444.2}{40.997} = 2.356$$

The t statistic shows that $\bar{y}_1$ and $\bar{y}_2$ differ by about 2.4 SEs.   ∎

How shall we judge whether our data are consistent with H_0? Of course, *perfect* agreement with H_0 would be expressed by a t statistic equal to zero ($t_s = 0$). But, even if the null hypothesis H_0 is exactly true, it is unlikely that t_s would be exactly zero; we expect the sample means to differ because of sampling variability. Fortunately, one can set limits on this sampling variability: in fact, the chance difference in the $\bar{y}$'s is not likely to exceed a few standard errors. To put this more precisely, it can be shown mathematically that

If H_0 is true, then the sampling distribution of t_s is a Student's t distribution with df $= n_1 + n_2 - 2$.

[*Note*: The preceding statement is true if certain conditions, to be discussed in Section 7.9, are met. Throughout Sections 7.4–7.8 we will assume that these conditions are satisfied.]

The essence of the _t_ test procedure is to locate the observed value t_s in the Student's _t_ distribution, as indicated in Figure 7.4. If t_s is near the center, as in Figure 7.4(a), then the data are regarded as compatible with H_0 because the observed deviation from H_0 (that is, the difference between $\bar{y}_1$ and $\bar{y}_2$) can be readily attributed to chance variation due to sampling. If, on the other hand, t_s falls in the far tail of the _t_ distribution, as in Figure 7.4(b), then the data are regarded as _incompatible_ with H_0, because the observed deviation _cannot_ be readily explained as chance variation. To put this another way, if H_0 is true, then a value of t_s in the far tails of the _t_ distribution is unlikely; consequently, a value of t_s in the far tails is interpreted as evidence against H_0.

FIGURE 7.4
Essence of the _t_ test.
(a) Data compatible with
H_0; **(b)** Data incompatible
with H_0.

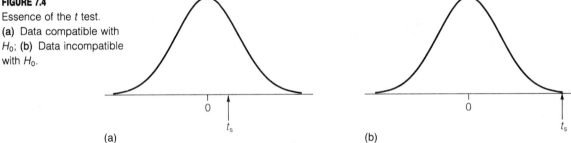

(a)   (b)

THE _P_-VALUE

To judge whether an observed value t_s is "far" in the tail of the _t_ distribution, we need a quantitative yardstick for locating t_s within the distribution. This yardstick is provided by the _P_-value, which can be defined (in the present context) as follows:

> The **_P_-value** of the data is the area under Student's _t_ curve in the double tails beyond $-t_s$ and $+t_s$.

Thus, the _P_-value is the shaded area in Figure 7.5. Note that we have defined the _P_-value as the total area in the _double_ tail; this is sometimes called the **two-tailed _P_-value**.

FIGURE 7.5
Definition of Two-Tailed
_P_-Value for the _t_ Test

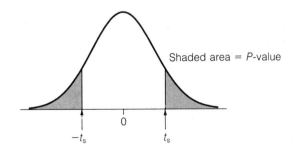

Shaded area = _P_-value

From the definition of P-value, it follows that **the P-value of the data is a measure of compatibility between the data and H_0**: a large P-value (close to 1) indicates a value of t_s near the center of the t distribution (compatible with H_0), whereas a small P-value (close to 0) indicates a value of t_s in the far tails of the t distribution (incompatible with H_0).

In analyzing data, how do we determine the P-value of the data? There are two answers to this question. If the samples are large enough, then the reference t distribution is approximately a normal distribution, so that a P-value can be approximately determined from Table 3, as illustrated in the following example.

EXAMPLE 7.9

Consider a set of data with $n_1 = 30$, $n_2 = 30$, and $t_s = 1.40$. The P-value of the data is (approximately) the two-tailed area in Figure 7.6; the curve in the figure is a Student's t curve with df $= 30 + 30 - 2 = 58$, which is approximately a normal curve. From Table 3 we determine the P-value of the data as approximately

$$P = 1 - (2)(.4192) = .16$$

FIGURE 7.6

P-Value for Example 7.9

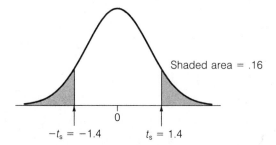

Shaded area $= .16$

$-t_s = -1.4$ $t_s = 1.4$

If the sample sizes are small, the Student's t curve with df $= n_1 + n_2 - 2$ is not well approximated by a normal curve, so that P-values cannot be determined from Table 3. Modern statistical computer programs, and some calculators, will provide exact P-values. In this book we will rely on the limited information in Table 4; we can use this information to *bracket* the P-value, rather than to determine it exactly. The following example illustrates the bracketing process.

EXAMPLE 7.10

TOLUENE AND THE BRAIN

For the brain NE data, the value of the t statistic (as determined in Example 7.8) is $t_s = 2.356$. The corresponding P-value is shaded in Figure 7.7; the t curve has df $= n_1 + n_2 - 2 = 9$. From Table 4 with df $= 9$, we find $t_{.05} = 2.262$ and $t_{.02} = 2.821$; because t_s is between these critical values, the P-value must be between .02 and .05. Thus, the P-value of the data is bracketed as

$$.02 < P < .05$$

(In fact, it can be determined by computer that the P-value for these data is $P = .043$.)

FIGURE 7.7

P-Value for Example 7.10

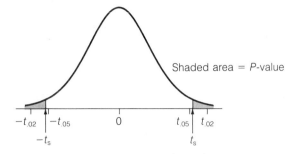

Shaded area = *P*-value

If the observed t_s is not within the boundaries of Table 4, then the *P*-value is bracketed on only one side. For example, if t_s is greater than $t_{.0001}$, then the *P*-value is bracketed as

$$P < .0001$$

DRAWING CONCLUSIONS FROM A *t* TEST

We have seen that the *P*-value is a measure of the compatibility between the data and H_0. But where does one draw the line between compatibility and incompatibility? Most people would agree that $P = .0001$ indicates incompatibility, and that $P = .80$ indicates compatibility, but what about intermediate values? For example, should $P = .10$ be regarded as compatible or incompatible with H_0? The answer is not intuitively obvious.

In much scientific research, it is not necessary to draw a sharp line. However, in many situations a *decision* must be reached. For example, the Food and Drug Administration (FDA) must decide whether the data submitted by a pharmaceutical manufacturer are sufficient to justify approval of a medication. As another example, a fertilizer manufacturer must decide whether the evidence favoring a new fertilizer is sufficient to justify the expense of further research.

Making a decision requires drawing a definite line between compatibility and incompatibility. The threshold value, on the *P*-value scale, is called the **significance level** of the test, and is denoted by the symbol α (alpha). The value of α is chosen by whoever is making the decision. Common choices are $\alpha = .10, .05$, and .01; the most standard choice is the 5% significance level, $\alpha = .05$. If the *P*-value of the data is less than α, the data are judged incompatible with H_0; in this case we say that H_0 is **rejected**, and that the data provide evidence in favor of H_A. If the *P*-value of the data is greater than α, we say that H_0 is **not rejected**, and that the data provide insufficient evidence to claim that H_A is true.

The following example illustrates the use of the *t* test to make a decision.

EXAMPLE 7.11

TOLUENE AND THE BRAIN

For the brain NE experiment of Example 7.6, the data are summarized in Table 7.7 (page 208).

TABLE 7.7

NE Concentration (ng/gm)

	TOLUENE	CONTROL
$\bar{y}$	540.8	444.2
s	66.1	69.6
n	6	5

In Example 7.10 we found that the P-value of these data is between .02 and .05. Suppose we choose to make a decision at the 5% significance level, $\alpha =$ .05. Because the P-value is less than .05 we reject H_0 and conclude that the data provide evidence in favor of H_A; specifically, the data provide evidence that $\mu_1 > \mu_2$. (The direction is determined by noting in Table 7.7 that $\bar{y}_1 > \bar{y}_2$.) The strength of the evidence is expressed by the statement that $.02 < P < .05$.

This completes the *formal* hypothesis test; but how should we *interpret* this result? Recall that the informal hypotheses were

H_0^*:   Toluene has no effect on NE concentration in rat medulla

H_A^*:   Toluene has some effect on NE concentration in rat medulla

Correspondingly, the informal conclusion from the t test may be stated as follows:

Conclusion: The data provide some evidence $(.02 < P < .05)$ that toluene *increases* NE concentration in the rat medulla.   ■

We emphasize that, strictly speaking, the t test addresses only the formal hypotheses H_0 and H_A. The t test does not *in itself* justify the verbal (informal) conclusion. For instance, the conclusion in Example 7.11 rests on a number of implicit assumptions—that the rats were randomly allocated to the toluene and control groups and that they were held in identical environments except for toluene exposure, that the measurements were made under comparable conditions, and so on. In Chapter 8 we will explicitly discuss some aspects of the design of experiments. Throughout this book, however, we must sometimes oversimplify scientific issues in order to clarify statistical issues. We will therefore take the liberty of stating conclusions in the language of the scientist, assuming that the experiment was adequately designed and carried out.

The next example illustrates a t test in which H_0 is *not* rejected.

EXAMPLE 7.12

WEIGHT GAIN OF COWS

In a study of the nutritional requirements of cattle, researchers measured the weight gains of cows during a 78-day period. For two breeds of cows, Hereford (HH) and Brown Swiss/Hereford (SH), the results are summarized in Table 7.8.[10]

TABLE 7.8

Weight Gain (kg)

	HH	SH
$\bar{y}$	18.3	13.9
s	17.8	19.1
n	33	51

The question of interest is whether one of the breeds has a tendency to gain more weight than the other under these conditions; informally, the hypotheses are

H_0^*: The mean weight gains of the two breeds are the same

H_A^*: The mean weight gains of the two breeds are different

The formal hypotheses are

H_0: $\mu_1 = \mu_2$

H_A: $\mu_1 \neq \mu_2$

The pooled method gives

$$SE_{(\bar{y}_1 - \bar{y}_2)} = 4.156$$

so that the *t* statistic is

$$t_s = \frac{18.3 - 13.9}{4.156} = 1.06$$

From Table 4 (with df = 60) we find that $t_{.20} = 1.296$, which is larger than t_s; consequently, the *P*-value of the data is greater than .20. We would not reject H_0 at $\alpha = .05$, at $\alpha = .10$, or even at $\alpha = .20$. To put it another way, the observed difference between the sample means (18.3 vs 13.9) can readily be explained as chance variation due to sampling. We may state the informal conclusion as follows:

Conclusion: There is insufficient evidence $(P > .20)$ to conclude that the breeds differ with respect to weight gain under these conditions. ∎

Note carefully the phrasing of the conclusion in Example 7.12. We do *not* say that there is evidence *for* the null hypothesis, but only that there is insufficient evidence *against* it. When we do not reject H_0, this indicates a lack of evidence that H_0 is false, which is *not* the same thing as evidence that H_0 is true. The astronomer Carl Sagan (in another context) summed up this principle of evidence in this succinct statement:[11]

Absence of evidence is not evidence of absence.

In other words, nonrejection of H_0 is *not* the same as *acceptance* of H_0. (To avoid confusion, it is best not to use the phrase "accept H_0" at all.)

Nonrejection of H_0 indicates that the data are compatible with H_0, but the data may *also* be quite compatible with H_A. For instance, in Example 7.12 we found that the observed difference between the sample means could easily be due to sampling variation, but this finding does not rule out the possibility that the observed difference is actually due to a real difference between the breeds. (Methods for such ruling out of possible alternatives will be discussed in Section 7.7 and optional Section 7.8.)

In testing a hypothesis, the researcher starts out with the assumption that H_0 is true and then asks whether the data contradict that assumption. This logic can make sense even if the researcher regards the null hypothesis as implausible, or indeed, impossible. For instance, in Example 7.12 it could be argued that there is almost certainly *some* difference (perhaps very small) between the breeds. The fact that we did not reject H_0 does not mean that we accept H_0; rather it means that *if* there is a difference between the breeds, we have little or no information as to the *direction* of the difference—that is, whether $\mu_1 > \mu_2$ or $\mu_1 < \mu_2$. As another illustration of the same idea, we might regard the data on NE concentration (Example 7.11) as giving information not primarily on the *existence* of an effect of toluene (it might be argued that any such treatment would have *some* effect on NE), but rather primarily on the *direction* of the effect (that is, to *increase* NE concentration).

REPORTING THE RESULTS OF A t TEST

In reporting the results of a t test, a researcher may choose to make a definite decision (to reject H_0 or not to reject H_0) at a specified significance level α, or he may choose simply to describe the results in phrases such as "There is very strong evidence that . . . " or "The evidence suggests that . . . " or "There is virtually no evidence that . . . ." In writing a report for publication, it is very desirable to state the P-value so that the reader can make a decision on his own.

The term *significant* is often used in reporting results. For instance, an observed difference is said to be "statistically significant at the 5% level" if it is large enough to justify rejection of H_0 at $\alpha = .05$. (When α is not specified, it is usually understood to be .05; we should emphasize, however, that α *is an arbitrarily chosen value and there is nothing "official" about .05*.) Unfortunately, the term "significant" is easily misunderstood and should be used with care; we will return to this point in Section 7.7.

Note: In this section we have used different notation for the formal (H_0, H_A) and informal (H_0^*, H_A^*) hypotheses in order to point out the logical distinction between them. For simplicity, in the remainder of this book we will abandon the notational distinction and simply use H_0 and H_A to denote both the formal and informal hypotheses.

EXERCISES 7.19–7.30

7.19 Use the normal approximation (Table 3) to find the two-tailed P-value of the following data as analyzed by the t test.

	SAMPLE 1	SAMPLE 2
$\bar{y}$	83	80
n	30	30

$$\text{SE}_{(\bar{y}_1 - \bar{y}_2)} = 2.1$$

7.20 Proceed as in Exercise 7.19 for the following data:

	SAMPLE 1	SAMPLE 2
$\bar{y}$	532	500
n	40	60
$SE_{(\bar{y}_1-\bar{y}_2)} = 10$		

7.21 Proceed as in Exercise 7.19 for the following data:

	SAMPLE 1	SAMPLE 2
$\bar{y}$	3.28	3.20
n	75	75
$SE_{(\bar{y}_1-\bar{y}_2)} = .042$		

7.22 For each of the following data sets, use Table 4 to bracket the two-tailed *P*-value of the data as analyzed by the *t* test.

a.

	SAMPLE 1	SAMPLE 2
$\bar{y}$	735	854
n	4	3
$SE_{(\bar{y}_1-\bar{y}_2)} = 38$		

b.

	SAMPLE 1	SAMPLE 2
$\bar{y}$	5.3	5.0
n	7	7
$SE_{(\bar{y}_1-\bar{y}_2)} = .24$		

c.

	SAMPLE 1	SAMPLE 2
$\bar{y}$	36	30
n	15	20
$SE_{(\bar{y}_1-\bar{y}_2)} = 1.3$		

7.23 For each of the following data sets, use Table 4 to bracket the two-tailed *P*-value of the data as analyzed by the *t* test.

a.

	SAMPLE 1	SAMPLE 2
$\bar{y}$	100.2	106.8
n	8	5
$SE_{(\bar{y}_1-\bar{y}_2)} = 5.7$		

b.

	SAMPLE 1	SAMPLE 2
$\bar{y}$	49.8	44.3
n	8	8
$SE_{(\bar{y}_1-\bar{y}_2)} = 1.9$		

c.

	SAMPLE 1	SAMPLE 2
$\bar{y}$	3.58	3.00
n	10	15
$SE_{(\bar{y}_1-\bar{y}_2)} = .12$		

7.24 Backfat thickness is a variable used in evaluating the meat quality of pigs. An animal scientist measured backfat thickness (cm) in pigs raised on two different diets, with the results given in the table.[12]

	DIET 1	DIET 2
$\bar{y}$	3.49	3.05
s	.40	.40

Suppose we use the pooled t test to compare the diets. Bracket the P-value, assuming that the number of pigs on each diet was
a. 5　　**b.** 10　　**c.** 15
[*Hint:* You need to calculate s_c only once.]

7.25 Heart disease patients often experience spasms of the coronary arteries. Because biological amines may play a role in these spasms, a research team measured amine levels in coronary arteries that were obtained postmortem from patients who had died of heart disease and also from a control group of patients who had died from other causes. The accompanying table summarizes the concentration of the amine serotonin.[13]

	SEROTONIN (ng/gm) Heart Disease	Controls
Mean	3,840	5,310
SE	850	640
n	8	12

a. For these data, the SE (pooled) of $(\bar{y}_1 - \bar{y}_2)$ is 1,046. Use a t test to compare the means at the 5% significance level.
b. Verify the value of $SE_{(\bar{y}_1-\bar{y}_2)}$ given in part a.

7.26 In a study of the periodical cicada (*Magicicada septendecim*), researchers measured the hind tibia lengths of the shed skins of 110 individuals. Results for males and females are shown in the accompanying table.[14]

Group	n	TIBIA LENGTH (micrometer units) Mean	Standard Deviation
Males	60	78.42	2.87
Females	50	80.44	3.52

a. Use an unpooled t test to investigate the dependence of tibia length on gender in this species. Use the 5% significance level.
b. Given the above data, if you were told the tibia length of an individual of this species, could you make a fairly confident prediction of its sex? Why or why not?
c. Repeat the t test of part a, assuming that the means and standard deviations were as given in the table, but that they were based on only one-tenth as many individuals (6 males and 5 females).

7.27 In a study of the development of the thymus gland, researchers weighed the glands of ten chick embryos. Five of the embryos had been incubated 14 days and five had been incubated 15 days. The thymus weights were as shown in the table.[15]

| | THYMUS WEIGHT (mg) | |
	14 days	15 days
	29.6	32.7
	21.5	40.3
	28.0	23.7
	34.6	25.2
	44.9	24.2
Mean	31.72	29.22
SS	304.79	206.71

   a. Use a *t* test to compare the means at $\alpha = .10$.
   b. Note that the chicks that were incubated longer had a smaller mean thymus weight. Is this "backward" result surprising, or could it easily be attributed to chance? Explain.

7.28 As part of an experiment on root metabolism, a plant physiologist grew birch tree seedlings in the greenhouse. He flooded four seedlings with water for one day, and kept four others as controls. He then harvested the seedlings and analyzed the roots for ATP content. The results (nmol ATP per mg tissue) are as follows:[16]

FLOODED	CONTROL
1.45	1.70
1.19	2.04
1.05	1.49
1.07	1.91

Use a *t* test to investigate the effect of flooding. Use $\alpha = .05$.

7.29 After surgery a patient's blood volume is often depleted. In one study, the total circulating volume of blood plasma was measured for each patient immediately after surgery. After infusion of a "plasma expander" into the bloodstream, the plasma volume was measured again and the increase in plasma volume (ml) was calculated. Two of the plasma expanders used were albumin (25 patients) and polygelatin (14 patients). The accompanying table reports the increase in plasma volume.[17] Use an unpooled *t* test to compare the mean increase in plasma volume under the two treatments. Let $\alpha = .01$.

	ALBUMIN	POLYGELATIN
n	25	14
Mean Increase	490	240
SE	60	30

7.30 Nutritional researchers conducted an investigation of two high-fiber diets intended to reduce serum cholesterol level. Twenty men with high serum cholesterol were randomly allocated to receive an "oat" diet or a "bean" diet for 21 days. The table summarizes the fall (before minus after) in serum cholesterol levels.[18] Use a t test to compare the diets at the 5% significance level.

		FALL IN CHOLESTEROL (mg/dl)	
Diet	n	Mean	SD
Oat	10	53.6	31.1
Bean	10	55.5	29.4

| | | | | | | | | | | | |

FURTHER DISCUSSION OF THE *t* TEST

In this section we discuss more fully the method and interpretation of the t test.

ANOTHER LOOK AT THE TEST PROCEDURE

Although we have described the t test in terms of P-value, the decision procedure can also be described in terms of critical values. To decide whether to reject H_0, one can simply compare the observed magnitude of t_s (symbolized as $|t_s|$) with the appropriate critical value from Table 4. For instance, the P-value will be less than .05 if and only if $|t_s|$ is greater than $t_{.05}$. This relationship is indicated in Figure 7.8. The t test decision procedure can be stated succinctly as

Reject H_0 if $|t_s| >$ critical value.

Thus, the critical value is a threshold value on the t scale.

FIGURE 7.8 Possible Outcomes of the t Test at $\alpha = .05$. **(a)** If $|t_s| < t_{.05}$, then $P > .05$ and H_0 is not rejected. **(b)** If $|t_s| > t_{.05}$, then $P < .05$ and H_0 is rejected.

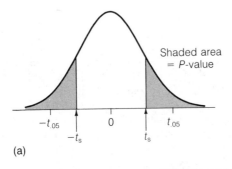

(a)

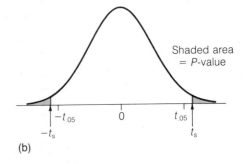

(b)

RELATIONSHIP BETWEEN TEST AND CONFIDENCE INTERVAL

There is a close connection between the confidence interval approach and the hypothesis testing approach to the comparison of μ_1 and μ_2. Consider, for example, a 95% confidence interval for $(\mu_1 - \mu_2)$ and its relationship to the t test at the 5% significance level. The t test and the confidence interval use the same three quantities—$(\bar{y}_1 - \bar{y}_2)$, $\mathrm{SE}_{(\bar{y}_1 - \bar{y}_2)}$, and $t_{.05}$—but manipulate them in different ways. In the t test, we reject H_0 if

$$\frac{|\bar{y}_1 - \bar{y}_2|}{\mathrm{SE}_{(\bar{y}_1 - \bar{y}_2)}} > t_{.05}$$

(See Figure 7.8.) To put this another way, we reject H_0 if

$$|\bar{y}_1 - \bar{y}_2| > t_{.05}\,\mathrm{SE}_{(\bar{y}_1 - \bar{y}_2)}$$

But the latter condition is equivalent to saying that the second term in the confidence interval expression

$$(\bar{y}_1 - \bar{y}_2) \pm t_{.05}\,\mathrm{SE}_{(\bar{y}_1 - \bar{y}_2)}$$

is smaller in magnitude than the first, in which case both the upper and lower confidence limits will have the same sign. It follows that the *t* test will reject H_0 at $\alpha = .05$ if and only if the 95% confidence interval for $(\mu_1 - \mu_2)$ does not contain 0, that is, if the confidence interval contains only positive values or only negative values. (The same relationship holds between the 99% confidence interval and the test at $\alpha = .01$, and so on.) We illustrate with an example.

EXAMPLE 7.13

WEIGHT GAIN OF COWS

For the cow weight gain data of Example 7.12, the pivotal quantities for a *t* test at $\alpha = .05$ are:

$$\bar{y}_1 - \bar{y}_2 = 18.3 - 13.9 = 4.4$$

$$\mathrm{SE}_{(\bar{y}_1 - \bar{y}_2)} = 4.156$$

$$t_{.05} = 2.000 \quad (\text{with df} = 60)$$

We can use these same three quantities to form a 95% confidence interval:

$$4.4 \pm (2.000)(4.156) \quad \text{or} \quad -3.9 \text{ kg} < \mu_1 - \mu_2 < +12.7 \text{ kg}$$

The confidence interval contains 0; it follows that the *t* test would not reject H_0 at the 5% significance level (and indeed we saw in Example 7.12 that it did not). Note that this equivalence between the test and the confidence interval makes common sense; according to the confidence interval, μ_1 may be as much as 3.9 kg *less*, or as much as 12.7 kg *more*, than μ_2; it is natural, then, to say that we are uncertain as to whether μ_1 is greater than (or less than, or equal to) μ_2. ∎

In the context of the Student's *t* method, the confidence interval approach and hypothesis testing approach are different ways of using the same basic information. The confidence interval has the advantage that it indicates the *magnitude* of the difference between μ_1 and μ_2. The testing approach has the advantage that the *P*-value describes on a continuous scale the *evidence* that μ_1 and μ_2 are really different. In Section 7.7 we will explore further the use of a confidence interval to supplement the interpretation of a *t* test. In later chapters we will encounter other hypothesis tests that cannot so readily be supplemented by a confidence interval.

INTERPRETATION OF α

In analyzing data or making a decision based on data, you will often need to choose a significance level α. How do you know whether to choose $\alpha = .05$ or $\alpha = .01$ or some other value? To make this judgment, it is helpful to have an *operational* interpretation of α. We now give such an interpretation.

Recall from Section 7.4 that the sampling distribution of t_s, if H_0 is true, is a Student's t distribution. Let us assume for definiteness that df = 60 and that α is chosen equal to .05. The critical value (from Table 4) is $t_{.05} = 2.000$. Figure 7.9 shows the Student's t distribution and the values ± 2.000. The total shaded area in the figure is .05; it is split into two equal parts of area .025 each. We can think of Figure 7.9 as a formal guide for deciding whether to reject H_0: If the observed value of t_s falls in the hatched regions of the t_s axis, then H_0 will be rejected. But the chance of this happening is 5%, if H_0 is true. Thus, we can say that

$$\text{Pr}\{\text{reject } H_0\} = .05 \quad \text{if } H_0 \text{ is true}$$

This probability has meaning in the context of a super meta-experiment (depicted in Figure 7.10) in which we repeatedly sample from two populations and calculate a value of t_s. It is important to realize that the probability refers to a situation in which H_0 is true. In order to concretely picture such a situation, you are invited to suspend disbelief for a moment and come on an imaginary trip in Example 7.14.

FIGURE 7.9

A t Test at $\alpha = .05$. H_0 is rejected if t_s falls in the hatched region.

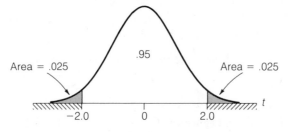

EXAMPLE 7.14

MUSIC AND MARIGOLDS

Imagine that the scientific community has developed great interest in the influence of music on the growth of marigolds. One school of investigation centers on whether music written by Bach or Mozart produces taller plants. Plants are randomly allocated to listen to Bach (treatment 1) or Mozart (treatment 2) and, after a suitable period of listening, data are collected on plant height. The null hypothesis is

H_0: Marigolds respond equally well to Bach or Mozart

or

H_0: $\mu_1 = \mu_2$

where

μ_1 = Mean height of marigolds if exposed to Bach
μ_2 = Mean height of marigolds if exposed to Mozart

FIGURE 7.10
Meta-Experiment
for the *t* Test

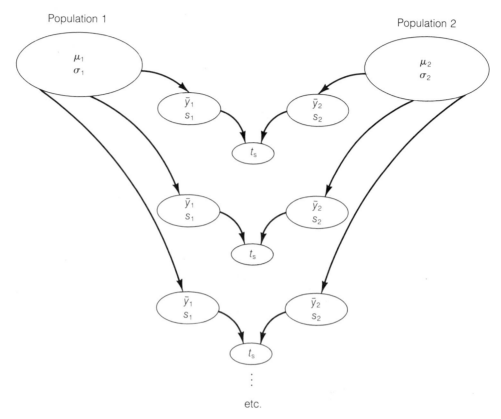

etc.

Assume for the sake of argument that H_0 is in fact true. Imagine now that many investigators perform the Bach vs Mozart experiment, and that each experiment uses 62 plants, so df = 60. Suppose each investigator analyzes his data with a *t* test at α = .05. What conclusions will the investigators reach? In the super meta-experiment of Figure 7.10, suppose each pair of samples represents a different investigator. Since we are assuming that μ_1 and μ_2 are actually equal, the values of t_s will deviate from 0 only because of chance sampling error. If all the investigators were to get together and make a frequency distribution of their t_s values, that distribution would follow a Student's *t* curve. The investigators would make their decisions as indicated by Figure 7.9, so we would expect them to have the following experiences:

95% of them would not reject H_0;

$2\frac{1}{2}$% of them would conclude that the plants prefer Bach;

$2\frac{1}{2}$% of them would conclude that the plants prefer Mozart.

Thus, a total of 5% of the investigators would reject the true null hypothesis.

■

Example 7.14 provides an image for interpreting α. Of course, in analyzing data, we are dealing not with a meta-experiment but with a single experiment. When we perform a t test at the 5% significance level, we are playing the role of *one* of the investigators in Example 7.14, and the others are imaginary. If we reject H_0, there are two possibilities:

1. H_0 is in fact false; or
2. H_0 is in fact true, but we are one of the unlucky 5% who rejected H_0 anyway.

We feel "confident" in our rejection of H_0 because the second possibility is unlikely (assuming that we regard 5% as a small percentage). Of course, we never know (unless someone replicates the experiment) whether or not we are one of the unlucky 5%.

We have seen that α can be interpreted as a probability:

$$\alpha = \Pr\{\text{reject } H_0\} \quad \text{if } H_0 \text{ is true}$$

The erroneous rejection of H_0 when H_0 is true is called a **Type I error**. In choosing α, we are choosing our level of protection against Type I error. Many researchers regard 5% as an acceptably small risk. If we do not regard 5% as small enough, we might choose to use a more conservative value of α such as $\alpha = .01$; in this case the percentage of true null hypotheses which we reject would be not 5% but 1%.

In practice, the choice of α may depend on the context of the particular experiment. For example, a regulatory agency might demand more exacting proof of efficacy for a toxic drug than for a relatively innocuous one. Also, a person's choice of α may be influenced by his prior opinion about the phenomenon under study. For instance, suppose an agronomist is skeptical of claims for a certain soil treatment; in evaluating a new study of the treatment, he might express his skepticism by choosing a very conservative significance level (say, $\alpha = .001$), thus indicating that it would take a lot of evidence to convince him that the treatment is effective. Note that, if a written report of an investigation includes a P-value, then each reader is free to choose his own value of α in evaluating the reported results.

SIGNIFICANCE LEVEL VERSUS P-VALUE   Students sometimes find it hard to distinguish between significance level (α) and P-value.* For the t test, both α and P are tail areas under Student's t curve. But α is an arbitrary prespecified value; it can be (and should be) chosen before looking at the data. By contrast, the P-value is determined from the data; indeed, giving the P-value is a way of describing the data. You may find it helpful at this point to compare Figure 7.5 with Figure 7.9. The shaded area represents P-value in the former and α in the latter figure.

---

*Unfortunately, the term "significance level" is not used consistently by all people who write about statistics. Some authors use the terms "significance level" or "significance probability" where we have used "P-value."

SECTION 7.6

ONE-TAILED *t* TESTS

The *t* test described in the preceding sections is called a **two-tailed *t* test** because the null hypothesis is rejected if t_s falls in either tail of the Student's *t* distribution and the *P*-value of the data is a two-tailed area under Student's *t* curve. A two-tailed *t* test is used to test the null hypothesis

$$H_0: \quad \mu_1 = \mu_2$$

against the alternative hypothesis

$$H_A: \quad \mu_1 \neq \mu_2$$

This alternative H_A is called a **nondirectional alternative**.

DIRECTIONAL ALTERNATIVE HYPOTHESES

In some studies it is apparent from the beginning that there is only one reasonable direction of deviation from H_0. In such situations it is appropriate to formulate a **directional alternative hypothesis**. The following is a directional alternative:

$$H_A: \quad \mu_1 < \mu_2$$

Another directional alternative is

$$H_A: \quad \mu_1 > \mu_2$$

The following two examples illustrate situations where directional alternatives are appropriate.

EXAMPLE 7.15
NIACIN
SUPPLEMENTATION

Consider a feeding experiment with lambs. The observation *Y* will be weight gain in a two-week trial. Ten animals will receive diet 1, and ten animals will receive diet 2, where

Diet 1 = Standard ration + Niacin

Diet 2 = Standard ration

On biological grounds it is expected that niacin may increase weight gain; there is no reason to suspect that it could possibly decrease weight gain. An appropriate formulation would be

H_0: Niacin is not effective in increasing weight gain $\quad (\mu_1 = \mu_2)$

H_A: Niacin is effective in increasing weight gain $\quad\quad (\mu_1 > \mu_2)$ ∎

EXAMPLE 7.16
HAIR DYE AND CANCER

Suppose a certain hair dye is to be tested to determine whether it is carcinogenic (cancer-causing). The dye will be painted on the skins of 20 mice (group 1), and an inert substance will be painted on the skins of 20 mice (group 2) who will serve as controls. The observation *Y* will be the number of tumors appearing on each mouse. An appropriate formulation is

H_0: The dye is not carcinogenic $\quad (\mu_1 = \mu_2)$

H_A: The dye is carcinogenic $\quad\quad (\mu_1 > \mu_2)$ ∎

THE ONE-TAILED TEST PROCEDURE

When the alternative hypothesis is directional, the t test procedure must be modified. The modified procedure is called a **one-tailed t test** and is carried out in two steps as follows:

STEP 1   Check directionality—see if the data deviate from H_0 in the *direction* specified by H_A:

a. If not, the P-value is $P > .50$.
b. If so, proceed to step 2.

STEP 2   The P-value of the data is the *one-tailed* area beyond t_s.

To conclude the test, one can make a decision at a prespecified significance level α: H_0 is rejected if $P < \alpha$.

The rationale of the two-step procedure is that the P-value measures deviation from H_0 *in the direction specified by* H_A. The one-tailed P-value is illustrated in Figure 7.11 and the two-step testing procedure is illustrated in Example 7.17.

FIGURE 7.11
One-Tailed P-Value for a t Test, **(a)** if the alternative is H_A: $\mu_1 < \mu_2$ and t_s is negative; **(b)** if the alternative is H_A: $\mu_1 > \mu_2$ and t_s is positive.

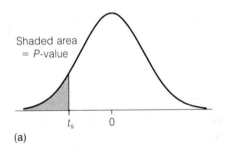
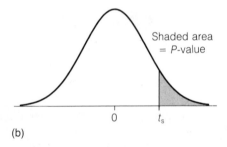

(a)   (b)

EXAMPLE 7.17

NIACIN
SUPPLEMENTATION

Consider the lamb feeding experiment of Example 7.15. The alternative hypothesis is

$$H_A: \quad \mu_1 > \mu_2$$

We will reject H_0 if $\bar{y}_1$ is sufficiently *greater* than $\bar{y}_2$. Because $n_1 = n_2 = 10$, we have df $= 18$. The critical values from Table 4 are reproduced in Table 7.9.

TABLE 7.9

Critical Values with df $= 18$

Two-Tailed Area	.20	.10	.05	.02	.01	.001	.0001
Critical Value	1.330	1.734	2.101	2.552	2.878	3.922	4.966

To illustrate the one-tailed test procedure, suppose that we have[19]

$$SE_{(\bar{y}_1 - \bar{y}_2)} = 2.2 \text{ lb}$$

and that we choose $\alpha = .05$. Let us consider various possibilities for the two sample means.
a. Suppose the data give $\bar{y}_1 = 10$ lb and $\bar{y}_2 = 13$ lb. This deviation from H_0 is opposite to the assertion of H_A: we have $\bar{y}_1 < \bar{y}_2$, but H_A asserts that $\mu_1 >$

μ_2. Consequently, $P > .5$, so we would not reject H_0 at any significance level. We conclude that the data provide no evidence that niacin is effective in increasing weight gain.

b. Suppose the data give $\bar{y}_1 = 14$ lb and $\bar{y}_2 = 10$ lb. This deviation from H_0 is in the direction of H_A (because $\bar{y}_1 > \bar{y}_2$), so we proceed to step 2. The value of t_s is

$$t_s = \frac{14 - 10}{2.2} = 1.82$$

From Table 7.9, we see that the two-tailed P-value would be bracketed as follows:

$.05 <$ two-tailed $P < .10$

Consequently, the one-tailed P-value is bracketed as

$.025 <$ one-tailed $P < .05$

Since $P < \alpha$, we reject H_0 and conclude that there is some evidence that niacin is effective.

c. Suppose the data give $\bar{y}_1 = 12$ lb and $\bar{y}_2 = 10$ lb. Then, proceeding as in part **b**, we compute the test statistic as $t_s = .91$, and bracket the one-tailed P-value as

$P > .10$

Since $P > \alpha$, we do not reject H_0; we conclude that there is insufficient evidence to claim that niacin is effective. Thus, although these data deviate from H_0 in the direction of H_A, the amount of deviation is not great enough to justify rejection of H_0.   ∎

Notice that what distinguishes a one-tailed t test from a two-tailed t test is the way in which P-value is determined, but *not* the directionality or nondirectionality of the conclusion. If we reject H_0, our conclusion is directional even if our H_A is nondirectional. (For instance, in Example 7.11 we concluded that toluene *increases* NE concentration.)

DIRECTIONAL VERSUS NONDIRECTIONAL ALTERNATIVES

The same data will give a different P-value depending on whether the alternative hypothesis is directional or nondirectional. It can happen that the same data will permit rejection of H_0 using the one-tailed procedure but not using the two-tailed procedure. For example, suppose for definiteness that df $= 60$, and that we are using the 5% significance level. From Table 4 with df $= 60$ we find $t_{.05} = 2.000$ and $t_{.10} = 1.671$. If the alternative hypothesis is nondirectional, as

H_A: $\mu_1 \neq \mu_2$

then we will reject H_0 if

$|t_s| > 2.000$

If the alternative hypothesis is directional, as

$$H_A: \quad \mu_1 > \mu_2$$

then we will reject H_0 if

$$t_s > 1.671$$

These two procedures are indicated in Figure 7.12. If, for instance, we observe $t_s = 1.8$, then the one-tailed procedure rejects H_0 but the two-tailed procedure does not. In this sense, it is "easier" to reject H_0 with the one-tailed procedure than with the two-tailed procedure.

FIGURE 7.12 Two-Tailed and One-Tailed t Tests at $\alpha = .05$ with df $= 60$. H_0 is rejected if t_s falls in the hatched region of the t-axis.

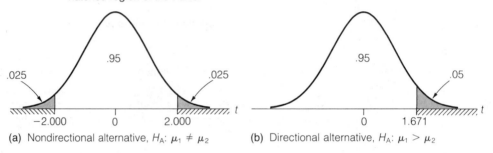

(a) Nondirectional alternative, H_A: $\mu_1 \neq \mu_2$ (b) Directional alternative, H_A: $\mu_1 > \mu_2$

Why do we use a different procedure when the alternative hypothesis is directional? The purpose of the modification in the procedure is to preserve the interpretation of significance level α as given in Section 7.5, that is,

$$\alpha = \Pr\{\text{reject } H_0\} \quad \text{if } H_0 \text{ is true}$$

For instance, consider the case $\alpha = .05$ and df $= 60$. In Figure 7.12, the total shaded area is equal to .05 in both (a) and (b). This means that, if a great many investigators were to test a true H_0, then 5% of them would reject H_0 and commit a Type I error; this statement is true whether the alternative H_A is directional or nondirectional.

The crucial point in justification of the modified procedure for testing against a directional H_A is that *if* the direction of deviation of the data from H_0 is *not* as specified by H_A, then we will not reject H_0. For example, in the carcinogenesis experiment of Example 7.16, if the mice exposed to the hair dye had *fewer* tumors than the control group, we might (1) simply conclude that the data do not indicate a carcinogenic effect, or (2) if the exposed group had *substantially* fewer tumors, so that the test statistic t_s was very far in the wrong tail of the t distribution, we might look for methodological errors in the experiment—for example, mistakes in lab technique or in recording the data, nonrandom allocation of the mice to the two groups, and so on.

A one-tailed t test is especially natural when only one direction of deviation from H_0 is believed to be plausible. However, one-tailed tests are also used in situations where deviation in both directions is possible, but only one direction is of primary interest. For instance, in the niacin experiment of Example 7.15, it is not crucial that the experimenter believe that it is *impossible* for niacin to reduce weight gain rather than increase it. In fact, some researchers would formulate the hypotheses for the niacin experiment as

$$H_0: \quad \mu_1 \leq \mu_2$$
$$H_A: \quad \mu_1 > \mu_2$$

The test procedure for the problem formulated in this way would be exactly the same as that illustrated in Example 7.17. Deviations in the wrong direction (less weight gain on niacin) would not lead to rejection of H_0, and thus to no claims about the effect of niacin;* this is the essential feature that distinguishes a directional from a nondirectional formulation.

CHOOSING THE FORM OF H_A

When is it legitimate to use a directional H_A, and so to perform a one-tailed *t* test? The answer to this question is linked to the directionality check—step 1 of the two-step test procedure given previously. Clearly such a check makes sense only if H_A was (or could have been) formulated before the data were inspected. (If we were to formulate a directional H_A that was "inspired" by the data, then of course the data would always deviate from H_0 in the "right" direction and the test procedure would *always* proceed to step 2.) This is the rationale for the following rule.

RULE FOR DIRECTIONAL ALTERNATIVES

It is legitimate to use a directional alternative H_A only if H_A is formulated before seeing the data.

In research, investigators often get more pleasure from rejecting a null hypothesis than from not rejecting one. In fact, research reports often contain phrases such as "we are unable to reject the null hypothesis" or "the results failed to reach statistical significance." Under these circumstances, one might wonder what the consequences would be if researchers succumbed to the natural temptation to ignore the above rule for using directional alternatives. After all, very often one can think of a rationale for an effect *ex post facto*—that is, after the effect has been observed. A return to the imaginary experiment on plants' musical tastes will illustrate this situation.

EXAMPLE 7.18
MUSIC AND MARIGOLDS

Recall the imaginary experiment of Example 7.14, in which investigators measure the heights of marigolds exposed to Bach or Mozart. Suppose, as before, that the null hypothesis is true, that df = 60, and that the investigators all perform *t* tests at $\alpha = .05$. Now suppose in addition that all of the investigators *violate* the rule

---

*It must be admitted, however, that many researchers, if faced with large deviations in the wrong direction which could not be explained as gross error or poor lab technique, would then revert to a nondirectional formulation so that they could reject H_0.

for use of directional alternatives, and that they formulate H_A *after* seeing the data. Half of the investigators would obtain data for which $\bar{y}_1 > \bar{y}_2$, and they would formulate the alternative

H_A: $\mu_1 > \mu_2$ (plants prefer Bach)

The other half would obtain data for which $\bar{y}_1 < \bar{y}_2$, and they would formulate the alternative

H_A: $\mu_1 < \mu_2$ (plants prefer Mozart)

Now envision what would happen. Since the investigators are using directional alternatives, they will all use the critical value 1.671 for the t test (see Figure 7.12). We would expect them to have the following experiences:

90% of them would not reject H_0;

5% of them would conclude that the plants prefer Bach;

5% of them would conclude that the plants prefer Mozart.

Thus, a total of 10% of the investigators would reject the true null hypothesis. Of course each investigator individually never realizes that the overall percentage of Type I errors is 10% rather than 5%. And the conclusions that plants prefer Bach or Mozart could be supported by *ex post facto* rationales that would be limited only by the imagination of the investigators.   ∎

As Example 7.18 illustrates, a researcher who uses a directional alternative when it is not justified pays the price of a doubled risk of Type I error. His protection against erroneous rejection of H_0 is not as great as he may believe.

It may seem strange that a rule of statistical analysis should specify the timing of a personal mental event—yet this is what the rule does by specifying that you must think of H_A before seeing the data. In later chapters we will encounter other statistical rules that are "psychological" in the same sense, and whose violation can exact a much higher price than doubling the chance of a Type I error.

EXERCISES 7.31–7.39

7.31 Use the normal approximation (Table 3) to find the one-tailed P-value of the following data as analyzed by the t test, assuming that the alternative hypothesis is H_A: $\mu_1 > \mu_2$.

	SAMPLE 1	SAMPLE 2
$\bar{y}$	10.8	10.5
n	50	50

$SE_{(\bar{y}_1-\bar{y}_2)} = .23$

7.32 Proceed as in Exercise 7.31 for the accompanying data.

	SAMPLE 1	SAMPLE 2
$\bar{y}$	750	730
n	100	100

$$SE_{(\bar{y}_1 - \bar{y}_2)} = 11$$

7.33 For each of the following data sets, use Table 4 to bracket the one-tailed *P*-value of the data as analyzed by the *t* test. Assume that the alternative hypothesis is H_A: $\mu_1 > \mu_2$.

a.

	SAMPLE 1	SAMPLE 2
$\bar{y}$	3.24	3.00
n	10	10

$$SE_{(\bar{y}_1 - \bar{y}_2)} = .061$$

b.

	SAMPLE 1	SAMPLE 2
$\bar{y}$	560	500
n	6	4

$$SE_{(\bar{y}_1 - \bar{y}_2)} = 45$$

c.

	SAMPLE 1	SAMPLE 2
$\bar{y}$	73	79
n	20	20

$$SE_{(\bar{y}_1 - \bar{y}_2)} = 2.8$$

7.34 Ecological researchers measured the concentration of red cells in the blood of 27 field-caught lizards (*Sceloporis occidentalis*). In addition, they examined each lizard for infection by the malarial parasite *Plasmodium*. The red cell counts ($10^{-3} \times$ cells per mm^3) were as reported in the table.[20]

	INFECTED ANIMALS	NONINFECTED ANIMALS	
$\bar{y}$	972.1	843.4	
s	245.1	251.2	$s_c = 248.5$
n	12	15	

One might expect that malaria would reduce the red cell count, and in fact previous research with another lizard species had shown such an effect. Do the data support this expectation? Test the null hypothesis of no difference against the alternative that the infected population has a lower red cell count. Use a pooled *t* test at
a. $\alpha = .05$ **b.** $\alpha = .10$

7.35 A study was undertaken to compare the respiratory responses of hypnotized and nonhypnotized subjects to certain instructions. The 16 male volunteers were allocated at random to an experimental group to be hypnotized and to a control group. Baseline measurements were taken at the start of the experiment. In analyzing the data, the researchers noticed that the baseline breathing patterns of the two groups were different; this was surprising, since all the subjects had been treated the same up to that time. One explanation proposed for this unexpected difference was that the experimental group were more excited in anticipation of the experience of being hypnotized. The accompanying table presents a summary of the baseline measurements of total ventilation (liters of air per minute per square meter of body area).[21]

	EXPERIMENTAL	CONTROL
$\bar{y}$	6.169	5.291
s	.621	.652
n	8	8

a. Use a t test to test the hypothesis of no difference against a nondirectional alternative. Let $\alpha = .05$.

b. Use a t test to test the hypothesis of no difference against the alternative that the experimental conditions produce a larger mean than the control conditions. Let $\alpha = .05$.

c. Which of the two tests, that of part a or part b, is more appropriate? Explain.

7.36 In a study of lettuce growth, 10 seedlings were randomly allocated to be grown in either standard nutrient solution or in a solution containing extra nitrogen. After 22 days of growth, the plants were harvested and weighed, with the results given in the table.[22] Are the data sufficient to conclude that the extra nitrogen enhances plant growth under these conditions? Use a t test at $\alpha = .05$ against a directional alternative.

Nutrient Solution	n	LEAF DRY WEIGHT (gm) Mean	SD
Standard	5	3.62	.54
Extra nitrogen	5	4.17	.67

7.37 An entomologist conducted an experiment to see if wounding a tomato plant would induce changes that improve its defense against insect attack. She grew larvae of the tobacco hornworm (*Manduca sexta*) on wounded plants or control plants. The accompanying table shows the weights (mg) of the larvae after 7 days of growth.[23] How strongly do the data support the researcher's expectation? Use a pooled t test at the 5% significance level. Let H_A be that wounding the plant tends to diminish larval growth.

	WOUNDED	CONTROL	
$\bar{y}$	28.66	37.96	
s	9.02	11.14	$s_c = 10.20$
n	16	18	

7.38 In a study of the effect of amphetamine on water consumption, a pharmacologist injected four rats with amphetamine and four with saline as controls. She measured the amount of water each rat consumed in 24 hours; the following are the results, expressed as ml water per kg body weight:[24]

AMPHETAMINE	CONTROL
118.4	122.9
124.4	162.1
169.4	184.1
105.3	154.9

Use a t test to compare the treatments at $\alpha = .05$. Let the alternative hypothesis be that amphetamine tends to suppress water consumption.

7.39 A pain-killing drug was tested for efficacy in 50 women who were experiencing uterine cramping pain following childbirth. Twenty-five of the women were randomly allocated to receive the drug, and the remaining 25 received a placebo (inert substance). Capsules of drug or placebo were given before breakfast and again at noon. A pain relief score, based on hourly questioning throughout the day, was computed for each woman. The possible pain relief scores ranged from 0 (no relief) to 56 (complete relief for 8 hours). Summary results are shown in the table.[25]

Treatment	Number of Women	PAIN RELIEF SCORE Mean	SD
Drug	25	31.96	12.05
Placebo	25	25.32	13.78

a. Test for evidence of efficacy using a t test. Use a directional alternative and $\alpha = .05$.

b. Use the normal approximation (Table 3) to find the P-value of the data.

c. If the alternative hypothesis were nondirectional, how would the answers to parts a and b change?

| | | | | | | | | | | | | |

SECTION 7.7

MORE ON INTERPRETATION OF STATISTICAL SIGNIFICANCE

Ideally, statistical analysis should aid the researcher by helping to clarify whatever message is contained in the data. For this purpose, it is not enough that the statistical calculations be correct; the results must also be correctly interpreted. In this section we explore some principles of interpretation that apply not only to the t test, but also to other statistical tests to be discussed later.

SIGNIFICANT DIFFERENCE VERSUS IMPORTANT DIFFERENCE

The term *significant* is often used in describing the results of a statistical analysis. For example, if an experiment to compare a drug against a placebo gave data with a very small P-value, then the conclusion might be stated as "The effect of the drug was highly significant." As another example, if two fertilizers for wheat gave a yield comparison with a large P-value, then the conclusion might be stated as "The wheat yields did not differ significantly between the two fertilizers" or "The difference between the fertilizers was not significant." As a third example, suppose a substance is tested for toxic effects by comparing exposed animals and control animals, and that the null hypothesis of no difference is not rejected. Then the conclusion might be stated as "No significant toxicity was found."

Clearly such phraseology using the term *significant* can be seriously misleading. After all, in ordinary English usage, the word *significant* connotes "substantial" or "important." In statistical jargon, however, the statement

"The difference was significant"

means nothing more or less than

"The null hypothesis of no difference was rejected."

By the same token, the statement

"The difference was not significant"

means

"The null hypothesis of no difference was not rejected."

This specialized usage of the word *significant* has become quite common in scientific writing, and understandably is the source of much confusion.

It is essential to recognize that a statistical test provides information about only one question: whether the difference observed in the data is large enough to infer that a difference in the same direction exists in the population. The question of whether a difference is *important*, as opposed to significant, cannot be decided on the basis of P-value. The following two examples illustrate this fact.

EXAMPLE 7.19
SERUM LD

Lactate dehydrogenase (LD) is an enzyme that may show elevated activity following damage to the heart muscle or other tissues. A large study of serum LD levels in healthy young people yielded the results shown in Table 7.10.[26]

TABLE 7.10
Seurm LD (U/l)

	MALES	FEMALES
$\bar{y}$	60	57
s	11	10
n	270	264

The difference between males and females is quite significant; in fact, $t_s = 3.3$, which gives a P-value of $P \approx .001$. However, this does not imply that the difference is large or important.   ∎

EXAMPLE 7.20
BODY WEIGHT

Imagine that we are studying the body weight of men and women, and we obtain the fictitious but realistic data shown in Table 7.11.[27]

TABLE 7.11
Body Weight (lb)

	MALES	FEMALES
$\bar{y}$	175	143
s	35	34
n	2	2

For these data, the t test gives $t_s = .93$, and we see from Table 4 that the P-value is $P > .20$. The observed difference between males and females is not small (it is $175 - 143 = 32$ lb), yet it is not statistically significant for any reasonable choice of α. The lack of statistical significance does not imply that the sex difference in body weight is small or unimportant. It means only that the data are inadequate to characterize the difference in the population means. ∎

CONFIDENCE INTERVALS TO ASSESS IMPORTANCE

The preceding examples show that the statistical significance or nonsignificance of a difference does not indicate whether the difference is important. Nevertheless, the question of "importance" can and should be addressed in most data analyses. To assess importance, one needs to consider the *magnitude* of the difference. A reasonable approach is to use the observed difference $(\bar{y}_1 - \bar{y}_2)$ to construct a confidence interval for the population difference $(\mu_1 - \mu_2)$. In interpreting the confidence interval, the judgment of what is "important" is made on the basis of experience with the particular practical situation. The following three examples illustrate this use of confidence intervals.

EXAMPLE 7.21
SERUM LD

For the LD data of Example 7.19, a 95% confidence interval for $(\mu_1 - \mu_2)$ is

$$3 \pm 1.8$$

or

$$1.2 \text{ U/l} < \mu_1 - \mu_2 < 4.8 \text{ U/l}$$

This interval implies (with 95% confidence) that the population mean difference between the sexes does not exceed 4.8 U/l. A physician evaluating this information would know that 4.8 U/l is less than the typical day-to-day fluctuation in a person's LD level, and that therefore the sex difference is negligible from the medical standpoint. For example, the physician might conclude that it is unnecessary to differentiate between the sexes in establishing clinical thresholds for diagnosis of illness. Thus, the sex difference in LD may be said to be statistically significant but medically unimportant. To put this another way, the data suggest that men do in fact tend to have higher levels than women, but not very much higher. ∎

EXAMPLE 7.22
BODY WEIGHT

For the body-weight data of Example 7.20, a 95% confidence interval for $(\mu_1 - \mu_2)$ is

$$32 \pm 148$$

or

$$-116 \text{ lb} < \mu_1 - \mu_2 < +180 \text{ lb}$$

From this confidence interval we cannot tell whether the true difference (between the population means) is large favoring females, is small, or is large favoring males. Because the confidence interval contains numbers of small magnitude and numbers of large magnitude, it does not tell us whether the difference between the sexes is important or unimportant. A definite statement about the importance of the difference would require more data. Suppose, for example, that the means and standard deviations were as given in Table 7.11, but that they were based on 2,000 rather than 2 people of each sex. Then the 95% confidence interval would be

$$32 \pm 2$$

or

$$30 \text{ lb} < \mu_1 - \mu_2 < 34 \text{ lb}$$

This interval would imply (with 95% confidence) that the difference is at least 30 lb, an amount which might reasonably be regarded as important, at least for some purposes.   ∎

EXAMPLE 7.23
YIELD OF TOMATOES

Suppose a horticulturist is comparing the yields of two varieties of tomatoes; yield is measured as pounds of tomatoes per plant. On the basis of practical considerations, the horticulturist has decided that a difference between the varieties is "important" only if it exceeds 1 pound per plant, on the average. That is, the difference is important if

$$|\mu_1 - \mu_2| > 1.0 \text{ lb}$$

Suppose the horticulturist's data give the following 95% confidence interval:

$$.2 \text{ lb} < \mu_1 - \mu_2 < .3 \text{ lb}$$

Because all values in the interval are less than 1.0 lb, the data support (with 95% confidence) the assertion that the difference is *not* important, using the horticulturist's criterion.   ∎

In many investigations, statistical significance and practical importance are both of interest. The following example shows how the relationship between these two concepts can be visualized using confidence intervals.

EXAMPLE 7.24

YIELD OF TOMATOES

Let us return to the tomato experiment of Example 7.23. The confidence interval was

$$.2 \text{ lb} < \mu_1 - \mu_2 < .3 \text{ lb}$$

Recall from Section 7.5 that the confidence interval can be interpreted in terms of a t test. Because all values within the confidence interval are positive, a t test (two-tailed) at $\alpha = .05$ would reject H_0. Thus, the difference between the two varieties is statistically significant, although it is not horticulturally important: The data indicate that variety 1 is better than variety 2, but also that it is not much better. The distinction between significance and importance for this example can be seen in Figure 7.13, which shows the confidence interval plotted on the $(\mu_1 - \mu_2)$-axis. Note that the confidence interval lies entirely to one side of zero and also entirely to one side of the "importance" threshold of 1.0.

FIGURE 7.13

Confidence Interval for Example 7.24

To further explore the relationship between significance and importance, let us consider other possible outcomes of the tomato experiment. Table 7.12 shows how the horticulturist would interpret various possible confidence intervals, still using the criterion that a difference must exceed 1.0 lb in order to be considered important.

TABLE 7.12

Interpretation of Confidence Intervals

95% CONFIDENCE INTERVAL	IS THE DIFFERENCE	
	Significant?	Important?
$.2 < \mu_1 - \mu_2 < .3$	Yes	No
$1.2 < \mu_1 - \mu_2 < 1.3$	Yes	Yes
$.2 < \mu_1 - \mu_2 < 1.3$	Yes	Cannot tell
$-.2 < \mu_1 - \mu_2 < .3$	No	No
$-1.2 < \mu_1 - \mu_2 < 1.3$	No	Cannot tell

Table 7.12 shows that a significant difference may or may not be important, and an important difference may or may not be significant. In practice, the assessment of importance using confidence intervals is a simple and extremely useful supplement to a test of hypothesis.

TYPE I AND TYPE II ERRORS

In Section 7.5 we defined a Type I error to be the rejection of a true null hypothesis. In order to fully appreciate the behavior of a hypothesis testing procedure, it is necessary to consider another kind of error, the Type II error. If

H_0 is false (and H_A is true), but we do not reject H_0, then we have made a **Type II error**. Table 7.13 displays the situations in which Type I and Type II errors can occur.

TABLE 7.13
Possible Outcomes
of Testing H_0

		TRUE SITUATION	
		H_0 True	H_0 False
OUR	Do not reject H_0	Correct	Type II error
DECISION	Reject H_0	Type I error	Correct

The consequences of Type I and Type II errors can be very different. The following two examples show some of the variety of these consequences.

EXAMPLE 7.25

MARIJUANA AND
THE PITUITARY

Cannabinoids, which are substances contained in marijuana, can be transmitted from mother to young through the placenta and through the milk. Suppose we conduct the following experiment on pregnant mice: We give one group of mice a dose of cannabinoids and keep another group as controls. We then evaluate the function of the pituitary gland in the offspring. The hypotheses would be

H_0:   Cannabinoids do not affect pituitary of offspring

H_A: Cannabinoids do affect pituitary of offspring

If in fact cannabinoids do not affect the pituitary of the offspring, but our data lead us to reject H_0, this would be a Type I error; the consequence might be unnecessary alarm if the conclusion were made public. On the other hand, if cannabinoids do affect the pituitary of the offspring, but our t test results in nonrejection of H_0, this would be a Type II error; one consequence might be unjustifiable complacency on the part of marijuana-smoking mothers.   ∎

EXAMPLE 7.26

IMMUNOTHERAPY

Chemotherapy is standard treatment for a certain cancer. Suppose we conduct a clinical trial to study the efficacy of supplementing the chemotherapy with immunotherapy (stimulation of the immune system). Patients are given either chemotherapy or chemotherapy plus immunotherapy. The hypotheses would be

H_0:   Immunotherapy is not effective in enhancing survival

H_A: Immunotherapy is effective in enhancing survival

If immunotherapy is actually not effective, but our data lead us to reject H_0 and conclude that immunotherapy is effective, then we have made a Type I error. The consequence, if this conclusion is acted on by the medical community, might be the widespread use of unpleasant, dangerous, and worthless immunotherapy. If, on the other hand, immunotherapy is actually effective, but our data do not enable us to detect that fact (perhaps because our samples are too small), then we have made a Type II error, with consequences quite different from those of a Type I error: The standard treatment will continue to be used until someone provides convincing evidence that supplementary immunotherapy is effective. If we still "believe" in immunotherapy, we can conduct another trial (perhaps with larger samples) to try again to establish its effectiveness.   ∎

As the foregoing examples illustrate, the consequences of a Type I error are usually quite different from those of a Type II error. The likelihoods of the two types of error may be very different, also. Recall from Section 7.5 that the significance level α can be interpreted as the probability of rejecting H_0 if H_0 is true. Because α is chosen at will, the hypothesis testing procedure "protects" you against Type I error by giving you control over the risk of such an error. This control is independent of the sample size and other factors. The chance of a Type II error, by contrast, depends on many factors, and may be large or small. In particular, an experiment with small sample sizes often has a high risk of Type II error.

We are now in a position to reexamine Carl Sagan's aphorism that "Absence of evidence is not evidence of absence." Because the risk of Type I error is controlled and that of Type II error is not, our state of knowledge is much stronger after rejection of a null hypothesis than after nonrejection. For example, suppose we are testing whether a certain soil additive is effective in increasing the yield of field corn. If we reject H_0, then either (1) we are right; or (2) we have made a Type I error; since the risk of a Type I error is controlled, we can be relatively confident of our conclusion that the additive is effective (although not necessarily *very* effective). Suppose, on the other hand, that the data are such that we do not reject H_0. Then either (1) we are right (that is, H_0 is true), or (2) we have made a Type II error. Since the risk of a Type II error may be quite high, we cannot say confidently that the additive is ineffective. In order to justify a claim that the additive is ineffective, we would need to supplement our test of hypothesis with further analysis, such as a confidence interval or an analysis of the chance of Type II error.

A THIRD TYPE OF ERROR In addition to Type I and Type II errors, a third type of error is possible: our conclusion might be in the wrong direction. For example, using a t test we might conclude that $\mu_1 < \mu_2$ when in fact $\mu_1 > \mu_2$ (suppose the males of a species are larger than the females but we wrongly conclude from our data that the females are larger). It can be shown that the risk for this third type of error is less than α, and so is under control just as the Type I error is.

POWER

As we have seen, Type II error is an important concept. For this reason special vocabulary is used to describe the chance of Type II error. The chance of *not* making a Type II error is called the **power** of a statistical test. The power can be expressed as

$$\text{Power} = \Pr\{\text{reject } H_0\} \quad \text{if } H_0 \text{ is false}$$

Thus, the power of a t test is a measure of the *sensitivity* of the test, or the ability of the test procedure to *detect* a difference between μ_1 and μ_2 when such a difference really does exist. In this way the power is analogous to the resolving power of a microscope.

The power of a statistical test depends on many factors in an investigation, including the sample sizes and the inherent variability of the observations. All other things being equal, using larger samples gives more information and thereby increases power. In addition, we will see that some statistical tests can be more powerful than others, and that some study designs can be more powerful than others.

The planning of a scientific investigation should always take power into consideration. No one wants to emerge from lengthy and perhaps expensive labor in the lab or the field, only to discover upon analyzing the data that the sample sizes were insufficient or the experimental material too variable, so that experimental effects that were considered important were not detected. Two techniques are available to aid the researcher in planning for adequate sample sizes. One technique is to decide how small each standard error ought to be and choose n using an analysis such as that of Section 6.4. A second technique is a quantitative analysis of the power of the statistical test. Such an analysis for the t test is discussed in the next section.

EXERCISES 7.40–7.48

7.40 (*Sampling exercise*)   Refer to the collection of 100 ellipses shown with Exercise 3.1, which can be thought of as representing a natural population of the organism *C. ellipticus*. Use random digits (from Table 1 or your calculator) to choose *two* random samples of five ellipses each. Use a metric ruler to measure the body length of each ellipse; measurements to the nearest millimeter will be adequate.
a. Compare the means of your two samples, using a two-tailed t test at $\alpha = .05$.
b. Did the analysis of part **a** lead you to a Type I error, a Type II error, or no error?

7.41 (*Sampling exercise*)   Simulate choosing random samples from two different populations, as follows. First, proceed as in Exercise 7.40 to choose two random samples of five ellipses each and measure their lengths. Then add 6 mm to *each* measurement in *one* of the samples.
a. Compare the means of your two samples, using a two-tailed t test at $\alpha = .05$.
b. Did the analysis of part **a** lead you to a Type I error, a Type II error, or no error?

7.42 (*Sampling exercise*)   Prepare simulated data as follows. First, proceed as in Exercise 7.40 to choose two random samples of five ellipses each and measure their lengths. Then, toss a coin. If the coin falls heads, add 6 mm to *each* measurement in *one* of the samples. If the coin falls tails, do not modify either sample.
a. Prepare two copies of the simulated data. On the Student Copy, show the data *only*; on the Instructor Copy, indicate also which sample (if any) was modified.
b. Give your Instructor Copy to the instructor and trade your Student Copy with another student when you are told to do so.
c. After you have received another student's paper, compare the means of his or her two samples using a two-tailed t test at $\alpha = .05$. If you reject H_0, decide which sample was modified.

7.43 A field trial was conducted to evaluate a new seed treatment that was supposed to increase soybean yield. When a statistician analyzed the data, he found that the mean yield from the treated seeds was 4 lb/acre greater than that from control plots planted with untreated seeds. However, the statistician declared the difference to be "not significant." Proponents of the treatment objected strenuously to the statistician's statement, pointing out that, at current market prices, 4 lb/acre would bring a tidy sum, which would be highly significant to the farmer. How would you answer this objection?[28]

7.44 In a clinical study of treatments for rheumatoid arthritis, patients were randomly allocated to receive either a standard medication or a newly designed medication. After a suitable period of observation, statistical analysis showed that there was no significant difference in the therapeutic response of the two groups, but that the incidence of undesirable side effects was significantly lower in the group receiving the new medication. The researchers concluded that the new medication should be regarded as preferable to the standard medication, because it had been shown to be equally effective therapeutically and to produce fewer side effects. In what respect is the researchers' reasoning faulty? (Assume that the term "significant" refers to rejection of H_0 at $\alpha = .05$.)

7.45 There is an old folk belief that the sex of a baby can be guessed before birth on the basis of its heart rate. In an investigation to test this theory, fetal heart rates were observed for mothers admitted to a maternity ward. The results (in beats per minute) are summarized in the table.

		HEART RATE (bpm)	
	n	Mean	SE
Males	250	137.21	.62
Females	250	137.18	.53

Construct a 95% confidence interval for the difference in population means. Does the confidence interval support the claim that the population mean sex difference (if any) in fetal heart rates is small and unimportant? (Use your own "expert" knowledge of heart rate to make a judgment of what is "unimportant.")

7.46 Coumaric acid is a compound that may play a role in disease resistance in corn. A botanist measured the concentration of coumaric acid in corn seedlings grown in the dark or in a light/dark photoperiod. The results (nmol acid per gm tissue) are given in the accompanying table.[30]

	DARK	PHOTOPERIOD
$\bar{y}$	106	102
s	21	27
n	4	4

Suppose the botanist considers the effect of lighting conditions to be "important" if the difference in means is 20%, that is, about 20 nmol/g. Based on a 95% confidence interval, do the above data indicate whether the true difference is "important"?

7.47 As part of a large study of serum chemistry in healthy people, the data in the table at the top of page 236. were obtained for the serum concentration of uric acid in men and women aged 18–55 years.[31]

	SERUM URIC ACID (mmol/l)	
	Men	Women
$\bar{y}$	.354	.263
s	.058	.051
n	530	420

Construct a 95% confidence interval for the true difference in population means (use the unpooled SE). Suppose the investigators feel that the difference in population means is "clinically important" if it exceeds .08 mmol/l. Does the confidence interval indicate whether the difference is "clinically important"?

7.48 Repeat Exercise 7.47, assuming that the means and standard deviations are as given in the table, but that the sample sizes are only one-tenth as large (that is, 53 men and 42 women).

S E C T I O N 7.8

PLANNING FOR ADEQUATE POWER (OPTIONAL)

We have defined the power of a statistical test as

Power = Pr{reject H_0}   if H_0 is false

To put this another way, the power of a test is the probability that it will yield a statistically significant result when it should (that is, when H_A is true).

Since the power is the probability of *not* making an error (Type II), high power is desirable. But power comes at a price. All other things being equal, more observations (larger samples) bring more power, and observations cost time and money. In this section we explain how a researcher can rationally plan an experiment to have adequate power for his purposes, and yet cost as little as possible.

Specifically, we will consider the power of the two-sample t test, conducted at significance level α. We assume that the populations are normal with equal SDs, and we denote the common value of the SD by σ (that is, $\sigma_1 = \sigma_2 = \sigma$). It can be shown that in this case the power is maximized if the sample sizes are equal; so we will assume that n_1 and n_2 are equal and denote the common value by n (that is, $n_1 = n_2 = n$).

Under the above assumptions, the power of the t test depends on the following factors: (a) α; (b) σ; (c) n; (d) $(\mu_1 - \mu_2)$. After briefly discussing each of these factors, we will address the all-important question of choosing the value of n.

DEPENDENCE OF POWER ON α

In choosing α, one chooses a level of protection against Type I error. However, this protection is traded for vulnerability to Type II error. If, for example, one chooses $\alpha = .01$ rather than $\alpha = .05$, then one is choosing to reject H_0 *less* readily, and so is (perhaps unwittingly) choosing to increase the risk of Type II error and reduce the power. Thus, there is an unavoidable trade-off between the risk of Type I error and the risk of Type II error.

DEPENDENCE ON σ

The larger σ, the smaller the power (all other things being equal). Recall from Chapter 5 that the reliability of a sample mean is determined by the quantity

$$\sigma_{\bar{Y}} = \frac{\sigma}{\sqrt{n}}$$

A larger σ tends to produce less reliable information about the population means, and so lower power to discern a difference between them. In order to increase power, then, a researcher usually tries to design the investigation so as to have σ as small as possible. For example, a botanist will try to hold light conditions constant throughout a greenhouse area, a pharmacologist will use genetically identical experimental animals, and so on. Usually, however, σ cannot be reduced to zero; there is still considerable variation in the observations.

DEPENDENCE ON n

The larger n, the higher the power (all other things being equal). If we increase n, we decrease $\sigma/\sqrt{n}$; this improves the precision of the sample means ($\bar{y}_1$ and $\bar{y}_2$). In addition, larger n gives more information about σ; this is reflected in a reduced critical value for the test (reduced because of more df). Thus, increasing n increases the power of the test in two ways.

DEPENDENCE ON $(\mu_1 - \mu_2)$

In addition to the factors we have discussed, the power of the t test also depends on the actual difference between the population means, that is, on $(\mu_1 - \mu_2)$. This dependence is very natural, as illustrated by the following example.

EXAMPLE 7.27
HEIGHTS OF PEOPLE

In order to clearly illustrate the concepts, we consider a familiar variable, body height of people. Imagine what would happen if an investigator were to measure the heights of two random samples of eleven people each ($n = 11$), and then conduct a two-tailed t test at $\alpha = .05$.

a. First, suppose that sample 1 consisted of 17-year-old males and sample 2 consisted of 17-year-old females. The two population means differ substantially; in fact, $(\mu_1 - \mu_2)$ is about 5 inches ($\mu_1 \approx 69.1$ and $\mu_2 \approx 64.1$ inches).[32] It can be shown (as we will see) that in this case the investigator has about a 99% chance of rejecting H_0 and correctly concluding that the males in the population of 17-year-olds are taller (on the average) than the females.

b. By contrast, suppose that sample 1 consisted of 17-year-old females and sample 2 consisted of 14-year-old females. The two population means differ, but by a modest amount; the difference is $(\mu_1 - \mu_2) = .6$ inch ($\mu_1 \approx 64.1$ and $\mu_2 \approx 63.5$ inches). It can be shown that in this case the investigator has less than a 10% chance of rejecting H_0; in other words, there is more than a 90% chance that he will fail to detect the fact that 17-year-old girls are taller than 14-year-old girls. (In fact, it can be shown that there is a 29% chance that $\bar{y}_1$ will be less than $\bar{y}_2$—that is, there is a 29% chance that eleven 17-year-old girls chosen at random will be shorter on the average than eleven 14-year-old girls chosen at random.)

The contrast between cases (a) and (b) is not due to any change in the SDs; in fact, for each of the three populations the value of σ is about 2.5 inches. Rather, the contrast is due to the simple fact that, with a fixed n and σ, it is easier to detect a large difference than a small difference.   ∎

PLANNING A STUDY

Suppose an investigator is planning a study for which the t test will be appropriate. How shall he take into account all the factors that influence the power of the test? First consider the choice of significance level α. A simple approach is to begin by determining the cost of an adequately powerful study using a liberal choice (say, $\alpha = .05$). If that cost is not high, the investigator can consider reducing α (say, to .01) and see if an adequately powerful study is still affordable.

Suppose, then, that the investigator has chosen a working value of α. Suppose also that the experiment has been designed to reduce σ as far as practicable, and that the investigator has available an estimate or guess of the value of σ.

At this point, the investigator needs to ask himself about the magnitude of the difference he wants to detect. As we saw in Example 7.27, a given sample size may be adequate to detect a large difference in population means, but entirely inadequate to detect a small difference. As a more realistic example, an experiment using 5 rats in a treatment group and 5 rats in a control group might be large enough to detect a substantial treatment effect, while detection of a subtle treatment effect would require more rats (perhaps 30) in each group.

The above discussion suggests that choosing a sample size for adequate power is somewhat analogous to choosing a microscope: We need high resolving power if we want to see a very tiny structure; for large structures a hand lens will do. In order to proceed with planning the experiment, the investigator needs to decide how large an effect he is looking for. How should the size of the effect be measured? It turns out that the right definition is

$$\text{Effect size} = \frac{|\mu_1 - \mu_2|}{\sigma}$$

That is, the **effect size** is the difference in population means expressed relative to the common population SD. Thus, the effect size is a kind of "signal to noise ratio," where $|\mu_1 - \mu_2|$ represents the signal we want to detect and σ represents the background noise that tends to obscure the signal.

If α and the effect size have been specified, then the power of the t test depends only on the sample sizes (n). Table 5 at the end of the book shows the value of n required in order to achieve a specified power against a specified effect size. (For a brief explanation of the calculations underlying Table 5, see Appendix 7.1.) Let us see how Table 5 applies to our familiar example of body height.

EXAMPLE 7.28

HEIGHTS OF PEOPLE

In Example 7.27, case (a), we considered samples of 17-year-old males and 17-year-old females. The effect size is

$$\frac{|\mu_1 - \mu_2|}{\sigma} = \frac{|69.1 - 64.1|}{2.5} = \frac{5}{2.5} = 2.0$$

For a two-tailed t test at $\alpha = .05$, Table 5 shows that the sample size required for a power of .99 is $n = 11$; this is the basis for the claim in Example 7.27 that the investigator has a 99% chance of detecting the difference between males and females. ∎

The definition of effect size that we are using is probably unfamiliar to the biologically oriented reader. It is more common in biology to "standardize" a difference of two quantities by expressing it as a percentage of one of them. For example, the height difference between the 17-year-old populations, expressed as a percentage of mean female height, is

$$\frac{\mu_1 - \mu_2}{\mu_2} = \frac{5''}{64''} = .08 \quad \text{or} \quad 8\%$$

Thus, males are only about 8% taller than females. However, from the viewpoint of power it is more relevant that the height distributions for males and females are 2 SDs apart. Two SDs is a substantial separation in the sense that the information contained in samples of only $n = 11$ is enough to discriminate reliably between the distributions.

As we have seen, in order to choose a sample size the researcher needs to specify not only the size of the effect he wishes to detect, but also how certain he wants to be of detecting it; that is, it is necessary to specify how much power is wanted. Since the power measures the protection against Type II error, the choice of a desired power level depends on the consequences that would result from a Type II error. If the consequences of a Type II error would be *very* unfortunate (for example, if a promising but risky cancer treatment is being tested on humans and a negative result would discredit the treatment so that it would never be tested again), then the researcher might specify a high power, say .95 or .99. But of course high power is expensive in terms of n. For much research, a Type II error is not a disaster, and a lower power such as .80 is considered adequate.

The following example illustrates a typical use of Table 5 in planning an experiment.

EXAMPLE 7.29

WEIGHT GAIN OF BEEF CATTLE

An animal scientist is planning an experiment to evaluate a new dietary supplement for beef cattle. One group of cattle will receive a standard diet, and another group will receive the standard diet plus the supplement. The response variable will be weight gain (kg) during a 100-day test period. In a previous experiment with the standard diet, the mean weight gain during a 100-day test was 115 kg, and the SD was 17 kg. The researcher will use a one-tailed t test at the 5% significance level. He now wants to decide how many cattle (n) to put in each group.

The researcher believes that a 10% effect of the supplement would be important to discover; more precisely, if the true (population) difference is at least

10% of 115 kg = 11.5 kg

then he wants to have a good chance of detecting that fact. He decides to define "good chance" as a power of .80. Using the historical value (17 kg) as a guess of σ, the effect size he wants to consider is

$$\frac{|\mu_1 - \mu_2|}{\sigma} = \frac{11.5}{17} \approx .7$$

For this effect size, and for a power of .80 with a one-tailed test at the 5% significance level, Table 5 yields $n = 26$. The researcher will need to put 26 animals on each diet.

At this point, the researcher might ask himself some questions, such as (1) Can he afford to use 52 animals for the study? If not, then (2) would he perhaps be willing to redefine an "important" effect as 20% rather than 10% in order to reduce the required n? With questions such as these, and repeated use of Table 5, he can finally decide on a firm value for n, or possibly decide to abandon the project because an adequate study would be too costly.   ∎

EXERCISES 7.49–7.55

7.49 One measure of the meat quality of pigs is backfat thickness. Suppose two researchers, Jones and Smith, are planning to measure backfat thickness in two groups of pigs raised on different diets. They have decided to use the same number (n) of pigs in each group, and to compare the mean backfat thickness using a two-tailed t test at the 5% significance level. Preliminary data indicate that the SD of backfat thickness is about .3 cm.

When the researchers approach a statistician for help in choosing n, she naturally asks how much difference they want to detect. Jones replies, "If the true difference is $\frac{1}{4}$ cm or more, I want to be reasonably sure of rejecting H_0." Smith replies, "If the true difference is $\frac{1}{2}$ cm or more, I want to be very sure of rejecting H_0."

If the statistician interprets "reasonably sure" as 80% power, and "very sure" as 95% power, what value of n will she recommend
a. to satisfy Jones' requirement?
b. to satisfy Smith's requirement?

7.50 Refer to the brain NE data of Example 7.6. Suppose you are planning a similar experiment; you will study the effect of LSD (rather than toluene) on brain NE. You anticipate using a two-tailed t test at $\alpha = .05$. Suppose you have decided that a 10% effect (increase or decrease in mean NE) of LSD would be important, and so you want to have good power (80%) to detect a difference of this magnitude.
a. Using the data of Example 7.6 as a "pilot study," determine how many rats you should have in each group. (The mean NE in the control group in Example 7.6 is 444.2 ng/g and the pooled SD is $s_c = 67.7$ ng/g.)
b. If you were planning to use a *one-tailed* t test, what would be the required number of rats?

7.51 Suppose you are planning a greenhouse experiment on growth of pepper plants. You will grow n individually potted seedlings in standard soil and another n seedlings in specially treated soil. After 21 days, you will measure Y = total stem length (cm) for each plant. If the effect of the soil treatment is to increase the population mean stem length by 2 cm, you would like to have a 90% chance of rejecting H_0 with a one-tailed t test. Data from a pilot study (such as the data in Exercise 2.16) on 15 plants grown in standard soil give $\bar{y}$ = 12.5 cm and s = .8 cm.

    a. Suppose you plan to test at α = .05. Use the pilot information to determine what value of n you should use.

    b. What assumptions are necessary for the validity of the calculation in part **a**? Which of these can be checked (roughly) from the data of the pilot study?

    c. Suppose you decide to adopt a more conservative posture and test at α = .01. What value of n should you use?

7.52 Diastolic blood pressure measurements on American men aged 18–44 years follow approximately a normal curve with μ = 81 mm Hg and σ = 11 mm Hg. The distribution for women aged 18–44 is also approximately normal with the same SD but with a lower mean: μ = 75 mm Hg.[33] Suppose we are going to measure the diastolic blood pressure of n randomly selected men and n randomly selected women in the age group 18–44 years. Let E be the event that the difference between men and women will be found statistically significant by a t test. How large must n be in order to have $\Pr\{E\}$ = .9,

    a. if we use a two-tailed test at α = .05?

    b. if we use a two-tailed test at α = .01?

    c. if we use a one-tailed test (in the correct direction) at α = .05?

7.53 Suppose you are planning an experiment to test the effect of a certain drug treatment on drinking behavior in the rat. You will use a two-tailed t test to compare a treated group of rats against a control group; the observed variable will be Y = one-hour water consumption after 23-hour deprivation. You have decided that, if the effect of the drug is to shift the population mean consumption by 2 ml or more, then you want to have at least an 80% chance of rejecting H_0 at the 5% significance level.

    a. Preliminary data (as in Example 2.6) indicate that the SD of Y under control conditions is approximately 2.5 ml. Using this as a guess of σ, determine how many rats you should have in each group.

    b. What assumptions are required for the calculation of part **a**? Which of these assumptions can be checked (roughly) from preliminary data such as those given in Example 2.6?

    c. Suppose that, because the calculation of part **a** indicates a rather large number of rats, you consider modifying the experiment so as to reduce σ. You find that, by switching to a better supplier of rats and by improving lab procedures, you could cut the SD in half; however, the cost of each observation would be doubled. Would these measures be cost-effective, that is, would the modified experiment be less costly?

7.54 Data from a large study indicate that the serum concentration of lactate dehydrogenase (LD) is higher in men than in women. (The data are summarized in Example 7.19.) Suppose Dr. Jones proposes to conduct his own study to replicate this finding; however, because of limited resources Jones can enlist only 35 men and 35 women for his study. Supposing that the true difference in population means is 4 U/l and each population SD is 10 U/l, what is the probability that Jones will be successful? Specifically, find the probability that Jones will reject H_0 with a one-tailed t test at the 5% significance level.

7.55 Refer to the painkiller study of Exercise 7.39. In that study, the evidence favoring the drug was marginally significant ($.025 < P < .05$). Suppose Dr. Smith is planning a new study on the same drug in order to try to replicate the original findings, that is, to show the drug to be effective. She will consider her study successful if she rejects H_0 with a one-tailed test at $\alpha = .05$. In the original study, the difference between the treatment means was about half a standard deviation [$(32 - 25)/13 \approx .5$]. Taking this as a provisional value for the effect size, determine how many patients Smith should have in each group in order for her chance of success to be
a. 80%   **b.** 90%
[*Note:* This problem illustrates that surprisingly large sample sizes may be required to make a replication study worthwhile, especially if the original findings were only marginally significant.]

SECTION 7.9

STUDENT'S *t*: ASSUMPTIONS AND SUMMARY

In the preceding sections we have discussed the comparison of two means using classical methods based on Student's t distribution. In this section we describe the assumptions on which these methods are based. In addition, we summarize the methods for convenient reference.

ASSUMPTIONS

The t test and confidence interval procedures we have described are appropriate if the following conditions are assumed to hold:

1. *Conditions on the design of the study*
    a. It must be reasonable to regard the data as random samples from their respective populations. The populations must be large. The observations within each sample must be independent.
    b. The two samples must be independent of each other.
2. *Conditions on the form of the population distributions*
    a. If the sample sizes are small, the population distributions must be approximately normal with approximately equal standard deviations ($\sigma_1 = \sigma_2$). In this case the pooled SE should be used.
    b. If the sample sizes are large, the population distributions need not be approximately normal. Also, σ_1 and σ_2 need not be approximately equal; if they are not, the unpooled SE should be used.

Condition 2(b) is based on an approximation theorem similar to the Central Limit Theorem. The required "largeness" in condition 2(b) depends on the degree of nonnormality of the populations (as in Section 6.5) and the degree of inequality of σ_1 and σ_2. In many practical situations, moderate sample sizes (say, $n_1 = 20$, $n_2 = 20$) are quite "large" enough.

VERIFICATION OF CONDITIONS

A check of the above conditions should be a part of every data analysis.

A check of condition 1(a) would proceed as for a confidence interval (Section 6.5), with the researcher looking for biases in the experimental design and verifying that there is no hierarchical structure within each sample.

Condition 1(b) means that there must be no pairing or dependency between the two samples. The full meaning of this condition will become clear in Chapters 8 and 9.

Sometimes it is known from previous studies whether the populations can be assumed to be approximately normal with equal standard deviations. In the absence of such information, the normality requirement can be checked by making histograms (or stem-and-leaf displays) for each sample separately, and the requirement of equal standard deviations ($\sigma_1 = \sigma_2$) can be checked by comparing the sample standard deviations, s_1 and s_2. But how do we decide whether the histograms are close enough to a normal form, and whether s_1 and s_2 are close enough in value?* These can be troublesome decisions.

If the sample sizes are large, then the decisions are not so crucial, because the approximate method (based on the unpooled SE) can be used. Furthermore, if the sample sizes are equal ($n_1 = n_2$), then the pooled and the unpooled SE are equal, so it is not necessary to decide between them.

If the sample sizes are small, then the assumptions of normality and of equality of SDs are more important, but unfortunately the samples give less reliable information about whether the populations satisfy these assumptions. For small or moderate sample sizes, the sample SDs (s_1 and s_2) can easily differ by as much as 30% just by chance. However, if s_1 and s_2 differ so much that one is twice as large as the other (or more), it is probably unwise to proceed on the assumption that $\sigma_1 = \sigma_2$.[†]

With respect to nonnormality, usually only a rather conspicuous departure from normality (outliers, or long straggly tails) should be cause for concern. Moderate skewness has very little effect on the *t* test, even for small samples.

CONSEQUENCES OF INAPPROPRIATE USE OF STUDENT'S *t*

Our discussion of the *t* test and confidence interval (in Sections 7.3–7.8) was based on the assumptions (1) and (2) listed above. Violation of the assumptions may render the methods inappropriate.

---

*When the SDs are very different, the comparison of means may not be appropriate at all (see Section 7.12).

[†]In this situation it is preferable to use the unpooled SE; an approximation to the critical value is obtained by entering Student's *t* table with df given by

$$df = \frac{(SE_1^2 + SE_2^2)^2}{SE_1^4/(n_1 - 1) + SE_2^4/(n_2 - 1)}$$

where $SE_1 = s_1/\sqrt{n_1}$ and $SE_2 = s_2/\sqrt{n_2}$. This approximation still assumes normality of the population distributions.

If the assumptions are not satisfied, then the *t* test may be inappropriate in two possible ways:

1. It may be **invalid** in the sense that the actual risk of Type I error is larger than the nominal significance level α. (To put this another way, the *P*-value yielded by the *t* test procedure may be inappropriately small.)
2. The *t* test may be valid but *less powerful* than a more appropriate test.

If the design includes hierarchical structures that are ignored in the analysis, the *t* test may be seriously invalid. If the samples are not independent of each other, the usual consequence is a loss of power.

One fairly common type of departure from the assumption of normality is for one or both populations to have long straggly tails. The effect of this form of nonnormality is to inflate the SE, and thus to rob the *t* test of power. Inequality of the population SDs can affect both power and validity.

Inappropriate use of confidence intervals is analogous to that for *t* tests. If the assumptions are violated, then the confidence interval may not be valid, or it may be valid but wider than necessary.

OTHER APPROACHES

Because methods based on Student's *t* distribution are not always the most appropriate, statisticians have devised other methods that serve similar purposes. One of these is the Mann–Whitney test, which we will describe in Section 7.11. Another approach to the difficulty is to transform the data, for instance to analyze (log *Y*) instead of *Y* itself; sometimes the transformation serves the dual purpose of reducing nonnormality and equalizing the SDs.

SUMMARY OF FORMULAS

For convenient reference, we summarize in the accompanying boxes the formulas for Student's *t* method for comparing the means of independent samples.

STANDARD ERROR OF $(\bar{y}_1 - \bar{y}_2)$

Unpooled method

$$\text{SE}_{(\bar{y}_1 - \bar{y}_2)} = \sqrt{\frac{s_1^2}{n_1} + \frac{s_2^2}{n_2}} = \sqrt{\text{SE}_1^2 + \text{SE}_2^2}$$

Pooled method

$$\text{SE}_{(\bar{y}_1 - \bar{y}_2)} = \sqrt{s_c^2 \left(\frac{1}{n_1} + \frac{1}{n_2} \right)}$$

where

$$s_c^2 = \frac{\text{SS}_1 + \text{SS}_2}{n_1 + n_2 - 2} = \frac{(n_1 - 1)s_1^2 + (n_2 - 1)s_2^2}{n_1 + n_2 - 2}$$

CONFIDENCE INTERVAL FOR $(\mu_1 - \mu_2)$

95% Confidence interval:

$$(\bar{y}_1 - \bar{y}_2) \pm t_{.05}\ \text{SE}_{(\bar{y}_1 - \bar{y}_2)}$$

Critical value $t_{.05}$ from Student's *t* distribution with

$$\text{df} = n_1 + n_2 - 2$$

Confidence intervals with other confidence coefficients (90%, 99%, etc.) are constructed analogously (using $t_{.10}$, $t_{.01}$, etc.)

---

t TEST

H_0: $\mu_1 = \mu_2$
H_A: $\mu_1 \neq \mu_2$ (nondirectional)
H_A: $\mu_1 < \mu_2$ (directional)
H_A: $\mu_1 > \mu_2$ (directional)

Test statistic: $t_s = \dfrac{\bar{y}_1 - \bar{y}_2}{\text{SE}_{(\bar{y}_1 - \bar{y}_2)}}$

P-value:

P = Tail area under Student's *t* curve with df = $n_1 + n_2 - 2$

Nondirectional H_A: P = Two-tailed area beyond t_s and $-t_s$
Directional H_A: Step 1. Check directionality.
                        Step 2. P = Single-tailed area beyond t_s
Decision: Reject H_0 if $P < \alpha$.

| | | | | | | | | | | | | |

EXERCISES 7.56–7.57

7.56 Refer to the sucrose consumption data analyzed in Exercise 7.17 and displayed in Figure 7.2.
   a. Does the assumption that the populations are normal with equal standard deviations appear to be reasonable for these data? Explain.
   b. In view of your answer to part **a**, on what grounds can you defend the application of Student's *t* method to these data?

7.57 Refer to the serotonin data of Exercise 7.25. On what grounds might an objection be raised to the use of the *t* test on these data?

Our study of the t test has illustrated some of the general principles of statistical tests of hypotheses. In the remainder of this book we will introduce several other types of tests besides the t test.

A GENERAL VIEW OF HYPOTHESIS TESTS

A typical statistical test involves a null hypothesis H_0, an alternative hypothesis H_A, and a test statistic that measures deviation or discrepancy of the data from H_0. The sampling distribution of the test statistic, under the assumption that H_0 is true, is called the **null distribution** of the test statistic. (For example, the null distribution of the t statistic t_s is—under certain assumptions—a Student's t distribution.) The null distribution indicates how much the test statistic can be expected to deviate from H_0 because of chance alone.

In testing a hypothesis, we assess the evidence against H_0 by locating the test statistic within the null distribution; the P-value is a measure of this location which indicates the degree of compatibility between the data and H_0. The dividing line between compatibility and incompatibility is specified by an arbitrarily chosen significance level α. The decision whether to reject the null hypothesis is made according to the following rule:

Reject H_0 if $P \le \alpha$.

In this book, we will not usually calculate the P-value exactly, but will bracket it using a table of critical values. If H_A is directional, the bracketing of P is a two-step procedure.

Every test of a null hypothesis H_0 has its associated risks of Type I error (rejecting H_0 when H_0 is true) and Type II error (not rejecting H_0 when H_0 is false). The risk of Type I error is always limited by the chosen significance level:

$\Pr\{\text{reject } H_0\} \le \alpha$ if H_0 is true

Thus, the hypothesis testing procedure treats the Type I error as the one to be most stringently guarded against. The risk of Type II error, by contrast, can be quite large if the samples are small.

ANOTHER LOOK AT P-VALUE

In order to place P-value in a general setting, let us consider some verbal interpretations of P-value.

First we revisit the t test. For a nondirectional H_A, we have defined the P-value of the data to be the two-tailed area under the Student's t curve beyond the observed value of t_s. Another way of defining the P-value is the following:

The **P-value** of the data is the probability (under H_0) of getting a result as extreme as, or more extreme than, the result that was actually observed.

To put this another way,

The **P-value** is the probability that, if H_0 were true, a result would be obtained which would deviate from H_0 as much as (or more than) the actual data do.

Actually, this description of P-value is a bit too limited. The P-value actually depends on the nature of the alternative hypothesis. When we are performing a t test against a *directional* alternative, the P-value of the data is (if the observed deviation is in the direction of H_A) only a *single-tailed* area beyond the observed value of t_s. The more general definition of P-value is the following:

> The **P-value** of the data is the probability (under H_0) of getting a result as deviant as, or more deviant than, the result actually observed—where deviance is measured as discrepancy from H_0 in the direction of H_A.

The P-value measures how easily the observed deviation could be explained as chance variation rather than by the alternative explanation provided by H_A. For example, if the t test yields a P-value of $P = .036$ for our data, then we may say that, if H_0 were true, we would expect data to deviate from H_0 as much as our data only 3.6% of the time (in the meta-experiment).

Another definition of P-value which is worth thinking about is the following:

> The **P-value** of the data is the value of α for which H_0 would just barely be rejected, using those data.

To interpret this definition, imagine that a research report which includes a P-value is read by a number of interested scientists. The scientists who are more skeptical of H_A might personally use a more conservative decision threshold, such as $\alpha = .001$; the scientists who are quite favorably disposed toward H_A might use a very liberal value such as $\alpha = .10$. The P-value of the data determines the point, within this spectrum of opinion, which separates those who find the data to be convincing in favor of H_A, and those who do not. Of course, if the P-value is large, for instance $P = .40$, then presumably no reasonable person would reject H_0 and be convinced of H_A.

As the preceding discussion shows, the P-value does not describe all facets of the data, but relates only to a test of a particular null hypothesis against a particular alternative. In fact, we will see that the P-value of the data also depends on which statistical test is used to test a given null hypothesis. For this reason, when describing in a scientific report the results of a statistical test, it is best to report the P-value (exact, if possible), the name of the statistical test, and whether the alternative hypothesis was directional or nondirectional.

We repeat here, because it applies to any statistical test, the principle expounded in Section 7.7: The P-value is a measure of the strength of the evidence against H_0, but the P-value does *not* reflect the *magnitude* of the discrepancy between the data and H_0. The data may deviate from H_0 only slightly, yet if the samples are large, the P-value may be quite small. By the same token, data that deviate substantially from H_0 can nevertheless yield a large P-value.

INTERPRETATION OF ERROR PROBABILITIES

A common mistake is to interpret the P-value as the probability that the null hypothesis is true. A related misconception is the belief that, if H_0 has been rejected at (for example) the 5% significance level, then the probability that H_0

is true is 5%. These interpretations are not correct. In fact, the probability that H_0 is true cannot be calculated at all. This point can be illustrated by an analogy with medical diagnosis.

In applying a diagnostic test for an illness, two types of error are possible: A healthy individual may be diagnosed as ill (false positive) or an ill individual may be diagnosed as healthy (false negative). Trying out a diagnostic test on individuals *known* to be healthy or ill will enable us to estimate the proportions of these groups who will be misdiagnosed; yet this information alone will not tell us what proportion of all positive diagnoses are false diagnoses. These ideas are illustrated numerically in the next example.

EXAMPLE 7.30
PKU SCREENING

At the present time, virtually all newborn babies in the United States are given a blood test for the congenital disease phenylketonuria (PKU). Using a widely accepted interpretation of the blood test, virtually 100% of babies who actually have PKU will be diagnosed as positive for PKU, while only $\frac{1}{2,000}$ or .05% of babies without the disease will be falsely diagnosed as positive. If a baby is diagnosed as positive, what is the probability that he actually has PKU? The answer is not 99.95%; in fact, the correct answer depends on the prevalence of the disease. Actually, about one in 10,000 newborns has PKU, so the results of screening 1 million babies might resemble those in Table 7.14.[34]

TABLE 7.14
Fictitious Results of
Screening 1 Million
Babies for PKU

| | | ACTUAL STATUS OF BABY | | |
		Non-PKU	PKU	Total
DIAGNOSIS	Negative	999,400	0	999,400
	Positive	500	100	600
Total		999,900	100	1,000,000

Notice in Table 7.14 that only 500 of 999,900 non-PKU babies, or .05%, give a false positive diagnosis. Nevertheless, the probability that a baby with a positive diagnosis has PKU is not 99.95% but rather is only $\frac{100}{600}$ or 17%. This startlingly low probability is due to the rarity of the disease. ∎

An analogy can be drawn between the medical diagnosis situation of Example 7.30 and a statistical test of a hypothesis. Table 7.14 is analogous to the table of Type I and Type II error (Table 7.13), with a Type I error corresponding to a false positive diagnosis and a Type II error to a false negative diagnosis. The analogy may clarify the nature of the probabilities involved in significance level and power. The risk of Type I error is a probability computed *under the assumption that H_0 is true*; similarly, the risk of a Type II error is computed assuming that H_A is true. If we have a well-designed study with adequate sample sizes, both of these probabilities will be small. We then have a good test procedure in the same sense that the PKU blood test is a good diagnostic procedure. But this does not in itself guarantee that most of the null hypotheses we reject are in fact false, or that most of those we do not reject are in fact true. The validity or nonvalidity of such guarantees would depend on an unknown and unknowable quantity —namely, the proportion of true null hypotheses among all null hypotheses that are tested (which is analogous to the incidence of PKU).

PERSPECTIVE

It seems only fair at this point to mention that the philosophy of statistical hypothesis testing which we have explained in this chapter is not shared by all statisticians. Our view, which is called the **frequentist view**, is widely held to be applicable to most problems of research. An alternative view, the **Bayesian view**, permits the quantitative evaluation of data to depend not only on the observed data but also on the researcher's (or consumer's) prior beliefs about the truth or falsity of H_0.

|||||||||||||

S E C T I O N 7.11

THE MANN–WHITNEY TEST

The **Mann–Whitney test** is used to compare two independent samples. It is a competitor to the t test, but unlike the t test, the Mann–Whitney test is valid no matter what the form of the population distributions. The Mann–Whitney test is therefore called a **distribution-free** type of test. In addition, the Mann–Whitney test does not focus on any particular parameter such as a mean or a median; for this reason it is called a **nonparametric** type of test.

STATEMENT OF H_0 AND H_A

Let us denote the observations in the two samples by Y_1 and Y_2. A general statement of the null hypothesis of a Mann–Whitney test is

H_0:   The population distributions of Y_1 and Y_2 are the same

In practice, it is more natural to state H_0 and H_A in words suitable to the particular application, as illustrated in Example 7.31.

EXAMPLE 7.31

BETA-ENDORPHIN

Human beta-endorphin (HBE) is a hormone secreted by the pituitary gland under conditions of stress. An exercise physiologist measured the resting (unstressed) blood concentration of HBE in two groups of men: Group 1 consisted of 11 men who had been jogging regularly for some time, and group 2 consisted of 15 men who had just entered a physical fitness program. The results are given in Table 7.15.[35]

TABLE 7.15

HBE (pg/ml) in Two Groups of Men

JOGGERS Group 1				FITNESS PROGRAM ENTRANTS Group 2			
39	40	32	60	70	47	54	27
19	52	41	32	31	42	37	41
13	37	28		9	18	33	23
				49	41	59	

As a step to understanding the possible effect of exercise on HBE, the researcher might compare the baseline HBE levels of the two groups. The null hypothesis could be stated as

H_0:   The populations from which the two groups were drawn have the same distribution of HBE level

or, more informally, as

H_0:   Joggers and fitness program entrants do not differ with respect to HBE level

A nondirectional alternative could be stated as

H_A: One of the two populations tends to have higher HBE levels than the other

or, the alternative hypothesis might be directional, for example,

H_A: Joggers tend to have higher HBE levels than fitness program entrants

■

METHOD

The Mann–Whitney test statistic, which is denoted U_s, measures the degree of separation or shift between the two samples. A large value of U_s indicates that the two samples are well separated, with relatively little overlap between them. Critical values for the Mann–Whitney test are given in Table 6 at the end of this book. The following example illustrates the Mann–Whitney test.

EXAMPLE 7.32
BETA-ENDORPHIN

Let us carry out a Mann–Whitney test on the HBE data of Example 7.31.

1. The value of U_s depends on the relative positions of the Y_1's and the Y_2's, but not on their actual values. The first step in determining U_s is to arrange the observations in increasing order, as follows:

					32									
Y_1:	13	19		28	32	37	39	40	41			52		60

Y_2:	9	18	23	27	31	33	37		41	42	47	49	54	59	70
									41						

(If you are doing this step by hand, a stem-and-leaf display can be labor-saving aid.)

2. We next determine two counts, K_1 and K_2, as follows:
   a. *The K_1 count*   For each observation in sample 1, we count the number of observations in sample 2 that are smaller in value (that is, to the left). Count $\frac{1}{2}$ for each tied observation. In the above data, there is one Y_2 less than the first Y_1; there are two Y_2's less than the second Y_1, four Y_2's less than the third Y_1, five Y_2's less than the fourth Y_1 (32), and five Y_2's less than the fifth Y_1 (also 32); there are six Y_2's less than the sixth Y_1 and one equal to

it, so we count $6\frac{1}{2}$. So far we have counts of 1, 2, 4, 5, 5, $6\frac{1}{2}$. Continuing in a similar way, we get further counts of 7, 7, 8, 12, and 14. All together there are eleven counts, one for each Y_1. The sum of all eleven counts is $K_1 = 71\frac{1}{2}$.

b. *The K_2 count* For each observation in sample 2, we count the number of observations in sample 1 that are smaller in value, counting $\frac{1}{2}$ for ties. This gives counts of 0, 1, 2, 2, 3, 5, $5\frac{1}{2}$, $8\frac{1}{2}$, $8\frac{1}{2}$, 9, 9, 9, 10, 10, 11. The sum of these counts is $K_2 = 93\frac{1}{2}$.

c. *Check* If the work is correct, the sum of K_1 and K_2 should be equal to the product of the sample sizes:

$$K_1 + K_2 \stackrel{?}{=} n_1 n_2$$
$$71\frac{1}{2} + 93\frac{1}{2} = 165 = (11)(15)$$

3. The test statistic U_s is the larger of K_1 and K_2. In this example, $U_s = 93.5$.

4. To determine critical values, we consult Table 6 with $n =$ the larger sample size, and $n' =$ the smaller sample size. In the present case, $n = 15$ and $n' = 11$. The critical values from Table 6 are reproduced in Table 7.16.

TABLE 7.16

Critical Values from Table 6 for $n = 15$, $n' = 11$

Nominal Tail Probability	.20	.10	.05	.02	.01	.002	.001
Critical Value	108	115	121	128	132	141	144

Let us test H_0 against a nondirectional alternative at significance level $\alpha = .05$. From Table 7.16, we note that $U_{.20} = 108$; since $93.5 < 108$, the P-value is $P > .20$ and H_0 is not rejected. There is insufficient evidence to conclude that HBE concentrations are different in joggers and fitness program entrants. ∎

As Example 7.32 illustrates, Table 6 is used to bracket the P-value for the Mann–Whitney test just as Table 4 is used for the t test. If H_A is nondirectional, one simply locates the critical values that bracket the observed U_s; one then brackets the P-value by the corresponding column headings. If U_s is exactly equal to a critical value, then the P-value is less than the column heading.* The following example illustrates the bracketing procedure.

EXAMPLE 7.33

BRACKETING THE P-VALUE

Suppose $n = 15$ and $n' = 11$, and H_A is nondirectional. Using the critical values shown in Table 7.16, we would bracket the P-value as follows:

If $U_s = 121$, then $.02 < P < .05$.

If $U_s = 125$, then $.02 < P < .05$.

If $U_s = 144$, then $P < .001$. ∎

*In a few cases, the P-value would be exactly equal to (rather than less than) the column heading. To simplify the presentation, we neglect this fine distinction.

DIRECTIONALITY For the t test, one determines the directionality of the data by seeing whether $\bar{y}_1 > \bar{y}_2$ or $\bar{y}_1 < \bar{y}_2$. Similarly, one can check directionality for the Mann–Whitney test by comparing K_1 and K_2: $K_1 > K_2$ indicates a trend for the Y_1's to be larger than the Y_2's, while $K_1 < K_2$ indicates the opposite trend. Often, however, this formal comparison is unnecessary; a glance at the data is enough.

DIRECTIONAL ALTERNATIVE If the alternative hypothesis H_A is directional rather than nondirectional, the Mann–Whitney procedure must be modified. As with the t test, the modified procedure has two steps and the second step involves halving the P-value.

STEP 1 Check directionality—see if the data deviate from H_0 in the direction specified by H_A.

a. If not, the P-value is $P > .5$.
b. If so, proceed to step 2.

STEP 2 The P-value of the data is *half* as much as it would be if H_A were nondirectional.

To make a decision at a prespecified significance level α, one rejects H_0 if $P \leq \alpha$.

The following example illustrates the two-step procedure.

EXAMPLE 7.34
DIRECTIONAL H_A

Suppose $n = 15$, $n' = 11$, and H_A is directional. Suppose further that the data do deviate from H_0 in the direction specified by H_A. The critical values shown in Table 7.16 can be used to bracket the P-value as follows:

If $U_s = 121$, then $.01 < P < .025$.
If $U_s = 125$, then $.01 < P < .025$.
If $U_s = 144$, then $P < .0005$.

Note that these P-values are half of those shown in Example 7.33. ∎

BLANK CRITICAL VALUES In some cases, certain entries in Table 6 are blank. The next example shows how the P-value is bracketed in such a case.

EXAMPLE 7.35
BLANK CRITICAL VALUES

If $n = 6$ and $n' = 5$, Table 6 reads as follows:

Nominal Tail Probability	.20	.10	.05	.02	.01	.002	.001
Critical Value	23	25	27	28	29		

Suppose H_A is nondirectional. Then the P-value would be bracketed as follows:

If $U_s = 29$, then $.002 < P < .01$.
If $U_s = 30$, then $.002 < P < .01$.

For these sample sizes, U_s cannot be larger than 30. The rationale for this bracketing procedure is explained in the next subsection. ∎

RATIONALE

Let us see why the Mann–Whitney test procedure makes sense. To take a specific case, suppose the sample sizes are $n_1 = 5$ and $n_2 = 4$. Then necessarily, regardless of what the data look like, we must have

$$K_1 + K_2 = 5 \times 4 = 20$$

The relative magnitudes of K_1 and K_2 indicate the amount of overlap of the Y_1's and the Y_2's. Figure 7.14 shows how this works. For the data of Figure 7.14(a), the two samples do not overlap at all; the data are *least* compatible with H_0, and U_s has its maximum value, $U_s = 20$. Similarly, $U_s = 20$ for Figure 7.14(b). On the other hand, the arrangement *most* compatible with H_0 is the one with maximal overlap, shown in Figure 7.14(c); for this arrangement $K_1 = 10$, $K_2 = 10$, and $U_s = 10$.

FIGURE 7.14

Three Data Arrays for a Mann–Whitney Test

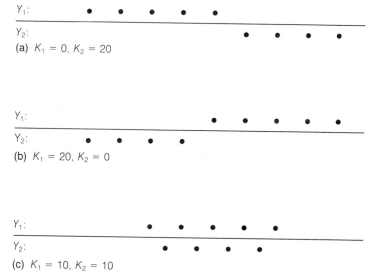

Y_1:

Y_2:

(a) $K_1 = 0$, $K_2 = 20$

Y_1:

Y_2:

(b) $K_1 = 20$, $K_2 = 0$

Y_1:

Y_2:

(c) $K_1 = 10$, $K_2 = 10$

All other possible arrangements of the data lie somewhere between the three arrangements shown in Figure 7.14; those with much overlap have U_s close to 10, and those with little overlap have U_s closer to 20. Thus, large values of U_s indicate incompatibility of the data with H_0.

We now briefly consider the null distribution of U_s and indicate how the critical values of Table 6 were determined. (Recall from Section 7.10 that, for any statistical test, the reference distribution for critical values is always the null distribution of the test statistic—that is, its sampling distribution under the assumption that H_0 is true.) To determine the null distribution of U_s, it is necessary to calculate the probabilities associated with various arrangements of the data, assuming that all the Y's were actually drawn from the same population.* (The method for calculating the probabilities is briefly described in Appendix 7.2.)

Figure 7.15(a) shows the null distribution of K_1 and K_2 for the case $n = 5$, $n' = 4$. For example, it can be shown that, if H_0 is true, then

$$\Pr\{K_1 = 0, K_2 = 20\} = .008$$

This is the first probability plotted in Figure 7.15(a). Note that Figure 7.15(a) is roughly analogous to a t distribution; large values of K_1 (right tail) represent evidence that the Y_1's tend to be larger than the Y_2's, and large values of K_2 (left tail) represent evidence that the Y_2's tend to be larger than the Y_1's.

FIGURE 7.15

Null Distributions for the Mann–Whitney Test When $n = 5$, $n' = 4$. **(a)** Null distribution of K_1 and K_2; **(b)** Null distribution of U_s.

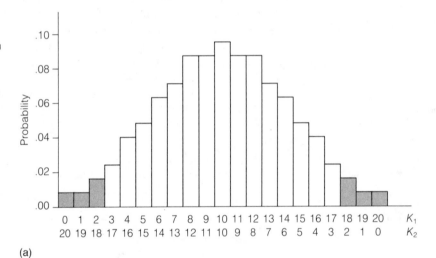

(a)

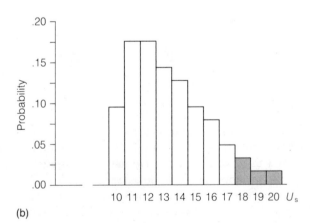

(b)

*In calculating the probabilities used in this section, it has been assumed that the chance of tied observations is negligible. This will be true for a continuous variable that is measured with high precision.

Figure 7.15(b) shows the null distribution of U_s, which is derived directly from the distribution in Figure 7.15(a). For instance, if H_0 is true then

$$\Pr\{K_1 = 0, K_2 = 20\} = .008$$

and

$$\Pr\{K_1 = 20, K_2 = 0\} = .008$$

so that

$$\Pr\{U_s = 20\} = .008 + .008 = .016$$

which is the rightmost probability plotted in Figure 7.15(b). Thus, both tails of the K distribution have been "folded" into the upper tail of the U distribution; for instance, the one-tailed shaded area in Figure 7.15(b) is equal to the two-tailed shaded area in Figure 7.15(a).

P-values for the Mann–Whitney test are upper-tail areas in the U_s distribution. For instance, it can be shown that the shaded area in Figure 7.15(b) is equal to .064; this means that if H_0 is true then

$$\Pr\{U_s \geq 18\} = .064$$

Thus, a data set that yielded $U_s = 18$ would have an associated P-value of $P = .064$ (assuming a nondirectional H_A). Note that, even when H_A is nondirectional, the Mann–Whitney test is, in a sense, a one-tailed test.

The critical values in Table 6 have been determined from the null distribution of U_s. For instance, here are the critical values for our specific case $n = 5$, $n' = 4$:

Nominal Tail Probability	.20	.10	.05	.02	.01	.002	.001
Critical Value	16	18	19	20			

Because the U distribution is discrete, the column headings do not correspond exactly to P-values as they do for the t distribution. Rather, the associated P-value is *less than or equal to* the column heading (this is why the columns are labeled *nominal* tail probabilities). For example, the critical value in the .10 column is 18, while we saw above that the exact P-value associated with $U_s = 18$ is .064 rather than .10; for these sample sizes (5 and 4) the value .064 is as close as the P-value can get to .10 without exceeding it. Now it should be clear why some of the critical value entries are blank: No value is given for $U_{.01}$, for instance, because the P-value cannot possibly be less than .01; the most extreme case is $U_s = 20$, which has a P-value equal to .016.

If one takes a decision making approach to the Mann–Whitney test, then one rejects H_0 if $P \leq \alpha$. Thus, when H_A is nondirectional, H_0 is rejected if U_s is greater than or equal to the critical value listed in Table 6 under column heading α. If the critical value does not exist, then H_0 can never be rejected. For instance, with sample sizes 5 and 4, a Mann–Whitney test against a nondirectional H_A at $\alpha = .01$ can never reject H_0.

APPLICABILITY OF THE MANN–WHITNEY TEST

In order for the Mann–Whitney test to be applicable, it must be reasonable to regard the data as random samples from their respective populations, with the observations within each sample being independent, and the two samples being independent of each other. Under these assumptions, the Mann–Whitney test is always valid no matter what the form of the population distributions, provided that the observed variable Y is continuous.[36]

The critical values given in Table 6 have been calculated assuming that ties do not occur. If the data contain only a few ties, then the critical values are approximately correct.

Actually, the Mann–Whitney test need not be restricted to continuous variables; it can be applied to any ordinal variable. However, if Y is discrete or categorical, then the data may contain many ties, and the test should not be used without appropriate modification of the critical values.

THE MANN–WHITNEY TEST VERSUS THE t TEST

The Mann–Whitney test and the t test are aimed at answering the same question, but they treat the data in very different ways. Unlike the t test, the Mann–Whitney test does not use the actual values of the Y's but only their relative positions in the rank ordering. This is both a strength and a weakness of the Mann–Whitney test. On the one hand, the test is distribution-free because the null distribution of U_s relates only to the various rankings of the Y's, and therefore does not depend on the form of the population distribution. On the other hand, the Mann–Whitney test can be inefficient: it can lack power because it does not use all the information in the data. This inefficiency is especially evident for small samples.

Neither of the competitors—the t test or the Mann–Whitney test—is clearly superior to the other. If the population distributions are not approximately normal, the t test may not even be valid. In addition, the Mann–Whitney test can be much more powerful than the t test, especially if the population distributions are highly skewed. If the population distributions are approximately normal with equal standard deviations, then the t test is better, but its advantage is not necessarily very great; for moderate sample sizes, the Mann–Whitney test can be nearly as powerful as the t test.

There is a confidence interval procedure that is associated with the Mann–Whitney test in the same way that the confidence interval for $(\mu_1 - \mu_2)$ is associated with the t test. The procedure is beyond the scope of this book.

EXERCISES 7.58–7.63

7.58 Consider two samples of sizes $n_1 = 5$, $n_2 = 7$. Use Table 6 to bracket the P-value, assuming that H_A is nondirectional and that
  a. $U_s = 26$ **b.** $U_s = 30$ **c.** $U_s = 35$

7.59 Consider two samples of sizes $n_1 = 4$, $n_2 = 8$. Use Table 6 to bracket the P-value, assuming that H_A is nondirectional and that
a. $U_s = 25$ **b.** $U_s = 31$ **c.** $U_s = 32$

7.60 In a pharmacological study, researchers measured the concentration of the brain chemical dopamine in six rats exposed to toluene and six control rats. (This is the same study described in Example 7.6.) The concentrations in the striatum region of the brain were as shown in the table.[37]

DOPAMINE (ng/gm)	
Toluene	Control
3,420	1,820
2,314	1,843
1,911	1,397
2,464	1,803
2,781	2,539
2,803	1,990

a. Use a Mann–Whitney test to compare the treatments at $\alpha = .05$. Use a nondirectional alternative.
b. Proceed as in part **a**, but let the alternative hypothesis be that toluene increases dopamine concentration.

7.61 In a study of hypnosis, breathing patterns were observed in an experimental group of subjects and in a control group. The measurements of total ventilation (liters of air per minute per square meter of body area) are shown below.[38] (These are the same data that were summarized in Exercise 7.35.) Use a Mann–Whitney test to compare the two groups at $\alpha = .05$. Use a nondirectional alternative.

EXPERIMENTAL	CONTROL
5.32	4.50
5.60	4.78
5.74	4.79
6.06	4.86
6.32	5.41
6.34	5.70
6.79	6.08
7.18	6.21

7.62 In an experiment to compare the effects of two different growing conditions on the heights of greenhouse chrysanthemums, all plants grown under condition 1 were found to be taller than any of those grown under condition 2 (that is, the two height distributions did not overlap). Calculate the value of U_s and bracket the P-value if the number of plants in each group was
a. 3   **b.** 4   **c.** 5
(Assume that H_A is nondirectional.)

7.63 In a study of preening behavior in the fruitfly *Drosophila melanogaster*, a single experimental fly was observed for three minutes while in a chamber with ten other flies of the same sex. The observer recorded the timing of each episode ("bout") of preening by the experimental fly. This experiment was replicated 15 times with male flies and 15 times with female flies (different flies each time). One question of interest was whether there is a sex difference in preening behavior. The observed preening times (average time per bout, in seconds) were as follows:[39]

> *Male:*   1.2, 1.2, 1.3, 1.9, 1.9, 2.0, 2.1, 2.2, 2.2, 2.3, 2.3, 2.4, 2.7, 2.9, 3.3
>
> $\bar{y} = 2.127$ $\sum(y - \bar{y})^2 = 4.969$
>
> *Female:*   2.0, 2.2, 2.4, 2.4, 2.4, 2.8, 2.8, 2.8, 2.9, 3.2, 3.7, 4.0, 5.4, 10.7, 11.7
>
> $\bar{y} = 4.093$ $\sum(y - \bar{y})^2 = 127.2$

a. For these data, the value of the Mann–Whitney statistic is $U_s = 189.5$. Use a Mann–Whitney test to investigate the sex difference in preening behavior. Let H_A be nondirectional and let $\alpha = .01$.

b. For these data, the standard error of $(\bar{y}_1 - \bar{y}_2)$ is SE $= .7933$ sec. Use a t test to investigate the sex difference in preening behavior. Let H_A be nondirectional and let $\alpha = .01$.

c. What assumptions are required for the validity of the t test but not for the Mann–Whitney test? What feature or features of the data suggest that these assumptions may not hold in this case?

d. Verify the value of U_s given in part **a**.

| | | | | | | | | | | |
SECTION 7.12
PERSPECTIVE

In this chapter we have discussed several techniques—confidence intervals and hypothesis tests—for comparing two independent samples when the observed variable is quantitative. In coming chapters we will introduce confidence interval and hypothesis testing techniques that are applicable in various other situations. Before proceeding, we pause to reconsider the methods of this chapter.

AN IMPLICIT ASSUMPTION

In discussing the tests of this chapter—the t test and the Mann–Whitney test—we have made an unspoken assumption, which we now bring to light. When interpreting the comparison of two distributions, we have assumed that the relationship between the two distributions is relatively simple—that if the distributions differ, then one of the two variables has a consistent tendency to be larger than the other. For instance, suppose we are comparing the effects of two diets on the weight gain of mice, with

> Y_1 = Weight gain of mice on diet 1
>
> Y_2 = Weight gain of mice on diet 2

Our implicit assumption has been that, if the two diets differ at all, then that difference is in a consistent direction for all individual mice. To appreciate the meaning of this assumption, suppose the two distributions are as pictured in Figure 7.16. In this case, even though the mean weight gain is higher on diet 1, it would be an oversimplification to say that mice tend to gain more weight on

FIGURE 7.16
Weight Gain Distributions
on Two Diets

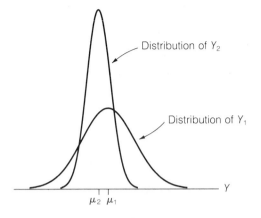

diet 1 than on diet 2; apparently *some* mice gain *less* on diet 1. Paradoxical situations of this kind do occasionally occur, and then the simple analysis typified by the *t* test and the Mann–Whitney test may be inadequate.

COMPARISON OF VARIABILITY

It sometimes happens that the variability of *Y*, rather than its average value, is of primary interest. For instance, in comparing two different lab techniques for measuring the concentration of an enzyme, a researcher might want primarily to know whether one of the techniques is more precise than the other, that is, whether its measurement error distribution has a smaller standard deviation. Techniques are available for testing the hypothesis $H_0: \sigma_1 = \sigma_2$, and for using a confidence interval to compare σ_1 and σ_2; these techniques are beyond the scope of this book.

**SUPPLEMENTARY
EXERCISES 7.64–7.78**

[*Note:* Exercises preceded by an asterisk refer to optional sections.]

7.64 For the following data, compute the standard error of $(\bar{y}_1 - \bar{y}_2)$ using
  a. the unpooled method.   **b.** the pooled method.

SAMPLE 1	SAMPLE 2
7	4
8	2
10	1
7	3
	4
	1
	6

7.65 To investigate the relationship between intracellular calcium and blood pressure, researchers measured the free calcium concentration in the blood platelets of 38 people with normal blood pressure and 45 people with high blood pressure. The results are given in the table.[40] Use the t test (unpooled) to compare the means. Let $\alpha = .01$ and let H_A be nondirectional.

Blood Pressure	PLATELET CALCIUM (nM)		
	Mean	SD	n
Normal	107.9	16.1	38
High	168.2	31.7	45

7.66 Refer to Exercise 7.65. Use the unpooled t method to construct a 95% confidence interval for the difference between the population means.

7.67 In a study of methods of producing sheep's milk for use in cheese manufacture, ewes were randomly allocated to either a mechanical or a manual milking method. The investigator suspected that the mechanical method might irritate the udder and thus produce a higher concentration of somatic cells in the milk. The accompanying data show the average somatic cell count for each animal.[41]

	SOMATIC CELL COUNT ($10^{-3} \times$ cells/ml)	
	Mechanical Milking	Manual Milking
	2,966	186
	269	107
	59	65
	1,887	126
	3,452	123
	189	164
	93	408
	618	324
	130	548
	2,493	139
Mean	1,215.6	219.0
SD	1,342.9	156.2
n	10	10

a. Do the data support the investigator's suspicion? Use an unpooled t test against a directional alternative at $\alpha = .05$. [The standard error of $(\bar{y}_1 - \bar{y}_2)$ is SE $= 427.54$.]

b. Do the data support the investigator's suspicion? Use a Mann–Whitney test against a directional alternative at $\alpha = .05$. (The value of the Mann–Whitney statistic is $U_s = 69$.) Compare with the result of part **a**.

c. What assumptions are required for the validity of the t test but not for the Mann–Whitney test? What features of the data cast doubt on these assumptions?

d. Verify the value of U_s given in part **b**.

7.68 A plant physiologist conducted an experiment to determine whether mechanical stress can retard the growth of soybean plants. Young plants were randomly allocated to two groups of 13 plants each. Plants in one group were mechanically agitated by shaking for 20 minutes twice daily, while plants in the other group were not agitated. After 16 days of growth, the total stem length (cm) of each plant was measured, with the results given in the accompanying table.[42] Use a t test to compare the treatments at $\alpha = .01$. Let the alternative hypothesis be that stress tends to retard growth.

	CONTROL	STRESS
$\bar{y}$	30.59	27.78
s	2.13	1.73
n	13	13

7.69 Refer to Exercise 7.68. Construct a 95% confidence interval for the population mean reduction in stem length. Does the confidence interval indicate whether the effect of stress is "horticulturally important," if "horticulturally important" is defined as a reduction in population mean stem length of at least
a. 1 cm **b.** 2 cm **c.** 5 cm

7.70 Refer to Exercise 7.68. The raw observations, in increasing order, are shown below. Compare the treatments using a Mann–Whitney test at $\alpha = .01$. Let the alternative hypothesis be that stress tends to retard growth.

CONTROL	STRESS
25.2	24.7
29.5	25.7
30.1	26.5
30.1	27.0
30.2	27.1
30.2	27.2
30.3	27.3
30.6	27.7
31.1	28.7
31.2	28.9
31.4	29.7
33.5	30.0
34.3	30.6

7.71 A developmental biologist removed the oocytes (developing egg cells) from the ovaries of 24 frogs (*Xenopus laevis*). For each frog the oocyte pH was determined. In addition, each frog was classified according to its response to a certain stimulus with the hormone progesterone. The pH values were as follows:[43]

Positive response: 7.06, 7.18, 7.30, 7.30, 7.31, 7.32, 7.33, 7.34, 7.36, 7.36, 7.40, 7.41, 7.43, 7.48, 7.49, 7.53, 7.55, 7.57

No response: 7.55, 7.70, 7.73, 7.75, 7.75, 7.77

Investigate the relationship of oocyte pH to progesterone response using a Mann–Whitney test at $\alpha = .05$. Use a nondirectional alternative.

7.72 Refer to Exercise 7.71. Summary statistics for the pH measurements are given in the table at the top of page 262. Investigate the relationship of oocyte pH to progesterone response using a pooled t test at $\alpha = .05$. Use a nondirectional alternative.

	POSITIVE RESPONSE	NO RESPONSE
$\bar{y}$	7.373	7.708
$\Sigma(y - \bar{y})^2$	.28120	.03288
n	18	6

7.73 A proposed new diet for beef cattle is less expensive than the standard diet. The proponents of the new diet have conducted a comparative study in which one group of cattle was fed the new diet and another group was fed the standard. They found that the mean weight gains in the two groups were not significantly different at the 5% significance level, and they stated that this finding supported the claim that the new cheaper diet was as good (for weight gain) as the standard diet. Criticize this statement.

*7.74 Refer to Exercise 7.73. Suppose you discover that the study used 25 animals on each of the two diets, and that the coefficient of variation of weight gain under the conditions of the study was about 20%. Using this additional information, write an expanded criticism of the proponents' claim, indicating how likely such a study would be to detect a 10% deficiency in weight gain on the cheaper diet (using a two-tailed t test at the 5% significance level).

7.75 (*Computer exercise*)   In an investigation of the possible influence of dietary chromium on diabetic symptoms, 14 rats were fed a low-chromium diet and 10 were fed a normal diet. One response variable was activity of the liver enzyme GITH, which was measured using a radioactively labelled molecule. The accompanying table shows the results, expressed as thousands of counts per minute per gram of liver.[44] Use a t test (pooled) to compare the diets at $\alpha = .05$. Use a nondirectional alternative.

LOW-CHROMIUM DIET		NORMAL DIET	
42.3	52.8	53.1	53.6
51.5	51.3	50.7	47.8
53.7	58.5	55.8	61.8
48.0	55.4	55.1	52.6
56.0	38.3	47.5	53.7
55.7	54.1		
54.8	52.1		

7.76 (*Computer exercise*)   Refer to Exercise 7.75. Use a Mann–Whitney test to compare the diets at $\alpha = .05$. Use a nondirectional alternative.

7.77 (*Computer exercise*)   Refer to Exercise 7.75.
   a. Construct a 95% confidence interval for the difference in population means.
   b. Suppose the investigators believe that the effect of the low-chromium diet is "unimportant" if it shifts mean GITH activity by less than 15%—that is, if the population mean difference is less than about 8 thousand cpm/gm. According to the confidence interval of part a, do the data support the conclusion that the difference is "unimportant"?
   c. How would you answer the question in part b if the criterion were 4 thousand rather than 8 thousand cpm/gm?

7.78 (*Computer exercise*)   In a study of the lizard *Sceloporis occidentalis*, researchers examined field-caught lizards for infection by the malarial parasite *Plasmodium*. To help assess the ecological impact of malarial infection, the researchers tested 15 infected and 15 noninfected lizards for stamina, as indicated by the distance each animal could run in two minutes. The distances (meters) are shown in the table.[45]

INFECTED ANIMALS		UNINFECTED ANIMALS	
16.4	36.7	22.2	18.4
29.4	28.7	34.8	27.5
37.1	30.2	42.1	45.5
23.0	21.8	32.9	34.0
24.1	37.1	26.4	45.5
24.5	20.3	30.6	24.5
16.4	28.3	32.9	28.7
29.1		37.5	

Do the data provide evidence that the infection is associated with decreased stamina? Investigate this question using
a. a *t* test.   **b.** a Mann–Whitney test.
Let H_A be directional and $\alpha = .05$.

| | | | | | | | | | | | | | | | | | | | | | | | | | | | | | | | | | | | | | | | | | | | | | | | | | | | | | | | | | | | | | | | | | | | | | |

C   H   A   P   T   E   R   8

CONTENTS

STATISTICAL PRINCIPLES OF DESIGN

SECTION 8.1

INTRODUCTION

In the previous chapters we have given our primary attention to the *analysis* of data, rather than to the *design* of the study which generated the data. Of course, the planning and design of a scientific investigation is a complex process that requires expert knowledge of the phenomenon under study. However, statistical considerations often play some role in the design process. In this chapter we will discuss some of those considerations, including

  selection of individuals for study,

  arrangement of experimental material in space and time,

  allocation of experimental units to treatment groups.

We have already discussed methods of data analysis in one-group and two-group studies in which the observed variable Y is quantitative. In the present chapter we broaden our perspective and consider studies with *any* number of groups, and with *any* kind of response variable (quantitative or categorical). In this chapter we concentrate our attention on design, not analysis. Some methods of analysis will be presented in later chapters; others are beyond the scope of this book.

ALLOCATION

In designing a biological experiment, researchers usually must face the question of **allocation**—that is, the assignment of experimental units to treatment groups. The following example is typical.

EXAMPLE 8.1

HORMONE ACTION IN MICE

A physiologist is planning to study the effect of a certain hormone on the uterus in young female mice whose ovaries have been removed. He will administer the hormone to ten mice and use another ten as controls. After 48 hours the animals will be sacrificed and their uteri weighed. In order to provide a meaningful comparison, the 20 animals will be treated exactly alike except for the administration of hormone. For instance, the hormone will be suspended in oil and administered by injection, and the controls will be injected with oil alone. In addition to such questions as the dosage and timing of the injections, the experimenter must also decide how to allocate the animals to the two treatment groups—that is, how to decide which individual mice should be in the hormone group and which in the control group.   ■

EXTRANEOUS VARIABLES

In most biological studies, the researcher must contend with **extraneous variables**—that is, variables which may affect the results but which are not themselves of direct interest. For instance, in the mouse experiment of Example 8.1 we can identify the following variables:

Response variable: Uterus weight

Manipulated variable: Hormone vs no hormone

Extraneous variables: Genetic differences among individual animals; environmental differences among individual animals such as nutrition, housing in the lab, and so on; variables affecting the measurement process such as dissection techniques, and so on.

An important consideration in the design of a biological study is how to take account of those extraneous variables which cannot simply be held constant. We will see that this can be done in a variety of ways.

OBSERVATIONAL VS EXPERIMENTAL STUDIES

A major consideration in interpreting the results of a biological study is whether the study was observational or experimental. In an **experimental study**, the researcher intervenes in or manipulates the experimental conditions. In an **observational study**, the researcher merely observes an existing situation. The distinction between observational and experimental studies was illustrated in Section 7.1; the following is another example.

EXAMPLE 8.2
CIGARETTE SMOKING

In studies of the effects of smoking cigarettes, both experimental and observational approaches have been used. Effects in animals can be studied experimentally, because animals (for instance, dogs) can be allocated to treatment groups and the groups can be given various doses of cigarette smoke.

Effects in humans are usually studied observationally. In one study, for example, pregnant women were questioned on their smoking habits, dietary habits, and so on.[1] When the babies were born, their physical and mental development was followed. One striking finding related to the babies' birthweights: the smokers tended to have smaller babies than the nonsmokers. The difference was not attributable to chance (the *P*-value was less than 10^{-5}). Nevertheless, it was far from clear that the difference was *caused* by smoking, because the women who smoked differed from the nonsmokers in many other aspects of their lifestyle besides smoking—for instance, they had very different dietary habits. (We will expand this example in Section 8.6.) ∎

As Example 8.2 illustrates, observational studies can be difficult to interpret. We will return to the subject of observational studies in Section 8.6. In Sections 8.2–8.5 we consider experimental studies, in which two or more experimental manipulations (*treatments*) are to be compared. In such studies the allocation of experimental units to treatment groups is under the control of the researcher. We will consider the two simplest schemes for allocation: the *completely randomized design* and the *randomized blocks design*.

SECTION 8.2

THE COMPLETELY RANDOMIZED DESIGN

A straightforward way to allocate experimental units to treatment groups is to use a **completely randomized design**, in which all possible allocations are equally likely.

HOW TO RANDOMIZE

A completely randomized allocation is simple to carry out. The first treatment group is filled by choosing a random sample from the available experimental units; then the second treatment group is filled by choosing a random sample from the remaining experimental units, and so on. The following example illustrates the procedure.

EXAMPLE 8.3

A DOSE–RESPONSE STUDY

An experiment to study the effects of drug A is to be conducted on dogs. Ten dogs are available, and they will be split into the following four treatment groups:

T_1:   No drug A (control)

T_2:   Low dose of drug A

T_3:   Moderate dose of drug A

T_4:   High dose of drug A

Suppose it has been decided* not to have equal-sized groups, but to put three dogs in each of T_1 and T_4, and two dogs in each of T_2 and T_3. The process of complete randomization is represented schematically in Figure 8.1.

FIGURE 8.1

Completely Randomized Allocation

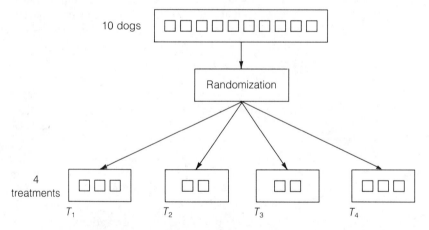

To carry out the completely randomized allocation, we can proceed as follows:

1. Choose a random sample of three dogs from the ten, and assign them to T_1.
2. Choose a random sample of two dogs from the remaining seven, and assign them to T_2.
3. Choose a random sample of two dogs from the remaining five, and assign them to T_3.
4. Assign the remaining three dogs to T_4.

The process of random sampling can be carried out as described in Section 3.2. For instance, suppose we assign the dogs identification numbers 0, 1, 2, 3, . . . ,

*In Section 13.5 we will explain why this choice of group sizes can be advantageous.

9, and we read random digits from Table 1 starting with row 06, column 16 and proceeding downward. Then the sequence of random digits would be

6, 4, 9, 3, 6, 1, 9, 5, 0, ...

and the treatment assignments would be as shown in Table 8.1. (Note that it is not necessary to read random digits to fill the last treatment group.)

TABLE 8.1
Randomized Allocation
for Example 8.3

T_1	T_2	T_3	T_4
Dog 6	Dog 3	Dog 5	Dog 2
Dog 4	Dog 1	Dog 0	Dog 7
Dog 9			Dog 8

The following two examples illustrate some of the variety of applications of the completely randomized design.

EXAMPLE 8.4
SOYBEAN GROWTH

For a study of plant growth, 60 one-week-old soybean seedlings will be set in individual pots in a greenhouse. The 60 pots will be divided into three groups of 20 each, and a different soil additive (A, B, or C) will be applied to each group. After 10 days of growth, 5 plants from each group will be harvested, dried, and weighed. Additional harvests will be made after 20, 30, and 40 days of growth. This experiment has 12 "treatments," as follows:

T_1 :   Additive A, harvest at 10 days

T_2 :   Additive B, harvest at 10 days

$\vdots$

T_{12}: Additive C, harvest at 40 days

A completely randomized allocation could be used, with 5 of the 60 pots assigned to each of the 12 treatment groups. ∎

EXAMPLE 8.5
WHITE BLOOD COUNT

In order to compare the effects of two drugs, A and B, on white blood count, 20 human volunteers will be divided into two groups, 10 to receive drug A and 10 to receive drug B. Blood specimens will be drawn at 0, 2, 4, 8, 16, and 24 hours after administration of the drug, and white blood cells will be counted in each specimen. For a completely randomized design, the 20 volunteers would be randomly allocated to the two treatment groups (A and B), 10 to each group. ∎

Notice that Example 8.4 and Example 8.5 both involve measures over time, but the structure of the two experiments is quite different. In Example 8.5, repeated measurements are made on the same individual; by contrast, in Example 8.4 each plant is measured only once. This is an important distinction, and we will see that the two structures generally require different methods of data analysis.

RANDOMIZATION AND THE RANDOM SAMPLING MODEL

Statistical methods for comparing two groups of observations (such as the t test) or several groups of observations are based on the random sampling model—the assumption that the groups to be compared can be regarded as random samples from their respective populations. The use of randomized allocation in an experiment helps to justify the application of the random sampling model in analyzing the data from the experiment. For instance, in Example 8.4 we might define conceptual populations as follows:

Population 1:   Weights of soybean plants grown with additive A for 10 days

Population 2:   Weights of soybean plants grown with additive B for 10 days

and so on.

In interpreting the experiment, it is crucial that the conceptual populations be identical in all respects except the experimental manipulations (soil additives and time of harvest). The physical act of randomization helps to justify this assumption. Thus, randomization is a kind of insurance against the possibility of hidden differences between the conceptual populations.

EXERCISES 8.1–8.4

[*Note:*   In several of these exercises you are asked to prepare a randomized allocation. For this purpose you can use either Table 1 or random digits from your calculator or a computer.]

8.1 For a medical experiment to investigate a certain new therapy, the investigators believe that the results for the new therapy will be of interest in themselves as well as in comparison to the standard therapy, and so they have decided to allocate twice as many patients to the new therapy. Suppose the experiment is to include 30 patients. Prepare a completely randomized allocation, assigning 10 patients to receive the standard therapy and 20 to receive the new therapy.

8.2 For a physiological study, eight monkeys are to be allocated to three treatment groups: groups 1 and 2 will contain three animals each, and group 3 will contain two animals. Prepare a completely randomized allocation.

8.3 For a greenhouse experiment on lettuce growth, 15 pots, each containing four plants, are to be allocated to five treatment groups, three pots to a group. (All plants in a pot receive the same treatment.) Prepare a completely randomized allocation of pots to treatments.

8.4 Canine parvovirus (CPV) is an intestinal disease that affects dogs. In a study to test a vaccine against CPV, 7-week-old beagle pups will be given either vaccine by injection, vaccine by nose drops, or placebo. The pups will be housed in pens of six pups each and watched for 5 weeks for signs of CPV. Because CPV is highly contagious, the unit of observation will be a pen rather than an individual pup; for instance, each pen will be classified as a success if all its pups remain free of CPV, and as a failure otherwise. All pups in a pen will receive the same treatment. Suppose 18 pens are available; prepare a completely randomized allocation of the 18 pens to the three treatment groups.[2]

S E C T I O N   8.3

**RESTRICTED
RANDOMIZATION:
BLOCKING AND
STRATIFICATION**

The completely randomized design makes no distinctions among the experimental units. Often an experiment can be improved by a more refined approach, one that takes advantage of known patterns of variability in the experimental units. We discuss two such improved designs.

THE RANDOMIZED BLOCKS DESIGN

In this approach, we first group the experimental units into sets, or **blocks**, of relatively similar units and then we randomly allocate treatments within each block. Here is an example.

EXAMPLE 8.6
BLOCKING BY LITTER

How does experience affect the anatomy of the brain? In a typical experiment to study this question, young rats are placed in one of three environments for 80 days:

T_1: *Standard environment.* The rat is housed with a single companion in a standard lab cage.

T_2: *Enriched environment.* The rat is housed with several companions in a large cage, furnished with various playthings.

T_3: *Impoverished environment.* The rat lives alone in a standard lab cage.

At the end of the 80-day experience, various anatomical measurements are made on the rats' brains.

Suppose a researcher plans to conduct the above experiment using 30 rats. To minimize variation in response, all 30 animals will be male, of the same age and strain. To reduce variation even further, the researcher can take advantage of the similarity of animals from the same litter. In this approach, the researcher would obtain three male rats from each of ten litters. The three littermates from each litter would be assigned at random: one to T_1, one to T_2, and one to T_3. This experimental design is indicated schematically in Figure 8.2.[3]

FIGURE 8.2
Blocking by Litter in
Rat Brain Study

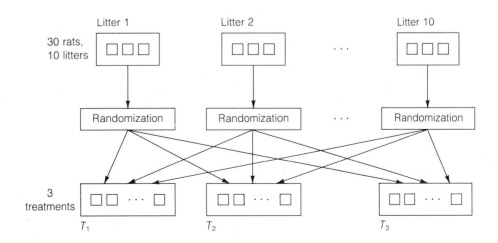

Another way to visualize the experimental design is in tabular form, as shown in Table 8.2. Each "Y" in the table represents an observation on one rat. Using the layout of Table 8.2, the experimenter can compare the responses of rats that received *different* treatments but are in the *same* litter. Such comparisons are, of course, not affected by any difference (genetic and other) that may exist between one litter and another.

TABLE 8.2
Format for Rat Brain Data

	TREATMENT		
	T_1	T_2	T_3
Litter 1	Y	Y	Y
Litter 2	Y	Y	Y
Litter 3	Y	Y	Y
.	.	.	.
.	.	.	.
Litter 10	Y	Y	Y

Example 8.6 is an illustration of a **randomized blocks design**. To carry out a randomized blocks design, the experimenter creates or identifies suitable blocks of experimental units and then randomly assigns treatments within each block in such a way that each treatment appears exactly once in each block.* In Example 8.6, the sets of three littermates serve as blocks. Here are two more examples of randomized blocks designs in biological experiments.

EXAMPLE 8.7
WITHIN-SUBJECT BLOCKING

A dermatologist is planning a study to compare two medicated lotions for their effectiveness in treating acne. Twenty patients are to participate in the study. Each patient will use lotion A on one side of his face and lotion B on the other; the dermatologist will observe the improvement on each side during a 3-month period. For each patient, the side of the face to receive lotion A is randomly selected; the other side receives lotion B. The bottles of medication have coded labels so that neither the patient nor the physician knows which bottle contains A and which contains B.[4]   ∎

EXAMPLE 8.8
BLOCKING IN AN AGRICULTURAL FIELD STUDY

When comparing several varieties of grain, an agronomist will generally plant several field plots of each variety and measure the yield of each plot. Differences in yields may reflect not only genuine differences among the varieties, but also differences among the plots in soil fertility, pH, water-holding capacity, and so on. Consequently, the spatial arrangement of the plots in the field is important. An efficient way to use the available field area is to divide the field into large regions called blocks, to subdivide each block into several plots, and then to randomly allocate the plots within each block to the various varieties. For

*Strictly speaking, the design we discuss is termed a **randomized *complete* blocks design** because every treatment appears in every block. In an incomplete blocks design, each block contains some, but not necessarily all, of the treatments.

instance, suppose we want to test four varieties of barley. Then each block would contain four plots. The resulting randomized allocation might look like Figure 8.3, which is a schematic map of the field. The "treatments" T_1, T_2, T_3, and T_4 are the four varieties of barley.

FIGURE 8.3
Layout of an Agricultural Randomized Blocks Design

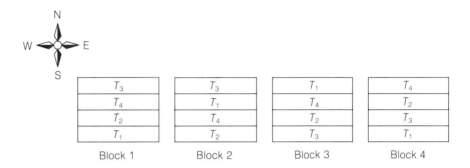

CREATING THE BLOCKS

As the preceding examples show, blocking is a way of *organizing* the inherent variation that exists among experimental units. Ideally, the blocking should be arranged so as to increase the information available from the experiment. To achieve this goal, **the experimenter should try to create blocks which are as homogeneous within themselves as possible, so that the inherent variation between experimental units becomes, as far as possible,** variation *between* **blocks rather than** *within* **blocks.** We saw one illustration of this principle in Example 8.6: in blocking by litter the experimenter is exploiting the fact that littermates are more similar to each other than to non-littermates. The following is another illustration.

EXAMPLE 8.9

AGRICULTURAL FIELD STUDY

For the barley experiment of Example 8.8, how would the agronomist determine the best arrangement or layout of blocks in a field? He would design the blocks to take advantage of any prior knowledge he may have of fertility patterns in the field. For instance, if he knows that an east–west fertility gradient exists in the field (perhaps the field slopes from east to west, with the result that the west end has a thicker layer of good soil), then he might choose blocks as in Figure 8.3; the layout maximizes soil differences between the blocks and minimizes differences between plots within each block. (But even if a field appears to be uniform, blocking is usually used in agronomic experiments, because plots closer together in the field are generally more similar than plots farther apart.)

To add solidity to this example, let us look at a set of data from a randomized blocks experiment on barley. Each entry in Table 8.3 (page 274) shows the yield (pounds of barley) of a plot 3.5 ft wide by 80 ft long.[5]

TABLE 8.3

Yield (lb) of Barley

	VARIETY			
	1	2	3	4
Block 1	28.8	31.7	20.7	26.6
Block 2	20.5	16.4	16.9	18.9
Block 3	15.6	14.6	15.4	15.6
Block 4	13.1	13.5	12.4	14.3
Mean	19.5	19.1	16.3	18.9

It appears from Table 8.3 that the yield potential of the blocks varies greatly; the data indicate a definite fertility gradient from block 1 to block 4. Because of the blocked design, comparison of the varieties is relatively unaffected by the fertility gradient. Of course, there also appears to be substantial variation within blocks. [You might find it an interesting exercise to peruse the data and ask yourself whether the observed differences between varieties are large enough to conclude that, for example, variety 1 is superior (in mean yield) to variety 3; use your intuition rather than a formal statistical analysis. The truth is revealed in Note 5.]   ∎

THE RANDOMIZATION PROCEDURE

Once the blocks have been created, the blocked allocation of experimental units is straightforward. Randomization is carried out for each block separately, as illustrated in the following example.

EXAMPLE 8.10

AGRICULTURAL FIELD
STUDY

Consider the agricultural field experiment of Example 8.8. In block 1, let us label the plots 1, 2, 3, 4, from north to south (see Figure 8.3); we will allocate one plot to each variety. The allocation proceeds as for the completely randomized design, by choosing plots at random from the four, and assigning the first plot chosen to T_1, the second to T_2, and so on. For instance, if we start reading Table 1 at row 06, column 16 and proceed downward, we get the following allocation:

Block 1:
T_1: Plot 4
T_2: Plot 3
T_3: Plot 1
T_4: Plot 2

This is in fact the assignment shown in Figure 8.3 for block 1. We can then repeat this procedure for block 2. There is no need to find a new starting place in the random digit table. If we continue reading consecutive digits from Table 1, we obtain for block 2:

Block 2:
T_1: Plot 2
T_2: Plot 4
T_3: Plot 1
T_4: Plot 3

which is the assignment shown in Figure 8.3. We can proceed in this way until all the blocks have been processed. ∎

STRATIFIED RANDOMIZATION

Stratification is a design strategy similar to blocking, in which discrete groups, or **strata**, of experimental units are identified, and random allocation to treatment groups is carried out separately in each stratum. Usually the number of strata is fairly small, so that each stratum contains a substantial number of experimental units. The following example is typical.

EXAMPLE 8.11

PROGNOSTIC STRATA IN MEDICAL EXPERIMENTS

Consider a medical experiment to compare several treatments for a certain disease. Suppose the investigators know in advance that the patient's overall outlook, or prognosis, depends on certain factors such as the patient's age, sex, and how far the disease has progressed at the time treatment is started. These prognostic factors can be used to define categories (strata) of patients. For instance, suppose we crudely divide age into two categories (young, old) and stage of disease into two categories (early, advanced). Then the joint criteria of sex, age, and stage of disease determine the following strata:

Stratum 1: Female, young, early

Stratum 2: Male, young, early

Stratum 3: Female, old, early

Stratum 4: Male, old, early

Stratum 5: Female, young, advanced

Stratum 6: Male, young, advanced

Stratum 7: Female, old, advanced

Stratum 8: Male, old, advanced

A **stratified randomization** would consist of randomly allocating patients to treatment groups, separately for each stratum. Note that, because of the nature of the strata, one would expect the different strata to contain different numbers of patients. ∎

SECTION 8.4

MORE ON RANDOMIZATION AND CONTROL

In this section we expand and relate the twin themes of randomization and control in biological experiments.

COMPLEMENTARITY OF RANDOMIZATION AND BLOCKING

Randomization and blocking (or stratification) are two complementary ways of dealing with the extraneous variables that may threaten an experiment. In biological experimentation, the experimental units may be similar but they are usually not identical; they differ in many respects, both visible and invisible. The advantage of randomization is that it tends to maintain balance among the treatment groups with respect to all extraneous variables—including those the experimenter has not even thought of. Of course, blocking gives a more definite assurance of balance, but blocking can control only a limited number of variables. No matter how carefully the blocks are constructed, the units within the blocks are still different. This is why randomization within blocks should always be used as a complement to blocking.

In many experiments, blocking is not practical at all. For one thing, blocking can be expensive. Another difficulty is that the experimenter may not have a good basis for forming blocks. Blocking on a variable that is not related to the response variable of interest is a waste of time and (as we will see) is inefficient.

Blocking actually has two related purposes. First, it improves the comparability of the various treatment groups by forcing them to be balanced with respect to the variables used in the blocking. Second, and often more important, is the effect of blocking on the *precision* of treatment comparisons in an experiment. By minimizing the variability among experimental units within a block, the experimenter enables treatment differences to stand out more clearly. Consequently, an experiment with well-constructed blocks can yield much more information than an unblocked experiment on similar experimental material; the blocked experiment is said to be more **efficient**. In Chapter 9 we will show in detail, for the specific case of comparing two treatments, how blocking can increase information, and how the added information can be extracted from the data by suitable analysis.

In many experiments randomization is used at several stages. For instance, in a greenhouse experiment a botanist may randomly allocate seedlings to pots, treatments to pots, and pots to positions on the greenhouse bench. Often, time as well as space must be considered in the design. For instance, suppose 20 mice are to be sacrificed and their livers assayed for a certain biochemical. If the same researcher is going to do all 20 assays, he may not be able to do them all simultaneously. The worst possible plan (although the simplest) would be to assay the animals systematically: all T_1 animals first, then all T_2 animals, and so on. A better plan would be to randomly allocate the 20 animals to the 20 time slots for assay.

STATISTICAL ADJUSTMENT FOR EXTRANEOUS VARIABLES

We have noted that it is desirable to minimize or control the influence of extraneous variables in an experiment. Randomization, blocking, and stratification are techniques for exerting this control through the design of the experiment. Control can also be imposed at the analysis stage. A battery of statistical techniques is available for taking account of extraneous variables in the analysis of data. These analytical techniques include post-stratification, regression, analysis of variance, and analysis of covariance. We will discuss some of these in subsequent chapters. Many of these analytical methods can be used in conjunction with the design techniques of randomization, blocking, and stratification.

WHY RANDOMIZE?

Although the process of randomized allocation is straightforward, it can be tedious and researchers may be tempted to skip it. A haphazard rather than random procedure might appear to be equally good. For instance, suppose we need to allocate rats to four treatment groups: T_1, T_2, T_3, and T_4. Rather than use random digits to allocate the rats, why not just reach in the cage and grab some rats for T_1, then some rats for T_2, and so on? This procedure would have a potentially serious bias: it would tend to assign the more sluggish (easier to catch) rats to T_1, and the most lively rats to T_4. The comparison of the treatments might then be distorted by the systematic differences in the rats. Of course, this particular source of bias is rather obvious. The advantage of randomization is that it guards against *all* biases, even those which are not at all obvious.* Note, however, that randomization is not magic; it cannot *guarantee* that the treatment groups are exactly comparable, but it provides a form of insurance that is far better than haphazard allocation. (In some situations statistical adjustment, as mentioned in the previous section, can be used to correct for imbalances remaining after randomization.)

HISTORICAL CONTROLS

Researchers may be particularly reluctant to use randomized allocation in medical experiments on human beings. Suppose, for instance, that researchers want to evaluate a promising new treatment for a certain illness. It can be argued that it would be unethical to withhold the treatment from any patients, and that therefore all current patients should receive the new treatment. But then who would serve as a control group? One possibility is to use **historical controls**—that is, previous patients with the same illness who were treated with another therapy. One difficulty with historical controls is that there is often a tendency for later patients to show a better response—even to the same therapy—than earlier patients with the same diagnosis. This tendency has been confirmed, for instance, by comparing experiments conducted at the same medical centers in different years.[6] One major reason for the tendency is that the overall characteristics of the patient population may change with time. For instance, because diagnostic techniques tend to improve, patients with a given diagnosis (say, breast cancer) in 1990 may have a better chance of recovery (even with the same treatment) than those with the same diagnosis in 1980, because they were diagnosed earlier in the course of the disease.

Medical researchers do not agree on the validity and value of historical controls. The following example illustrates the importance of this controversial issue.

*An additional argument in favor of randomization is as a direct justification for certain test procedures. For instance, the null distribution of the Mann–Whitney test can be derived by using (rather than the random sampling model) a mathematical model that explicitly incorporates the physical act of randomization. A similar development can be used to justify the t test as approximately correct.

EXAMPLE 8.12
CORONARY ARTERY
DISEASE

Disease of the coronary arteries is often treated by surgery (such as bypass surgery), but it can also be treated with drugs only. Many studies have attempted to evaluate the effectiveness of surgical treatment for this common disease. In a review of 29 of these studies, each study was classified as to whether it used randomized controls or historical controls; the conclusions of the 29 studies are summarized in Table 8.4.[7]

TABLE 8.4
Coronary Artery Disease
Studies

TYPE OF CONTROLS	CONCLUSION ABOUT EFFECTIVENESS OF SURGERY		Total Number of Studies
	Effective	Not Effective	
Randomized	1	7	8
Historical	16	5	21

It would appear from Table 8.4 that enthusiasm for surgery is much more common among researchers who use historical controls than among those who use randomized controls. ∎

Proponents of the use of historical controls argue that statistical adjustment can provide meaningful comparison between a current group of patients and a group of historical controls; for instance, if the current patients are younger than the historical controls, then the data can be analyzed in a way that adjusts, or corrects, for the effect of age. Critics reply that such adjustment may be grossly inadequate.

The concept of historical controls is not limited to medical studies. The issue arises whenever a researcher compares current data with past data. Whether the data are from the lab, the field, or the clinic, the researcher must confront the question: Can the past and current results be meaningfully compared? One should always at least ask whether the experimental material, and/or the environmental conditions, may have changed enough over time to distort the comparison.

BLINDING

In many experiments, some or all of the researchers are kept **blind**; that is, during the experiment they are kept ignorant of the treatment assignment. Consider, for instance, the following:

In a study to compare two treatments for lung cancer, a radiologist reads X-rays to evaluate each patient's progress. The X-ray films are coded so that the radiologist cannot tell which treatment each patient received.

Mice are fed one of three diets, and the effects on their liver are assayed by a researcher who does not know which diet each mouse received.

The most obvious reason for blinding is to reduce the possibility of subjective bias influencing the observation process itself: a researcher who *expects* or *wants* certain results may unconsciously influence those results. Such bias can enter even apparently "objective" measurements, through subtle variation in dissection techniques, titration procedures, and so on.

In medical studies of human beings, blinding often serves additional purposes. For one thing, a patient must be asked whether he consents to participate in a medical study. If the physician who asks the question already knows which treatment the patient would receive, then by discouraging certain patients and encouraging others, he can (consciously or unconsciously) create noncomparable treatment groups. The effect of such biased assignment can be surprisingly large, and it has been noted that it generally favors the "new" or "experimental" treatment.[8] Another reason for blinding in medical studies is that a physician may (consciously or unconsciously) provide more psychological encouragment, or even better care, to the patients who are receiving the treatment which the physician regards as superior.

Note that blinding may be difficult or impossible if nonrandomized allocation is used. This is another argument against the use of historical controls.

In experiments on humans, **double-blinding** is often used. This means that the treatment assignment is kept secret from the experimental subject as well as from the experimenter. The purpose of blinding the subject is to minimize the extent to which his expectations influence the results of the experiment. It is well known that people often exhibit a *placebo response*; that is, they tend to respond favorably to *any* treatment, even if it is only a *placebo* such as an unmedicated lotion or a sugar pill. If the patient is unaware of which treatment he is receiving, then the placebo effect may not be eliminated, but at least it will not favor one treatment more than another.

Similar to placebo treatment is *sham* treatment, which is used on animals as well as humans. An example of sham treatment is injecting control animals with an inert substance such as saline. In some studies of surgical treatments, control animals (even, occasionally, humans) are given a "mock" surgery.

MISSING DATA

Sometimes the best-laid plans go awry, and an experimenter is faced with the vexing problem of **missing data**—that is, observations which were planned but cannot be made. This can happen, for instance, because experimental animals or plants die, because equipment malfunctions, or because human subjects fail to return for a follow-up observation.

A common approach to the problem of missing data is to simply use the remaining data and ignore the fact that some observations are missing. This approach is temptingly simple, but must be used with extreme caution, because comparisons based on the remaining data may be seriously biased. For instance, if observations on some experimental mice are missing because the mice died of causes related to the treatment they received, it is obviously not valid to simply compare the mice who survived. As another example, if patients drop out of a medical study because they think their treatment is not working, then analysis of the remaining patients could produce a greatly distorted picture.

Naturally, it is best to make every effort to avoid missing data. But if data are missing, it is crucial that the possible reasons for the omissions be considered in interpreting and reporting the results.

[*Note:* In several of these exercises you are asked to prepare a randomized allocation. For this purpose you can use either Table 1 or random digits from your calculator or a computer.]

8.5 In an experiment to compare six different fertilizers for tomatoes, 36 individually potted seedlings are to be used, six to receive each fertilizer. The tomato plants will be grown in a greenhouse, and the total yield of tomatoes will be observed for each plant. The experimenter has decided to use a randomized blocks design: the pots are to be arranged in six blocks of six plants each on the greenhouse bench. Two possible arrangements of the blocks are shown in the accompanying figure.

Arrangement I:

Steam pipe

Arrangement II:

Steam pipe

One factor that affects tomato yield is temperature, which cannot be held exactly constant thoughout the greenhouse. In fact, a temperature gradient across the bench is likely. Heat for the greenhouse is provided by a steam pipe which runs lengthwise under one edge of the bench, and so the side of the bench near the steam pipe is likely to be warmer.

a. Which arrangement of blocks (I or II) is better? Why?

b. Prepare a randomized allocation of treatments to the pots within each block. (Assume that the assignments of seedlings to pots and of pots to positions within the block have already been made.)

8.6 An experiment on vitamin supplements is to be conducted on young piglets, using litters as blocks in a randomized blocks design. There will be five treatments: four types of supplement and a control. Thus, five piglets from each litter will be used. The experiment will include five litters. Prepare a randomized blocks allocation of piglets to treatments.

8.7 Refer to the vitamin experiment of Exercise 8.6. Suppose a colleague of the experimenter proposes an alternative design—that all pigs in a given litter receive the same treatment, with the five litters being randomly allocated to the five treatments. He points out that his proposal would save labor and greatly simplify the record-keeping. If you were the experimenter, how would you reply to this proposal?

8.8 In a pharmacological experiment on eating behavior in rats, 18 rats are to be randomly allocated to three treatment groups: T_1, T_2, and T_3. While under observation, the animals will be kept in individual cages in a rack. The rack has three tiers with six cages per tier. In spite of efforts to keep the lighting uniform, the lighting conditions vary somewhat from one tier to another (the bottom tier is darkest), and the experimenter is concerned about this because lighting is thought to influence eating behavior in rats. The following three plans are proposed for allocating the rats to positions in the rack (to be done after the allocation of rats to treatment groups):

Plan I.   Randomly allocate the 18 rats to the 18 positions in the rack.

Plan II.   Put all T_1 rats on the first tier, all T_2 rats on the second, and all T_3 rats on the third tier.

Plan III.   On each tier, put two T_1 rats, two T_2 rats, and two T_3 rats.

Of these three plans, which is the *worst*? Why?

8.9 An experimenter is planning an agricultural field experiment to compare the yields of 25 varieties of corn. He will use a randomized blocks design with six blocks; thus, there will be 150 plots, and the yield of each plot must be measured. The experimenter realizes that the time required to harvest and weigh all the plots is so long that rain might interrupt the operation. If rain should intervene, there could be a yield difference between the harvests before and after the rain. The experimenter is considering the following plans.

Plan I.   Harvest all plots of variety 1 first, all of variety 2 next, and so on.

Plan II.   Harvest all plots of block 1 first, all of block 2 next, and so on.

Which plan is better? Why?

8.10 For an experiment to compare two methods of artificial insemination in cattle, the following cows are available:

Heifers (14–15 months old): 8 animals

Young cows (2–3 years old): 8 animals

Mature cows (4–8 years old): 10 animals

The animals are to be randomly allocated to the two treatment groups, using the three age groups as strata. Prepare a suitable allocation, randomly dividing each stratum into two equal groups.

SECTION 8.5

LEVELS OF REPLICATION

Replication is central to the statistical analysis of experimental data. By considering the variation among experimental units treated alike, the researcher obtains a benchmark for assessing differences between treatments. Increased replication tends to provide increased information (all other things being equal). For instance, we saw in Chapter 6 how the information in an experiment, as measured by the standard error of the mean, depends on the number of replicate observations, through the formula

$$SE_{\bar{y}} = \frac{s}{\sqrt{n}}$$

In Section 6.4 we discussed the important topic of planning an experiment to include sufficient replication (that is, large enough n) to achieve a desired SE.

In this section we consider an issue that can cause confusion in both the planning and the analysis of biological experiments. Sometimes variation arises at several different hierarchical levels in an experiment, and it can be a challenge to sort these out, and particularly, to correctly identify the quantity n. Example 8.13 illustrates this issue.

EXAMPLE 8.13

GERMINATION OF SPORES

In a study of the fungus that causes the anthracnose disease of corn, interest focused on the survival of the fungal spores.[9] Batches of spores, all prepared from a single culture of the fungus, were stored in chambers under various environmental conditions, and then assayed for their ability to germinate, as follows. Each batch of spores was suspended in water, and then plated on agar in a petri dish. Ten "plugs" of 3-mm diameter were cut from each petri dish, and were incubated at 25°C for 12 hours. Each plug was then examined with a microscope for germinated and ungerminated spores. The environmental conditions of storage (the "treatments") included the following:

T_1:   Storage at 70% relative humidity for one week

T_2:   Storage at 60% relative humidity for one week

T_3:   Storage at 60% relative humidity for two weeks

and so on . . .

All together there were 43 treatments.

The design of the experiment is indicated schematically in Figure 8.4. There were 129 batches of spores, which were randomly allocated to the 43 treatments, three batches to each treatment. Each batch of spores resulted in one petri dish, and each petri dish resulted in ten plugs.

FIGURE 8.4

Design of Spore
Germination Experiment

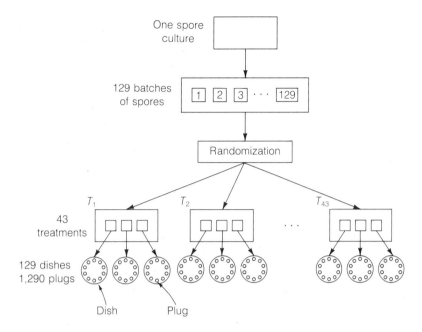

To get a feeling for the issues raised by this design, let us look at some of the raw data. Table 8.5 shows the percentage of the spores that had germinated for each plug asssayed for treatment 1.

TABLE 8.5

Percentage Germination
Under Treatment 1

	DISH I	DISH II	DISH III
	49	66	49
	58	84	60
	48	83	54
	69	69	72
	45	72	57
	43	85	70
	60	59	65
	44	60	68
	44	75	66
	68	68	60
Mean	52.8	72.1	62.1
SD	10.1	9.5	7.4

Table 8.5 shows that there is considerable variability both *within* each petri dish and *between* the dishes. The variability within the dishes reflects local variation in the percent germination, perhaps due largely to differences among the spores themselves (some of the spores were more mature than others). The variability between dishes is even larger, because it includes not only local vari-

ation but also larger-scale variation such as the variability among the original batches of spores, and temperature and relative humidity variations within the storage chambers.

Now consider the problem of comparing treatment 1 against the other treatments. Would it be legitimate to take the point of view that we have 30 observations for each treatment? To focus this question, let us consider the matter of calculating the standard error for the mean of treatment 1. The mean and SD of all 30 observations are

Mean $= 62.33$

SD $= 11.88$

Is it legitimate to calculate the SE of the mean as

$$SE_{\bar{y}} = \frac{s}{\sqrt{n}} = \frac{11.88}{\sqrt{30}} = 2.2 \quad ?$$

As you may suspect, **this is not legitimate**. There is a hierarchical structure in the data (see Section 6.5), and so we cannot apply the SE formula so naively. An acceptable way to calculate the SE is to consider the mean for each dish as an observation; thus, we obtain the following:*

Observations: 52.8,   72.1,   62.1

$n = 3$

Mean $= 62.33$

SD $= 9.65$

$$SE_{\bar{y}} = \frac{s}{\sqrt{n}} = \frac{9.65}{\sqrt{3}} = 5.6$$

Notice that the incorrect analysis gave the same mean (62.33) as this analysis, but an inappropriately small SE (2.2 rather than 5.6). If we were comparing several treatments, the same pattern would tend to hold; the incorrect analysis would tend to produce SEs which were (individually and pooled) too small, which might cause us to "overinterpret" the data, in the sense of thinking we see treatment differences where none exists.

We should emphasize that, even though the correct analysis requires combining the measurements on the ten plugs in a dish into a single observation for that dish, nevertheless the experimenter was not wasting his effort by measuring ten plugs per dish instead of, say, only one plug per dish. The mean of ten plugs is a much better estimate of the average for the entire dish than is a measurement on one plug; the improved precision for measuring ten plugs is reflected in a smaller between-dish SD. For instance, for treatment 1 the SD was 9.65; if fewer plugs per dish had been measured, this SD would probably have been larger. ■

---

*An alternative way to aggregate the data from the ten plugs in a dish would be to combine the raw counts of germinated and ungerminated spores for the whole dish, and express these as an overall percent germination.

The pitfall illustrated by Example 8.13 has trapped many an unwary researcher. When hierarchical structures result from repeated measurements on the same individual organism (as illustrated in Section 6.5), they are relatively easy to recognize. But the hierarchical structure in Example 8.13 has a different origin; it is due to the fact that the unit of observation is an individual plug, but individual plugs are not randomly allocated to the treatment groups. Rather, the unit that is randomly allocated to treatment is a batch of spores, which later is plated in a petri dish, which then gives rise to ten plugs. In the language of experimental design, plugs are *nested* within petri dishes. *Whenever observational units are nested within the units that are randomly allocated to treatments, a hierarchical structure may potentially exist in the data.* Note that the difficulty is only "potential"; in some cases a nonhierarchical analysis may be acceptable. For instance, if experience had shown that the differences between petri dishes were negligible, then we might ignore the hierarchical structure in analyzing the data. The decision can be a difficult one, and may require expert statistical advice.

The issue of hierarchical data structures has important implications for the design of an experiment as well as its analysis. The sample size (n) must be appropriately identified in order to determine whether the experiment includes enough replication. As a simple example, suppose it is proposed to do a spore germination experiment such as that of Example 8.13, but with only *one* dish per treatment, rather than three. To see the flaw in this proposal, suppose for simplicity that the proposed experiment is to include only three treatments, with one dish per treatment. How, then, would we distinguish treatment differences from inherent differences between the dishes? The answer is that we could not. The intertreatment differences and the interdish differences would be mutually entangled, or **confounded**. You can easily visualize this situation if you look at the data in Table 8.5 and pretend that those data came from the proposed experiment; that is, pretend that dishes I, II, and III had received different treatments, and that we had no other data. It would be difficult to extract meaningful information about intertreatment differences unless we knew *for certain* that interdish variation was negligible. (This is not to say that a proposal to use one dish per treatment is irredeemably flawed. Such a design can succeed if the treatment effects are sufficiently regular; we will see an application of this idea in Chapter 13.)

DETERMINATION OF SAMPLE SIZE We saw in Section 6.4 how to use a preliminary estimate of the SD to determine the sample size (n) required to attain a desired degree of precision, as expressed by the SE. (A similar development in optional Section 7.8 expressed desired precision in terms of statistical power.) These ideas carry over to experiments involving hierarchical data structures. For example, suppose a botanist is planning a spore germination experiment such as that of Example 8.13. If he has already decided to use ten plugs per dish, the remaining problem would be to decide on the number of dishes per treatment. This question could be approached as in Section 6.4, considering the dish as the experimental unit, and using a preliminary estimate of the SD between dishes (which was 9.65

in Example 8.13). If, however, he wants to choose optimal values for *both* the number of plugs per dish *and* the number of dishes per treatment, he may wish to consult a statistician.

8.11 Four treatments were compared for their effect on the growth of spinach cells in cell culture flasks. Using a completely randomized design, the experimenter allocated two flasks to each treatment. After a certain time on treatment, he randomly drew three aliquots (1 cc each) from each flask and measured the cell density in each aliquot; thus, he had six cell density measurements for each treatment. In calculating the standard error of a treatment mean, the experimenter calculated the standard deviation of the six measurements and divided by $\sqrt{6}$. On what grounds might an objection be raised to this method of calculating the SE?

8.12 In an experiment on soybean varieties, individually potted soybean plants were grown in a greenhouse, using a completely randomized design with 10 plants of each variety. From the harvest of each plant, five seeds were chosen at random and individually analyzed for their percentage of oil. This gave a total of 50 measurements for each variety. To calculate the standard error of the mean for a variety, the experimenter calculated the standard deviation of the 50 observations and divided by $\sqrt{50}$. Why would this calculation be of doubtful validity?

SECTION 8.6

OBSERVATIONAL STUDIES

In this section we discuss some issues that pertain especially to observational, rather than experimental, studies. The difficulties in interpreting observational studies arise from two primary sources:

Nonrandom selection from populations

Uncontrolled extraneous variables

The following example illustrates both of these.

EXAMPLE 8.14
RACE AND BRAIN SIZE

In the 19th century, much effort was expended in the attempt to show "scientifically" that certain human races were inferior to others. A leading researcher on this subject was the American physician S. G. Morton, who won widespread admiration for his studies of human brain size. Throughout his life, Morton collected human skulls from various sources, and he carefully measured the cranial capacities of hundreds of these skulls. His data appeared to suggest that (as he suspected) the "inferior" races had smaller cranial capacities. Table 8.6 gives a summary of Morton's data comparing Caucasian skulls to those of Indians (Native Americans).[10] According to a *t* test, the difference between these two samples is "statistically significant" ($P < .001$). But is it *meaningful*?

TABLE 8.6
Cranial Capacity (in³)

	CAUCASIAN	INDIAN
Mean	87	82
SD	8	10
n	52	144

In the first place, the notion that cranial capacity is a measure of intelligence is no longer taken seriously. Leaving that question aside, one can still ask whether it is true that the mean cranial capacity of Native American Indians is less than that of Caucasians. Such an inference beyond the actual data requires that the data be viewed as random samples from their respective populations. Of course, in actuality Morton's data are not random samples but "samples of convenience," because Morton measured those skulls which he happened to obtain. But might the data be viewed "as if" they were generated by random sampling? One way to approach this question is to look for sources of bias. In 1977 the noted biologist Stephen Jay Gould reexamined Morton's data with this goal in mind, and indeed Gould found several sources of bias. For instance, the 144 Indian skulls represent many different groups of Indians; as it happens, 25% of the skulls (that is, 36 of them) were from Inca Peruvians, who were a small-boned people with small skulls, while relatively few were from large-skulled tribes such as the Iroquois. Clearly a comparison between Indians and Caucasians is meaningless unless somehow adjusted for such imbalances. When Gould made such an adjustment, he found that the difference between Indians and Caucasians vanished. ∎

Even though the story of Morton's skulls is more than 100 years old, it can still serve to alert us to the pitfalls of inference. Morton was a conscientious researcher and took great care to make accurate measurements; Gould's re-examination did not reveal any suggestion of conscious fraud on Morton's part. Morton may have overlooked the biases in his data because they were *invisible* biases; that is, they related to aspects of the selection process rather than aspects of the measurements themselves.

When we look at a set of observational data, we can sometimes become so hypnotized by its apparent *solidity* and *objectivity* that we forget to ask how the observational units were selected. The question should always be asked. If the selection was haphazard rather than truly random, the results can be severely distorted.

CAUSATION

Many observational studies are aimed at discovering some kind of causal relationship. Such discovery can be very difficult because of extraneous variables that enter in an uncontrolled (and perhaps unknown) way. The investigator must be guided by the maxim:

Association is not causation.

For instance, it is known that some populations whose diets are high in fiber enjoy a reduced incidence of colon cancer. But this observation does not in itself show that it is the high-fiber diet, rather than some other factor, which provides the protection against colon cancer.

We do not mean to say that an observed association *cannot* be causally interpreted, but only that such interpretation requires particular caution.

The following example shows how uncontrolled extraneous variables can cloud an observational study, and what kinds of steps can be taken to clarify the picture.

EXAMPLE 8.15

SMOKING AND
BIRTHWEIGHT

In a large observational study of pregnant women, it was found that the women who smoked cigarettes tended to have smaller babies than the nonsmokers.[1] (This study was mentioned in Example 8.2.) It is plausible that smoking could cause a reduction in birthweight, for instance by interfering with the flow of oxygen and nutrients across the placenta. But of course plausibility is not proof. In fact, the investigators found that the smokers differed from the nonsmokers with respect to many extraneous variables. For instance, the smokers drank more coffee than the nonsmokers; they also drank more whiskey. Either of these might plausibly be linked to a deficit in growth. In addition—and this is especially puzzling—it was found that the smokers began to menstruate at younger ages than the nonsmokers. This phenomenon (early onset of menstruation) could not possibly have been *caused* by smoking, because it occurred (in almost all instances) *before* the woman began to smoke. One interpretation that has been proposed is that the two populations—women who choose to smoke and those who do not—are different in some biological way; thus, it has been suggested that the reduced birthweight is due "to the *smoker*, not the *smoking*."[11]

A number of more recent studies have attempted to shed some light on the relationship between maternal smoking and infant development. Researchers in one study[12] observed, in addition to smoking habits, about 50 extraneous variables, including the mother's age, weight, height, blood type, upper arm circumference, religion, education, income, and so on. After applying complex statistical methods of adjustment, they concluded that birthweight varies with smoking even when all the extraneous factors are held constant. In another study[13] of pregnant women, researchers measured various quantities related to the functioning of the placenta. They found that, compared to nonsmokers, women who smoked had more abnormalities of the placenta, and that their infants had very much higher blood levels of cotinine, a substance derived from nicotine. They also found evidence that, in the women who smoked, the circulation of blood in the placenta was notably improved by abstaining from smoking for 3 hours.

A third study[14] used a matched design to try to isolate the effect of smoking behavior. The investigators identified 159 women who had smoked during one pregnancy, but quit smoking before the next pregnancy. These women were individually matched with 159 women who smoked during two consecutive pregnancies; pairs were matched with respect to the birthweight of the first child, amount of smoking during the first pregnancy, and several other factors. Thus, the members of a pair were believed to have identical "reproductive potential." The researchers then considered the birthweight of the second child; they found that the women who had quit smoking gave birth to infants who weighed more than the infants of their matched controls who continued to smoke. Of course, we cannot rule out the possibility that the women who quit smoking also quit other harmful habits, such as drinking too much alcohol, and that the increased birthweight was not really caused by giving up smoking. ∎

Example 8.15 shows that observational studies *can* provide information about causality, but must be interpreted cautiously. Researchers generally agree that a causal interpretation of an observed association requires extra support—for instance, that the association be observed consistently in observational studies conducted under various conditions and taking various extraneous factors into account, and also, ideally, that the causal link be supported by experimental evidence.

Against this background, you can appreciate the enormous scientific value of randomized allocation. For instance, if we could randomly allocate women to smoking and nonsmoking groups, the effect of smoking on birthweight could be explored much more directly. Of course, such an experiment on humans is not ethically or socially defensible.

CASE–CONTROL STUDIES

Observational investigation of the possible causes of a disease often takes the form of a **case–control study**, in which cases of the disease are compared against controls who do not have the disease. Here is an example.

EXAMPLE 8.16
DIET AND STOMACH CANCER

It is suspected that certain foods might increase the risk of stomach cancer. In order to investigate this, suppose that we plan to interview some stomach cancer patients to learn about their dietary habits, and that we will also interview some controls who do not have stomach cancer. Then we might compare the diets of the cases and the controls to see if any differences emerge. But of course many factors, especially social and economic ones, influence dietary habits, and some of these factors may themselves be linked with cancer. How can controls be chosen who are comparable to the cases with respect to these extraneous factors?

Here is how one group of investigators approached the problem.[15] The cases were 220 Japanese stomach cancer patients from several cooperating hospitals in Hawaii. For each case, two controls were chosen according to an unambiguous rule: the controls were the next older and the next younger Japanese of the same sex in the same section of the hospital at the time of the interview, subject to the condition that the control did not have stomach cancer, other diseases of the stomach, or other cancers of the digestive system. Because of this selection rule, the controls were matched to the cases on age, sex, and—most important—on many unmeasurable extraneous variables (such as socioeconomic status) that were indirectly expressed by the location of the hospital. Note that all the cases and controls were Japanese; this factor was not matched, but rather was constant.

∎

In a case–control study, the cases are usually a sample of convenience—say, all the available patients with the given disease. A primary design consideration is how to choose the controls. Often (as in Example 8.16) the cases are individually matched to the controls with respect to several important extraneous variables. Unfortunately, reports of case–control studies sometimes do not specify how the controls were chosen; this is a serious omission that can undermine the entire report.

8.13 For an early study of the relationship between diet and heart disease, the investigator obtained data on heart disease mortality in various countries and on national average dietary compositions in the same countries. The accompanying graph shows, for six countries, the 1948–1949 death rate from degenerative heart disease (among men aged 55–59 years) plotted against the amount of fat in the diet.[16] The countries are:

1. United States   4. England and Wales
2. Canada   5. Italy
3. Australia   6. Japan

In what ways might this graph be misleading? Which extraneous variables might be relevant here? Discuss.

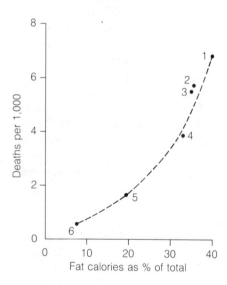

In this chapter we have introduced some of the statistical principles that can guide the scientist in designing an investigation. We have considered several types of observational study and two basic designs using randomized allocation.

DESIGN OF EXPERIMENTS

There is much more to the subject of randomized allocation than we have space to present. For instance, in many situations it is possible to retrieve much information from an experiment in which replication is absent or incomplete, if the experiment has been designed correctly. The statistical discipline called **design**

of experiments is devoted to the investigation of schemes for designing experiments which are informative and make efficient use of the available resources. These include methods of allocation of experimental units to treatments, how and where to incorporate replication, and how to choose which treatments to include (for instance, which doses of a drug, or which combinations of light and temperature).

ONLY STATISTICAL?

The term "statistical" is sometimes used—or, rather, misused—as an epithet. For instance, some people say that the evidence linking dietary cholesterol and heart disease is "only statistical." What they really mean is "only observational." Statistical evidence can be very strong indeed, if it flows from a randomized study rather than an observational study. As we saw in Section 8.6, statistical evidence from an observational study must be interpreted with great care, because of potential distortions caused by extraneous variables. For this reason, serious observational research today frequently involves collaboration with a statistician.

ANALYSIS AND DESIGN

Techniques for data analysis must be appropriately tailored to the design of the study that generated the data. In Chapter 7 we described techniques for comparing data from two *independent* samples. In observational studies, the requirement of independence means that the observational units were selected independently from the two populations, with no matching or stratification in the selection process. In randomized studies the requirement of independence means that the design was completely randomized, with no blocking or stratification in the randomization process.

In Chapter 9 we will describe techniques for analyzing a randomized blocks experiment with two treatments. The same techniques can be used to compare two matched observational samples.

In Chapters 11 and 12 we will expand the comparison of independent samples to include analysis of categorical data and the comparison of three or more treatments. Analysis of randomized blocks experiments with three or more treatments is beyond the scope of this book.

SUPPLEMENTARY EXERCISES 8.14–8.19

[*Note:* In several of these exercises you are asked to prepare a randomized allocation. For this purpose you can use either Table 1 or random digits from your calculator or a computer.]

8.14 It is known that alcohol consumption during pregancy can harm the fetus. To study this phenomenon, 20 pregnant mice are to be allocated to three treatment groups. Group 1 (ten mice) will receive no alcohol, group 2 (five mice) will receive a low dose of alcohol, and group 3 (five mice) will receive a high dose. Prepare a completely randomized allocation.

8.15 Refer to the alcohol study of Exercise 8.14. When each mouse gives birth, the birthweight of each pup will be measured. Suppose the mice in group 1 give birth to a total of 85 pups, so the experimenter has 85 observations of Y = birthweight. To calculate the standard error of the mean of these 85 observations, the experimenter could calculate the standard deviation of the 85 observations and divide by $\sqrt{85}$. On what grounds might an objection be raised to this method of calculating the SE?

8.16 The following 24 subjects are available for a medical study:

SUBJECT	SEX	AGE	SUBJECT	SEX	AGE
C.M.	F	19	K.N.	M	43
A.F.	F	23	C.O.	M	45
E.G.	F	32	T.K.	M	48
M.A.	F	39	T.W.	M	50
D.N.	F	46	A.J.	M	52
B.B.	F	51	D.S.	M	56
A.S.	M	22	G.D.	M	56
M.L.	M	27	W.H.	M	59
P.C.	M	34	M.S.	M	61
L.W.	M	38	J.H.	M	63
J.P.	M	40	C.F.	M	66
C.R.	M	40	R.V.	M	72

Suppose the subjects are to be allocated to two treatment groups of 12 subjects each, using a completely randomized design. Prepare a suitable allocation.

8.17 Refer to Exercise 8.16. Suppose a randomized blocks design is to be used to compare the two treatments. Use the given information to form 12 pairs of subjects, matched by sex and (as closely as possible) age. Randomly allocate the members of each pair to the two treatment groups.

8.18 Refer to Exercise 8.16. Suppose a stratified design is to be used to compare the two treatments. Randomly allocate the 24 subjects to two groups of 12 each, but stratify by sex—that is, randomly divide the females into two groups of three each and the males into two groups of nine each.

8.19 Refer to Exercise 8.16. Suppose these 24 subjects are to participate in a study to compare *three* treatments, using a randomized blocks design. Use the given information to form eight triplets of subjects, matched by sex and (as closely as possible) age. Randomly allocate the members of each triplet to the three treatment groups.

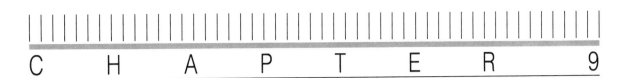

C H A P T E R 9

CONTENTS

COMPARISON OF TWO PAIRED SAMPLES

INTRODUCTION

Chapter 7 dealt with the comparison of two independent samples with respect to a quantitative variable Y. In the present chapter we consider the comparison of two samples which are not independent but are paired. In **a paired design**, the observations (Y_1, Y_2) occur in pairs; the observational units in a pair are linked in some way, so that they have more in common with each other than with members of another pair.

The following is an example of a paired design.

EXAMPLE 9.1

NERVE CELL
REGENERATION

Certain types of nerve cells have the ability to regenerate a part of the cell that has been amputated. In an early study of this process, measurements were made on the nerves in the spinal cord in rhesus monkeys. Nerves emanating from the left side of the cord were cut, while nerves from the right side were kept intact. During the regeneration process, the content of creatine phosphate (CP) was measured in the left and the right portion of the spinal cord. Table 9.1 shows the data for the right (control) side (Y_1), and for the left (regenerating) side (Y_2). The units of measurement are mg CP per 100 gm tissue.[1]

TABLE 9.1

CP in Monkey Spinal
Cord (mg/100 gm)

ANIMAL	RIGHT SIDE (CONTROL)	LEFT SIDE (REGENERATING)
1	16.3	11.5
2	4.8	3.6
3	10.9	12.5
4	14.2	6.3
5	16.3	15.2
6	9.9	8.1
7	29.2	16.6
8	22.4	13.1
Mean	15.50	10.86
SD	7.61	4.49

In Example 9.1, the observational units (sides of the spinal cord) occur in pairs, and the members of a pair are linked by virtue of being in the same animal. A suitable analysis of the data should take advantage of this pairing.

In Section 9.2 we show how to analyze paired data using methods based on Student's t distribution. In Section 9.4 we describe a nonparametric test for paired data. Sections 9.3, 9.5, and 9.6 contain more examples and discussion of the paired design.

**THE PAIRED-SAMPLE
t TEST AND
CONFIDENCE
INTERVAL**

In this section we discuss the use of Student's t distribution to obtain tests and confidence intervals for paired data.

STANDARD ERROR OF $(\bar{y}_1 - \bar{y}_2)$

To compare the means of two paired samples, we need to calculate a standard error for the difference between the means, that is,

$$SE_{(\bar{y}_1 - \bar{y}_2)}$$

To calculate this SE, we make a simple shift of viewpoint: instead of considering Y_1 and Y_2 separately, we consider the difference d, defined as

$$d = Y_1 - Y_2$$

Let us denote the mean of the d's as $\bar{d}$. The quantity $\bar{d}$ is related to the individual sample means as follows:

$$\bar{d} = \bar{y}_1 - \bar{y}_2$$

The relationship between population means is analogous:

$$\mu_d = \mu_1 - \mu_2$$

Thus, we may say that *the mean of the difference is equal to the difference of the means.* Because of this simple relationship, a comparison of the means can be carried out by concentrating entirely on the d's.

A standard error for $\bar{d}$ is easy to calculate. Because $\bar{d}$ is just the mean of a single sample, we can apply the SE formula of Chapter 6 to obtain the following formula:

$$SE_{(\bar{y}_1 - \bar{y}_2)} = SE_{\bar{d}} = \frac{s_d}{\sqrt{n_d}}$$

where s_d is the standard deviation of the d's and n_d is the number of d's. The following example illustrates the calculation.

EXAMPLE 9.2

NERVE CELL
REGENERATION

Table 9.2 shows the spinal cord CP data of Example 9.1 and the differences d.

TABLE 9.2

CP in Monkey Spinal
Cord (mg/100 gm)

ANIMAL	CONTROL SIDE Y_1	REGENERATING SIDE Y_2	$d = Y_1 - Y_2$
1	16.3	11.5	4.8
2	4.8	3.6	1.2
3	10.9	12.5	−1.6
4	14.2	6.3	7.9
5	16.3	15.2	1.1
6	9.9	8.1	1.8
7	29.2	16.6	12.6
8	22.4	13.1	9.3
Mean	15.50	10.86	4.64
SD	7.61	4.49	4.89

Note that the mean of the difference is indeed the difference of the means:

$$\bar{d} = 4.64 = 15.50 - 10.86$$

We calculate the standard error of this difference as follows:

$$s_d = 4.8858$$

$$n_d = 8$$

$$SE_{(\bar{y}_1 - \bar{y}_2)} = \frac{4.8858}{\sqrt{8}} = 1.727$$

∎

CONFIDENCE INTERVAL AND TEST OF HYPOTHESIS

The standard error described above is the basis for the **paired-sample t method** of analysis, which can take the form of a confidence interval or a test of hypothesis.

A 95% confidence interval for $(\mu_1 - \mu_2)$ is constructed as

$$(\bar{y}_1 - \bar{y}_2) \pm t_{.05} \, SE_{(\bar{y}_1 - \bar{y}_2)}$$

that is,

$$\bar{d} \pm t_{.05} \, SE_{\bar{d}}$$

where the constant $t_{.05}$ is determined from Student's t distribution with

$$df = n_d - 1$$

Intervals with other confidence coefficients (such as 90%, 99%, etc.) are constructed analogously (using $t_{.10}$, $t_{.01}$, etc.). The following example illustrates the confidence interval.

EXAMPLE 9.3

NERVE CELL
REGENERATION

For the spinal cord CP data, we have df $= 8 - 1 = 7$. From Table 4 we find that $t_{.05} = 2.365$; thus, the 95% confidence interval for $(\mu_1 - \mu_2)$ is

$$15.50 - 10.86 \pm (2.365)(1.727)$$

$$4.64 \pm 4.08$$

$$.6 < \mu_1 - \mu_2 < 8.7 \text{ mg/100 gm}$$

∎

We can also compare the two means with a t test. To test the null hypothesis

$$H_0: \quad \mu_1 = \mu_2$$

we use the test statistic

$$t_s = \frac{\bar{y}_1 - \bar{y}_2}{SE_{(\bar{y}_1 - \bar{y}_2)}}$$

that is,

$$t_s = \frac{\bar{d}}{SE_{\bar{d}}}$$

Critical values are obtained from Student's t distribution (Table 4) with df $=$ $n_d - 1$. The following example illustrates the t test.

EXAMPLE 9.4

NERVE CELL
REGENERATION

For the spinal cord CP data, let us formulate the null hypothesis and nondirectional alternative:

H_0: Mean CP is the same in regenerating and in normal tissue

H_A: Mean CP is different in regenerating and in normal tissue

or, in symbols,

H_0: $\mu_1 = \mu_2$

H_A: $\mu_1 \neq \mu_2$

Let us test H_0 against H_A at significance level $\alpha = .05$. The test statistic is

$$t_s = \frac{4.64}{1.727} = 2.69$$

From Table 4, $t_{.05} = 2.365$ and $t_{.02} = 2.998$. We reject H_0 and find that there is sufficient evidence ($.02 < P < .05$) to conclude that mean CP is lower in regenerating than in normal tissue. ∎

The confidence interval and t test formulas for paired data are directly analogous to those given in Chapter 7 for analysis of independent samples. The formula for a 95% confidence interval is

$$(\bar{y}_1 - \bar{y}_2) \pm t_{.05} \, SE_{(\bar{y}_1 - \bar{y}_2)}$$

and for the t statistic is

$$t_s = \frac{\bar{y}_1 - \bar{y}_2}{SE_{(\bar{y}_1 - \bar{y}_2)}}$$

Notice that these formulas appear identical to those given in Chapter 7. However, the paired-sample analysis is different from the independent-samples analysis in two respects: the SE is calculated differently and the number of degrees of freedom is different.

RESULT OF IGNORING PAIRING

Suppose that a study is conducted using a paired design, but that the pairing is ignored in the analysis of the data. Such an analysis is not valid because it assumes that the samples are independent when in fact they are not. The incorrect analysis can be grossly inadequate, as the following example illustrates.

EXAMPLE 9.5

NERVE CELL
REGENERATION

For the spinal cord CP data, the two sample SDs are $s_1 = 7.609$ and $s_2 = 4.488$. If we proceed as if the samples were independent and apply the SE formula of Chapter 7, we obtain

$$
\begin{aligned}
\mathrm{SE}_{(\bar{y}_1 - \bar{y}_2)} &= \sqrt{\frac{s_1^2}{n_1} + \frac{s_2^2}{n_2}} \\[2mm]
&= \sqrt{\frac{(7.609)^2}{8} + \frac{(4.488)^2}{8}} \\[2mm]
&= 3.123
\end{aligned}
$$

This SE is much larger than the value (SE = 1.727) that we calculated using the pairing.

Continuing to proceed as if the samples were independent, we calculate df = $n_1 + n_2 - 2 = 14$. If we conduct a t test, our test statistic is

$$
t_s = \frac{4.64}{3.123} = 1.49
$$

which does not approach significance, even at $\alpha = .10$. When we used a paired-sample analysis of these data (Example 9.4), we found evidence of a treatment effect; the present analysis, ignoring the pairing, fails to discover this evidence.

To further compare the paired and unpaired analysis, let us consider the 95% confidence interval for $(\mu_1 - \mu_2)$. For the unpaired analysis, we find $t_{.05} = 2.145$ with df = 14; this yields the confidence interval

4.64 ± (2.145)(3.123)

4.64 ± 6.70

The paired analysis (Example 9.3) yielded the narrower interval

4.64 ± (2.365)(1.727)

4.64 ± 4.08

The paired-sample interval is narrower because it uses a smaller SE; this effect is slightly offset by a larger value of $t_{.05}$ (2.365 vs 2.145).

Why is the paired-sample SE so much smaller than the independent-samples SE calculated from the same data (SE = 1.7 vs SE = 3.1)? Table 9.1 reveals the reason. The data show that there is large variation from one animal to the next. For instance, animal 2 has very low CP values (on both the right and left sides) and animal 7 has very high values. The independent-samples SE formula incorporates all of this variation (expressed through s_1 and s_2); in the paired-sample approach, interanimal variation in CP level has no influence on the calculations because only the d's are used. By using each animal as its own control, the experimenter has greatly increased the precision of the experiment. But if the pairing is ignored in the analysis, the extra precision is wasted.   ■

The preceding example illustrates the gain in precision that can result from a paired design coupled with a paired analysis. The choice between a paired and an unpaired design will be discussed in Section 9.3.

CONDITIONS FOR VALIDITY OF STUDENT'S *t* ANALYSIS

The conditions for validity of the paired-sample *t* test and confidence interval are as follows:

1. It must be reasonable to regard the *differences* (the *d*'s) as a random sample from some large population.
2. The population distribution of the *d*'s must be normal. The methods are approximately valid if the population distribution is approximately normal or if the sample size (n_d) is large.

The above conditions are the same as those given in Chapter 6; in the present case the conditions apply to the *d*'s because the analysis is based on the *d*'s. Verification of the conditions can proceed as described in Chapter 6. First, the design should be checked to assure that the *d*'s are independent of each other, and especially, that there is no hierarchical structure within the *d*'s. (Note, however, that the Y_1's are not independent of the Y_2's, because of the pairing.) Second, a histogram or stem-and-leaf display of the *d*'s can provide a rough check for approximate normality.

Notice that normality of the Y_1's and Y_2's is not required, because the analysis depends only on the *d*'s. For the same reason, equality of the standard deviations σ_1 and σ_2 is not required.[2]

SUMMARY OF FORMULAS

For convenient reference, we summarize in the boxes the formulas for the paired-sample methods based on Student's *t*.

BASIC RELATIONSHIP

$$\bar{y}_1 - \bar{y}_2 = \bar{d}$$

STANDARD ERROR OF $(\bar{y}_1 - \bar{y}_2)$

$$SE_{(\bar{y}_1 - \bar{y}_2)} = SE_{\bar{d}} = \frac{s_d}{\sqrt{n_d}}$$

CONFIDENCE INTERVAL FOR $(\mu_1 - \mu_2)$

95% confidence interval:

$$(\bar{y}_1 - \bar{y}_2) \pm t_{.05}\, SE_{(\bar{y}_1 - \bar{y}_2)}$$

Intervals with other confidence coefficients (e.g., 90%, 99%) are constructed analogously (e.g., using $t_{.10}$, $t_{.01}$).

---

t TEST OF H_0: $\mu_1 = \mu_2$

Test statistic:

$$t_s = \frac{\bar{y}_1 - \bar{y}_2}{SE_{(\bar{y}_1 - \bar{y}_2)}}$$

Critical values from Student's t distribution with

$$df = n_d - 1$$

EXERCISES 9.1–9.7

9.1 In an agronomic field experiment, blocks of land were subdivided into two plots of 346 square feet each. The plots were planted with two varieties of wheat, using a randomized blocks design. The plot yields (lb) of wheat are given in the table.[3]

	VARIETY	
BLOCK	1	2
1	32.1	34.5
2	30.6	32.6
3	33.7	34.6
4	29.7	31.0
Mean	31.52	33.17
SD	1.76	1.72

a. Calculate the standard error of the difference between the variety means.
b. Test for a difference between the varieties using a paired t test at $\alpha = .05$. Use a nondirectional alternative.
c. Test for a difference between the varieties the wrong way, using an independent-samples t test. Compare with the result of part **b**.

9.2 In an investigation of possible brain damage due to alcoholism, an X-ray procedure known as a computerized tomography (CT) scan was used to measure brain densities in eleven chronic alcoholics. For each alcoholic, a nonalcoholic control was selected who matched the alcoholic on age, sex, education, and other factors. The brain density measurements on the alcoholics and the matched controls are reported in the accompanying table.[4]

a. Calculate the standard error of the difference between the means.
b. Use a t test to test the null hypothesis of no difference against the alternative that alcoholism reduces brain density. Let $\alpha = .02$.
c. Calculate the standard error of the difference between the means the wrong way, using the independent-samples method; compare with the result of part a. If a t test were based on this erroneous SE, what would be the conclusion?

PAIR	ALCOHOLIC	CONTROL	DIFFERENCE
1	40.1	41.3	−1.2
2	38.5	40.2	−1.7
3	36.9	37.4	−.5
4	41.4	46.1	−4.7
5	40.6	43.9	−3.3
6	42.3	41.9	.4
7	37.2	39.9	−2.7
8	38.6	40.4	−1.8
9	38.5	38.6	−.1
10	38.4	38.1	.3
11	38.1	39.5	−1.4
$\bar{y}$	39.14	40.66	−1.52
s	1.72	2.56	1.58

9.3 Cyclic adenosine monophosphate (cAMP) is a substance that can mediate cellular response to hormones. In a study of maturation of egg cells in the frog *Xenopus laevis*, oocytes from each of four females were divided into two batches; one batch was exposed to progesterone and the other was not. After two minutes, each batch was assayed for its cAMP content, with the results given in the table.[5] Use a *t* test to investigate the effect of progesterone on cAMP. Let H_A be nondirectional and $\alpha = .05$.

FROG	cAMP (pmol/oocyte)	
	Control	Progesterone
1	6.01	5.23
2	2.28	1.21
3	1.51	1.40
4	2.12	1.38

9.4 Under certain conditions, electrical stimulation of a beef carcass will improve the tenderness of the meat. In one study of this effect, beef carcasses were split in half; one side (half) was subjected to a brief electrical current and the other side was an untreated control. For each side, a steak was cut and tested in various ways for tenderness. In one test, the experimenter obtained a specimen of connective tissue (collagen) from the steak and determined the temperature at which the tissue would shrink; a tender piece of meat tends to yield a low collagen shrinkage temperature. The data are given in the table at the top of page 302.[6]

a. Construct a 95% confidence interval for the mean difference between the treated side and the control side.

b. Construct a 95% confidence interval the wrong way, using the independent-samples method. How does this interval differ from the one you obtained in part **a**?

CARCASS	COLLAGEN SHRINKAGE TEMPERATURE (°C)		
	Treated Side	Control Side	Difference
1	69.50	70.00	−.50
2	67.00	69.00	−2.00
3	70.75	69.50	1.25
4	68.50	69.25	−.75
5	66.75	67.75	−1.00
6	68.50	66.50	2.00
7	69.50	68.75	.75
8	69.00	70.00	−1.00
9	66.75	66.75	.00
10	69.00	68.50	.50
11	69.50	69.00	.50
12	69.00	69.75	−.75
13	70.50	70.25	.25
14	68.00	66.25	1.75
15	69.00	68.25	.75
Mean	68.750	68.633	.117
SD	1.217	1.302	1.118

9.5 Refer to Exercise 9.4. Use a t test to test the null hypothesis of no effect against the alternative hypothesis that the electrical treatment tends to reduce the collagen shrinkage temperature. Let $\alpha = .10$.

9.6 Women who smoke cigarettes tend to have smaller babies than nonsmokers. In an investigation of this phenomenon, two groups of women were identified: one group (SS) had smoked during two consecutive pregnancies; another group (SN) had smoked during one pregnancy but quit smoking before the next pregnancy. From these two groups, 159 matched pairs of women were formed. The members of a pair—one SS and one SN—were matched with respect to several factors, including the birthweight of the first child and the amount of smoking at the time of the first birth. The data on birthweight (gm) of the second child are reported in the accompanying table.[7] Use a t test to compare the two groups. Use a directional alternative and let $\alpha = .01$.

GROUP	MEAN BIRTHWEIGHT	MEAN DIFFERENCE	SD OF DIFFERENCE
SS	3,212	169	617
SN	3,381		

9.7 Invent a paired data set, consisting of five pairs of observations, for which $\bar{y}_1$ and $\bar{y}_2$ are not equal, and $SE_{\bar{y}_1} > 0$ and $SE_{\bar{y}_2} > 0$, but $SE_{(\bar{y}_1 - \bar{y}_2)} = 0$.

S E C T I O N   9.3

THE PAIRED DESIGN

Ideally, in a paired design the members of a pair are relatively similar to each other—that is, more similar to each other than to members of other pairs—with respect to extraneous variables. The advantage of this arrangement is that, when members of a pair are compared, the comparison is free of the extraneous variation that originates in between-pair differences. We will expand on this theme after giving some examples.

EXAMPLES OF PAIRED DESIGNS

Paired designs can arise in a variety of ways, including the following:

Randomized blocks experiments with two experimental units per block

Observational studies with individually matched controls

Repeated measurements on the same individual at two different times

Blocking by time

RANDOMIZED BLOCKS EXPERIMENTS A randomized blocks design (Chapter 8) is a paired design if there are only two treatments. Each block would then contain two experimental units, one to receive each treatment. The following is an example.

EXAMPLE 9.6

FERTILIZERS FOR EGGPLANTS

In a greenhouse experiment to compare two fertilizer treatments for eggplants, individually potted plants are arranged on the greenhouse bench in blocks of two (that is, pairs). Within each pair, one (randomly chosen) plant will receive treatment 1 and the other will receive treatment 2. ∎

OBSERVATIONAL STUDIES As noted in Chapter 8, matching in observational studies is an important method of controlling extraneous variables. A **matched-pair design** arises if members of two groups, matched with respect to various extraneous variables, are to be compared. Here is an example.

EXAMPLE 9.7

SMOKING AND LUNG CANCER

In a case–control study of lung cancer, 100 lung cancer patients were identified. For each case, a control was chosen who was individually matched to the case with respect to age, sex, and education level. The smoking habits of the cases and the controls were compared. ∎

REPEATED MEASUREMENTS Many biological investigations involve repeated measurements made on the same individual at different times. These include studies of growth and development, studies of biological processes, and studies in which measurements are made before and after application of a certain treatment. When only two times are involved, the measurements are paired, as in the following example.

EXAMPLE 9.8

EXERCISE AND SERUM TRIGLYCERIDES

Triglycerides are blood constituents that are thought to play a role in coronary artery disease. To see whether regular exercise could reduce triglyceride levels, researchers measured the concentration of triglycerides in the blood serum of seven male volunteers, before and after participation in a 10-week exercise program. The results are shown in Table 9.3 (page 304).[8] Note that there is considerable variation from one participant to another. For instance, participant 1 had relatively low triglyceride levels both before and after, while participant 3 had relatively high levels.

TABLE 9.3

Serum Triglycerides
(mmol/l)

PARTICIPANT	BEFORE	AFTER
1	.87	.57
2	1.13	1.03
3	3.14	1.47
4	2.14	1.43
5	2.98	1.20
6	1.18	1.09
7	1.60	1.51

BLOCKING BY TIME   In some situations, blocks or pairs are formed implicitly when replicate measurements are made at different times. The following is an example.

EXAMPLE 9.9

GROWTH OF VIRUSES

In a series of experiments on a certain virus (mengovirus), a microbiologist measured the growth of two strains of the virus—a mutant strain and a nonmutant strain—on mouse cells in petri dishes. Replicate experiments were run on 19 different days. The data are shown in Table 9.4. Each number represents the total growth in 24 hours of the viruses in a single dish.[9]

TABLE 9.4

Virus Yield at 24 Hours

RUN	NONMUTANT STRAIN	MUTANT STRAIN	RUN	NONMUTANT STRAIN	MUTANT STRAIN
1	160	97	11	61	15
2	36	55	12	14	10
3	82	31	13	140	150
4	100	95	14	68	44
5	140	80	15	110	31
6	73	110	16	37	14
7	110	100	17	95	57
8	180	100	18	64	70
9	62	6	19	58	45
10	43	7			

Note that there is considerable variation from one run to another. For instance, run 1 gave relatively large values (160 and 97), whereas run 2 gave relatively small values (36 and 55). This variation between runs arises from unavoidable small variations in the experimental conditions. For instance, both the growth of the viruses and the measurement technique are highly sensitive to environmental conditions such as the temperature and CO_2 concentration in the incubator. Slight fluctuations in the environmental conditions cannot be prevented, and these fluctuations cause the variation that is reflected in the data. In this kind of situation the advantage of running the two strains concurrently (that is, in pairs) is particularly striking.

Examples 9.8 and 9.9 both involve measurements at different times. But notice that the pairing structure in the two examples is entirely different. In Example 9.8 the members of a pair are measurements on the same individual at two times, whereas in Example 9.9 the members of a pair are measurements on two petri dishes at the same time. Nevertheless, in both examples the principle of pairing is the same: members of a pair are similar to each other with respect to extraneous variables. In Example 9.9 time is an extraneous variable, whereas in Example 9.8 the comparison between two times (before and after) is of primary interest and interperson variation is extraneous.

PURPOSES OF PAIRING

Pairing in an experimental design can serve to reduce bias, to increase precision, or both.

We noted in Chapter 8 that blocking or matching can reduce bias by controlling variation due to extraneous variables. The variables used in the matching are necessarily balanced in the two groups to be compared, and therefore cannot distort the comparison. For instance, if two groups are composed of age-matched pairs of people, then a comparison between the two groups is free of any bias due to a difference in age distribution.

In observational studies, where randomization is impossible, the control of bias is often the primary reason for using a pair-matched design. In randomized experiments, where bias can be controlled by randomized allocation, a major reason for pairing is to increase precision.

Effective pairing increases precision by increasing the information available in an experiment. An appropriate analysis, which extracts this extra information, leads to more powerful tests and narrower confidence intervals. Thus, an effectively paired experiment is more efficient; it yields more information than an unpaired experiment with the same number of observations.

We saw an instance of effective pairing in the spinal cord CP experiment of Example 9.1. The pairing was effective because much of the variation in the measurements was interanimal variation, which did not enter the comparison between the treatments. As a result, the experiment yielded more precise information about the treatment difference than would a comparable unpaired experiment—that is, an experiment which would compare measurements on regenerating tissue in eight animals versus measurements on normal (control) tissue in eight *different* animals.

The effectiveness of a given pairing can be displayed visually in a *scatterplot* of Y_1 against Y_2; each point in the scatterplot represents a single pair (Y_1, Y_2). Figure 9.1 (page 306) shows a scatterplot for the spinal cord CP data; each point represents a single animal. Notice that the points in the scatterplot show a definite upward trend. This upward trend indicates the effectiveness of the pairing: measurements on the same animal have more in common than measurements on different animals, so that an animal with a relatively high value of Y_1 tends to have a relatively high value of Y_2, and similarly for low values.

FIGURE 9.1

Scatterplot for the Spinal
Cord CP Data

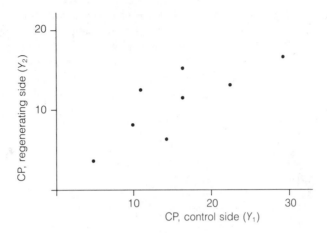

Note that pairing is a strategy of *design*, not of analysis, and is therefore carried out *before* the Y's are observed. It is not correct to use the observations themselves to form pairs. Such a data manipulation could distort the experimental results as severely as outright fakery.

RANDOMIZED PAIRS DESIGN VS COMPLETELY RANDOMIZED DESIGN

In planning a randomized experiment, the experimenter may need to decide between a paired design and a completely randomized design. We have said that effective pairing can greatly enhance the precision of an experiment. On the other hand, pairing in an experiment may *not* be effective, if the observed variable Y is not related to the factors used in the pairing. For instance, suppose pairs were matched on age only, but in fact Y turned out not to be age-related. It can be shown that ineffective pairing actually yields less precision than no pairing at all. For instance, in relation to a *t* test, ineffective pairing would not tend to reduce the SE, but it would reduce the degrees of freedom, and the net result would be a loss of power.

The choice whether to use a paired design depends on practical considerations (pairing may be expensive or unwieldy) and on precision considerations. With respect to precision, the choice depends on how effective the pairing is expected to be. The following example illustrates this issue.

EXAMPLE 9.10

FERTILIZERS FOR
EGGPLANTS

A horticulturist is planning a greenhouse experiment with individually potted eggplants. Two fertilizer treatments are to be compared, and the observed variable is to be Y = yield of eggplants (pounds). The experimenter knows that Y is influenced by such factors as light and temperature, which vary somewhat from place to place on the greenhouse bench. The allocation of pots to positions on the bench could be carried out according to a completely randomized design, or according to a randomized blocks (paired) design, as in Example 9.6. In deciding between these options, the experimenter must use his knowledge of how effective the pairing would be—that is, whether two pots sitting adjacent on the bench would be very much more similar in yield than pots farther apart. If he judges that the pairing would not be very effective, he may opt for the completely randomized design. ■

Note that effective pairing is *not* the same as simply holding experimental conditions constant. Pairing is a way of *organizing* the unavoidable variation that still remains after experimental conditions have been made as constant as possible. The ideal pairing organizes the variation in such a way that the variation within each pair is minimal and the variation between pairs is maximal.

CHOICE OF ANALYSIS

Generally speaking, the analysis of data should fit the design of the study. If the design is paired, a paired-sample analysis should be used; if the design is unpaired, an independent-samples analysis (as in Chapter 7) should be used.

Note that the extra information made available by an effectively paired design is *entirely wasted* if an unpaired analysis is used. (We saw an illustration of this in Example 9.5.) Thus, the paired design does not increase efficiency unless it is accompanied by a paired analysis.

EXERCISES 9.8–9.11

9.8 (*Sampling exercise*) This exercise illustrates the application of a matched-pairs design to the population of 100 ellipses (shown with Exercise 3.1). The accompanying table shows a grouping of the 100 ellipses into 50 pairs.

PAIR	ELLIPSE ID NUMBERS		PAIR	ELLIPSE ID NUMBERS		PAIR	ELLIPSE ID NUMBERS	
01	75	39	18	92	16	35	29	46
02	9	31	19	47	80	36	88	76
03	6	41	20	15	53	37	64	90
04	22	27	21	95	72	38	20	7
05	60	28	22	44	56	39	51	54
06	97	61	23	18	83	40	96	62
07	33	36	24	12	93	41	00	81
08	32	66	25	57	55	42	77	98
09	10	13	26	24	50	43	14	38
10	25	11	27	70	35	44	82	91
11	94	43	28	5	73	45	37	59
12	26	40	29	71	8	46	4	30
13	48	42	30	49	23	47	79	21
14	87	2	31	65	63	48	19	68
15	67	17	32	78	58	49	99	3
16	1	74	33	86	89	50	52	69
17	84	34	34	45	85			

To better appreciate this exercise, imagine the following experimental setting. We want to investigate the effect of a certain treatment, T, on the organism *C. ellipticus*. We will observe the variable Y = length. We can measure each individual only once, and so we will compare n treated individuals with n untreated controls. We know that the individuals available for the experiment are of various ages, and we know that age is related to length, so we have formed 50 age-matched pairs, some of which will be used in the experiment. The purpose of the pairing is to increase the power of the experiment by eliminating the random variation due to age. (Of course, the ellipses do not actually have ages, but the pairing shown in the table has been constructed in a way that *simulates* age-matching.)

a. Use random digits (from Table 1 or your calculator) to choose a random sample of five *pairs* from the list.

b. For each pair, use random digits (or toss a coin) to randomly allocate one member to treatment (T) and the other to control (C).

c. Measure the lengths of all ten ellipses. Then, to simulate a treatment effect, add 6 mm to each length in the T group.

d. Apply a paired-sample t test to the data. Use a nondirectional alternative and let $\alpha = .05$.

e. Did the analysis of part d lead you to a Type II error?

9.9 (*Continuation of Exercise 9.8*)   Apply an independent-samples t test to your data. Use a nondirectional alternative and let $\alpha = .05$. Does this analysis lead you to a Type II error?

9.10 (*Sampling exercise*)   Refer to Exercise 9.8. Imagine that a matched-pairs experiment is not practical (perhaps because the ages of the individuals cannot be measured), so we decide to use a completely randomized design to evaluate the treatment T.

a. Use random digits (from Table 1 or your calculator) to choose a random sample of ten individuals from the ellipse population (shown with Exercise 3.1). From these ten, randomly allocate five to T and five to C. (Or, equivalently, just randomly select five from the population to receive T and five to receive C.)

b. Measure the lengths of all ten ellipses. Then, to simulate a treatment effect, add 6 mm to each length in the T group.

c. Apply an independent-samples t test to the data. Use a nondirectional alternative and let $\alpha = .05$.

d. Did the analysis of part c lead you to a Type II error?

9.11 Refer to each exercise indicated below. Construct a scatterplot of the data. Does the appearance of the scatterplot indicate that the pairing was effective?

a. Exercise 9.1   **b.** Exercise 9.2   **c.** Exercise 9.4

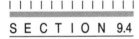

SECTION 9.4

THE SIGN TEST

The **sign test** is a nonparametric test that can be used to compare two paired samples. It is not particularly powerful, but it is very flexible in application and is especially simple to use and understand—a blunt but handy tool.

METHOD

Like the paired-sample t test, the sign test is based on the differences

$$d = Y_1 - Y_2$$

The only information used by the sign test is the sign (positive or negative) of each difference. If the differences are preponderantly of one sign, this is taken as evidence against the null hypothesis. Critical values for the sign test are given in Table 7 (at the end of the book). The following example illustrates the sign test.

EXAMPLE 9.11
GROWTH OF VIRUSES

Table 9.5 shows the virus growth data of Example 9.9, together with the signs of the differences.

TABLE 9.5 Virus Growth at 24 Hours

RUN	NONMUTANT STRAIN Y_1	MUTANT STRAIN Y_2	SIGN OF $d = Y_1 - Y_2$	RUN	NONMUTANT STRAIN Y_1	MUTANT STRAIN Y_2	SIGN OF $d = Y_1 - Y_2$
1	160	97	+	11	61	15	+
2	36	55	−	12	14	10	+
3	82	31	+	13	140	150	−
4	100	95	+	14	68	44	+
5	140	80	+	15	110	31	+
6	73	110	−	16	37	14	+
7	110	100	+	17	95	57	+
8	180	100	+	18	64	70	−
9	62	6	+	19	58	45	+
10	43	7	+				

Let us carry out a sign test to compare the growth of the two strains. We choose $\alpha = .05$. The null hypothesis and nondirectional alternative are

H_0: The two strains of virus grow equally well

H_A: One of the strains grows better than the other

The first step is to determine the following counts:

$N_+ =$ Number of positive differences
$N_- =$ Number of negative differences

The test statistic B_s is defined as

$B_s =$ Larger of N_+ and N_-

For the present data, we have

$N_+ = 15$
$N_- = 4$
$B_s = 15$

The next step is to bracket the P-value. Using Table 7 with $n_d = 19$, we obtain the following critical values:

Nominal Tail Probability	.20	.10	.05	.02	.01	.002	.001
Critical Value	13	14	15	15	16	17	17

To bracket the P-value, we find the rightmost column in the table with critical value less than or equal to B_s; then the P-value is bracketed between that column heading and the next one. In the present case the result is

$.01 < P < .02$

We reject H_0 and find that the data provide sufficient evidence to conclude that the nonmutant strain grows better (at 24 hours) than the mutant strain of virus.

∎

BRACKETING THE P-VALUE   Like the Mann–Whitney test, the sign test has a discrete null distribution. The sign test statistic B_s may be exactly equal to a column heading in Table 7, and in such a case the P-value is less than the column heading.* Also, certain critical value entries in Table 7 are blank. Both these situations are already familiar from our study of the Mann–Whitney test. Table 7 has another peculiarity that is not shared by the Mann–Whitney test: some critical values appear more than once in the same row. This feature is also due to the discreteness of the null distribution, and does not cause any particular difficulty; to bracket the P-value, we move to the right in Table 7 as far as possible, stopping when the next critical value in the table is *larger* than the observed value B_s.

DIRECTIONAL ALTERNATIVE   If the alternative hypothesis is directional, the sign test is carried out using the familiar two-step procedure—first checking for directionality and then cutting the P-value in half if the data deviate from H_0 in the direction specified by H_A.

A CAUTION   Note that Table 7, for the sign test, and Table 4, for the t test, are organized differently: Table 7 is entered with n_d, while Table 4 is entered with $(n_d - 1)$.

TREATMENT OF ZEROS   It may happen that some of the differences $(Y_1 - Y_2)$ are equal to zero. Should these be counted as positive or negative in determining B_s? A recommended procedure is to drop the corresponding pairs from the analysis and reduce the sample size n_d accordingly. In other words, each pair whose difference is zero is ignored entirely; such pairs are regarded as providing

---

*In a few cases the P-value would be exactly equal to (rather than less than) the column heading. To simplify the presentation, we neglect this fine distinction.

no evidence against H_0 in either direction. Notice that this procedure has no parallel in the t test; the t test treats differences of zero the same as any other value.

RATIONALE

The rationale for the sign test is very simple. Each difference (d) is either positive or negative (ignoring zeros); let π represent the probability that a difference will be positive. Then the null hypothesis for the sign test is

$$H_0: \quad \pi = .5$$

For instance, in the virus growth experiment of Example 9.11, if the two strains grow equally well then the observed difference between the two strains in any run is due only to chance, and is equally likely to be positive or negative; in this case, the probability of a positive difference would be $\frac{1}{2}$ or .5 for each run.

The null distribution for the sign test is based on the binomial distribution (Chapter 3). If H_0 is true, then the probability distribution of N_+ (the number of positive d's) will be a binomial distribution with

$$n = n_d \quad \text{and} \quad \pi = .5$$

The following is an example.

EXAMPLE 9.12

NULL DISTRIBUTION

Consider an experiment with ten pairs, so that $n_d = 10$. If H_0 is true, then the probability distribution of N_+ is a binomial distribution with $n = 10$ and $\pi = .5$. Figure 9.2(a) shows this binomial distribution, together with the associated values of N_+, N_-, and B_s. Figure 9.2(b) shows the null distribution of B_s, which is a "folded" version of Figure 9.2(a). (We saw a similar relationship between parts (a) and (b) of Figure 7.15.)

FIGURE 9.2 Null Distributions for the Sign Test When $n_d = 10$. **(a)** Distribution of N_+ and N_-; **(b)** Distribution of B_s.

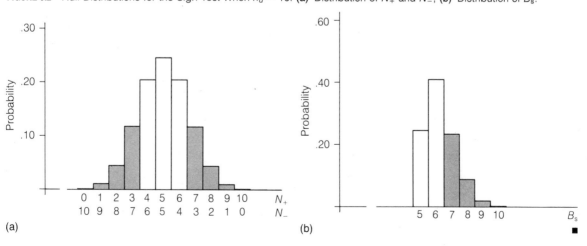

(a)   (b)

The following example shows how the binomial distribution can be used to calculate exact P-values for the sign test.

EXAMPLE 9.13
EXACT P-VALUES

Suppose $n_d = 10$, so that the null distribution is the one shown in Figure 9.2. Suppose further that H_A is nondirectional, and that the observed test statistic is $B_s = 7$. If H_0 is true, the probability of 7 or more plus (+) signs is calculated from the binomial formula as

$$C_7(.5)^7(.5)^3 + C_8(.5)^8(.5)^2 + C_9(.5)^9(.5)^1 + C_{10}(.5)^{10}(.5)^0$$
$$= 120(.5)^{10} + 45(.5)^{10} + 10(.5)^{10} + 1(.5)^{10}$$
$$= (176)(.5)^{10}$$
$$= .1719$$

This value (.1719) is the sum of the shaded bars in the right-hand tail in Figure 9.2(a). The sum of the shaded bars in the left-hand tail is also equal to .1719; consequently, the total shaded area, which is the P-value of the data, is

$$P = 2(.1719) = .3438 \approx .34$$

In terms of the null distribution of B_s, the P-value is an upper-tail probability; thus, the sum of the shaded bars in Figure 9.2(b) is equal to .34.   ∎

HOW TABLE 7 IS CALCULATED   Throughout your study of statistics you are asked to take on faith the critical values given in various tables. Table 7 is an exception; the following example shows how you could (if you wished to) calculate the critical values yourself. Understanding the example will help you to appreciate how the other tables of critical values have been obtained.

EXAMPLE 9.14

Suppose $n_d = 10$. We saw in Example 9.13 that

If $B_s = 7$ the P-value of the data is .3438.

Similar calculations using the binomial formula show that

If $B_s = 8$, the P-value of the data is .1094.
If $B_s = 9$, the P-value of the data is .0215.
If $B_s = 10$, the P-value of the data is .00195.

For $n_d = 10$, the critical values are given in Table 7 as follows:

Nominal Tail Probability	.20	.10	.05	.02	.01	.002	.001
Critical Value	8	9	9	10	10	10	

These critical values have been determined from the above P-values, using the principle that the P-value corresponding to each entry should be as close as possible to the column heading without exceeding it. Thus, for instance, the critical value in the .05 column is equal to 9 because the P-value for $B_s = 9$ (.0215) is less than .05, but the next larger P-value (.1094) is greater than .05. In fact, .1094 is also greater than .10, and this is why 9 (rather than 8) is also listed in the .10 column. Look now at the .02 column. The only possible P-value less than .02 is .00195, which is also less than .01 and less than .002; this accounts for the three critical values listed as 10. On the other hand, .00195 is the smallest possible P-value and yet is greater than .001; for this reason the .001 column is left blank. ∎

APPLICABILITY OF THE SIGN TEST

The sign test is valid in any situation where the d's are independent of each other and the null hypothesis can be appropriately translated as

$$H_0: \quad \Pr\{d \text{ is positive}\} = .5$$

Thus, the sign test is distribution-free; its validity does not depend on any assumptions about the form of the population distribution of the d's. This broad validity is bought at a price: if the population distribution of the d's is indeed normal, then the sign test is much less powerful than the t test.*

The sign test is useful because it can be applied quickly and in a wide variety of settings. In fact, sometimes the sign test can be applied to data which do not permit a t test at all, as in the following example.

EXAMPLE 9.15
THC AND
CHEMOTHERAPY

Chemotherapy for cancer often produces nausea and vomiting. The effectiveness of THC (the active ingredient of marijuana) in preventing these side effects was compared with the standard drug Compazine. Of the 46 patients who tried both drugs (but were not told which was which), 21 expressed no preference, while 20 preferred THC and 5 preferred Compazine. Since "preference" indicates a *sign* for the difference, but not a *magnitude*, a t test is impossible in this situation. For a sign test, we have $n_d = 25$ and $B_s = 20$, so that $.002 < P < .01$; even at $\alpha = .01$ we would reject H_0 and find that the data provide sufficient evidence to conclude that THC is preferred to Compazine.[10] ∎

EXERCISES 9.12–9.23

9.12 Use Table 7 to bracket the P-value for a sign test (against a nondirectional alternative), assuming that $n_d = 9$ and
    a. $B_s = 6$ **b.** $B_s = 7$ **c.** $B_s = 8$ **d.** $B_s = 9$

9.13 Use Table 7 to bracket the P-value for a sign test (against a nondirectional alternative), assuming that $n_d = 15$ and,
    a. $B_s = 10$ **b.** $B_s = 11$ **c.** $B_s = 12$
    d. $B_s = 13$ **e.** $B_s = 14$ **f.** $B_s = 15$

---

*Actually, there is another simple test for paired data, called the Wilcoxon signed-ranks test, which is generally more powerful than the sign test and yet is distribution-free.

9.14 Skin from cadavers can be used to provide temporary skin grafts for severely burned patients. The longer such a graft survives before its inevitable rejection by the immune system, the more the patient benefits. A medical team investigated the usefulness of matching graft to patient with respect to the HL-A antigen system. Each patient received two grafts, one with close HL-A compatibility, and the other with poor compatibility. The survival times (in days) of the skin grafts are shown in the accompanying table.[11] Use a sign test to compare the two types of grafts. Use a directional alternative and let $\alpha = .05$. (Notice that a t test could not be applied here because two of the observations are incomplete; patient 3 died with a graft still surviving and the observation on patient 10 was incomplete for an unspecified reason.)

PATIENT	HL-A COMPATIBILITY	
	Close	Poor
1	37	29
2	19	13
3	57+	15
4	93	26
5	16	11
6	23	18
7	20	26
8	63	43
9	29	18
10	60+	42
11	18	19

9.15 Refer to Exercise 9.14. Calculate the exact P-value of the data as analyzed by the sign test. (Note that H_A is directional.)

9.16 Can mental exercise build "mental muscle"? In one study of this question, twelve littermate pairs of young male rats were used; one member of each pair, chosen at random, was raised in an "enriched" environment with toys and companions, while its littermate was raised alone in an "impoverished" environment. (See Example 8.6.) After 80 days, the animals were sacrificed and their brains were dissected by a researcher who did not know which treatment each rat had received. One variable of interest was the weight of the cerebral cortex, expressed relative to total brain weight. For 10 of the 12 pairs, the relative cortex weight was greater for the "enriched" rat than for his "impoverished" littermate; in the other 2 pairs, the "impoverished" rat had the larger cortex. Use a sign test to compare the environments at $\alpha = .05$; let the alternative hypothesis be that environmental enrichment tends to increase the relative size of the cortex.[12]

9.17 Refer to Exercise 9.16. Calculate the exact P-value of the data as analyzed by the sign test. (Note that H_A is directional.)

9.18 Twenty institutionalized epileptic patients participated in a study of a new anticonvulsant drug, valproate. Ten of the patients (chosen at random) were started on daily valproate and the remaining 10 received an identical placebo pill. During an eight-week observation period, the numbers of major and minor epileptic seizures were counted for each patient. After this, all patients were "crossed over" to the other treatment, and seizure counts were made during a second eight-week observation period. The numbers of minor seizures are given in the accompanying table.[13] Test for efficacy of valproate using the sign test at $\alpha = .05$. Use a directional alternative. (Note that this analysis ignores the possible effect of time—that is, first versus second observation period.)

PATIENT NUMBER	PLACEBO PERIOD	VALPROATE PERIOD	PATIENT NUMBER	PLACEBO PERIOD	VALPROATE PERIOD
1	37	5	11	7	8
2	52	22	12	9	8
3	63	41	13	65	30
4	2	4	14	52	22
5	25	32	15	6	11
6	29	20	16	17	1
7	15	10	17	54	31
8	52	25	18	27	15
9	19	17	19	36	13
10	12	14	20	5	5

*9.19 (*This exercise is based on material from optional Section 5.5.*)   Refer to Exercise 9.18. Use the normal approximation to the binomial distribution (with the continuity correction) to calculate the P-value of the data as analyzed by the sign test. (Note that H_A is directional.)

9.20 An ecological researcher studied the interaction between birds of two subspecies, the Carolina Junco and the Northern Junco. He placed a Carolina male and a Northern male, matched by size, together in an aviary and observed their behavior for 45 minutes beginning at dawn. This was repeated on different days with different pairs of birds. The table shows counts of the episodes in which one bird displayed dominance over the other—for instance, by chasing it or displacing it from its perch.[14] Use a sign test to compare the subspecies. Use a nondirectional alternative and let $\alpha = .01$.

PAIR	NUMBER OF EPISODES IN WHICH	
	Northern Was Dominant	Carolina Was Dominant
1	0	9
2	0	6
3	0	22
4	2	16
5	0	17
6	2	33
7	1	24
8	0	40

9.21 Refer to Exercise 9.20. Calculate the exact P-value of the data as analyzed by the sign test. (Note that H_A is nondirectional.)

9.22 a. Suppose a paired data set has $n_d = 7$ and $B_s = 7$. Calculate the exact *P*-value of the data as analyzed by the sign test (against a nondirectional alternative).

b. Explain why, in Table 7 with $n_d = 7$, no critical value is given in the .01 column.

9.23 a. Suppose a paired data set has $n_d = 15$. Calculate the exact *P*-value of the data as analyzed by the sign test (against a nondirectional alternative) if (i) $B_s = 13$; (ii) $B_s = 14$; (iii) $B_s = 15$.

b. Explain why, in Table 7 with $n_d = 15$, the critical value in the .002 column is $B_{.002} = 14$.

c. If Table 7 had a .005 column, what would be the entry (that is, $B_{.005}$) for $n_d = 15$?

SECTION 9.5

FURTHER CONSIDERATIONS IN PAIRED EXPERIMENTS

In this section we discuss two additional topics: the interpretation of before–after studies, and the reporting of paired data.

BEFORE–AFTER STUDIES

Many studies in the life sciences compare measurements before and after some experimental intervention. These studies can be difficult to interpret, because the effect of the experimental intervention may be confounded with other changes over time. One way to protect against this difficulty is to use randomized concurrent controls, as in the following example.

EXAMPLE 9.16

BIOFEEDBACK AND BLOOD PRESSURE

A medical research team investigated the effectiveness of a biofeedback training program designed to reduce high blood pressure. Volunteers were randomly allocated to a biofeedback group or a control group. All volunteers received health education literature and a brief lecture. In addition, the biofeedback group received 8 weeks of relaxation training, aided by biofeedback, meditation, and breathing exercises. The results for systolic blood pressure, before and after the 8 weeks, are shown in Table 9.6.[15]

TABLE 9.6

Results of Biofeedback Experiment

GROUP	n	SYSTOLIC BLOOD PRESSURE (mm Hg)			
		BEFORE Mean	AFTER Mean	DIFFERENCE Mean	SE
Biofeedback	99	145.2	131.4	13.8	1.34
Control	93	144.2	140.2	4.0	1.30

Let us analyze the before–after changes by paired *t* tests at $\alpha = .05$. In the biofeedback group, the mean systolic blood pressure fell by 13.8 mm Hg. To evaluate the statistical significance of this drop, the test statistic is

$$t_s = \frac{13.8}{1.34} = 10.3$$

which is highly significant ($P \ll .0001$). However, this result alone does not demonstrate the effectiveness of the biofeedback training; the drop in blood pressure might be partly or entirely due to other factors, such as the health education literature or the special attention received by all the participants. Indeed, a paired t test applied to the control group gives

$$t_s = \frac{4.0}{1.30} = 3.08 \qquad .001 < P < .01$$

Thus, the people who received *no* biofeedback training *also* experienced a statistically significant drop in blood pressure.

To isolate the effect of the biofeedback training, we can compare the experience of the two treatment groups, using an independent-samples t test. We again choose $\alpha = .05$. The difference between the mean changes in the two groups is

$$13.8 - 4.0 = 9.8 \text{ mm Hg}$$

and the standard error (unpooled) of this difference is

$$\sqrt{(1.34)^2 + (1.30)^2} = 1.87$$

Thus, the t statistic is

$$t_s = \frac{9.8}{1.87} = 5.24$$

This test provides strong evidence ($P < .0001$) that the biofeedback program is effective. If the experimental design had not included the control group, then this last crucial comparison would not have been possible, and the support for efficacy of biofeedback would have been shaky indeed. ∎

REPORTING OF DATA

In communicating experimental results, it is desirable to choose a form of reporting that conveys the extra information provided by pairing. With small samples, a graphical approach can be used, as in the following example.

EXAMPLE 9.17

PLASMA ALDOSTERONE IN DOGS

Aldosterone is a hormone involved in maintaining fluid balance in the body. In a veterinary study, six dogs with heart failure were treated with the drug Captopril, and plasma concentrations of aldosterone were measured before and after the treatment. The results are displayed in Figure 9.3 (page 318).[16] The experience of each dog is represented by two points joined by a line. Note that the lines carry crucial information. For instance, all the lines slope downward, which indicates that all six dogs experienced a fall (rather than a rise) in plasma aldosterone. Also, lines that are parallel represent falls of equal magnitude; in Figure 9.3 four of the lines are approximately parallel. If the lines were omitted from the plot, the reader would be unable to assess either the statistical or the practical significance of the before–after change.

FIGURE 9.3

Plasma Aldosterone in Six
Dogs Before and After
Treatment with Captopril

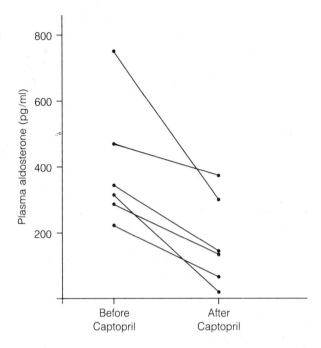

In published reports of biological research, the crucial information related to pairing is often omitted. For instance, a common practice is to report the means and standard deviations of Y_1 and Y_2, but to omit the standard deviation of the difference, d. It is best to report some description of d, using either a display like Figure 9.3, or a histogram of the d's, or at least the standard deviation of the d's.

EXERCISES 9.24–9.25

9.24 Thirty-three men with high serum cholesterol, all regular coffee drinkers, participated in a study to see whether abstaining from coffee would affect their cholesterol level. Twenty-five of the men (chosen at random) drank no coffee for 5 weeks, while the remaining eight men drank coffee as usual. The accompanying table shows the serum cholesterol levels (in mg/dl) at baseline (at the beginning of the study) and the change from baseline after 5 weeks.[17]

	NO COFFEE ($n = 25$)		USUAL COFFEE ($n = 8$)	
	Mean	SD	Mean	SD
Baseline	341	37	331	30
Change from Baseline	−35	27	+26	56

For the following t tests, use nondirectional alternatives and let $\alpha = .05$.
a. The no-coffee group experienced a 35 mg/dl drop in mean cholesterol level. Use a t test to assess the statistical significance of this drop.

b. The usual-coffee group experienced a 26 mg/dl rise in mean cholesterol level. Use a *t* test to assess the statistical significance of this rise.

c. Use a *t* test (unpooled) to compare the no-coffee mean change (-35) to the usual-coffee mean change ($+26$).

9.25 Eight young women participated in a study to investigate the relationship between the menstrual cycle and food intake. Dietary information was obtained every day by interview; the study was double-blind in the sense that the participants did not know its purpose and the interviewer did not know the timing of their menstrual cycles. The table shows, for each participant, the average caloric intake for the 10 days preceding and the 10 days following the onset of the menstrual period (these data are for one cycle only). For these data, prepare a display like that of Figure 9.3.[18]

PARTICIPANT	FOOD INTAKE (Cal)	
	Premenstrual	Postmenstrual
1	2,378	1,706
2	1,393	958
3	1,519	1,194
4	2,414	1,682
5	2,008	1,652
6	2,092	1,260
7	1,710	1,239
8	1,967	1,758

SECTION 9.6

PERSPECTIVE

We have discussed several statistical methods of comparing two samples. In analyzing real data, it is wise to keep in mind that these statistical methods address only limited questions.

The paired *t* test is limited in two ways:

1. It is limited to questions concerning $\bar{d}$.
2. It is limited to questions about *aggregate* differences.

The second limitation is very broad; it applies not only to the methods of this chapter but also to those of Chapter 7 and to many other elementary statistical techniques. We will discuss these two limitations separately.

LIMITATION OF $\bar{d}$

One limitation of the paired *t* test and confidence interval is simple but too often overlooked: When some of the *d*'s are positive and some are negative, the magnitude of $\bar{d}$ does not reflect the "typical" magnitude of the *d*'s. The following example shows how misleading $\bar{d}$ can be.

EXAMPLE 9.18

MEASURING SERUM
CHOLESTEROL

Suppose a clinical chemist wants to compare two methods of measuring serum cholesterol; she is interested in how closely the two methods agree with each other. She takes blood specimens from 400 patients, splits each specimen in half, and assays one half by method A and the other by method B. Table 9.7 (page 320) shows fictitious data, exaggerated to clarify the issue.

TABLE 9.7
Serum Cholesterol
(mg/dl)

SPECIMEN	METHOD A	METHOD B	$d = A - B$
1	200	234	-34
2	284	272	$+12$
3	146	153	-7
4	263	250	$+13$
5	258	232	$+26$
$\vdots$	$\vdots$	$\vdots$	$\vdots$
400	176	190	-14
Mean	215.2	214.5	.7
SD	45.6	59.8	18.8

In Table 9.7, the sample mean difference is small ($\bar{d} = .7$). Furthermore, the data indicate that the population mean difference is small (a 95% confidence interval is -1.1 mg/dl $< \mu_d < 2.5$ mg/dl). But such discussion of $\bar{d}$ or μ_d does not address the central question, which is: How closely do the methods agree? In fact, Table 9.7 indicates that the two methods do not agree well; the individual differences between method A and method B are not small. The mean $\bar{d}$ is small because the positive and negative differences tend to cancel each other. To assess agreement between the methods, it would be far more relevant to analyze the *absolute* (unsigned) magnitudes of the d's (that is, 34, 12, 7, 13, 26, and so on). These magnitudes could be analyzed in various ways: we could average them, we could count how many are "large" (say, more than 10 mg/dl), and so on. ∎

LIMITATION OF THE AGGREGATE VIEWPOINT

Consider a paired experiment in which two treatments, say A and B, are applied to the same person. If we apply a t test or a sign test, we are viewing the people as an ensemble rather than individually. This is appropriate if we are willing to assume that the difference (if any) between A and B is in a consistent direction for all people—or, at least, that the important features of the difference are preserved even when the people are viewed *en masse*. The following example illustrates the issue.

EXAMPLE 9.19
TREATMENT OF ACNE

Consider a clinical study to compare two medicated lotions for treating acne. Twenty patients participate. Each patient uses lotion A on one side of his face and lotion B on the other side. After 3 weeks, each side of the face is scored for total improvement.

First, suppose that the A side improves more than the B side in ten patients, while in the other ten the B side improves more. According to a sign test, this result is in perfect agreement with the null hypothesis. And yet, two very different interpretations are logically possible:

Interpretation 1: Treatments A and B are in fact completely equivalent; their action is indistinguishable. The observed differences between A and B sides of the face were entirely due to chance variation.

Interpretation 2: Treatments A and B are in fact completely different. For some people (about 50% of the population), treatment A is more effective than treatment B, whereas in the remaining half of the population treatment B is more effective. The observed differences between A and B sides of the face were biologically meaningful.*

The same ambiguity of interpretation arises if the results favor one treatment over another. For instance, suppose the A side improved more than the B side in 18 of the 20 cases, while B was favored in 2 patients. This result, which is statistically significant ($P < .001$), could again be interpreted in two ways. It could mean that treatment A is in fact superior to B for everybody, but chance variation obscured its superiority in two of the patients; or it could mean that A is superior to B for most people, but for about 10% of the population ($\frac{2}{20} = .10$) B is superior to A. ∎

The difficulty illustrated by Example 9.19 is not confined to randomized blocks experiments. In fact, it is particularly clear in another type of paired experiment—the measurement of change over time. Consider, for instance, the blood pressure data of Example 9.16. Our discussion of that study hinged on an aggregate measure of blood pressure—the mean. If some patients' pressures rose as a result of biofeedback and others fell, these details were ignored in the analysis based on Student's *t*; only the average change was analyzed.

Neither is the difficulty confined to human experiments. Suppose, for instance, that two fertilizers, A and B, are to be compared in an agronomic field experiment using a randomized blocks design, with the data to be analyzed by a paired *t* test. If treatment A is superior to B on acid soils, but B is better than A on alkaline soils, this fact would be obscured in an experiment that included soils of both types.

The issue raised by the preceding examples is a very general one. Simple statistical methods such as the sign test and the *t* test are designed to evaluate treatment effects *in the aggregate*—that is, *collectively*—for a population of people, or of mice, or of plots of ground. The segregation of differential treatment effects in subpopulations requires more delicate handling, both in design and analysis.

This confinement to the aggregate point of view applies to Chapter 7 (independent samples) even more forcefully than to the present chapter. For instance, if treatment A is given to one group of mice and treatment B to another, it is quite impossible to know how a mouse in group A would have responded *if* it had received treatment B; the only possible comparison is an aggregate one. In Section 7.12 we stated that the statistical comparison of independent samples depends on an "implicit assumption"; essentially, the assumption is that the phenomenon under study can be adequately perceived from an aggregate viewpoint.

*This may seem farfetched, but phenomena of this kind do occur; as an obvious example, consider the response of patients to blood transfusions of type A or type B blood.

In many, perhaps most, biological investigations the phenomena of interest are reasonably universal, so that this issue of submerging the individual in the aggregate does not cause a serious problem. Nevertheless, one should not lose sight of the fact that aggregation may obscure important individual detail.

EXERCISES 9.26–9.27

9.26 For each of 29 healthy dogs, a veterinarian measured the glucose concentration in the anterior chamber of the left eye and the right eye, with the results shown in the table.[19]

ANIMAL NUMBER	GLUCOSE (mg/dl) Right eye	GLUCOSE (mg/dl) Left eye	ANIMAL NUMBER	GLUCOSE (mg/dl) Right eye	GLUCOSE (mg/dl) Left eye
1	79	79	16	80	80
2	81	82	17	78	78
3	87	91	18	112	110
4	85	86	19	89	91
5	87	92	20	87	91
6	73	74	21	71	69
7	72	74	22	92	93
8	70	66	23	91	87
9	67	67	24	102	101
10	69	69	25	116	113
11	77	78	26	84	80
12	77	77	27	78	80
13	84	83	28	94	95
14	83	82	29	100	102
15	74	75			

Using the paired t method, a 95% confidence interval for the mean difference is -1.1 mg/dl $< \mu_d < .7$ mg/dl. Does this result suggest that, for the average dog in the population, the difference in glucose concentration between the two eyes is less than 1.1 mg/dl? Explain.

9.27 Tobramycin is a powerful antibiotic. To minimize its toxic side effects, the dose can be individualized for each patient. Thirty patients participated in a study of the accuracy of this individualized dosing. For each patient, the predicted peak concentration of Tobramycin in the blood serum was calculated, based on the patient's age, sex, weight, and other characteristics. Then Tobramycin was administered and the actual peak concentration (μg/ml) was measured. The results were reported as in the table.[20]

	PREDICTED	ACTUAL
Mean	4.52	4.40
SD	.90	.85
n	30	30

Does the reported summary give enough information for you to judge whether the individualized dosing is, on the whole, accurate in its prediction of peak concentration? If so, describe how you would make this judgment. If not, describe what additional information you would need and why.

SUPPLEMENTARY EXERCISES 9.28–9.37

9.28 For each of nine horses, a veterinary anatomist measured the density of nerve cells at specified sites in the intestine. The results for site I (midregion of jejunum) and site II (mesenteric region of jejunum) are given in the accompanying table.[21] Each density value is the average of counts of nerve cells in five equal sections of tissue.

ANIMAL	SITE I	SITE II	DIFFERENCE
1	50.6	38.0	12.6
2	39.2	18.6	20.6
3	35.2	23.2	12.0
4	17.0	19.0	−2.0
5	11.2	6.6	4.6
6	14.2	16.4	−2.2
7	24.2	14.4	9.8
8	37.4	37.6	−.2
9	35.2	24.4	10.8
Mean	29.36	22.02	7.33
SD	13.32	10.33	7.79

a. Construct a 95% confidence interval for the difference in mean density between the sites.
b. Construct a 95% confidence interval the wrong way, using the independent-samples method. How does this interval differ from the one obtained in part **a**?

9.29 Refer to Exercise 9.28. Compare the sites using a t test at $\alpha = .05$. Use a nondirectional alternative.

9.30 Refer to Exercise 9.28.
a. Compare the sites using a sign test at $\alpha = .05$. Use a nondirectional alternative.
b. Calculate the exact P-value for the analysis of part **a**.

9.31 Refer to Exercise 9.28. Construct a scatterplot of the data. Does the appearance of the scatterplot indicate that the pairing was effective? Explain.

9.32 As part of a study of the physiology of wheat maturation, an agronomist selected six wheat plants at random from a field plot. For each plant, she measured the moisture content in two batches of seeds: one batch from the "central" portion of the wheat head, and one batch from the "top" portion, with the results shown in the table at the top of page 324.[22] Construct a 95% confidence interval for the mean difference in moisture content of the two regions of the wheat head.

PLANT	PERCENT MOISTURE	
	Central	Top
1	62.7	59.7
2	63.6	61.6
3	60.9	58.2
4	63.0	60.5
5	62.7	60.6
6	63.7	60.8

9.33 In a study of the effect of caffeine on muscle metabolism, nine male volunteers underwent arm exercise tests on two separate occasions. On one occasion, the volunteer took a placebo capsule an hour before the test; on the other occasion he received a capsule containing pure caffeine. (The time order of the two occasions was randomly determined.) During each exercise test, the subject's respiratory exchange ratio (RER) was measured. The RER is the ratio of carbon dioxide produced to oxygen consumed, and is an indicator of whether energy is being obtained from carbohydrates or from fats. The results are presented in the accompanying table.[23] Use a t test to assess the effect of caffeine. Use a nondirectional alternative and let $\alpha = .05$.

SUBJECT	RER (%)	
	Placebo	Caffeine
1	105	96
2	119	99
3	92	89
4	97	95
5	96	88
6	101	95
7	94	88
8	95	93
9	98	88

9.34 For the data of Exercise 9.33, construct a display like that of Figure 9.3.

9.35 In an experiment to compare two diets for fattening beef steers, nine pairs of animals were chosen from the herd; members of each pair were matched as closely as possible with respect to hereditary factors. The members of each pair were randomly allocated, one to each diet. The table shows the weight gains (lb) of the animals over a 140-day test period.[24] Use a t test to compare the diets at $\alpha = .05$. Use a nondirectional alternative.

PAIR	DIET 1	DIET 2	DIFFERENCE
1	596	498	98
2	422	460	−38
3	524	468	56
4	454	458	−4
5	538	530	8
6	552	482	70
7	478	528	−50
8	564	598	−34
9	556	456	100
Mean	520.4	497.6	22.9
SD	57.1	47.3	59.3

9.36 (*Computer exercise*) For an investigation of the mechanism of wound healing, a biologist chose a paired design, using the left and right hindlimbs of the salamander *Notophthalmus viridescens*. After amputating each limb, she made a small wound in the skin and then kept the limb for 4 hours in either a solution containing benzamil or a control solution. She theorized that the benzamil would impair the healing. The accompanying table shows the amount of healing, expressed as the area (mm^2) covered with new skin after 4 hours.[25]

ANIMAL	CONTROL LIMB	BENZAMIL LIMB	ANIMAL	CONTROL LIMB	BENZAMIL LIMB
1	.55	.14	10	.42	.21
2	.15	.08	11	.49	.11
3	.00	.00	12	.08	.03
4	.13	.13	13	.32	.14
5	.26	.10	14	.18	.37
6	.07	.08	15	.35	.25
7	.20	.11	16	.03	.05
8	.16	.00	17	.24	.16
9	.03	.05			

  a. Assess the effect of benzamil using a t test at $\alpha = .05$. Let the alternative hypothesis be that the researcher's expectation is correct.
  b. Proceed as in part **a** but use a sign test.
  c. Construct a 95% confidence interval for the mean effect of benzamil.
  d. Construct a scatterplot of the data. Does the appearance of the scatterplot indicate that the pairing was effective? Explain.

9.37 (*Computer exercise*) In a study of hypnotic suggestion, 16 male volunteers were randomly allocated to an experimental group and a control group. Each subject participated in a two-phase experimental session. In the first phase, respiration was measured while the subject was awake and at rest. (These measurements were also described in Exercises 7.35 and 7.61.) In the second phase, the subject was told to imagine that he was performing muscular work, and respiration was measured again.

For subjects in the experimental group, hypnosis was induced between the first and second phases; thus, the suggestion to imagine muscular work was "hypnotic suggestion" for experimental subjects and "waking suggestion" for control subjects. The accompanying table shows the measurements of total ventilation (liters of air per minute per square meter of body area) for all 16 subjects.[26]

EXPERIMENTAL GROUP			CONTROL GROUP		
Subject	Rest	Work	Subject	Rest	Work
1	5.74	6.24	9	6.21	5.50
2	6.79	9.07	10	4.50	4.64
3	5.32	7.77	11	4.86	4.61
4	7.18	16.46	12	4.78	3.78
5	5.60	6.95	13	4.79	5.41
6	6.06	8.14	14	5.70	5.32
7	6.32	11.72	15	5.41	4.54
8	6.34	8.06	16	6.08	5.98

a. Use a t test to compare the mean resting values in the two groups. Use a nondirectional alternative and let $\alpha = .05$. [This is the same as Exercise 7.35(a).]

b. Use suitable paired and unpaired t tests to investigate (i) the response of the experimental group to suggestion; (ii) the response of the control group to suggestion; (iii) the difference between the responses of the experimental and control groups. Use directional alternatives (suggestion increases ventilation, and hypnotic suggestion increases it more than waking suggestion) and let $\alpha = .05$ for each test.

c. Repeat the investigations of part **b** using suitable nonparametric tests (sign and Mann–Whitney tests).

d. Use suitable histograms or stem-and-leaf displays to investigate the reasonableness of the normality assumptions underlying the t tests of part **b**. How does this investigation shed light on the discrepancies between the results of parts **b** and **c**?

C H A P T E R 10

CONTENTS

ANALYSIS OF CATEGORICAL DATA. I. CONFIDENCE INTERVALS AND GOODNESS-OF-FIT TESTS

SECTION 10.1

INTRODUCTION

In Chapters 6, 7, and 9 we have described methods for analysis of data sets in which the observed variable is quantitative. In Chapters 10 and 11 we will consider methods of analysis when the observed variable is **categorical** rather than quantitative. Recall from Chapter 2 that, with a categorical variable, each observation is a category, rather than a number; each observational unit belongs to one and only one category. For instance, human blood type is a categorical variable; each person's type is either A, B, AB, or O.

In Chapter 10 we will consider analysis of a single sample of categorical data. We assume that the data can be regarded as a random sample from some population. The population distribution of a categorical variable can be described in terms of the population proportion, or probability, of each category. In Section 10.2 we discuss construction of a confidence interval for a population proportion, and in Section 10.3 we show how to test certain kinds of hypotheses about the proportions.

SECTION 10.2

CONFIDENCE INTERVAL FOR A POPULATION PROPORTION

Consider a random sample of n categorical observations, and let us fix attention on *one* of the categories. For instance, suppose a geneticist observes n guinea pigs whose coat color can be either black, sepia, cream, or albino; let us fix attention on the category "black." Let π denote the population proportion of the category, and let $\hat{\pi}$ denote the corresponding sample proportion. (This is the same notation used in Section 2.9, in Chapter 3, and in Section 5.2.). The notation is schematically represented in Figure 10.1.

FIGURE 10.1

Notation for Population and Sample Proportion

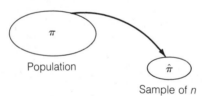

Under the random sampling model, a natural estimate of the population proportion (π) is the sample proportion ($\hat{\pi}$). How close to π is $\hat{\pi}$ likely to be? Recall from Chapter 5 that this question can be answered in terms of the sampling distribution of $\hat{\pi}$, which in turn is computed from the binomial distribution.

In Chapter 6 we showed how to use sample data on a quantitative variable to construct a confidence interval for the population mean, μ; the rationale for the method was based on the sampling distribution of $\bar{Y}$. In a similar way, sample data on the relative abundance of a category can be used to construct a confidence interval for the population proportion, π. The rationale for the method is based on the sampling distribution of $\hat{\pi}$.

A confidence interval for π can be constructed directly from the binomial distribution. However, for many practical situations a simple approximate method can be used instead. When the sample size (n) is large, the sampling distribution of $\hat{\pi}$ is approximately normal; this approximation is related to the Central Limit

Theorem. If you review Figure 5.4, you will see that the sampling distributions resemble normal curves, especially the distribution with $n = 80$. (The approximation is described in detail in optional Section 5.5.) We now describe an approximate confidence interval method based on this normal approximation.

CONFIDENCE INTERVAL FOR π: METHOD

Recall from Chapter 6 that a 95% confidence interval for a population mean μ is constructed as

$$\bar{y} \pm t_{.05}\, SE_{\bar{y}}$$

An approximate confidence interval for a population proportion π is constructed analogously. The first step is to calculate a standard error for $\hat{\pi}$, using the formula given in the box.

STANDARD ERROR OF A PROPORTION

$$SE_{\hat{\pi}} = \sqrt{\frac{\hat{\pi}(1 - \hat{\pi})}{n}}$$

Critical values for the confidence interval are obtained from the normal distribution; these can be found most easily from Table 4 with df $= \infty$. (Recall from Chapter 6 that the t distribution with df $= \infty$ is a normal distribution.)

For a 95% confidence interval, we find from Table 4 with df $= \infty$ that $t_{.05} = 1.960$. Thus, the approximate 95% confidence interval for a population proportion π is constructed as shown in the box.

APPROXIMATE CONFIDENCE INTERVAL FOR π

95% confidence interval: $\quad \hat{\pi} \pm (1.960)SE_{\hat{\pi}}$

Intervals with other confidence coefficients are constructed analogously; for instance, for a 90% confidence interval one would use 1.645 instead of 1.960.

The following example illustrates the confidence interval method.

EXAMPLE 10.1

LEFT-HANDEDNESS

In a survey of English and Scottish college students, 40 of 400 male students were found to be left-handed. Let us construct a 95% confidence interval for the proportion, π, of left-handed individuals in the population.[1]

The sample proportion is

$$\hat{\pi} = \frac{40}{400} = .100$$

and the SE is

$$SE_{\hat{\pi}} = \sqrt{\frac{(.100)(.900)}{400}} = .015$$

An approximate 95% confidence interval for π is

$$.100 \pm (1.960)(.015)$$

or

$$.07 < \pi < .13$$

With 95% confidence, we conclude that between 7% and 13% of the sampled population are left-handed.   ∎

STRUCTURE OF THE SE   Recall from Chapter 6 that the standard error of a sample mean is given by the formula

$$SE_{\bar{y}} = \frac{s}{\sqrt{n}}$$

To compare this with the standard error of a proportion, let us write $SE_{\hat{\pi}}$ as

$$SE_{\hat{\pi}} = \frac{\sqrt{\hat{\pi}(1 - \hat{\pi})}}{\sqrt{n}}$$

The numerator of this expression, namely,

$$\sqrt{\hat{\pi}(1 - \hat{\pi})}$$

is analogous to s in $SE_{\bar{y}}$. It is dissimilar to s in one respect: it depends on $\hat{\pi}$ itself, whereas the value of s is not constrained by the value of $\bar{y}$. And yet, in a more fundamental respect, this numerator is similar to s: it is unchanged if we increase the sample size while retaining the sample composition.* Thus, $SE_{\hat{\pi}}$ is inversely proportional to $\sqrt{n}$ in a way analogous to $SE_{\bar{y}}$. The following example illustrates this idea.

EXAMPLE 10.2
LEFT-HANDEDNESS

Suppose, as in Example 10.1, that a sample of n individuals contains 10% left-handers. Then $\hat{\pi} = .10$ and

$$SE_{\hat{\pi}} = \sqrt{\frac{(.10)(.90)}{n}} = \frac{.30}{\sqrt{n}}$$

If $n = 400$, then

$$SE_{\hat{\pi}} = .015$$

and if $n = 1,600$, then

$$SE_{\hat{\pi}} = .0075$$

Thus, a sample with the same composition (that is, 10% left-handers) but 4 times as large, would yield twice as much precision in the estimation of π.   ∎

*We noted in Section 2.8 that this is approximately true for s.

PLANNING A STUDY TO ESTIMATE π

In Section 6.4 we discussed a method for choosing the sample size n so that a proposed study would have sufficient precision for its intended purpose. The approach depended on two elements: (1) a specification of the desired $SE_{\bar{y}}$; and (2) a preliminary guess of the SD. In the present context, when the observed variable is categorical, a similar approach can be used. If a desired value of $SE_{\hat{\pi}}$ is specified, and if a rough informed guess of $\hat{\pi}$ is available, then the required sample size n can be determined from the following equation:

$$\text{Desired SE} = \sqrt{\frac{(\text{Guessed } \hat{\pi})(1 - \text{Guessed } \hat{\pi})}{n}}$$

The following example illustrates the use of the method.

EXAMPLE 10.3

LEFT-HANDEDNESS

Suppose we regard the left-handedness data of Example 10.1 as a pilot study, and we now wish to plan a new study large enough to estimate π with a standard error of 1 percentage point, that is, .01. Our guessed value of $\hat{\pi}$ from the pilot study is .10, so the required n must satisfy the following relation:

$$.01 = \sqrt{\frac{(.10)(.90)}{n}}$$

This equation is easily solved to give $n = 900$. We should plan to examine 900 men.   ■

PLANNING IN IGNORANCE   Suppose no preliminary informed guess of $\hat{\pi}$ is available. Remarkably, in this situation it is still possible to plan an experiment to achieve a desired value of $SE_{\hat{\pi}}$.* Such a "blind" plan depends on the fact that the crucial quantity $\sqrt{\hat{\pi}(1 - \hat{\pi})}$ is *largest* when $\hat{\pi} = .5$; you can see this in the graph of Figure 10.2. It follows that a value of n calculated using "guessed $\hat{\pi}$" = .5 will be *conservative*—that is, it will certainly be large enough. (Of course, it will be much larger than necessary if $\hat{\pi}$ is really very different from .5.) The following example shows how such "worst-case" planning is used.

FIGURE 10.2

How $\sqrt{\hat{\pi}(1 - \hat{\pi})}$ Depends on $\hat{\pi}$

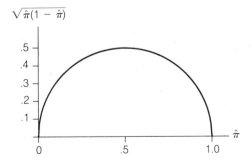

*By contrast, it would *not* be possible if we were planning a study to estimate a population mean μ, and we had no information whatsoever about the value of the SD.

EXAMPLE 10.4
LEFT-HANDEDNESS

Suppose, as in Example 10.3, that we are planning a study of left-handedness and we want $SE_{\hat{\pi}}$ to be .01, but suppose that we have no preliminary information whatsoever. We can proceed as in Example 10.3, but using "guessed $\hat{\pi}$" = .5; the calculation yields $n = 2,500$. Thus, a sample of 2,500 people would be adequate to estimate π with a standard error of .01, regardless of the actual value of $\hat{\pi}$. (Of course, if $\hat{\pi} = .1$, this value of n is much larger than necessary.)   ∎

EXERCISES 10.1–10.13

10.1 A series of patients with bacterial wound infections were treated with the antibiotic Cefotaxime. Bacteriologic response (disappearance of the bacteria from the wound) was considered "satisfactory" in 84% of the patients.[2] Determine the standard error of the observed proportion of "satisfactory" responses if the series contained
a. 50 patients   **b.** 200 patients

10.2 In an experiment with a certain mutation in the fruitfly *Drosophila*, n individuals were examined; of these, 20% were found to be mutants. Determine the standard error of the proportion of mutants in the population if
a. $n = 100$ **b.** $n = 400$

10.3 Refer to Exercise 10.2. In each case ($n = 100$ and $n = 400$) construct a 95% confidence interval for the population proportion of mutants.

10.4 In a natural population of mice (*Mus musculus*) near Ann Arbor, Michigan, the coats of some individuals are white-spotted on the belly. In a sample of 580 mice from the population, 28 individuals were found to have white-spotted bellies.[3] Construct a 95% confidence interval for the population proportion of this trait.

10.5 To evaluate the policy of routine vaccination of infants for whooping cough, adverse reactions were monitored in 339 infants who received their first injection of vaccine. Reactions were noted in 69 of the infants.[4] Construct a 95% confidence interval for the probability of a reaction to the vaccine.

10.6 In a study of human blood types in nonhuman primates, a sample of 71 orangutans were tested and 14 were found to be blood type B.[5] Construct a 95% confidence interval for the relative frequency of blood type B in the orangutan population.

10.7 In populations of the snail *Cepaea*, the shells of some individuals have dark bands, while other individuals have unbanded shells.[6] Suppose that a biologist is planning a study to estimate the percentage of banded individuals in a certain natural population, and that he wants to estimate the percentage—which he anticipates will be in the neighborhood of 60%—with a standard error not to exceed 4 percentage points. How many snails should he plan to collect?

10.8 (*Continuation of Exercise* 10.7)   What would the answer be if the anticipated percentage of banded snails were 50% rather than 60%?

10.9 The ability to taste the compound phenylthiocarbamide (PTC) is a genetically controlled trait in humans. In Europe and Asia, about 70% of people are tasters.[7] Suppose a study is being planned to estimate the relative frequency of tasters in a certain Asian population, and it is desired that the standard error of the estimated relative frequency should be .01. How many people should be included in the study?

10.10 Refer to Exercise 10.9. Calculate a conservative value for the required sample size, assuming that the preliminary guess (70%) is not available.

10.11 Refer to Exercise 10.9. Suppose the SE requirement is relaxed by a factor of 2—from .01 to .02. Would this reduce the required sample size by a factor of 2? Explain.

10.12 The "Luso" variety of wheat is resistant to the Hessian fly. In order to understand the genetic mechanism controlling this resistance, an agronomist plans to examine the progeny of a certain cross involving Luso and a nonresistant variety. Each progeny plant will be classified as resistant or susceptible and the agronomist will estimate the proportion of progeny that are resistant.[8] How many progeny does he need to classify in order to *guarantee* that the standard error of his estimate of this proportion will not exceed .05?

10.13 (*Continuation of Exercise 10.12*)   Suppose the agronomist is considering two possible genetic mechanisms for the inheritance of resistance; the population ratio of resistant to susceptible progeny would be 1 : 1 under one mechanism and 3 : 1 under the other. If the agronomist uses the sample size determined in Exercise 10.12, can he be sure that a 95% confidence interval will exclude at least one of the mechanisms? Explain.

| | | | | | | | | | | | |

SECTION 10.3

THE CHI-SQUARE GOODNESS-OF-FIT TEST

In analyzing a set of categorical observations, it is sometimes appropriate to test a null hypothesis H_0 that specifies the population proportions, or probabilities, of the various categories. Here is an example.

EXAMPLE 10.5
SNAPDRAGON COLORS

In the snapdragon (*Antirrhinum majus*), individual plants can be red-flowered, pink-flowered, or white-flowered. According to a certain Mendelian genetic model, self-pollination of pink-flowered plants should produce progeny that are red, pink, and white in the ratio 1 : 2 : 1. This Mendelian prediction can be formulated as the following null hypothesis:

$$H_0: \quad \Pr\{Red\} = .25, \quad \Pr\{Pink\} = .50, \quad \Pr\{White\} = .25$$

This hypothesis asserts that each progeny plant has a 25% chance of being red, a 50% chance of being pink, and a 25% chance of being white. Equivalently, H_0 asserts that, in the conceptual population of all potential progeny, 25% of the individuals are red, 50% are pink, and 25% are white.   ■

THE CHI-SQUARE STATISTIC

Given a random sample of n categorical observations, how can one judge whether they agree with a null hypothesis H_0 which specifies the probabilities of the categories? There are two complementary approaches to this question. First, as a way of describing the data, one can calculate the estimated probability—that is, the observed relative frequency—of each category. The following notation for relative frequencies is sometimes useful: when a probability $Pr\{E\}$ is estimated from observed data, the estimate is denoted by a hat ("⌢"), thus:

$$\widehat{Pr}\{E\}$$

The second approach is to use a statistical test, called a **goodness-of-fit test**, to assess the compatibility of the data with H_0. The most widely used goodness-of-fit test is the **chi-square test** or χ^2 test (χ is the Greek letter "chi").

We will describe the calculation of the chi-square test statistic in terms of the absolute, rather than the relative, frequencies of the categories. For each category, let O represent the **observed frequency** of the category and let E represent the **expected frequency**—that is, the frequency that would be expected according to H_0. The E's are calculated by multiplying each probability specified in H_0 by n. The test statistic for the chi-square goodness-of-fit test is then calculated from the O's and the E's using the formula given in the accompanying box. Example 10.6 illustrates the calculation of estimated probabilities and of the chi-square statistic.

THE CHI-SQUARE STATISTIC

$$\chi_s^2 = \sum \frac{(O - E)^2}{E}$$

where the summation is over all the categories.

EXAMPLE 10.6

SNAPDRAGON COLORS

Suppose a geneticist, investigating the Mendelian prediction of Example 10.5, self-pollinates pink-flowered snapdragon plants and produces 209 progeny with the following colors:

    Red: 43 plants

    Pink: 119 plants

    White: 47 plants

The estimated category probabilities are

$$\widehat{Pr}\{Red\} = \frac{43}{209} = .21$$

$$\widehat{Pr}\{Pink\} = \frac{119}{209} = .57$$

$$\widehat{Pr}\{White\} = \frac{47}{209} = .22$$

These estimated probabilities agree fairly well, but not exactly, with those specified by H_0.

We proceed to the calculation of the chi-square statistic. The observed frequencies are

Red: $O = 43$
Pink: $O = 119$
White: $O = 47$

The expected frequencies are calculated from H_0:

Red: $E = (.25)(209) = 52.25$
Pink: $E = (.50)(209) = 104.50$
White: $E = (.25)(209) = 52.25$

For calculating χ_s^2 it is convenient to arrange the observed and expected frequencies together, with the expected frequencies in parentheses, as follows:

RED	PINK	WHITE	TOTAL
43 (52.25)	119 (104.50)	47 (52.25)	209

Note that the sum of the expected frequencies is the same as the sum of the observed frequencies (209). The χ^2 statistic is

$$\chi_s^2 = \frac{(43 - 52.25)^2}{52.25} + \frac{(119 - 104.50)^2}{104.50} + \frac{(47 - 52.25)^2}{52.25}$$

$$= 4.18 \qquad\qquad\qquad \blacksquare$$

COMPUTATIONAL NOTES   The following tips are helpful in calculating a chi-square statistic.

1. The table of observed frequencies must include *all* categories, so that the sum of the O's is equal to the total number of observations.
2. The O's must be absolute, rather than relative, frequencies.
3. It is convenient to add each term $(O - E)^2/E$ to memory after calculating it. If you prefer to write down and reenter the expected frequencies, you may round them to two decimal places.

THE χ^2 DISTRIBUTION

From the way in which χ_s^2 is defined, it is clear that small values of χ_s^2 would indicate that the data agree with H_0, while large values of χ_s^2 would indicate disagreement. In order to base a statistical test on this agreement or disagreement, we need to know how much χ_s^2 may be affected by sampling variation.

We consider the null distribution of χ_s^2—that is, the sampling distribution which χ_s^2 follows if H_0 is true. It can be shown (using the methods of mathematical statistics) that, if the sample size is large enough, then the null distribution of χ_s^2 can be approximated by a distribution known as a χ^2 **distribution**. The form of a χ^2 distribution depends on a parameter called "degrees of freedom" (df). Figure 10.3 (page 336) shows the χ^2 distribution with df = 5.

FIGURE 10.3
The χ^2 Distribution
with df = 5

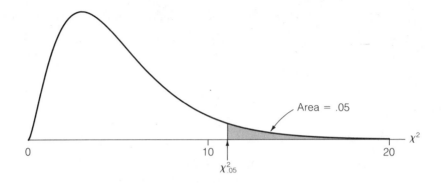

Table 8 (at the end of this book) gives critical values for the χ^2 distribution. For instance, for df = 5, the 5% critical value is $\chi^2_{.05} = 11.07$. This critical value corresponds to an area of .05 in the upper tail of the χ^2 distribution, as shown in Figure 10.3.

THE GOODNESS-OF-FIT TEST

For a chi-square goodness-of-fit test, the null distribution of χ^2_s is approximately a χ^2 distribution with

    df = (Number of categories) − 1

The test of H_0 is carried out using critical values from Table 8, as illustrated in the following example.

EXAMPLE 10.7

SNAPDRAGON COLORS

For the snapdragon data of Example 10.6, the observed chi-square statistic was $\chi^2_s = 4.18$. Because there are three color categories, the df for the null distribution are calculated as

    df = 3 − 1 = 2

From Table 8 with df = 2 we find that $\chi^2_{.20} = 3.22$ and $\chi^2_{.10} = 4.61$. The P-value is bracketed as $(.10 < P < .20)$. We would not reject H_0 even at $\alpha = .10$. Thus, we conclude that the data are reasonably consistent with the Mendelian model.

■

Note that the critical values for the chi-square test do not depend on the sample size, n. However, the test procedure *is* affected by n, through the value of the chi-square statistic. If we change the size of a sample while keeping its percentage composition fixed, then χ^2_s varies directly as the sample size, n. For instance, imagine appending to a sample a replicate of itself. Then the expanded sample would have twice as many observations, the value of each O would be doubled, the value of each E would be doubled, and so the value of χ^2_s would be doubled [because in each term of χ^2_s the numerator $(O - E)^2$ would be multiplied by 4, and the denominator E would be multiplied by 2].

COMPOUND HYPOTHESES AND DIRECTIONALITY

Let us examine the goodness-of-fit null hypothesis more closely. In a two-sample comparison such as a t test, the null hypothesis contains exactly one assertion—for instance, that two population means are equal. By contrast, a goodness-of-fit null hypothesis can contain more than one assertion. Such a null hypothesis may be called a **compound null hypothesis**. The following is an example.

EXAMPLE 10.8

SNAPDRAGON COLORS

The Mendelian null hypothesis of Example 10.5 is

$$H_0: \quad \text{Pr\{Red\}} = .25, \quad \text{Pr\{Pink\}} = .50, \quad \text{Pr\{White\}} = .25$$

This is a compound hypothesis because it makes two independent assertions, namely

$$\text{Pr\{Red\}} = .25 \quad \text{and} \quad \text{Pr\{Pink\}} = .50.$$

Note that the third assertion ($\text{Pr\{White\}} = .25$) is not an independent assertion because it follows from the other two.   ∎

When the null hypothesis is compound, the chi-square test has two special features. First, the alternative hypothesis is necessarily nondirectional. Second, if H_0 is rejected the test does not yield a directional conclusion. The following example illustrates these points.

EXAMPLE 10.9

SNAPDRAGON COLORS

In Example 10.5, the alternative hypothesis (which we did not state explicitly) was as follows:

$$H_A: \quad \text{At least one of the probabilities specified in } H_0 \text{ is incorrect}$$

or, in other words,

$$H_A: \quad \text{Pr\{Red\}} \neq .25, \quad \text{and/or} \quad \text{Pr\{Pink\}} \neq .50, \quad \text{and/or} \quad \text{Pr\{White\}} \neq .25$$

This alternative hypothesis is nondirectional. (Perhaps "omnidirectional" would be a better term.)

Suppose a geneticist were to obtain the following data on 209 plants:

Red: 52 plants
Pink: 128 plants
White: 29 plants

For these data, $\chi_s^2 = 15.63$; from Table 8 we find that $.0001 < P < .001$. Thus, H_0 would be rejected, even at $\alpha = .001$. The conclusion from this test would be that the Mendelian prediction does not hold in this situation. However, the test would *not* yield a directional conclusion such as $\text{Pr\{Red\}} < .25$, $\text{Pr\{Pink\}} > .50$, $\text{Pr\{White\}} < .25$. Indeed, for this particular data set the observed relative frequency of red plants is $\frac{52}{209} = .2488$, which (you can see intuitively) provides little or no evidence that $\text{Pr\{Red\}} < .25$.   ∎

When H_0 is compound, the chi-square test is nondirectional in nature because the chi-square statistic measures deviations from H_0 in all directions. Statistical methods are available which do yield directional conclusions and which can handle directional alternatives, but such methods are beyond the scope of this book.

DICHOTOMOUS VARIABLES

If the categorical variable analyzed by a goodness-of-fit test is dichotomous, then the null hypothesis is not compound, and directional alternatives and directional conclusions do not pose any particular difficulty.

DIRECTIONAL CONCLUSION The following example illustrates the directional conclusion.

EXAMPLE 10.10
SEXES OF BIRDS

In an ecological study of the Carolina Junco, 53 birds were captured from a certain population; of these, 40 were male.[9] Is this evidence that males outnumber females in the population? An appropriate null hypothesis is

H_0: Population is 50% male and 50% female

Equivalently, H_0 can be restated in terms of the probability that a randomly chosen bird will be male or female:

H_0: Pr{Male} = .50, Pr{Female} = .50

This hypothesis is not compound because it contains only one independent assertion. (Note that the second assertion is redundant; it follows from the first.)
   Let us test H_0 against the nondirectional alternative

H_A: Pr{Male} ≠ .50

The data yield $\chi_s^2 = 13.8$, and from Table 8 we find that $.0001 < P < .001$. Even at $\alpha = .001$ we would reject H_0 and find that there is sufficient evidence to conclude that the population contains more males than females. ∎

   To recapitulate, the directional conclusion in Example 10.10 is legitimate because we know that if H_0 is false then necessarily either Pr{Male} < .5 or Pr{Male} > .5. By contrast, in Example 10.9 H_0 may be false but Pr{Red} may still be equal to .25; the chi-square analysis does not determine *which* of the probabilities are not as specified by H_0.

DIRECTIONAL ALTERNATIVE A chi-square goodness-of-fit test against a directional alternative (when the observed variable is dichotomous) uses the familiar two-step procedure: First check for directionality and then cut the P-value in half if the data deviate from H_0 in the direction specified by H_A. The following example illustrates the procedure.

EXAMPLE 10.11
COLOR VISION

A study of color vision in squirrels used an apparatus containing three small translucent panels that could be separately illuminated. The animals were trained to choose, by pressing a lever, the panel that appeared different from the other two. (During these "training" trials, the panels differed in brightness, rather than color.) Then the animals were tested for their ability to discriminate between various colors.

In one series of "testing" trials on a single animal, one of the panels was red and the other two were white; the location of the red panel was varied randomly from trial to trial. In 75 trials, the animal chose correctly 45 times and incorrectly 30 times.[10] How strongly does this support the interpretation that the animal can discriminate between the two colors?

We may formulate null and alternative hypotheses as follows:

H_0: The animal cannot discriminate red from white

H_A: The animal can discriminate red from white

These hypotheses can be translated as

H_0: $\Pr\{\text{correct}\} = \frac{1}{3}$

H_A: $\Pr\{\text{correct}\} > \frac{1}{3}$

Notice that in this formulation even an animal that can discriminate is not expected to choose correctly all (or even most) of the time, but it is expected to choose correctly more often than would be predicted by chance alone.

From the data on the 75 trials, we first note that the data do deviate from H_0 in the direction specified by H_A, because the observed relative frequency of correct answers is $\frac{45}{75}$, which is greater than $\frac{1}{3}$. The value of the chi-square statistic is $\chi_s^2 = 24.0$; from Table 8 we bracket the P-value as $P < .00005$, and conclude that the evidence is very strong that the animal can discriminate red from white. Note that this conclusion applies only to the particular animal; no inference is made to other individuals of the species. ∎

**EXERCISES
10.14–10.20**

10.14 A cross between white and yellow summer squash gave progeny of the following colors:[11]

Color	White	Yellow	Green
Number of Progeny	155	40	10

Are these data consistent with the $12:3:1$ ratio predicted by a certain genetic model? Use a chi-square test at $\alpha = .10$.

10.15 Refer to Exercise 10.14. Suppose the sample had the same composition but was 10 times as large: 1,550 white, 400 yellow, and 100 green progeny. Would the data be consistent with the $12:3:1$ model?

10.16 How do bees recognize flowers? As part of a study of this question, researchers used the following two artificial "flowers":[12]

Flower 1   Flower 2

The experiment was conducted as a series of trials on individual bees; each trial consisted of presenting a bee with both flowers and observing which flower it landed on first. (Flower 1 was sometimes on the left, and sometimes on the right.) During the "training" trials, flower 1 contained a sucrose solution and flower 2 did not; thus, the bee was trained to prefer flower 1. During the "testing" trials, neither flower contained sucrose. In 25 testing trials with a particular bee, the bee chose flower 1 twenty times and flower 2 five times. Use a goodness-of-fit test to assess the evidence that the bee could remember and distinguish the flower patterns. Use a directional alternative and let $\alpha = .05$.

10.17 At a midwestern hospital there were a total of 932 births in 20 consecutive weeks. Of these births, 216 occurred on weekends.[13] Do these data reveal more than chance deviation from random timing of the births? (Test for goodness of fit, with two categories of births—weekday and weekend. Use a nondirectional alternative and let $\alpha = .05$.)

10.18 In a breeding experiment, white chickens with small combs were mated and produced 190 offspring, of the types shown in the accompanying table.[14] Are these data consistent with the Mendelian expected ratios of $9:3:3:1$ for the four types? Use a chi-square test at $\alpha = .10$.

TYPE	NUMBER OF OFFSPRING
White feathers, small comb	111
White feathers, large comb	37
Dark feathers, small comb	34
Dark feathers, large comb	8
Total	190

10.19 Among n babies born in a certain hospital, 52% were boys.[15] Suppose we want to test the hypothesis that the true probability of a boy is $\frac{1}{2}$. Calculate the value of χ_s^2, and bracket the P-value for testing against a nondirectional alternative, if
a. $n = 1,000$ **b.** $n = 3,000$ **c.** $n = 6,000$

10.20 An experimental design using litter-matching was employed to test a certain drug for cancer-causing potential. From each of 50 litters of rats, three females were selected; one of these three, chosen at random, received the test drug, and the other two were kept as controls. During a 2-year observation period, the time of occurrence of a tumor, and/or death from various causes, was recorded for each animal. One way to analyze the data is to note simply which rat (in each triplet) developed a tumor first. Some triplets were uninformative on this point because either (a) none of the three littermates developed a tumor, or (b) a rat developed a tumor after its littermate had died from some other cause. The results for the 50 triplets are shown in the table.[16] Use a goodness-of-fit test to evaluate the evidence that the drug causes cancer. Use a directional alternative and let $\alpha = .01$. [*Hint:* Use only the 20 triplets that provide complete information.]

	NUMBER OF TRIPLETS
Tumor first in the treated rat	12
Tumor first in one of the two control rats	8
No tumor	23
Death from another cause	7
Total	50

SECTION 10.4

APPLICABILITY OF METHODS

We have described two techniques for analyzing categorical data: the confidence interval in Section 10.2 and the chi-square goodness-of-fit test in Section 10.3. We now give guidelines for deciding when to apply these techniques.

CONDITIONS FOR VALIDITY

The confidence interval for π and the chi-square goodness-of-fit test are valid under the following conditions:

1. *Design conditions*   It must be reasonable to regard the data as a random sample of categorical observations from a large population. The observations must be independent of each other.
2. *Sample size conditions*   The sample size must be large enough. Both the confidence interval and the test are approximate, and the approximation is best for large samples.
3. *Form of* H_0 (for the test only)   The null hypothesis must specify numerical values for the category probabilities.*

VERIFICATION OF DESIGN CONDITIONS

To verify the design conditions, we need to identify a population from which the data may be viewed as a random sample. Sometimes the "population" is rather abstract. For instance, in applications to genetics such as Example 10.5, we may

---

*A slightly modified form of the chi-square test can be used to test a hypothesis which merely constrains the probabilities rather than specifying them exactly. An example would be testing the fit of data to a binomial distribution (see optional Section 3.5).

think of the progeny of a given mating or cross as a random sample from the conceptual population of all *potential* progeny of that mating or cross. Another instance is Example 10.11, which involves a sequence of independent trials with unknown probability of success; the "population" is the conceptual population of potential responses for the particular animal. (Note that the same approach could not be used to analyze combined data from more than one individual; such data would be hierarchical in structure and the observations would not be independent.)

To verify that the data may be viewed as a random sample, the following must be ruled out: (1) bias in the sampling procedure; (2) complex random sampling schemes such as stratified random sampling (recall from Chapter 2 that this book deals only with *simple* random sampling); (3) hierarchical structure or dependency of the observations.

The importance of checking for bias cannot be overemphasized. The following example shows that a large sample size is no cure for bias in sampling; even a very large sample can be grossly unrepresentative if the sampling procedure is biased.

EXAMPLE 10.12
OPINION POLLING

In 1976 Ann Landers, an advice columnist, conducted a survey of parents through her nationally syndicated column. She asked her readers to respond by postcard to the question: "If you had it to do over again, would you have children?" She received more than 10,000 responses, of which 70% were "No." In an attempt to replicate this unsettling result, two newspapers commissioned their own surveys, with the following results:[17]

> *Kansas City Star:*   94% "Yes"   ($n = 409$)
> *Newsday:*   91% "Yes"   ($n = 1,373$)

The *Star* and *Newsday* polls were based on samples chosen by a random method.* The Ann Landers poll, although much larger, produced a result that was seriously biased because the respondents were self-selected rather than randomly selected; people who want to air a complaint are far more likely to send in a postcard than contented people are.   ∎

The following example shows the relevance of checking for dependency in the observations.

EXAMPLE 10.13
FOOD CHOICE BY
INSECT LARVAE

In a behavioral study of the clover root curculio *Sitona hispidulus*, 20 larvae were released into each of six petri dishes. Each dish contained nodulated and non-nodulated alfalfa roots, arranged in a symmetric pattern. (This experiment was more fully described in Example 1.5.) After 24 hours the location of each larva was noted, with the results shown in Table 10.1.[18]

*The sampling methods used in opinion polls are, however, often more complex than the simple random sampling method treated in this book.

TABLE 10.1

Food Choice by
Sitona Larvae

DISH	NUMBER OF LARVAE		
	Nodulated Roots	Nonnodulated Roots	Other (died, lost, etc.)
1	5	3	12
2	9	1	10
3	6	3	11
4	7	1	12
5	5	1	14
6	14	3	3
Total	46	12	62

Suppose the following analysis is proposed. A total of 58 larvae made a choice; the observed frequencies of choosing nodulated and nonnodulated roots were 46 and 12, and the corresponding expected frequencies (assuming random choice) would be 29 and 29; these data yield $\chi_s^2 = 19.93$, from which (using a directional alternative) we find from Table 8 that $P < .00005$. The validity of this proposed analysis is highly doubtful because it depends on the assumption that all the observations in a given dish are independent of each other; this assumption would certainly be false if (as is biologically plausible) the larvae tend to follow each other in their search for food.

How, then, should the data be analyzed? One approach is to make the reasonable assumption that the observations in one dish are independent of those in another dish. Under this assumption one could use a paired analysis on the six dishes ($n_d = 6$); a paired t test yields $P \approx .005$ and a sign test yields $P \approx .02$. Note that the questionable assumption of independence within dishes led to a P-value that was much too small.   ∎

GUIDELINES FOR SAMPLE SIZE CONDITIONS

For the chi-square test   The critical values given in Table 8 are only approximately correct for determining the P-value associated with χ_s^2. As a rule of thumb, the approximation is considered adequate if each expected frequency (E) is at least equal to 5.

For the confidence interval   The confidence interval based on $\mathrm{SE}_{\hat{\pi}}$ is only approximate. Table 10.2 (page 344) shows how large n must be for the approximation to be reasonably good.[19] Note that the required n depends on $\hat{\pi}$. If $\hat{\pi}$ is close to 0 or 1, the requirement can be easily stated in terms of the number of observations in the *smaller* category—that is, the category containing less than half of the observations.

TABLE 10.2

Sample Size Required for Good Agreement Between Approximate and Exact Confidence Intervals for π

$\hat{\pi}$	n	NUMBER IN THE SMALLER CATEGORY
<.01	—	25
.01	2,400	24
.05	440	22
.10	180	18
.20	60	12
.30	20	6
.40	15	6
.50	10	5
.60	15	6
.70	20	6
.80	60	12
.90	180	18
.95	440	22
.99	2,400	24
>.99	—	25

The following examples illustrate the application of the sample size guidelines.

EXAMPLE 10.14

SNAPDRAGON COLORS

The snapdragon data of Example 10.6 satisfy the sample size guidelines for the chi-square test because the expected frequencies (52.25, 104.50, and 52.25) all exceed 5. ∎

EXAMPLE 10.15

LEFT-HANDEDNESS

The left-handedness data of Example 10.1 satisfy the requirement of Table 10.2 because $\hat{\pi} = .10$ and $n = 400$, which is larger than 180. Another way to check the requirement is to note that the sample contains 40 observations in the smaller category (left-handed), and 40 is larger than 18. ∎

Example 10.16 shows how neglecting the sample size requirements can lead to seriously wrong conclusions.

EXAMPLE 10.16

SIDE EFFECTS

Of 100 patients who were given an experimental drug, none developed adverse side effects. Thus, the sample proportion of side effects is

$$\hat{\pi} = \frac{0}{100} = 0$$

and the standard error formula gives

$$SE_{\hat{\pi}} = \sqrt{\frac{(0)(1)}{100}} = 0$$

Of course, it would be ridiculous to conclude from this that there is *no* sampling error in $\hat{\pi}$! The standard error is meaningless in this case because the requirement of Table 10.2 has been grossly violated; the sample contains 0 patients with side effects and 0 is less than 25. ∎

When the sample size is not large enough to use the approximate chi-square test or approximate confidence interval, exact methods can be used instead.* The exact methods are beyond the scope of this book.

CONFIDENCE INTERVAL AND TEST

The confidence interval and the goodness-of-fit test are two closely linked techniques for analyzing categorical data. A confidence interval for a category probability can be a useful adjunct to a goodness-of-fit test. The following is an example.

EXAMPLE 10.17
SEXES OF BIRDS

The ecologists of Example 10.10 observed that 40 of 53 birds were male. Letting π represent the proportion of male birds in the population, we can calculate $\hat{\pi} = \frac{40}{53} \approx .75$, and construct the following 95% confidence interval for π:

$$.64 < \pi < .87$$

The width of the confidence interval shows that the data provide only rather weak information about the value of π; nevertheless the confidence interval indicates that the data are incompatible with H_0 because the interval does not include the value $\pi = .5$. ∎

It is natural to wonder if, when the observed variable is dichotomous, the χ^2 goodness-of-fit test is equivalent to simply noting (as in Example 10.17) whether a confidence interval contains the hypothesized probability. In fact the two procedures are approximately, but not exactly, equivalent.

CONFIRMING A NULL HYPOTHESIS

In some applications of the goodness-of-fit test, notably in genetics, the experimenter wishes to *confirm* the null hypothesis. As with any other test of a hypothesis, nonrejection of H_0 with a chi-square test means that the data are compatible with H_0, but this does not in itself confirm that H_0 is true. The same data may also be compatible with *other* plausible hypotheses.

In order to establish positive support of a null hypothesis H_0, it is necessary to go beyond the chi-square test of H_0. One way to do this is to show, using confidence intervals, that each category probability is *close* (in some practical sense) to the probability specified by H_0. Another way, particularly applicable to genetic ratios, is to use additional chi-square tests to exclude other competing genetic models. The following example illustrates the latter approach.

*For the data of Example 10.16, it can be shown that the exact method yields the 95% confidence interval $0 \leq \pi \leq .036$.

EXAMPLE 10.18
SNAPDRAGON COLORS

According to the chi-square test in Example 10.7, the snapdragon data of Example 10.6 are quite compatible with the Mendelian model which predicts a $1:2:1$ ratio. Suppose, however, that the experimenter was also considering another Mendelian model which predicted a $3:9:4$ ratio. Testing the same data for goodness of fit to a $3:9:4$ ratio yields $\chi_s^2 = .92$, indicating a high degree of compatibility with that model also. To distinguish between the two models, the experimenter would need a larger sample of snapdragons.   ■

**EXERCISES
10.21–10.24**

10.21 Refer to the data indicated below. Is the sample size large enough for the chi-square test (using the critical values in Table 8) to be valid?
a. The squash color data of Exercise 10.14
b. The bee behavior data of Exercise 10.16
c. The chicken breeding data of Exercise 10.18

10.22 Refer to the data indicated below. Is the sample size large enough for the approximate confidence interval method to be valid?
a. The mouse coat color data of Exercise 10.4
b. The vaccination data of Exercise 10.5
c. The orangutan blood type data of Exercise 10.6

10.23 Are mice right-handed or left-handed? In a study of this quesion, 320 mice of a highly inbred strain were tested for paw preference by observing which forepaw— right or left—they used to retrieve food from a narrow tube. Each animal was tested 50 times, for a total of $320 \times 50 = 16{,}000$ observations. The results were as follows:[20]

	RIGHT	LEFT
Number of observations	7,871	8,129

Suppose we assign an expected frequency of 8,000 to each category and perform a goodness-of-fit test; we find that $\chi_s^2 = 4.16$, so that at $\alpha = .05$ we would reject the hypothesis of a $1:1$ ratio and find that there is sufficient evidence to conclude that mice of this strain are (slightly) biased toward use of the left paw. This analysis contains a fatal flaw; what is it?

10.24 Ragweed is a plant that can spread by means of underground stems. In a study of habitat selection by a particular ragweed clone, six individual stems were rooted in the centers of six long narrow pots, and then one end of each pot was kept saline by repeatedly watering it with salt water. After four months, 216 new shoots had emerged in the six pots; of these, 86 had emerged in the saline end of their pot and 130 had emerged in the nonsaline end. From these observations, a chi-square statistic can be calculated as follows:[21]

$$\chi_s^2 = \frac{(130 - 108)^2}{108} + \frac{(86 - 108)^2}{108} = 8.96$$

Even at $\alpha = .01$, we would reject H_0 and find that there is sufficient evidence to conclude that plants of this clone preferentially migrate to nonsaline soil. On what grounds can the above analysis be criticized? [*Hint:* Are the conditions for validity of the test satisfied?]

SUPPLEMENTARY
EXERCISES
10.25–10.35

10.25 A U.S. government publication reported the results of a national health survey in which 16,780 people had been questioned. The percentage of Americans who "considered themselves overweight" was estimated to be 46%, with a standard error of .42 percentage point.[22] If this were a simple random sample of size $n = 16,780$, what would the SE be? (Your answer will not agree exactly with the reported SE, because the survey design was actually more complex than the simple random sampling design considered in this book.)

10.26 When male mice are grouped, one of them usually becomes dominant over the others. In order to see how a parasitic infection might affect the competition for dominance, male mice were housed in groups, three mice to a cage; two mice in each cage received a mild dose of the parasitic worm *H. polygyrus*. Two weeks later, criteria such as the relative absence of tail wounds were used to identify the dominant mouse in each cage. It was found that the uninfected mouse had become dominant in 15 of 30 cages.[23] Is this evidence that the parasitic infection tends to inhibit the development of dominant behavior? Use a goodness-of-fit test against a directional alternative. Let $\alpha = .05$. [*Hint:* The observational unit in this experiment is not an individual mouse, but a cage of three mice.]

10.27 In a study of environmental effects upon reproduction, 123 adult female white-tailed deer from the central Adirondack area were captured and 97 were found to be pregnant.[24] Construct a 95% confidence interval for the proportion of females pregnant in this deer population.

10.28 Refer to Exercise 10.27.
  a. Is the sample size large enough for the approximate confidence interval method to be valid?
  b. Which of the conditions for validity of the confidence interval might have been violated in this study?

10.29 Refer to the cortex-weight data of Exercise 9.16.
  a. Use a goodness-of-fit test to test the hypothesis that the environmental manipulation has no effect. As in Exercise 9.16, use a directional alternative and let $\alpha = .05$. (This exercise shows how, by a shift of viewpoint, the sign test can be reinterpreted as a goodness-of-fit test. Of course, the chi-square goodness-of-fit test described in this chapter can be used only if the number of observations is large enough.)
  b. Is the number of observations large enough for the test in part **a** to be valid?

10.30 A researcher is planning a study of cancer induction in which mice will be injected with a radioactive substance and then observed for 20 months for the appearance of bone tumors. Past experience suggests that about 15% of the mice will develop tumors. If the researcher wants to estimate the percentage with a standard error of 3 percentage points, how many mice should she include in the experiment?[25]

10.31 People who harvest wild mushrooms sometimes accidentally eat the toxic "death cap" mushroom, *Amanita phalloides*. In reviewing 205 European cases of death-cap poisoning from 1971 through 1980, researchers found that 45 of the victims had died.[26] Use a goodness-of-fit test to compare this mortality to the 30% mortality that was recorded before 1970. Let the alternative hypothesis be that mortality has decreased with time, and let $\alpha = .05$.

10.32 It has been proposed that elevated blood levels of alpha$_1$-fetoprotein (AFP) are an indicator of liver cancer. In one study, 90 of 107 liver cancer patients were found to have elevated AFP.[27]
  a. Construct a 95% confidence interval for the prevalence of elevated AFP among liver cancer patients.
  b. Does the result of part **a** imply (with high confidence) that a majority of patients with liver cancer have elevated AFP? Explain.
  c. Does the result of part **a** imply (with high confidence) that a majority of patients with elevated AFP have liver cancer? Explain.

10.33 At a certain university there are 20,000 students. Suppose you want to estimate, based on examining a random sample, the prevalence of nearsightedness in the student body. The prevalence in the general population is about 45%.[28] Using this as a preliminary guess, determine how many students you will need to examine if you want the standard error of your estimate not to exceed 2 percentage points.

10.34 The appearance of leaf pigment glands in the seedling stage of cotton plants is genetically controlled. According to one theory of the control mechanism, the population ratio of glandular to glandless plants resulting from a certain cross should be $11:5$; according to another theory it should be $13:3$. In one experiment, the cross produced 89 glandular and 36 glandless plants.[29] Use goodness-of-fit tests (at $\alpha = .10$) to determine whether these data are consistent with
  a. the $11:5$ theory   **b.** the $13:3$ theory

10.35 When fleeing a predator, the minnow *Fundulus notti* will often head for shore and jump onto the bank. In a study of spatial orientation in this fish, individuals were caught at various locations and later tested in an artificial pool to see which direction they would choose when released: would they swim in a direction which, at their place of capture, would have led toward shore? The following are the directional choices ($\pm45°$) of 50 fish tested under cloudy skies:[30]

Toward shore	18
Away from shore	12
Along shore to the right	13
Along shore to the left	7

Use chi-square tests at $\alpha = .05$ to test the hypothesis that directional choice under cloudy skies is random,
  a. using the four categories listed above.
  b. collapsing to two categories—"toward shore" and "away from or along shore"— and using a directional H_A.

[*Note:* Although the chi-square test is valid in this setting, it should be noted that more powerful tests are available for analysis of orientation data.[31]]

C H A P T E R 11

CONTENTS

ANALYSIS OF CATEGORICAL DATA. II. CONTINGENCY TABLES

| | | | | | | | | | | |

SECTION 11.1

INTRODUCTION

In Chapter 10 we considered the analysis of a single sample of categorical data. The basic techniques were estimation of category probabilities and comparison of category frequencies with frequencies "expected" according to a null hypothesis. In Chapter 11 these basic techniques will be extended to more complicated situations. To set the stage, here are two examples.

EXAMPLE 11.1
TREATMENT OF ANGINA

Angina pectoris is a chronic heart condition in which the sufferer has periodic attacks of chest pain. In a study to evaluate the effectiveness of the drug Timolol in preventing angina attacks, patients were randomly allocated to receive a daily dosage of either Timolol or placebo for 28 weeks. The numbers of patients who became completely free of angina attacks are shown in Table 11.1.[1]

TABLE 11.1
Response to Angina
Treatment

	TREATMENT	
	Timolol	Placebo
Angina-free	44	19
Not Angina-free	116	128
TOTAL	160	147

A natural way to express the results is in terms of percentages, as follows:

Of Timolol patients, $\frac{44}{160}$ or 28% were angina-free.

Of placebo patients, $\frac{19}{147}$ or 13% were angina-free.

In this study, the angina-free response was more common among the Timolol-treated patients than among the placebo-treated patients—28% vs 13%. ■

EXAMPLE 11.2
COLOR-BLINDNESS AND GENDER

In an early study of red–green color-blindness, researchers examined a large number of Norwegian schoolchildren, with the results shown in Table 11.2.[2]

TABLE 11.2
Color-Blindness
and Gender

	Males	Females
Color-blind	725	40
Not Color-blind	8,324	9,032
TOTAL	9,049	9,072

To compare the estimated probabilities of color-blindness in the two sexes, we note the following:

Among males, $\frac{725}{9,049}$ or 8.0% were color-blind.

Among females, $\frac{40}{9,072}$ or .44% were color-blind.

The comparison shows that, among these 18,121 schoolchildren, color-blindness was much more common—nearly 20 times as common—among males than among females.   ∎

Tables such as Tables 11.1 and 11.2 are called **contingency tables**. The focus of interest in a contingency table is the dependence or association between the row variable and the column variable—for instance, between treatment and response in Table 11.1, and between color-blindness and gender in Table 11.2. (The word *contingent* means *dependent*.) In particular, Tables 11.1 and 11.2 are called **2 × 2** ("two-by-two") **contingency tables**, because they consist of two rows (excluding the "total" row) and two columns. Each category in the contingency table is called a **cell**; thus, a 2 × 2 contingency table has four cells.

Sections 11.2 and 11.3 will be concerned with the analysis and interpretation of 2 × 2 contingency tables. In Section 11.4 the discussion will be extended to contingency tables larger than 2 × 2.

| | | | | | | | | | | | | |

S E C T I O N 11.2

THE CHI-SQUARE TEST FOR THE 2 × 2 CONTINGENCY TABLE

In Chapter 10 we considered the estimate $\hat{\pi}$ of a population proportion, or probability, π. In analyzing a 2 × 2 contingency table, it is natural to think in terms of two probabilities, π_1 and π_2, which are to be compared. One form of comparison is through a statistical test of the null hypothesis that the probabilities are equal:

H_0: $\pi_1 = \pi_2$

The following example illustrates this null hypothesis.

EXAMPLE 11.3

TREATMENT OF ANGINA

For the angina study of Example 11.1, the probabilities of interest are

π_1 = Probability that a patient will become angina-free if given Timolol

π_2 = Probability that a patient will become angina-free if given a placebo

The estimated probabilities from the data are:

$$\hat{\pi}_1 = \frac{44}{160} = .28$$

$$\hat{\pi}_2 = \frac{19}{147} = .13$$

The null hypothesis asserts that the corresponding true (population) probabilities are equal.   ∎

THE CHI-SQUARE STATISTIC

Clearly, a natural way to test the above null hypothesis would be to reject H_0 if $\hat{\pi}_1$ and $\hat{\pi}_2$ are different by a sufficient amount. We describe a test procedure that compares $\hat{\pi}_1$ and $\hat{\pi}_2$ indirectly, rather than directly. The procedure is a chi-square test, based on a test statistic calculated as follows:

THE CHI-SQUARE STATISTIC

$$\chi_s^2 = \sum \frac{(O - E)^2}{E}$$

In the formula, the sum is taken over all four cells in the contingency table. Each O represents an observed frequency and each E represents the corresponding expected frequency according to H_0. We now describe how to calculate the E's.

The first step in determining the E's for a contingency table is to calculate the row and column total frequencies (these are called the **marginal frequencies**) and also the grand total of all the cell frequencies. The E's then follow from a simple rationale, as illustrated in Example 11.4.

EXAMPLE 11.4

TREATMENT OF ANGINA

Table 11.3 shows the angina data of Example 11.1, together with the marginal frequencies.

TABLE 11.3

Observed Frequencies for Angina Study

	TREATMENT		TOTAL
	Timolol	Placebo	
Angina-free	44	19	63
Not Angina-free	116	128	244
TOTAL	160	147	307

The E's should agree exactly with the null hypothesis. Because H_0 asserts that the probability of an angina-free response does not depend on the treatment, we can generate an estimate of this probability by pooling the two treatment groups; from Table 11.3, the pooled estimate is $\frac{63}{307}$. We can then apply this estimate to each treatment group to yield the number of angina-free patients expected according to H_0, as follows:

Timolol group: $\left(\frac{63}{307}\right) \times 160 = 32.83$

Placebo group: $\left(\frac{63}{307}\right) \times 147 = 30.17$

After applying a similar rationale to the expected frequencies of a not angina-free response, we obtain the expected frequencies shown in parentheses in Table 11.4. Note that the marginal totals for the E's are the same as for the O's.

TABLE 11.4
Observed and Expected
Frequencies for Angina
Study

	TREATMENT		TOTAL
	Timolol	Placebo	
Angina-free	44 (32.83)	19 (30.17)	63
Not Angina-free	116 (127.17)	128 (116.83)	244
TOTAL	160	147	307

■

In practice, it is not necessary to proceed through a chain of reasoning to obtain the expected frequencies for a contingency table. The procedure for calculating the E's can be condensed into a simple formula. The expected frequency for each cell is calculated from the marginal total frequencies for the same row and column, as follows:

EXPECTED FREQUENCIES IN CONTINGENCY TABLE

$$E = \frac{(\text{Row total}) \times (\text{Column total})}{\text{Grand total}}$$

The formula produces the same calculation as does the rationale given in Example 11.4, as the following example shows.

EXAMPLE 11.5
TREATMENT OF ANGINA

We will apply the above formula to the angina data of Example 11.1. The expected frequency of angina-free Timolol patients is calculated from the marginal totals as

$$E = \frac{(63)(160)}{307} = 32.83$$

Note that this is the same answer obtained in Example 11.4. Proceeding similarly for each cell in the contingency table, we would obtain all the E's shown in Table 11.4. ■

Note that, although the formula for χ_s^2 for contingency tables is the same as given for goodness-of-fit tests in Chapter 10, the method of calculating the E's is quite different for contingency tables because the null hypothesis is different.

THE TEST PROCEDURE

The chi-square test for a contingency table is carried out similarly to the chi-square goodness-of-fit test. Large values of χ_s^2 indicate evidence against H_0. Critical values are determined from Table 8; the number of degrees of freedom for a 2×2 contingency table is

$$df = 1$$

For a 2×2 contingency table, the alternative hypothesis can be directional or nondirectional. Directional alternatives are handled by the familiar two-step procedure, cutting the P-value in half if the data deviate from H_0 in the direction specified by H_A. Note that χ_s^2 itself does not express directionality; to determine the directionality of the data, one must calculate and compare the estimated probabilities.

The following example illustrates the chi-square test.

EXAMPLE 11.6

TREATMENT OF ANGINA

For the angina experiment of Example 11.1, let us apply a chi-square test at $\alpha = .01$. We may state the null hypothesis and a directional alternative informally as follows:

H_0:   Timolol is no better than placebo for preventing angina

H_A: Timolol is better than placebo for preventing angina

Symbolically, the statements are

H_0: $\pi_1 = \pi_2$

H_A: $\pi_1 > \pi_2$

To check the directionality of the data, we calculate the estimated probabilities of response:

$$\hat{\pi}_1 = \frac{44}{160} = .28$$

$$\hat{\pi}_2 = \frac{19}{147} = .13$$

and we note that

$$\hat{\pi}_1 > \hat{\pi}_2$$

Thus, the data do deviate from H_0 in the direction specified by H_A. We proceed to calculate the chi-square statistic from Table 11.4 as

$$\chi_s^2 = \frac{(44 - 32.83)^2}{32.83} + \frac{(116 - 127.17)^2}{127.17} + \frac{(19 - 30.17)^2}{30.17} + \frac{(128 - 116.83)^2}{116.83}$$

$$= 10.0$$

From Table 8 with df $= 1$, we find that $\chi_{.01}^2 = 6.63$ and $\chi_{.001}^2 = 10.83$, and so we have $.0005 < P < .005$. Thus, we reject H_0 and find that the data provide sufficient evidence to conclude that Timolol is better than placebo for producing an angina-free response.   ∎

Note that, even though $\hat{\pi}_1$ and $\hat{\pi}_2$ do not enter into the calculation of χ_s^2, *the calculation of $\hat{\pi}_1$ and $\hat{\pi}_2$ is an important part of the test procedure*; the information provided by the quantities $\hat{\pi}_1$ and $\hat{\pi}_2$ is essential for meaningful interpretation of the results.*

COMPUTATIONAL NOTES   The following tips are helpful in analyzing a 2 × 2 contingency table:

1. The contingency table format is convenient for computations. For presenting the data in a report, however, it is usually better to use a more readable form of display; some examples are shown in the exercises.
2. For calculating χ_s^2, the observed frequencies (O's) must be absolute, rather than relative, frequencies; also, *the table must contain all four cells*, so that the sum of the O's is equal to the total number of observations.
3. For a 2 × 2 contingency table, the four terms that are added to produce χ_s^2 all have the same numerator (but different denominators). You can use this fact to simplify the calculation.

ILLUSTRATION OF THE NULL HYPOTHESIS

The chi-square statistic measures discrepancy between the data and the null hypothesis in an indirect way; the quantities $\hat{\pi}_1$ and $\hat{\pi}_2$ are involved indirectly in the calculation of the expected frequencies. If $\hat{\pi}_1$ and $\hat{\pi}_2$ are equal, then the value of χ_s^2 is zero. Here is an example.

EXAMPLE 11.7

Table 11.5 shows fictitious data for an angina study similar to that described in Example 11.1.

TABLE 11.5
Fictitious Data for
Angina Study

	TREATMENT	
	Drug	Placebo
Angina-free	30	20
Not Angina-free	120	80
TOTAL	150	100

For the data of Table 11.5, the estimated probabilities of an angina-free response are *equal*:

$$\hat{\pi}_1 = \frac{30}{150} = .20$$

$$\hat{\pi}_2 = \frac{20}{100} = .20$$

*It is natural to wonder why we do not use a more direct comparison of $\hat{\pi}_1$ and $\hat{\pi}_2$. In fact, there is a test procedure based on a t-type statistic, calculated by dividing $(\hat{\pi}_1 - \hat{\pi}_2)$ by its standard error. This t-type procedure is equivalent to the chi-square test. We have chosen to present the chi-square test instead, for two reasons: (1) it can be extended to contingency tables larger than 2 × 2; (2) in certain applications the chi-square statistic is more natural than the t-type statistic; some of these applications appear in Section 11.3.

You can easily verify that, for Table 11.5, the expected frequencies are equal to observed frequencies, so that the value of χ_s^2 is zero. Also notice that the columns of the table are proportional to each other:

$$\frac{30}{120} = \frac{20}{80}$$

∎

As the preceding example suggests, an "eyeball" analysis of a contingency table is based on checking for proportionality of the columns. If the columns are nearly proportional, then the data agree fairly well with H_0; if they are highly nonproportional, then the data disagree with H_0. Note, however, that the actual value of χ_s^2 depends on the sample sizes as well as the degree of nonproportionality; as discussed in Section 10.3, the value of χ_s^2 varies directly with the number of observations if the percentage composition of the data is kept fixed and the number of observations is varied. This reflects the fact that a given percentage deviation from H_0 is less likely to occur by chance with a larger number of observations.

EXERCISES 11.1–11.10

11.1 The accompanying partially complete contingency table shows the responses to two treatments.

		TREATMENT	
		1	2
RESPONSE	Success	70	
	Failure		
TOTAL		100	200

a. Invent a fictitious data set that agrees with the table and for which $\chi_s^2 = 0$.
b. Calculate the estimated probabilities of success ($\hat{\pi}_1$ and $\hat{\pi}_2$) for your data set. Are they equal?

11.2 Proceed as in Exercise 11.1 for the following contingency table:

		TREATMENT	
		1	2
RESPONSE	Success	30	
	Failure		
TOTAL		300	100

11.3 Proceed as in Exercise 11.1 for the following contingency table:

		TREATMENT	
		1	2
RESPONSE	Success	5	20
	Failure	10	

11.4 Proceed as in Exercise 11.1 for the following contingency table:

		TREATMENT	
		1	2
RESPONSE	Success	20	10
	Failure	80	

11.5 Most salamanders of the species *P. cinereus* are red-striped, but some individuals are all red. The all-red form is thought to be a mimic of the salamander *N. viridescens*, which is toxic to birds. In order to test whether the mimic form actually survives more successfully, 163 striped and 41 red individuals of *P. cinereus* were exposed to predation by a natural bird population. After two hours, 65 of the striped and 23 of the red individuals were still alive.[3] Use a chi-square test to assess the evidence that the mimic form survives more successfully. Use a directional alternative and let α = .05.

11.6 Can attack of a plant by one organism induce resistance to subsequent attack by a different organism? In a study of this question, individually potted cotton (*Gossypium*) plants were randomly allocated to two groups. Each plant in one group received an infestation of spider mites (*Tetranychus*); the other group were kept as controls. After two weeks the mites were removed and all plants were inoculated with *Verticillium*, a fungus that causes wilt disease. The accompanying table shows the numbers of plants that developed symptoms of wilt disease.[4] Do the data provide sufficient evidence to conclude that infestation with mites induces resistance to wilt disease? Use a chi-square test against a directional alternative. Let α = .01.

		TREATMENT	
		Mites	No Mites
RESPONSE	Wilt Disease	11	17
TO FUNGUS	No Wilt Disease	15	4
TOTAL		26	21

11.7 Phenytoin is a standard anticonvulsant drug which unfortunately has many toxic side effects. A study was undertaken to compare phenytoin with valproate, a new drug, in the treatment of epilepsy. Patients were randomly allocated to receive either phenytoin or valproate for 12 months. Of 20 patients receiving valproate, 6 were free of seizures for the 12 months, while 6 of 17 patients receiving phenytoin were seizure-free.[5]

a. Use a chi-square test to compare the seizure-free response rates for the two drugs. Let H_A be nondirectional and α = .10.

b. Does the test in part **a** provide evidence that valproate and phenytoin are equally effective in preventing seizures? Discuss.

11.8 Estrus synchronization products are used to bring cows into heat at a predictable time so that they can be reliably impregnated by artificial insemination. In a study of two estrus synchronization products, 42 mature cows (aged 4–8 years) were randomly allocated to receive either product A or product B, and then all cows were bred by artificial insemination. The table shows how many of the inseminations resulted in pregnancy.[6] Use a chi-square test to compare the effectiveness of the two products in producing pregnancy. Use a nondirectional alternative and let $\alpha = .05$.

	TREATMENT	
	Product A	Product B
Total number of cows	21	21
Number of cows pregnant	8	15

11.9 Experimental studies of cancer often use strains of animals which have a naturally high incidence of tumors. In one such experiment, tumor-prone mice were kept in a sterile environment; one group of mice were maintained entirely germ-free, while another group were exposed to the intestinal bacterium *Escherichia coli*. The accompanying table shows the incidence of liver tumors.[7]

TREATMENT	TOTAL NUMBER OF MICE	MICE WITH LIVER TUMORS	
		Number	Percent
Germ-free	49	19	39
E. Coli	13	8	62

a. How strong is the evidence that tumor incidence is higher in mice exposed to *E. coli*? Use a chi-square test against a directional alternative. Let $\alpha = .05$.

b. How would the result of part **a** change if the percentages (39% and 62%) of mice with tumors were the same, but the sample sizes were (i) doubled (98 and 26)? (ii) tripled (147 and 39)? [*Hint:* Part **b** requires almost no calculation.]

11.10 In a randomized clinical trial to determine the most effective timing of administration of chemotherapeutic drugs to lung cancer patients, 16 patients were given four drugs simultaneously and 11 patients were given the same drugs sequentially. Objective response to the treatment (defined as shrinkage of the tumor by at least 50%) was observed in 11 of the patients treated simultaneously and in 3 of the patients treated sequentially.[8] Do the data provide evidence as to which timing is superior? Use a chi-square test against a nondirectional alternative. Let $\alpha = .05$.

S E C T I O N 11.3

INDEPENDENCE AND ASSOCIATION IN THE 2 × 2 CONTINGENCY TABLE

The 2 × 2 contingency table is deceptively simple. In this section we explore further the relationships that it can express.

TWO CONTEXTS FOR CONTINGENCY TABLES

A 2 × 2 contingency table can arise in two contexts, namely:

1. Two independent samples, dichotomous observed variable
2. One sample, two dichotomous observed variables

The first context is illustrated by the angina data of Example 11.1, which can be viewed as two independent samples—the Timolol group and the placebo group—of sizes $n_1 = 160$ and $n_2 = 147$. The observed variable is angina-free status. Any study involving a dichotomous observed variable and completely randomized allocation to two treatments can be viewed this way.

The second context is illustrated by the color-blindness data of Example 11.2, which can be viewed as a single sample of $n = 18{,}121$ children, observed with respect to two dichotomous variables—color vision (color-blind or not) and gender (male or female).

The two contexts are not always sharply differentiated. For instance, the color-blindness data of Example 11.2 could also be viewed as two samples—9,049 males and 9,072 females—observed with respect to one dichotomous variable (color vision).

The arithmetic of the chi-square test is the same in both contexts, but the statement and interpretation of hypotheses and conclusions can be very different. To describe relationships in the second context, it is useful to extend the language of probability to include a new concept, conditional probability.

CONDITIONAL PROBABILITY

Recall that the probability of an event predicts how often the event will occur. A **conditional probability** predicts how often an event will occur *under specified conditions*. The notation for a conditional probability is

$$\Pr\{E|C\}$$

which is read "probability of E, *given* C." When a conditional probability is estimated from observed data, the estimate is denoted by a hat ("⌢"), thus:

$$\widehat{\Pr}\{E|C\}$$

The following example illustrates these ideas.

EXAMPLE 11.8
COLOR-BLINDNESS
AND GENDER

Suitable conditional probabilities to describe the relation between color-blindness and gender (Example 11.2) would be as follows:

$\Pr\{CB|M\}$ = Probability that a person is color-blind, *given* that the person is male

$\Pr\{CB|F\}$ = Probability that a person is color-blind, *given* that the person is female

where CB denotes color-blind, and M and F denote male and female. The estimates of these conditional probabilities from the data of Table 11.2 are

$$\widehat{\Pr}\{CB|M\} = \frac{725}{9{,}049} = .080$$

and

$$\widehat{\Pr}\{CB|F\} = \frac{40}{9{,}072} = .0044$$

∎

Note that the conditional probability notation is a substitute for the π notation of Section 11.2. For instance, in Example 11.8 we can make the identification

$$\pi_1 = \Pr\{CB|M\}$$
$$\pi_2 = \Pr\{CB|F\}$$

It may seem unnecessary to introduce a new and complicated notation when a simpler notation was already available. However, we will find that we sometimes need the greater flexibility of the conditional probability notation.

INDEPENDENCE AND ASSOCIATION

In many contingency tables, the columns of the table play a different role than the rows. For instance, in the angina data of Example 11.1, the columns represent treatments and the rows represent responses. Also, in Example 11.8 it seems more natural to define the columnwise conditional probabilities $\Pr\{CB|M\}$ and $\Pr\{CB|F\}$ rather than the rowwise conditional probabilities $\Pr\{M|CB\}$ and $\Pr\{F|CB\}$.

On the other hand, in some cases it is natural to think of the rows and the columns of the contingency table as playing interchangeable roles. In such a case, conditional probabilities may be calculated either rowwise or columnwise, and the null hypothesis for the chi-square test may be expressed either rowwise or columnwise. The following is an example.

EXAMPLE 11.9
HAIR COLOR AND EYE COLOR

To study the relationship between hair color and eye color in a German population, an anthropologist observed a sample of 6,800 men, with the results shown in Table 11.6.[9]

TABLE 11.6
Hair Color and Eye Color

		HAIR COLOR		TOTAL
		Dark	Light	
EYE	Dark	726	131	857
COLOR	Light	3,129	2,814	5,943
TOTAL		3,855	2,945	6,800

The data of Table 11.6 would be naturally viewed as a single sample of size $n = 6,800$ with two dichotomous observed variables—hair color and eye color. To describe the data, let us denote dark and light eyes by DE and LE, and dark and light hair by DH and LH. We may calculate estimated columnwise conditional probabilities as follows:

$$\widehat{\Pr}\{DE|DH\} = \frac{726}{3,855} \approx .19$$

$$\widehat{\Pr}\{DE|LH\} = \frac{131}{2,945} \approx .04$$

A natural way to analyze the data is to compare these values: .19 vs .04. On the other hand, it is just as natural to calculate and compare estimated rowwise conditional probabilities:

$$\widehat{\Pr}\{DH|DE\} = \frac{726}{857} \approx .85$$

$$\widehat{\Pr}\{DH|LE\} = \frac{3,129}{5,943} \approx .53$$

Corresponding to these two views of the contingency table, the null hypothesis for the chi-square test can be stated columnwise as

H_0: $\Pr\{DE|DH\} = \Pr\{DE|LH\}$

or rowwise as

H_0: $\Pr\{DH|DE\} = \Pr\{DH|LE\}$

As we shall see, these two hypotheses are equivalent—that is, any population satisfying one of them must satisfy the other. ■

When a data set is viewed as a single sample with two observed variables, the relationship expressed by H_0 is called **statistical independence** of the row variable and the column variable. Variables that are not independent are called **dependent** or **associated**. Thus, the chi-square test is sometimes called a "test of independence" or a "test for association."

EXAMPLE 11.10

HAIR COLOR AND
EYE COLOR

The null hypothesis of Example 11.9 can be stated verbally as

H_0: Eye color is independent of hair color

or

H_0: Hair color is independent of eye color

or, more symmetrically,

H_0: Hair color and eye color are independent ■

The null hypothesis of independence can be stated generically as follows. Two groups, G_1 and G_2, are to be compared with respect to the probability of a characteristic C. The null hypothesis is

H_0: $\Pr\{C|G_1\} = \Pr\{C|G_2\}$

Note that each of the two statements of H_0 in Example 11.9 is of this form.

To further clarify the meaning of the null hypothesis of independence, in the following example we examine a data set that agrees *exactly* with H_0.

EXAMPLE 11.11

PLANT HEIGHT AND
DISEASE RESISTANCE

Consider a (fictitious) species of plant that can be categorized as short (S) or tall (T) and as resistant (R) or nonresistant (NR) to a certain disease. Consider the following null hypothesis:

H_0:   Plant height and disease resistance are independent

Each of the following is a valid statement of H_0:

(1) H_0: $\Pr\{R|S\}$ $= \Pr\{R|T\}$
(2) H_0: $\Pr\{NR|S\} = \Pr\{NR|T\}$
(3) H_0: $\Pr\{S|R\}$ $= \Pr\{S|NR\}$
(4) H_0: $\Pr\{T|R\}$ $= \Pr\{T|NR\}$

The following is *not* a statement of H_0:

(5) $\Pr\{R|S\} = \Pr\{NR|S\}$

Note the difference between (5) and (1). Statement (1) compares *two* groups (short and tall plants) with respect to disease resistance, whereas (5) is a statement about the distribution of disease resistance in only *one* group (short plants); statement (5) merely asserts that half (50%) of short plants are resistant and half are nonresistant.

Suppose, now, that we choose a random sample of 100 plants from the population and we obtain the data in Table 11.7.

TABLE 11.7
Plant Height and Disease Resistance

		HEIGHT		
		S	T	
RESISTANCE	R	12	18	30
	NR	28	42	70
		40	60	100

The data in Table 11.7 agree exactly with H_0; this agreement can be checked in four different ways, corresponding to the four different symbolic statements of H_0:

(1) $\widehat{\Pr}\{R|S\} = \widehat{\Pr}\{R|T\}$

$$\frac{12}{40} = .30 = \frac{18}{60}$$

(2) $\widehat{\Pr}\{NR|S\} = \widehat{\Pr}\{NR|T\}$

$$\frac{28}{40} = .70 = \frac{42}{60}$$

(3) $\widehat{\Pr}\{S|R\} = \widehat{\Pr}\{S|NR\}$

$$\frac{12}{30} = .40 = \frac{28}{70}$$

(4) $\widehat{\Pr}\{T|R\} = \widehat{\Pr}\{T|NR\}$

$$\frac{18}{30} = .60 = \frac{42}{70}$$

Note that the data in Table 11.7 do *not* agree with statement (5):

$$\widehat{\Pr}\{R|S\} = \frac{12}{40} = .30 \quad \text{and} \quad \widehat{\Pr}\{NR|S\} = \frac{28}{40} = .70$$

$$.30 \neq .70$$

∎

FACTS ABOUT ROWS AND COLUMNS

The data in Table 11.7 display independence whether viewed rowwise or columnwise. This is no accident, as the following fact shows.

Fact 11.1 The columns of a 2 × 2 table are proportional if and only if the rows are proportional. Specifically, suppose that $a, b, c,$ and d are any positive numbers, arranged as in Table 11.8.

TABLE 11.8
A General 2 × 2
Contingency Table

			TOTAL
	a	b	$a + b$
	c	d	$c + d$
TOTAL	$a + c$	$b + d$	

Then

$$\frac{a}{c} = \frac{b}{d} \quad \text{if and only if} \quad \frac{a}{b} = \frac{c}{d}$$

Another way to express this is:

$$\frac{a}{a + c} = \frac{b}{b + d} \quad \text{if and only if} \quad \frac{a}{a + b} = \frac{c}{c + d}$$

∎

You can easily show that Fact 11.1 is true; just use simple algebra. Because of Fact 11.1, the relationship of independence in a 2 × 2 contingency table is the same whether the table is viewed rowwise or columnwise. Note also that the expected frequencies, and therefore the value of χ_s^2, would remain the same if the rows and columns of the contingency table were interchanged. The following fact shows that the *direction* of dependence is also the same whether viewed rowwise or columnwise.

Fact 11.2 Suppose that $a, b, c,$ and d are any positive numbers, arranged as in Table 11.8. Then

$$\frac{a}{a + c} > \frac{b}{b + d} \quad \text{if and only if} \quad \frac{a}{a + b} > \frac{c}{c + d}$$

Also

$$\frac{a}{a+c} < \frac{b}{b+d} \quad \text{if and only if} \quad \frac{a}{a+b} < \frac{c}{c+d}$$ ∎

Supplementary discussion   The usual textbook treatment of conditional probability and independence is more formal than we have given here. Formal definitions and more discussion are given in Appendix 11.1.

VERBAL DESCRIPTION OF ASSOCIATION

Ideas of logical implication are expressed in everyday English in subtle ways. The following excerpt is from *Alice in Wonderland*, by Lewis Carroll:

> ". . . you should say what you mean," the March Hare went on.
>
> "I do," Alice hastily replied; "at least—at least I mean what I say—that's the same thing, you know."
>
> "Not the same thing a bit!" said the Hatter. "Why, you might just as well say that 'I see what I eat' is the same thing as 'I eat what I see'!"
>
> . . . "You might just as well say," added the Dormouse . . . , "That 'I breathe when I sleep' is the same thing as 'I sleep when I breathe'!"
>
> "It *is* the same thing with you," said the Hatter . . .

We also use ordinary language to express ideas of probability, conditional probability, and association. For instance, consider the following four statements:

Color-blindness is more common among males than among females.

Maleness is more common among color-blind people than femaleness.

Most color-blind people are male.

Most males are color-blind.

The first three statements are all true; are they actually just different ways of saying the same thing? Notice that the last statement is false.

In interpreting contingency tables, it is often necessary to describe probabilistic relationships in words. This can be quite a challenge. If you become fluent in such description, then you can always "say what you mean" *and* "mean what you say." The following two examples illustrate some of the issues.

EXAMPLE 11.12

PLANT HEIGHT AND
DISEASE RESISTANCE

For the plant height and disease resistance study of Example 11.11, we considered the null hypothesis

H_0:   Height and resistance are independent

This hypothesis could also be expressed verbally in various other ways, such as:

H_0:   Short and tall plants are equally likely to be resistant

H_0:   Resistant and nonresistant plants are equally likely to be tall

H_0:   Resistance is equally common among short and tall plants   ∎

EXAMPLE 11.13

HAIR COLOR AND
EYE COLOR

Let us consider the interpretation of Table 11.6. The chi-square statistic is $\chi_s^2 = 314$; from Table 8 we see that the P-value is tiny, so that the null hypothesis of independence is overwhelmingly rejected. We might state our conclusion in various ways. For instance, suppose we focus on the incidence of dark eyes. From the data we found that

$$\widehat{\text{Pr}}\{DE|DH\} > \widehat{\text{Pr}}\{DE|LH\}$$

that is,

$$\frac{726}{3,855} = .19 > \frac{131}{2,945} = .04$$

A natural conclusion from this comparison would be

Conclusion 1:   There is sufficient evidence to conclude that dark-haired men have a greater tendency to be dark-eyed than light-haired men do.

This statement is carefully phrased, because the statement

"Dark-haired men have a greater tendency to be dark-eyed"

is ambiguous by itself; it could mean

"Dark-haired men have a greater tendency to be dark-eyed than light-haired men do"

or

"Dark-haired men have a greater tendency to be dark-eyed than to be light-eyed"

The first of these statements says that

$$\text{Pr}\{DE|DH\} > \text{Pr}\{DE|LH\}$$

whereas the second says that

$$\text{Pr}\{DE|DH\} > \text{Pr}\{LE|DH\}$$

The second statement asserts that more than half of dark-haired men have dark eyes. Note that the data do *not* support this assertion; of the 3,855 dark-haired men, only 19% have dark eyes.

Conclusion 1 is only one of several possible wordings of the conclusion from the contingency table analysis. For instance, one might focus on dark hair and find

Conclusion 2:   There is sufficient evidence to conclude that dark-eyed men have a greater tendency to be dark-haired than light-eyed men do.

A more symmetrical phrasing would be

Conclusion 3:   There is sufficient evidence to conclude that dark hair is associated with dark eyes.

However, the phrasing in conclusion 3 is easily misinterpreted; it may suggest something like

> "There is sufficient evidence to conclude that most dark-haired men are dark-eyed"

which is not a correct interpretation.   ∎

We emphasize once again the principle that we stated in Section 11.2: *The calculation and comparison of appropriate conditional probabilities or $\hat{\pi}$'s is an essential part of the chi-square test.* Example 11.13 provides ample illustration of this point.

INDEPENDENCE AND CONCORDANCE

A common misconception is that traits which usually occur together cannot be independent. The following example shows that this intuitively appealing idea is incorrect.

EXAMPLE 11.14
INHERITANCE OF LEFT-HANDEDNESS

Suppose pairs of fraternal twins are examined, and the handedness of each twin is determined. For clarity, let us assume that all the twins are brother–sister pairs. Consider the following null hypothesis:

H_0:   Twin brother and sister are independent with respect to handedness

or

H_0:   Handedness of twin sister is independent of handedness of twin brother

Suppose data are collected for 1,000 twin pairs, with the results shown in Table 11.9.[10]

TABLE 11.9
Handedness of Twin Pairs

		SISTER		
		Left	Right	
BROTHER	Left	15	85	100
	Right	135	765	900
		150	850	1,000

Note that these data agree exactly with H_0; we may verify this by noting, for instance, that

$$\widehat{\Pr}\{BL|SL\} = \widehat{\Pr}\{BL|SR\}$$
$$\frac{15}{150} = \frac{85}{850}$$

where *BL* represents the event "brother is left-handed," and similarly for *SL* and *SR*.

Now let us consider some additional statements that also describe Table 11.9. (As a convention, let us agree that "most" means "75% or more.")

(1) Most of the brothers have the same handedness as their sisters.

(2) Most of the sisters have the same handedness as their brothers.

(3) Most of the twin pairs are concordant with respect to handedness (that is, they are either both right-handed or both left-handed).

Statements (1), (2), and (3) all say the same thing; the following calculation shows that the statements are true for the data of Table 11.9:

$$\text{Proportion of concordant pairs} = \frac{15 + 765}{1,000}$$
$$= .78 \quad \text{or} \quad 78\%$$
$$= \text{"most"}$$

At first glance, it may appear that statements (1), (2), and (3) actually *contradict* H_0. However, this is an illusion; the data of Table 11.8 agree with H_0 and with the three statements! The fact is that the twin pairs are no more concordant than randomly chosen unrelated male–female pairs, but—surprisingly—even unrelated pairs would usually be concordant.

As a further illustration of how everyday language can be misleading in describing conditional probability relationships, consider the following statements:

(4) Most left-handed sisters have right-handed brothers.

(5) Most left-handed brothers have right-handed sisters.

At first glance it may appear that these statements contradict the previous ones; for instance, statement (4) may appear to contradict statement (1). In fact there is no contradiction; as you can easily verify, statements (4) and (5) are also true for the data of Table 11.9.   ∎

EXERCISES
11.11–11.22

11.11 Consider a fictitious population of mice. Each animal's coat is either black (B) or grey (G) in color, and is either wavy (W) or smooth (S) in texture. Express each of the following relationships in terms of probabilities or conditional probabilities relating to the population of animals.
  a. Smooth coats are more common among black mice than among grey mice.
  b. Smooth coats are more common among black mice than wavy coats are.
  c. Smooth coats are more often black than are wavy coats.
  d. Smooth coats are more often black than grey.
  e. Smooth coats are more common than wavy coats.

11.12 Consider a fictitious population of mice in which each animal's coat is either black (B) or grey (G) in color, and is either wavy (W) or smooth (S) in texture (as in Exercise 11.11). Suppose a random sample of mice is selected from the population and the coat color and texture are observed; consider the accompanying partially complete contingency table (top of page 368) for the data.

		COLOR	
		B	G
TEXTURE	W		50
	S		
		60	150

a. Invent fictitious data sets that agree with the table and for which
(i) $\widehat{\Pr}\{W|B\} > \widehat{\Pr}\{W|G\}$;   (ii) $\widehat{\Pr}\{W|B\} = \widehat{\Pr}\{W|G\}$.
In each case, verify your answer by calculating the estimated conditional probabilities.
b. For each of the two data sets you invented in part **a**, calculate $\widehat{\Pr}\{B|W\}$ and $\widehat{\Pr}\{B|S\}$.
c. Which of the data sets of part **a** has $\widehat{\Pr}\{B|W\} > \widehat{\Pr}\{B|S\}$? Can you invent a data set for which

$$\widehat{\Pr}\{W|B\} > \widehat{\Pr}\{W|G\} \quad \text{but} \quad \widehat{\Pr}\{B|W\} < \widehat{\Pr}\{B|S\}?$$

If so, do it. If not, explain why not.

11.13 A medical team investigated the relation between immunological factors and survival after a heart attack. Blood specimens from 213 male heart-attack patients were tested for presence of antibody to milk protein. The patients were followed to determine whether they lived for 6 months following their heart attack. The results are given in the table.[11]

		ANTIBODY TEST		TOTAL
		Positive	Negative	
SURVIVAL	Died	29	10	39
	Alive	80	94	174
TOTAL		109	104	213

a. Let D and A represent died and alive, respectively, and let P and N represent positive and negative antibody tests. Calculate $\widehat{\Pr}\{D|P\}$, $\widehat{\Pr}\{D|N\}$, $\widehat{\Pr}\{P|D\}$, and $\widehat{\Pr}\{P|A\}$.
b. The value of the contingency-table chi-square statistic for these data is $\chi_s^2 = 10.27$. Test for a relationship between the antibody and survival. Use a nondirectional alternative and let $\alpha = .05$.

11.14 Refer to Exercise 11.13. Is the antibody test a good predictor of survival? To answer this question, imagine trying to predict survival solely on the basis of the antibody test. Use the data to estimate the probability that such a prediction would be correct (that is, the percentage of heart attack patients for whom the prediction would be correct).

11.15 In a study of behavioral asymmetries, 2,391 women were asked which hand they preferred to use (for instance, to write) and which foot they preferred to use (for instance, to kick a ball). The results are reported in the table.[12]

PREFERRED HAND	PREFERRED FOOT	NUMBER OF WOMEN
Right	Right	2,012
Right	Left	142
Left	Right	121
Left	Left	116
Total		2,391

a. Estimate the conditional probability that a woman is right-footed, given that she is right-handed.

b. Estimate the conditional probability that a woman is right-footed, given that she is left-handed.

c. Suppose we want to test the null hypothesis that hand preference and foot preference are independent. Calculate the chi-square statistic for this hypothesis.

d. Suppose we want to test the null hypothesis that right-handed women are equally likely to be right-footed or left-footed. Calculate the chi-square statistic for this hypothesis.

11.16 Consider a study to investigate a certain suspected disease-causing agent. One thousand people are to be chosen at random from the population, and each individual is to be classified as diseased or not diseased, and as exposed or not exposed to the agent. The results are to be cast in the following contingency table:

		EXPOSURE	
		Yes	No
DISEASE	Yes		
	No		

Let EY and EN denote exposure and nonexposure, and let DY and DN denote presence and absence of the disease. Express each of the following statements in terms of conditional probabilities. (Note that "a majority" means "more than half.")

a. The disease is more common among exposed than among nonexposed people.

b. Exposure is more common among diseased people than among nondiseased people.

c. Exposure is more common among diseased people than is nonexposure.

d. A majority of diseased people are exposed.

e. A majority of exposed people are diseased.

f. Exposed people are more likely to be diseased than are nonexposed people.

g. Exposed people are more likely to be diseased than to be nondiseased.

11.17 Refer to Exercise 11.16. Which of the statements express the assertion that occurrence of the disease is associated with exposure to the agent? (There may be more than one.)

11.18 Refer to Exercise 11.16. Invent fictitious data sets as specified below, and verify your answer by calculating appropriate estimated conditional probabilities. (Your data need not be statistically significant.)

a. Invent a data set for which

$$\widehat{\Pr}\{DY|EY\} > \widehat{\Pr}\{DY|EN\} \quad \text{but} \quad \widehat{\Pr}\{EY|DY\} < \widehat{\Pr}\{EN|DY\}$$

or explain why it is not possible.

b. Invent a data set that agrees with statement **a** of Exercise 11.16 but with neither **d** nor **e**; or, explain why it is not possible.

c. Invent a data set for which

$$\widehat{\Pr}\{DY|EY\} > \widehat{\Pr}\{DY|EN\} \quad \text{but} \quad \widehat{\Pr}\{EY|DY\} < \widehat{\Pr}\{EY|DN\}$$

or explain why it is not possible.

11.19 An ecologist studied the spatial distribution of tree species in a wooded area. From a total area of 21 acres, he randomly selected 144 quadrats (plots), each 38 feet square, and noted the presence or absence of maples and hickories in each quadrat. The results are shown in the table.[13]

		MAPLES	
		Present	Absent
HICKORIES	Present	26	63
	Absent	29	26

The value of the chi-square statistic for this contingency table is $\chi_s^2 = 7.96$. Test the null hypothesis that the two species are distributed independently of each other. Use a nondirectional alternative and let $\alpha = .01$. In stating your conclusion, indicate whether the data suggest attraction between the species, or repulsion. Support your interpretation with estimated conditional probabilities from the data.

11.20 Refer to Exercise 11.19. Suppose the data for fictitious tree species, A and B, were as presented in the accompanying table. The value of the chi-square statistic for this contingency table is $\chi_s^2 = 9.07$. As in Exercise 11.19, test the null hypothesis of independence and interpret your conclusion in terms of attraction or repulsion between the species.

		SPECIES A	
		Present	Absent
SPECIES B	Present	30	10
	Absent	49	55

11.21 In a study of the effects of smoking, 9,793 pregnant women were questioned about their smoking habits. (This study was mentioned in Examples 8.2 and 8.15.) The accompanying table shows the incidence of low birthweight (2,500 gm or less) among their infants.[14]

		SMOKING STATUS		TOTAL
		Smoker	Nonsmoker	
BIRTHWEIGHT	Low	237	197	434
	Normal	3,489	5,870	9,359
TOTAL		3,726	6,067	9,793

Let L and N represent low and normal birthweight, respectively, and let S and NS represent smoker and nonsmoker.

a. Calculate $\widehat{\text{Pr}}\{L|S\}$ and $\widehat{\text{Pr}}\{L|NS\}$.   **b.** Calculate $\widehat{\text{Pr}}\{S|L\}$ and $\widehat{\text{Pr}}\{S|N\}$.

11.22 Refer to Exercise 11.21. Invent a fictitious data set on smoking and birthweight for 10,000 women, for which $\widehat{\text{Pr}}\{L|S\}$ is twice as great as $\widehat{\text{Pr}}\{L|NS\}$, but nevertheless the majority of women who have low-birthweight babies are nonsmokers.

SECTION 11.4

THE *r* × *k* CONTINGENCY TABLE

The ideas of Sections 11.2 and 11.3 extend readily to contingency tables that are larger than 2 × 2. We now consider a contingency table with r rows and k columns, which is termed an **$r \times k$ contingency table**. Here is an example.

EXAMPLE 11.15
DISTRIBUTION OF BLOOD TYPE

Table 11.10 shows the observed distribution of ABO blood type in three samples of black Americans living in different locations.[15]

TABLE 11.10
Frequency Distributions of Blood Type

		LOCATION		
		I (Florida)	II (Iowa)	III (Missouri)
	A	122	1,781	353
BLOOD	B	117	1,351	269
TYPE	AB	19	289	60
	O	244	3,301	713
TOTAL		502	6,722	1,395

To compare the distributions in the three locations, one can calculate the columnwise percentages, as displayed in Table 11.11. (For instance, of the Florida sample, $\frac{122}{502}$ or 24.3% are Type A.) Inspection of Table 11.11 shows that the three percentage distributions (columns) are fairly similar.

TABLE 11.11
Percentage Distributions of Blood Type

		LOCATION		
		I (Florida)	II (Iowa)	III (Missouri)
	A	24.3	26.5	25.3
BLOOD	B	23.3	20.1	19.3
TYPE	AB	3.8	4.3	4.3
	O	48.6	49.1	51.1
TOTAL		100.0	100.0	100.0

THE CHI-SQUARE TEST FOR THE $r \times k$ TABLE

The goal of statistical analysis of an $r \times k$ contingency table is to investigate the relationship between the row variable and the column variable. Such an investigation can begin with an inspection of the columnwise or rowwise percentages, as in Table 11.11. One route to further analysis is to ask whether the discrepancies in percentages are too large to be explained as sampling error. This question can be answered by a chi-square test. The chi-square statistic is calculated from the familiar formula

$$\chi^2_s = \sum \frac{(O - E)^2}{E}$$

where the sum is over all cells of the contingency table, and the expected frequencies (E's) are calculated as

$$E = \frac{(\text{Row total}) \times (\text{Column total})}{\text{Grand total}}$$

This method of calculating the E's can be justified by a simple extension of the rationale given in Section 11.2. Critical values for the chi-square test are obtained from Table 8 with

$$df = (r - 1)(k - 1)$$

The following example illustrates the chi-square test.

EXAMPLE 11.16

DISTRIBUTION OF
BLOOD TYPE

Let us apply the chi-square test to the blood type data of Example 11.15. The null hypothesis is

H_0: The distribution of blood type is the same in the three populations

This hypothesis can be stated symbolically in conditional probability notation as follows:

$$H_0: \begin{cases} \Pr\{A|I\} = \Pr\{A|II\} = \Pr\{A|III\} \\ \Pr\{B|I\} = \Pr\{B|II\} = \Pr\{B|III\} \\ \Pr\{AB|I\} = \Pr\{AB|II\} = \Pr\{AB|III\} \\ \Pr\{O|I\} = \Pr\{O|II\} = \Pr\{O|III\} \end{cases}$$

Note that the percentages in Table 11.11 are the estimated conditional probabilities; that is,

$$\widehat{\Pr}\{A|I\} = .243$$
$$\widehat{\Pr}\{A|II\} = .265$$

and so on. We test H_0 against the nondirectional alternative hypothesis

H_A: The distribution of blood type is not the same in the three populations

Table 11.12 shows the observed and expected frequencies.

TABLE 11.12

Observed and Expected Frequencies of Blood Types

		LOCATION			TOTAL
		I	II	III	
BLOOD TYPE	A	122 (131.40)	1,781 (1,759.47)	353 (365.14)	2,256
	B	117 (101.17)	1,351 (1,354.69)	269 (281.14)	1,737
	AB	19 (21.43)	289 (287.00)	60 (59.56)	368
	O	244 (248.00)	3,301 (3,320.84)	713 (689.16)	4,258
TOTAL		502	6,722	1,395	8,619

From Table 11.12, we can calculate the test statistic as

$$\chi_s^2 = \frac{(122 - 131.40)^2}{131.40} + \frac{(117 - 101.17)^2}{101.17} + \cdots + \frac{(713 - 689.16)^2}{689.16}$$

$$= 5.65$$

For these data, $r = 4$ and $k = 3$, so that

$$df = (4 - 1)(3 - 1) = 6$$

From Table 8 with $df = 6$, we find that $\chi_{.20}^2 = 8.56$, so that $P > .20$. The null hypothesis would not be rejected at any reasonable significance level. Thus, the chi-square test shows that the observed differences among the three blood type distributions are no more than would be expected from sampling variation. ∎

Note that H_0 in Example 11.16 is a compound null hypothesis in the sense defined in Section 10.3—that is, H_0 contains more than one independent assertion. This will always be true for contingency tables larger than 2×2, and consequently for such tables the alternative hypothesis for the chi-square test will always be nondirectional and the conclusion, if H_0 is rejected, will be nondirectional. Thus, the chi-square test will often not represent a complete analysis of an $r \times k$ contingency table.

TWO CONTEXTS FOR $r \times k$ CONTINGENCY TABLES

We noted in Section 11.3 that a 2×2 contingency table can arise in two different contexts. Similarly, an $r \times k$ contingency table can arise in the following two contexts:

1. k independent samples, categorical observed variable with r categories
2. One sample, two categorical observed variables—one with k categories and one with r categories

As with the 2×2 table, the calculation of the chi-square statistic is the same for both contexts, but the statement of hypotheses and conclusions can differ. The following example illustrates the second context.

EXAMPLE 11.17
HAIR COLOR AND
EYE COLOR

Table 11.13 shows the relationship between hair color and eye color for 6,800 German men.[9] (This is the same study as in Example 11.9.)

TABLE 11.13
Hair Color and Eye Color

		HAIR COLOR			
		Brown	Black	Fair	Red
EYE COLOR	Brown	438	288	115	16
	Grey or Green	1,387	746	946	53
	Blue	807	189	1,768	47

Let us use a chi-square test to test the hypothesis

H_0: Hair color and eye color are independent

For the data of Table 11.13, one can calculate $\chi_s^2 = 1,074$. From Table 8 with df $= (3 - 1)(4 - 1) = 6$ we find $\chi_{.0001}^2 = 27.86$. Thus, H_0 is overwhelmingly rejected and we conclude that there is extremely strong evidence that hair color and eye color are associated.   ■

**EXERCISES
11.23–11.26**

11.23 It is thought that psychological states can influence the progress of cancer; feelings of helplessness and loss of control seem to reduce the body's ability to reject a tumor. To study this phenomenon, researchers injected 93 rats with tumor cells and then randomly allocated the animals to three treatment groups. Animals in group 1 received periodic electric shocks, but were able to terminate each shock by pressing a lever. Group 2 animals received the same amount of shock as group 1 animals, but they had no control over the shocks. Group 3 animals received no shocks. After one month, each animal was examined to determine whether it had rejected the tumor. The results are given in the table.[16]

TREATMENT	TOTAL NUMBER OF ANIMALS	TUMOR REJECTION Number	TUMOR REJECTION Percent
1. Escapable shock	30	19	63
2. Inescapable shock	30	8	27
3. No shock	33	18	55

a. Use a chi-square test to compare the treatments at $\alpha = .05$. (The value of the chi-square statistic is $\chi_s^2 = 8.85$.)

b. Verify the value of χ_s^2 given in part a.

11.24 For a study of free-living populations of the fruitfly *Drosophila subobscura*, researchers placed baited traps in two woodland sites and one open-ground area. The numbers of male and female flies trapped in a single day are given in the table.[17]

	WOODLAND SITE I	WOODLAND SITE II	OPEN GROUND
Males	89	34	74
Females	31	20	136
Total	120	54	210

a. Use a chi-square test to compare the sex ratios at the three sites. Let $\alpha = .05$.

b. Construct a table that displays the data in a more readable format, such as the one in Exercise 11.23.

11.25 In a classic study of peptic ulcer, blood types were determined for 1,655 ulcer patients. The accompanying table shows the data for these patients and for an independently chosen group of 10,000 healthy controls from the same city.[18]

BLOOD TYPE	ULCER PATIENTS	CONTROLS
O	911	4,578
A	579	4,219
B	124	890
AB	41	313
Total	1,655	10,000

a. The value of the chi-square statistic for this contingency table is $\chi_s^2 = 49.0$. Carry out the chi-square test at $\alpha = .01$.

b. Construct a table showing the percentage distributions of blood type for patients and for controls.

c. Verify the value of χ_s^2 given in part a.

11.26 The two claws of the lobster (*Homarus americanus*) are identical in the juvenile stages. By adulthood, however, the two claws normally have differentiated into a stout claw called a "crusher" and a slender claw called a "cutter." In a study of the differentiation process, 26 juvenile animals were reared in smooth plastic trays, and 18 were reared in trays containing oyster chips (which they could use to exercise their claws). Another 23 animals were reared in trays containing only one oyster chip. The claw configurations of all the animals as adults are summarized in the table.[19]

		CLAW CONFIGURATION		
		Right crusher, Left cutter	Right cutter, Left crusher	Right cutter, Left cutter
TREATMENT	Oyster chips	8	9	1
	Smooth plastic	2	4	20
	One oyster chip	7	9	7

a. The value of the contingency-table chi-square statistic for these data is $\chi_s^2 = 24.35$. Carry out the chi-square test at $\alpha = .01$.

b. Verify the value of χ_s^2 given in part a.

   c. Construct a table showing the percentage distribution of claw configurations for each of the three treatments.

APPLICABILITY OF METHODS

In this section we discuss guidelines for deciding when to use contingency table analysis and the associated chi-square test.

CONDITIONS FOR VALIDITY

The contingency-table chi-square test is valid under the following conditions:

1. *Design conditions*   It must be appropriate to view the data in one of the following ways:
   a. As two or more independent random samples, observed with respect to a categorical variable; or
   b. As one random sample, observed with respect to two categorical variables.
2. *Sample size conditions*   The sample size must be large enough. The critical values given in Table 8 are only approximately correct for determining the P-value associated with χ_s^2. As a rule of thumb, the approximation is considered adequate if each expected frequency (E) is at least equal to 5.
3. *Form of H_0*   The null hypothesis must be appropriate. A generic form of the null hypothesis for the contingency-table chi-square test may be stated as follows:

   H_0:   The row variable and the column variable are independent

VERIFICATION OF DESIGN CONDITIONS

It is all too easy to overlook the design conditions and to apply the chi-square test inappropriately. The conditions should be checked by considering the way in which samples were drawn and treatments (if any) were allocated to experimental units.

Only simple random sampling is allowed. This excludes the more complex schemes such as cluster sampling and stratified sampling that are often used in surveys.

If the data consist of several samples [situation (b) above], then the samples are required to be independent of each other. Failure to observe this restriction may result in a loss of power. If the design includes any pairing, blocking, or matching of experimental units, then the samples would not be independent. Methods of analysis for dependent samples are beyond the scope of this book.*

In either context—situation (a) or situation (b)—it is required that observations on different experimental units be independent of each other. There must be no dependency or hierarchical structure in the design. Failure to observe this restriction can result in a vastly inflated chance of Type I error (which is usually much more serious than a loss of power). The following example illustrates the situation.

*However, one method is described in Exercise 11.42.

EXAMPLE 11.18

POLLINATION OF
FLOWERS

A study was conducted to determine the adaptive significance of flower color in the scarlet gilia (*Ipomopsis aggregata*). Six red-flowered plants and six white-flowered plants were chosen for observation in field conditions; hummingbirds were permitted to visit the flowers, but the other major pollinator, a moth, was excluded by covering the plants at night. Table 11.14 shows, for each plant, the total number of flowers at the end of the season and the number that had set fruit.[20]

TABLE 11.14

Fruit Set in Scarlet Gilia

	RED-FLOWERED PLANTS			WHITE-FLOWERED PLANTS		
	Number of Flowers	Number Setting Fruit	Percent Setting Fruit	Number of Flowers	Number Setting Fruit	Percent Setting Fruit
	140	26	19	125	21	17
	116	11	9	134	17	13
	34	0	0	273	81	30
	79	9	11	146	38	26
	185	28	15	103	17	17
	106	11	10	82	24	29
Sum	660	85		863	198	

The question of interest is whether the percentage of fruit-set is higher for red-flowered than for white-flowered plants. Suppose this question is approached by regarding the individual flower as the observational unit; then the data could be cast in the contingency table format of Table 11.15.

TABLE 11.15

Fruit Set in Scarlet
Gilia Flowers

		FLOWER COLOR	
		Red	White
FRUIT SET	Yes	85	198
	No	575	665
Total		660	863
Percent Setting Fruit		13	23

Table 11.15 yields $\chi_s^2 = 25.0$, for which Table 8 gives $P < .0001$. However, this analysis is not correct, because the observations on flowers on the same plant are not independent of each other; they are dependent because the pollinator (the hummingbird) tends to visit flowers in groups, and perhaps also because the flowers on the same plant are physiologically and genetically related. The chi-square test is invalidated by the hierarchical structure in the data.

A better approach would be to treat the entire plant as the observational unit. For instance, one could take the "Percentage Setting Fruit" column of Table 11.14

as the basic observations; applying a t test to the values yields $t_s = 2.88$ (with $.01 < P < .02$), and applying a Mann–Whitney test yields $U_s = 32$ (with $.02 < P < .05$). Thus, the P-value from the inappropriate chi-square analysis is much too small.   ∎

POWER CONSIDERATIONS

In many studies the chi-square test is valid but is not as powerful as a more appropriate test. Specifically, consider a situation in which the rows or the columns (or both) of the contingency table correspond to a *rankable* categorical variable with more than two categories. The following is an example.

EXAMPLE 11.19
PAIN MEDICATION

In a completely randomized study to compare two pain medications, A and B, each patient rated the amount of pain relief on a 4-point subjective scale. The results are shown in Table 11.16.

TABLE 11.16
Response to Pain
Medication

		TREATMENT	
		Drug A	Drug B
PAIN RELIEF	None	3	7
	Some	7	11
	Substantial	10	5
	Complete	5	2
TOTAL		25	25

A contingency-table chi-square test would be valid to compare drugs A and B, but the test would lack power because it does not use the information contained in the *ordering* of the pain relief categories (none, some, substantial, complete). A related weakness of the chi-square test is that, even if H_0 is rejected, the test does not yield a directional conclusion such as "drug A relieves pain better than drug B."   ∎

Methods are available to analyze contingency tables with rankable row and/or column variables; such methods, however, are beyond the scope of this book.

EXERCISES
11.27–11.29

11.27 Refer to the chemotherapy data of Exercise 11.10. Are the sample sizes large enough for the approximate validity of the chi-square test?

11.28 In a study of prenatal influences on susceptibility to seizures in mice, pregnant females were randomly allocated to a control group or a "handled" group. Handled mice were given sham injections three times during gestation, while control mice were not touched. The offspring were tested for their susceptibility to seizures induced by a loud noise. The investigators noted that the response varied considerably from litter to litter. The accompanying table summarizes the results.[21]

TREATMENT	NUMBER OF LITTERS	NUMBER OF MICE	RESPONSE TO LOUD NOISE No Response	Wild Running	Seizure
Handled	19	104	23	10	71
Control	20	120	47	13	60

If these data are analyzed as a 2×3 contingency table, the chi-square statistic is $\chi_s^2 = 8.45$ and Table 8 gives $.01 < P < .02$. Is this an appropriate analysis for this experiment? Explain. [*Hint:* Does the design meet the conditions for validity of the chi-square test?]

11.29 In control of diabetes it is important to know how blood glucose levels change after eating various foods. Ten volunteers participated in a study to compare the effects of two foods—a sugar and a starch. A blood specimen was drawn before each volunteer consumed a measured amount of food; then additional blood specimens were drawn at eleven times during the next 4 hours. Each volunteer repeated the entire test on another occasion with the other food. Of particular concern were blood glucose levels that dropped below the initial level; the accompanying table shows the number of such values.[22]

FOOD	NO. OF VALUES LESS THAN INITIAL VALUE	TOTAL NUMBER OF OBSERVATIONS
Sugar	26	110
Starch	14	110

Suppose we analyze the given data as a contingency table. The test statistic would be

$$\chi_s^2 = \frac{(26 - 20)^2}{20} + \frac{(14 - 20)^2}{20} + \frac{(84 - 90)^2}{90} + \frac{(96 - 90)^2}{90} = 4.40$$

At $\alpha = .05$ we would reject H_0 and find that there is sufficient evidence to conclude that blood glucose values below the initial value occur more often after ingestion of sugar than after ingestion of starch. This analysis contains two flaws. What are they? [*Hint:* Are the conditions for validity of the test satisfied?]

S E C T I O N   11.6

CONFIDENCE
INTERVAL FOR
DIFFERENCE
BETWEEN
PROBABILITIES
(OPTIONAL)

The chi-square test for a 2×2 contingency table answers only a limited question: Do the estimated probabilities $\hat{\pi}_1$ and $\hat{\pi}_2$ differ enough to conclude that the true probabilities π_1 and π_2 are not equal? A complementary mode of analysis is to use the values $\hat{\pi}_1$ and $\hat{\pi}_2$ to describe the *magnitude* of dependence in the table. In this section we show how to construct a confidence interval for the difference, $(\pi_1 - \pi_2)$.

Consider a 2×2 contingency table which can be viewed as a comparison of two samples with respect to a dichotomous response variable. The two estimated probabilities are $\hat{\pi}_1$ and $\hat{\pi}_2$ and the sample sizes are n_1 and n_2. Recall from

Chapter 7 that, for a quantitative response variable, a natural measure of the degree of difference between two samples is the difference between the sample means $\bar{y}_1$ and $\bar{y}_2$. Analogously, in the present situation a natural measure is the difference between $\hat{\pi}_1$ and $\hat{\pi}_2$.

Like all quantities calculated from samples, the quantity $(\hat{\pi}_1 - \hat{\pi}_2)$ is subject to sampling error. The magnitude of the sampling error can be expressed by the standard error of $(\hat{\pi}_1 - \hat{\pi}_2)$, which is calculated from the following formula:

$$SE_{(\hat{\pi}_1-\hat{\pi}_2)} = \sqrt{\frac{\hat{\pi}_1(1 - \hat{\pi}_1)}{n_1} + \frac{\hat{\pi}_2(1 - \hat{\pi}_2)}{n_2}}$$

Notice that this SE has the form

$$SE_{(\hat{\pi}_1-\hat{\pi}_2)} = \sqrt{SE_1^2 + SE_2^2}$$

where SE_1 and SE_2 are the separate SEs calculated as described in Section 10.2. Thus, $SE_{(\hat{\pi}_1-\hat{\pi}_2)}$ is analogous to the unpooled form of $SE_{(\bar{y}_1-\bar{y}_2)}$ as described in Section 7.2.

An approximate confidence interval can be based on $SE_{(\hat{\pi}_1-\hat{\pi}_2)}$; for instance, a 95% confidence interval is

$$(\hat{\pi}_1 - \hat{\pi}_2) \pm (1.960)SE_{(\hat{\pi}_1-\hat{\pi}_2)}$$

The confidence interval is approximately valid for large sample sizes n_1 and n_2. The following example illustrates the construction of the confidence interval.

EXAMPLE 11.20
TREATMENT OF ANGINA

For the angina data of Example 11.1, the sample sizes are $n_1 = 160$ and $n_2 = 147$, and the estimated probabilities of the angina-free response are:

$$\hat{\pi}_1 = \frac{44}{160} = .2750$$

$$\hat{\pi}_2 = \frac{19}{147} = .1293$$

The difference between these is

$$\hat{\pi}_1 - \hat{\pi}_2 = .2750 - .1293$$
$$= .1457$$
$$\approx .15$$

Thus, we estimate that treatment with Timolol increases the probability of the angina-free response by .15, compared to placebo. To set confidence limits on this estimate, we calculate the standard error as

$$SE_{(\hat{\pi}_1-\hat{\pi}_2)} = \sqrt{\frac{(.2750)(.7250)}{160} + \frac{(.1293)(.8707)}{147}}$$
$$= .04485$$

The 95% confidence interval is

$$.1457 \pm (1.960)(.04485)$$
$$.1457 \pm .08791$$
$$.06 < \pi_1 - \pi_2 < .23$$

We are 95% confident that the effect of Timolol, expressed as an increase in the probability of the angina-free response, is between .06 and .23. ∎

RELATIONSHIP TO TEST The chi-square test for a 2×2 contingency table (Section 11.2) is approximately, but not exactly, equivalent to checking whether a confidence interval for $(\pi_1 - \pi_2)$ includes zero. [Recall from Section 7.5 that there is an exact equivalence between a t test and a confidence interval for $(\mu_1 - \mu_2)$.]

EXERCISES
11.30–11.34

11.30 Refer to the estrus synchronization data of Exercise 11.8. Let π_1 and π_2 represent the probabilities of pregnancy using products A and B, respectively. Construct a 95% confidence interval for $(\pi_1 - \pi_2)$.

11.31 Refer to the liver tumor data of Exercise 11.9. Let π_1 and π_2 represent the probabilities of liver tumors under the germ-free and the *E. coli* conditions, respectively. Construct a 95% confidence interval for $(\pi_1 - \pi_2)$.

11.32 For women who are pregnant with twins, complete bed rest in late pregnancy is commonly prescribed in order to reduce the risk of premature delivery. To test the value of this practice, 212 women with twin pregnancies were randomly allocated to a bed-rest group or a control group. The accompanying table shows the incidence of preterm delivery (less than 37 weeks of gestation).[23]

	BED REST	CONTROLS
No. of preterm deliveries	32	20
No. of women	105	107

Let π_1 and π_2 represent the probabilities of preterm delivery in the two conditions. Construct a 95% confidence interval for $(\pi_1 - \pi_2)$. Does the confidence interval suggest that bed rest is beneficial?

11.33 Refer to Exercise 11.32. The numbers of infants with low birthweight (2,500 gm or less) born to the women are shown in the table.

	BED REST	CONTROLS
No. of low-birthweight babies	76	92
Total no. of babies	210	214

Let π_1 and π_2 represent the probabilities of a low-birthweight baby in the two conditions. Explain why the above information is not sufficient to construct a confidence interval for $(\pi_1 - \pi_2)$.

11.34 Refer to the blood type data of Exercise 11.25. Let π_1 and π_2 represent the probabilities of Type O blood in the patient population and the control population, respectively. Construct a 95% confidence interval for $(\pi_1 - \pi_2)$.

SECTION 11.7

SUMMARY OF CHI-SQUARE TESTS

We have discussed two types of chi-square tests—goodness-of-fit tests in Chapter 10 and contingency table tests in Chapter 11. These tests are similar but are used for different purposes. The following summary should serve as a convenient reference for both tests and as a guide for distinguishing between them.

SUMMARY OF CHI-SQUARE TESTS

GOODNESS-OF-FIT TEST

Null hypothesis:

H_0 specifies the probability of each category

Calculation of expected frequencies:

$E = n \times$ Probability specified by H_0

Test statistic:

$$\chi_s^2 = \sum \frac{(O - E)^2}{E}$$

Null distribution (approximate): χ^2 distribution with

df = (Number of categories) $-$ 1

This approximation is adequate if $E \geq 5$ for every category.

CONTINGENCY TABLE

Null hypothesis:

H_0: Row variable and column variable are independent

Calculation of expected frequencies:

$$E = \frac{(\text{Row total}) \times (\text{Column total})}{\text{Grand total}}$$

Test statistic:

$$\chi_s^2 = \sum \frac{(O - E)^2}{E}$$

Null distribution (approximate): χ^2 distribution with

df = $(r - 1)(k - 1)$

where r is the number of rows and k is the number of columns in the contingency table. This approximation is adequate if $E \geq 5$ for every cell.

| | | | | | | | | | | | | |

**SUPPLEMENTARY
EXERCISES
11.35–11.46**

11.35 One explanation for the widespread incidence of the hereditary condition known as sickle-cell trait is that the trait confers some protection against malarial infection. In one investigation, 543 African children were checked for the trait and for malaria. The results are shown in the table.[24] Do the data provide evidence in favor of the explanation? The value of the chi-square statistic for this contingency table is $\chi_s^2 = 5.33$. Carry out the chi-square test against a directional alternative at $\alpha = .05$.

		MALARIA		
		Heavy infection	Noninfected or lightly infected	
SICKLE-CELL	Yes	36	100	136
TRAIT	No	152	255	407
		188	355	543

11.36 As part of a study of environmental influences on sex determination in the fish *Menidia*, eggs from a single mating were divided into two groups and raised in either a warm or a cold environment. It was found that 73 of 141 offspring in the warm environment and 107 of 169 offspring in the cold environment were females.[25] In each of the following chi-square tests, use a nondirectional alternative and let $\alpha = .05$.
  a. Test the hypothesis that the population sex ratio is 1:1 in the warm environment.
  b. Test the hypothesis that the population sex ratio is 1:1 in the cold environment.
  c. Test the hypothesis that the population sex ratio is the same in the warm as in the cold environment.
  d. Define the population to which the conclusions reached in parts **a–c** apply. (Is it the entire genus *Menidia*?)

11.37 The cilia are hair-like structures that line the nose and help to protect the respiratory tract from dust and foreign particles. A medical team obtained specimens of nasal tissue from nursery school children who had viral upper respiratory infections, and also from healthy children in the same classroom. The tissue was sectioned and the cilia were examined with a microscope for specific defects, with the results shown in the accompanying table.[26] The data show that the percentage of defective cilia was much higher in the tissue from infected children (15.7% vs 3.1%). Would it be valid to apply a chi-square test to compare these percentages? If so, do it. If not, explain why not.

	NUMBER OF CHILDREN	TOTAL NUMBER OF CILIA COUNTED	CILIA WITH DEFECTS	
			Number	Percent
Control	7	556	17	3.1
Respiratory infection	22	1,493	235	15.7

11.38 A group of mountain climbers participated in a trial to investigate the usefulness of the drug acetazolamide in preventing altitude sickness. The climbers were randomly assigned to receive either drug or placebo during an ascent of Mt. Rainier. The experiment was supposed to be double-blind, but the question arose whether some of the climbers might have received clues (perhaps from the presence or absence of side effects or from a perceived therapeutic effect or lack of it) as to which treatment they were receiving. To investigate this possibility, the climbers were asked (after the trial was over) to guess which treatment they had received.[27] The results can be cast in the following contingency table, for which $\chi_s^2 = 5.07$:

		TREATMENT RECEIVED	
		Drug	Placebo
GUESS	Correct	20	12
	Incorrect	11	21

Alternatively, the same results can be rearranged in the following contingency table, for which $\chi_s^2 = .01$:

		TREATMENT RECEIVED	
		Drug	Placebo
GUESS	Drug	20	21
	Placebo	11	12

Consider the null hypothesis

H_0: The blinding was perfect (the climbers received no clues)

Carry out the chi-square test of H_0 against the alternative that the climbers did receive clues. Let $\alpha = .05$. (You must decide which contingency table is relevant to this question.) [*Hint:* To clarify the issue for yourself, try inventing a fictitious data set in which most of the climbers *have* received strong clues; then arrange your fictitious data in each of the two contingency table formats and note which table would yield a large value of χ_s^2.]

11.39 Samples of 1,522 English and 1,029 Chinese people were classified according to the M–N system of blood type, with the results displayed in the accompanying table.[28] (Percentages do not add to exactly 100 because of round-off error.)

BLOOD TYPE	NUMBER		PERCENT	
	English	Chinese	English	Chinese
M	464	342	30	33
N	325	187	21	18
MN	733	500	48	49
Total	1,522	1,029	99	100

a. The value of the contingency-table chi-square statistic is $\chi_s^2 = 4.59$. Carry out the chi-square test. Let $\alpha = .05$.

b. If the sample sizes were increased by a factor of 10 (15,220 English and 10,290 Chinese), but the percentage blood type distributions were the same, how would the conclusion of part **a** be changed? [*Hint:* Part **b** requires almost no calculation.]

11.40 Consider a fictitious drug that can cause two adverse side effects—headache and rash. Suppose the drug is tried on 500 patients, and their reports of side effects are cast in the accompanying contingency table.

		HEADACHE		
		Yes	No	
RASH	Yes			200
	No			300
		100	400	

Consider the following possible situations:

(i)   Headache and rash occur independently of each other.

(ii) A person who experiences headache is more likely (than a person without headache) to experience a rash.

(iii) Headache and rash occur more often separately than together.

(iv) Headache and rash occur more often together than separately.

(v)   The occurrence of headache tends to be associated with the occurrence of rash.

Invent fictitious data sets as specified below, and verify each answer by calculating the estimated conditional probabilities.

a. A data set that agrees with (i)

b. A data set that agrees with (ii)

c. A data set that agrees with both (ii) and (iii)

Do you think that (iv) and (v) mean the same thing? (This may be a matter of opinion.)

11.41 Desert lizards (*Dipsosaurus dorsalis*) regulate their body temperature by basking in the sun or moving into the shade, as required. Normally the lizards will maintain a daytime temperature of about 38°C. When they are sick, however, they maintain a temperature about 2° to 4° higher—that is, a "fever." In an experiment to see whether this fever might be beneficial, lizards were given a bacterial infection; then 36 of the animals were prevented from developing a fever by keeping them in a 38° enclosure, while 12 animals were kept at a temperature of 40°. The table at the top of page 386 describes the mortality after 24 hours.[29] How strongly do these results support the hypothesis that fever has survival value? Use a chi-square test against a directional alternative. Let $\alpha = .05$.

	TEMPERATURE	
	38°	40°
Number of deaths	18	2
Number of animals	36	12
Percent mortality	50	17

11.42 As part of a study of risk factors for stroke, 155 women who had experienced a hemorrhagic stroke (cases) were interviewed. For each case, a control was chosen who had not experienced a stroke; the control was matched to the case by neighborhood of residence, age, and race. Each woman was asked whether she used oral contraceptives. The data for the 155 pairs are displayed in the table. "Yes" and "No" refer to use of oral contraceptives.[30]

		CASE	
		No	Yes
CONTROL	No	107	30
	Yes	13	5

To test for association between oral contraceptive use and stroke, one can consider only the 43 discordant pairs (pairs who answered differently), and test the hypothesis that a discordant pair is equally likely to be Yes/No or No/Yes. Use a chi-square goodness-of-fit test to test this hypothesis against a nondirectional alternative at $\alpha = .05$. (This test is known as **McNemar's test**; it can be regarded as a form of the sign test.)

11.43 The habitat selection behavior of the fruitfly *Drosophila subobscura* was studied by capturing flies from two different habitat sites. The flies were marked with colored fluorescent dust to indicate the site of capture and then released at a point midway between the original sites. On the following two days, flies were recaptured at the two sites. The results are summarized in the table.[31] The value of the chi-square statistic for this contingency table is $\chi_s^2 = 10.44$. Test the null hypothesis of independence against the alternative that the flies preferentially tend to return to their site of capture. Let $\alpha = .01$.

		SITE OF RECAPTURE	
		I	II
SITE OF	I	78	56
ORIGINAL CAPTURE	II	33	58

11.44 A study was conducted to evaluate the effectiveness of an oral antibiotic in the treatment of acne. The 42 patients who participated were randomly allocated to receive either antibiotic or placebo capsules. After 2 weeks of treatment, the subjective impressions of patient and physician (neither of whom knew which medication was being administered) were used to categorize the patient as "improved" or "not improved." The results were as summarized in the table.[32] How strongly do the results support the claim that the antibiotic is effective? Use a chi-square test against a directional alternative. Let $\alpha = .05$.

TREATMENT	NUMBER OF PATIENTS	IMPROVED	NO IMPROVEMENT
Antibiotic	23	18 (78%)	5 (22%)
Placebo	19	9 (47%)	10 (53%)

11.45 In the garden pea *Pisum sativum*, seed color can be yellow (Y) or green (G), and seed shape can be round (R) or wrinkled (W). Consider the following three hypotheses describing a population of plants:

$$H_0^{(1)}: \quad \Pr\{Y\} = \tfrac{3}{4}$$
$$H_0^{(2)}: \quad \Pr\{R\} = \tfrac{3}{4}$$
$$H_0^{(3)}: \quad \Pr\{R \mid Y\} = \Pr\{R \mid G\}$$

The first hypothesis asserts that yellow and green plants occur in a 3:1 ratio; the second hypothesis asserts that round and wrinkled plants occur in a 3:1 ratio, and the third hypothesis asserts that color and shape are independent. (In fact, for a population of plants produced by a certain cross—the dihybrid cross—all three hypotheses are known to be true.)

Suppose a random sample of 1,600 plants is to be observed, with the data to be arranged in the following contingency table:

		COLOR		
		Y	G	
SHAPE	R			
	W			
				1,600

Invent fictitious data sets as specified below, and verify each answer by calculating the estimated conditional probabilities. [*Hint:* In each case, begin with the marginal frequencies.]
a. A data set that agrees perfectly with $H_0^{(1)}$, $H_0^{(2)}$, and $H_0^{(3)}$
b. A data set that agrees perfectly with $H_0^{(1)}$ and $H_0^{(2)}$ but not with $H_0^{(3)}$
c. A data set that agrees perfectly with $H_0^{(3)}$ but not with $H_0^{(1)}$ or $H_0^{(2)}$

11.46 (*Computer exercise*)   In a study of the effects of smoking cigarettes during pregnancy, researchers examined the placenta from each of 58 women after childbirth. They noted the presence or absence (P or A) of a particular placental abnormality—atrophied villi. In addition, each woman was categorized as a nonsmoker (N), moderate smoker (M), or heavy smoker (H). The table on page 388 shows, for each woman, an ID number (#) and the results for smoking (S) and atrophied villi (V).[33]

#	S	V	#	S	V	#	S	V	#	S	V
1	N	A	16	H	P	31	M	A	46	M	A
2	M	A	17	H	P	32	M	A	47	H	P
3	N	A	18	N	A	33	N	A	48	H	P
4	M	A	19	M	P	34	N	A	49	H	A
5	M	A	20	N	P	35	N	A	50	N	P
6	M	P	21	M	A	36	H	P	51	N	A
7	H	P	22	H	A	37	N	A	52	M	P
8	N	A	23	M	P	38	H	P	53	M	A
9	N	A	24	N	A	39	H	P	54	H	P
10	M	P	25	N	P	40	N	A	55	H	A
11	N	A	26	N	A	41	M	A	56	M	P
12	N	P	27	N	A	42	N	A	57	H	P
13	H	P	28	M	P	43	H	A	58	H	P
14	M	A	29	N	A	44	M	A			
15	M	P	30	N	A	45	M	P			

a. Test for a relationship between smoking status and atrophied villi. Use a chi-square test at $\alpha = .05$.

b. Prepare a table which shows the total number of women in each smoking category, and the number and percentage in each category who had atrophied villi.

c. What pattern appears in the table of part **b** that is not used by the test of part **a**?

CONTENTS

COMPARING THE MEANS OF k INDEPENDENT SAMPLES

SECTION 12.1

INTRODUCTION

In Chapter 7 we considered the comparison of two independent samples with respect to a quantitative variable Y. The classical techniques for comparing the two sample means $\bar{y}_1$ and $\bar{y}_2$ are the test and the confidence interval based on Student's t distribution. In the present chapter we consider the comparison of the means of k independent samples, where k may be greater than 2. The following example illustrates an experiment with $k = 4$.

EXAMPLE 12.1

GROWTH OF SOYBEANS

A plant physiologist investigated the effect of mechanical stress on the growth of soybean plants. Individually potted seedlings were randomly allocated to four treatment groups of 13 seedlings each. Seedlings in two groups were stressed by shaking for 20 minutes twice daily, while two control groups were not stressed. Also, plants were grown in either low or moderate light. Thus, the treatments were:

Treatment 1:   Low light, control

Treatment 2:   Low light, stress

Treatment 3:   Moderate light, control

Treatment 4:   Moderate light, stress

After 16 days of growth, the plants were harvested, and the total leaf area (cm^2) of each plant was measured. The results are given in Table 12.1 and plotted in Figure 12.1.[1] Note that, although there is large variation within each treatment group, there are also considerable differences between the treatment means.

TABLE 12.1

Leaf Area (cm^2) of Soybean Plants

	TREATMENT			
	1	2	3	4
	264	235	314	283
	200	188	320	312
	225	195	310	291
	268	205	340	259
	215	212	299	216
	241	214	268	201
	232	182	345	267
	256	215	271	326
	229	272	285	241
	288	163	309	291
	253	230	337	269
	288	255	282	282
	230	202	273	257
Mean	245.3	212.9	304.1	268.8
SD	27.0	29.7	26.9	35.2
n	13	13	13	13

FIGURE 12.1
Leaf Area of Soybean
Plants Receiving Four
Different Treatments

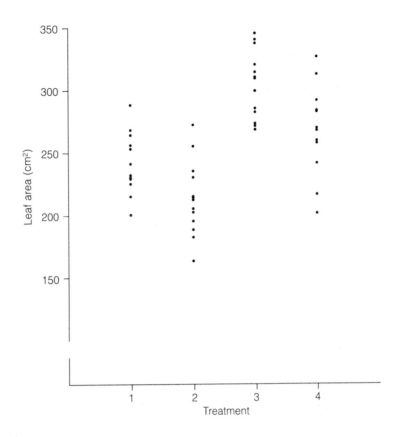

We will discuss the classical method of analyzing data from k independent samples. The method is called an **analysis of variance**, or **ANOVA**. This name is somewhat misleading, because the analysis is aimed at comparing *means* rather than variances; in Section 12.2 we will explain the origin of the name.

In applying analysis of variance, the data are regarded as random samples from k populations. We will denote the means of these populations as $\mu_1, \mu_2, \ldots, \mu_k$, and the standard deviations as $\sigma_1, \sigma_2, \ldots, \sigma_k$.

WHY NOT REPEATED t TESTS?

It is natural to wonder why the comparison of the means of k samples requires any new methods. For instance, why not just use a two-sample t test on each pair of samples? There are three reasons why this is not a good idea.

1. *The problem of multiple comparisons* The most serious difficulty with a naive "repeated t tests" procedure concerns Type I error: the probability of false rejection of a null hypothesis may be much higher than it appears to be. For

instance, suppose $k = 6$. Among six means there are 15 pairs of means, so that 15 hypotheses can be considered; these hypotheses are

(1) H_0: $\mu_1 = \mu_2$

(2) H_0: $\mu_1 = \mu_3$

(3) H_0: $\mu_2 = \mu_3$

$$\vdots$$

(15) H_0: $\mu_5 = \mu_6$

If t tests are used to test each of these hypotheses at $\alpha = .05$, then there is a 5% risk of Type I error for *each* of the 15 tests, but the *overall* risk, of *at least one* Type I error, is much higher than 5%.

The consequences of using repeated t tests are indicated by Table 12.2. For t tests at $\alpha = .05$, Table 12.2 shows the overall risk of Type I error,* that is,

> Overall risk = Probability that at least one of the t tests will reject its null hypothesis, when in fact $\mu_1 = \mu_2 = \cdots = \mu_k$

If $k = 2$, then the overall risk is .05, as it should be, but with larger k the risk increases rapidly; for $k = 6$ it is .37. It is clear from Table 12.2 that the researcher who uses repeated t tests is highly vulnerable to Type I error unless k is quite small.

TABLE 12.2

Overall Risk of Type I Error in Using Repeated t Tests at $\alpha = .05$

k	OVERALL RISK
2	.05
3	.12
4	.20
6	.37
8	.51
10	.63

The difficulties illustrated by Table 12.2 are due to **multiple comparisons**— that is, many comparisons on the same set of data. These difficulties can be reduced when the comparison of several groups is approached through ANOVA.

2. *Estimation of the standard deviation*   A pooled t test uses a standard deviation that is calculated by pooling the information from two samples. The ANOVA technique uses a pooled SD that is based on *all* of the samples simultaneously. This global sharing of information can yield improved precision in the analysis.

3. *Structure in the groups*   In many studies the logical structure of the treatments or groups to be compared may inspire questions that cannot be answered by

---

*Table 12.2 was computed assuming that the sample sizes are large and equal and that the population distributions are normal with equal standard deviations.

simple pairwise comparisons. For instance, for the soybean experiment of Example 12.1 we may ask whether the effect of stress on soybean growth is the same in low light as it is in moderate light. This question can be addressed using ANOVA (see optional Section 12.5).

A LOOK AHEAD

When data are analyzed by analysis of variance, the usual first step is to test the following **global null hypothesis**:

$$H_0: \quad \mu_1 = \mu_2 = \cdots = \mu_k$$

which asserts that all of the population means are equal. A statistical test of H_0 will be described in Section 12.3. (In some situations it is appropriate to use more powerful tests of the global null hypothesis; we will describe one of these in Chapter 13.)

If the global null hypothesis is rejected, then the data provide sufficient evidence to conclude that *some* of the μ's are unequal; the researcher would usually proceed to detailed comparisons to determine the *pattern* of differences among the μ's. If the global null hypothesis is not rejected, then the researcher might choose to construct one or more confidence intervals to characterize the lack of difference among the μ's.

All of the statistical procedures of this chapter—the test of the global null hypothesis and various methods of making detailed comparisons among the means—depend on the same basic calculations. These calculations are presented in Section 12.2.

| | | | | | | | | | | |

SECTION 12.2

THE BASIC ANALYSIS OF VARIANCE

In this section we present the basic ANOVA calculations that are used to describe the data and to facilitate further analysis. Recall from Chapter 7 that the pooled Student's t method uses two basic quantities from the data: (1) the quantity $(\bar{y}_1 - \bar{y}_2)$, which describes the difference between the sample means, and (2) the quantity s_c^2, which describes the variability within the samples. Analogously, the analysis of variance of k samples, or groups, begins with the calculation of quantities that describe the variability of the data *between* the groups and *within* the groups.* (For clarity, in this chapter we will often refer to the samples as "groups" of observations.)

NOTATION

To describe several groups of quantitative observations, we will use the following notation:

---

*Grammatically speaking, the word *among* should be used rather than *between* when referring to three or more groups; however, we will use "between" because it more clearly suggests that the groups are being compared against each other.

k = Number of groups

n = Number of observations in a group

$\bar{y}$ = Group mean

Sometimes, for clarity, we will use subscripts on the $\bar{y}$'s and the n's, thus:

$$\bar{y}_1, \quad \bar{y}_2, \quad \ldots, \quad \bar{y}_k,$$
$$n_1, \quad n_2, \quad \ldots, \quad n_k;$$

often we will omit them.

We will use two notations for summation: Σ will denote summation within a group and $\overset{*}{\Sigma}$ will denote summation across groups. Thus, a group mean is

$$\bar{y} = \frac{\Sigma y}{n}$$

The total number of observations is

$$n^* = \overset{*}{\Sigma} n$$

and the **grand mean**, the mean of all the observations, is

$$\bar{\bar{y}} = \frac{\overset{*}{\Sigma} \Sigma y}{n^*} = \text{Grand mean}$$

The following example illustrates this notation.

EXAMPLE 12.2

WEIGHT GAIN OF LAMBS

Table 12.3 shows the weight gains (in 2 weeks) of young lambs on three different diets. (These data are fictitious, but are realistic in all respects except for the fact that the group means are whole numbers.[2])

TABLE 12.3

Weight Gains
of Lambs (lb)

	DIET 1	DIET 2	DIET 3
	8	9	15
	16	16	10
	9	21	17
		11	6
		18	
n	3	5	4
Sum = Σy	33	75	48
Mean = $\bar{y}$	11	15	12

The total number of observations is

$$n^* = 3 + 5 + 4 = 12$$

and the total of all the observations is

$$\overset{*}{\sum}\sum y = 33 + 75 + 48 = 156$$

The grand mean is

$$\bar{\bar{y}} = \frac{156}{12} = 13 \text{ lb} \qquad \blacksquare$$

If the sample sizes (n's) are all equal, then the grand mean $\bar{\bar{y}}$ is just the simple average of the group means (the $\bar{y}$'s); but if the sample sizes are unequal this is not the case. For instance, in Example 12.2 note that

$$\frac{12 + 15 + 11}{3} \neq 13$$

VARIATION WITHIN GROUPS

A combined measure of variation within the k groups is the **sum of squares within groups**, or **SS(within)**, defined as follows:

DEFINITION OF SUM OF SQUARES WITHIN GROUPS

$$SS(\text{within}) = \overset{*}{\sum}\sum(y - \bar{y})^2$$

The double sum ($\overset{*}{\sum}\sum$) is calculated by first adding within each group ($\sum$) and then adding across groups ($\overset{*}{\sum}$). The following example illustrates the calculation of SS(within).

EXAMPLE 12.3
WEIGHT GAIN OF LAMBS

Table 12.4 shows the lamb weight-gain data, together with the group means and sums of squares.

TABLE 12.4
Calculation of SS(within) for Lamb Weight Gains

	DIET 1	DIET 2	DIET 3
	8	9	15
	16	16	10
	9	21	17
		11	6
		18	
n	3	5	4
Mean = $\bar{y}$	11	15	12
SS = $\Sigma(y - \bar{y})^2$	38	98	74

To calculate SS(within), we first calculate the quantity $\Sigma(y - \bar{y})^2$ for each group, as shown in Table 12.4; for instance:

$$(8 - 11)^2 + (16 - 11)^2 + (9 - 11)^2 = 38$$

Then we add across groups:

$$SS(within) = 38 + 98 + 74 = 210$$ ∎

Associated with SS(within) is a quantity called **degrees of freedom within groups**, or **df(within)**, which is defined as follows:

---
df WITHIN GROUPS

$$df(within) = n^* - k$$

Finally, we define the **mean square within groups**, or **MS(within)**, as follows:

MEAN SQUARE WITHIN GROUPS

$$MS(within) = \frac{SS(within)}{df(within)}$$

The MS(within) is actually an extension of a concept familiar from Chapter 7. Recall from Section 7.2 that the pooled variance s_c^2 from two samples is calculated as

$$s_c^2 = \frac{SS_1 + SS_2}{n_1 + n_2 - 2}$$

where each SS denotes the quantity $\Sigma(y - \bar{y})^2$ for a group. The MS(within) can be written analogously:

$$MS(within) = \frac{SS_1 + SS_2 + \cdots + SS_k}{n_1 + n_2 + \cdots + n_k - k}$$

Because of this relationship, MS(within) can be interpreted as a pooled variance, and we may write

$$s_c^2 = MS(within)$$

That is, s_c^2 is an alternate notation for MS(within). Thus, the pooled standard deviation s_c is given as follows:

POOLED STANDARD DEVIATION

$$s_c = \sqrt{MS(within)}$$

The pooled standard deviation s_c can be expressed in terms of the individual standard deviations $s_1, s_2, \ldots, s_k$, as follows:

$$s_c^2 = \frac{(n_1 - 1)s_1^2 + (n_2 - 1)s_2^2 + \cdots + (n_k - 1)s_k^2}{(n_1 - 1) + (n_2 - 1) + \cdots + (n_k - 1)}$$

(In Section 7.2 we discussed this relationship for $k = 2$.) The number of degrees of freedom associated with s_c (that is, the denominator of s_c^2) is the sum of the df associated with each sample SD:

$$\text{df(within)} = n^* - k = (n_1 - 1) + (n_2 - 1) + \cdots + (n_k - 1)$$

These relationships have a simple interpretation. Recall from Section 6.2 that the df associated with a sample SD is the number of independent pieces of information (about variability) upon which the SD is based. If we assume that the population SD is the same in all k populations, then s_c is an estimate of the population SD, and df(within) expresses the total amount of information upon which the estimate is based.

The following example illustrates the calculation and interpretation of MS(within) and s_c.

EXAMPLE 12.4

WEIGHT GAIN OF LAMBS

For the lamb growth data of Example 12.2, $k = 3$ and $n^* = 12$, so that

$$\text{df(within)} = 12 - 3 = 9$$

We found in Example 12.3 that SS(within) = 210; thus,

$$\text{MS(within)} = \frac{210}{9} = 23.333$$

and

$$s_c = \sqrt{23.333} = 4.83 \text{ lb}$$

If we assume that the population standard deviation of weight gains is the same for all three diets, then we estimate that standard deviation to be 4.83 lb.

Table 12.5 shows the individual sample SDs and their associated df. (Notice that s_c is within the range of the individual SDs; this will be true for any data set.) The individual SDs are estimates of the corresponding population SDs, and the df reflect the precision of the estimates. The pooled estimate s_c is based on $2 + 4 + 3 = 9$ df, and is related to the individual SDs as follows:

$$s_c^2 = \frac{2}{9} s_1^2 + \frac{4}{9} s_2^2 + \frac{3}{9} s_3^2$$

TABLE 12.5

Sample SDs and df for Lamb Weight Gains

	DIET 1	DIET 2	DIET 3
SD	4.36	4.95	4.97
df = $n - 1$	2	4	3

■

VARIATION BETWEEN GROUPS

For two groups, the difference between the groups is simply described by $(\bar{y}_1 - \bar{y}_2)$. How can we describe between-group variability for more than two groups? It turns out that a convenient measure is the **sum of squares between groups**, or **SS(between)**, defined as follows:

DEFINITION OF SUM OF SQUARES BETWEEN GROUPS

$$\text{SS(between)} = \overset{*}{\sum} n(\bar{y} - \bar{\bar{y}})^2$$

Each term in SS(between) is the square of a difference between the group mean $\bar{y}$ and the grand mean $\bar{\bar{y}}$, multiplied by the group size, n; this can be written explicitly as:

$$\text{SS(between)} = n_1(\bar{y}_1 - \bar{\bar{y}})^2 + n_2(\bar{y}_2 - \bar{\bar{y}})^2 + \cdots + n_k(\bar{y}_k - \bar{\bar{y}})^2$$

Associated with SS(between) is the **degrees of freedom between groups**, or **df(between)**, defined as follows:

---
df BETWEEN GROUPS

$$\text{df(between)} = k - 1$$

The **mean square between groups**, or **MS(between)**, is defined as follows:

MEAN SQUARE BETWEEN GROUPS

$$\text{MS(between)} = \frac{\text{SS(between)}}{\text{df(between)}}$$

The following example illustrates these definitions.

EXAMPLE 12.5
WEIGHT GAIN OF LAMBS

For the data of Example 12.2, the quantities that enter SS(between) are shown in Table 12.6.

TABLE 12.6
Calculation of
SS(between) for
Lamb Weight Gains

	DIET 1	DIET 2	DIET 3
n	3	5	4
Mean $\bar{y}$	11	15	12

Grand mean $\bar{\bar{y}} = 13$

From Table 12.6 we calculate

$$SS(\text{between}) = 3(11 - 13)^2 + 5(15 - 13)^2 + 4(12 - 13)^2 = 36$$

Since $k = 3$, we have

$$df(\text{between}) = 3 - 1 = 2$$

so that

$$MS(\text{between}) = \frac{36}{2} = 18$$

■

A FUNDAMENTAL RELATIONSHIP OF ANOVA

The name "analysis of variance" derives from a fundamental relationship involving SS(between) and SS(within). Suppose y denotes an individual observation; then it is obviously true that

$$y - \bar{\bar{y}} = (\bar{y} - \bar{\bar{y}}) + (y - \bar{y})$$

This equation expresses the deviation of an observation from the grand mean as the sum of two parts: a between-group deviation $(\bar{y} - \bar{\bar{y}})$ and a within-group deviation $(y - \bar{y})$. It is also true (but not at all obvious) that the analogous relationship holds for the corresponding sums of squares; that is,

$$\overset{*}{\sum}\sum(y - \bar{\bar{y}})^2 = \overset{*}{\sum}n(\bar{y} - \bar{\bar{y}})^2 + \overset{*}{\sum}\sum(y - \bar{y})^2 \qquad (12.1)$$

The quantity on the left-hand side of (12.1) is called the **total sum of squares**, or **SS(total)**:

DEFINITION OF TOTAL SUM OF SQUARES

$$SS(\text{total}) = \overset{*}{\sum}\sum(y - \bar{\bar{y}})^2$$

Note that SS(total) measures variability among all n^* observations in the k groups. The relationship (12.1) can be written as

---

RELATIONSHIP BETWEEN SUMS OF SQUARES

$$SS(\text{total}) = SS(\text{between}) + SS(\text{within})$$

The above fundamental relationship shows how the total variation in the data set can be *analyzed*, or broken down, into two interpretable components: between-sample variation and within-sample variation. This partition is an "analysis of variance."

The **total degrees of freedom**, or **df(total)**, is defined as follows:

TOTAL df
$$df(total) = n^* - 1$$

With this definition, the degrees of freeedom add, just as the sums of squares do; that is,

$$df(total) = df(between) + df(within)$$
$$n^* - 1 = (k - 1) + (n^* - k)$$

Notice that, if we were to consider all n^* observations as a single sample, then the SS for that sample (that is, the numerator of the variance) would be SS(total) and the associated df (that is, the denominator of the variance) would be df(total).

The following example illustrates the fundamental relationships between the sums of squares and degrees of freedom.

EXAMPLE 12.6
WEIGHT GAIN OF LAMBS

For the data of Table 12.3, we found $\bar{\bar{y}} = 13$; we calculate SS(total) as

$$SS(total) = \overset{*}{\sum}\sum(y - \bar{\bar{y}})^2$$
$$= [(8 - 13)^2 + (16 - 13)^2 + (9 - 13)^2]$$
$$+ [(9 - 13)^2 + (16 - 13)^2 + (21 - 13)^2 + (11 - 13)^2 + (18 - 13)^2]$$
$$+ [(15 - 13)^2 + (10 - 13)^2 + (17 - 13)^2 + (6 - 13)^2]$$
$$= 246$$

For these data, we found that SS(between) = 36 and SS(within) = 210. We verify that

$$246 = 36 + 210$$

Also, we found that df(between) = 2 and df(within) = 9. We verify that

$$df(total) = 12 - 1 = 11 = 2 + 9$$ ∎

THE ANOVA TABLE

When working with the ANOVA quantities, it is customary to arrange them in a table. The following example shows a typical format for the ANOVA table.

EXAMPLE 12.7
WEIGHT GAIN OF LAMBS

Table 12.7 shows the ANOVA for the lamb weight gain data. Notice that the ANOVA table clearly shows the additivity of the sums of squares and the degrees of freedom.

TABLE 12.7
ANOVA Table for Lamb
Weight Gains

SOURCE	SS	df	MS
Between diets	36	2	18
Within diets	210	9	23.333
Total	246	11	

∎

COMPUTATIONAL FORMULAS (OPTIONAL)

The formulas given above for the ANOVA sums of squares are *definitional* formulas. Recall that in Chapter 2 we gave a computational formula as well as the definitional formula for the standard deviation. Similarly, there are computational formulas for the ANOVA sums of squares. These computational formulas can be used to shorten the calculations when data analysis is done on a hand calculator. (In practice, however, there is a strong trend toward using computers to calculate ANOVAs.)

The computational formulas for the ANOVA sums of squares are as follows:

COMPUTATIONAL FORMULAS

$$SS(\text{between}) = \overset{*}{\sum}\left[\frac{\left(\sum y\right)^2}{n}\right] - \frac{\left(\overset{*}{\sum}\sum y\right)^2}{n^*}$$

$$SS(\text{within}) = \overset{*}{\sum}\sum y^2 - \overset{*}{\sum}\left[\frac{\left(\sum y\right)^2}{n}\right]$$

The following example illustrates the application of the computational formulas.

EXAMPLE 12.8
WEIGHT GAIN OF LAMBS

To apply the computational formulas to the lamb data of Example 12.2, we begin with the calculations shown in Table 12.8.

TABLE 12.8
Application of
Computational Formulas
to the Lamb Data

	DIET 1	DIET 2	DIET 3	
	8	9	15	
	16	16	10	
	9	21	17	
		11	6	
		18		
n	3	5	4	
$\sum y$	33	75	48	$\overset{*}{\sum}\sum y = 156$
$\sum y^2$	401	1,223	650	

We then calculate the following:

$$\overset{*}{\sum}\left[\frac{\left(\sum y\right)^2}{n}\right] = \frac{(33)^2}{3} + \frac{(75)^2}{5} + \frac{(48)^2}{4} = 2{,}064$$

$$\frac{\left(\overset{*}{\sum}\sum y\right)^2}{n^*} = \frac{(156)^2}{12} = 2{,}028$$

$$\overset{*}{\sum}\sum y^2 = 401 + 1{,}223 + 650 = 2{,}274$$

Finally, the sums of squares are calculated as

SS(between) = 2,064 − 2,028 = 36

SS(within) = 2,274 − 2,064 = 210

which are the same as the answers we obtained from the definitional formulas in Examples 12.3 and 12.5. ∎

SUMMARY OF DEFINITIONAL FORMULAS

For convenient reference, we display in the box the definitional formulas for the basic ANOVA quantities.

ANOVA QUANTITIES WITH DEFINITIONAL FORMULAS

SOURCE	SS (SUM OF SQUARES)	df	MS (MEAN SQUARE)
Between groups	$\overset{*}{\sum} n(\bar{y} - \bar{\bar{y}})^2$	$k - 1$	SS/df
Within groups	$\overset{*}{\sum}\sum (y - \bar{y})^2$	$n^* - k$	SS/df
Total	$\overset{*}{\sum}\sum (y - \bar{\bar{y}})^2$	$n^* - 1$	

EXERCISES 12.1–12.8

12.1 The accompanying table shows fictitious data for three samples.

	SAMPLE	
1	2	3
48	40	39
39	48	30
42	44	32
43		35
Mean 43	44	34

a. Compute SS(between) and SS(within), using the definitional formulas.

b. Use the definitional formula to compute SS(total), and verify the relationship between SS(between), SS(within), and SS(total).

c. Compute MS(between), MS(within), and s_c.

12.2 Proceed as in Exercise 12.1 for the following data:

	SAMPLE		
	1	2	3
	23	18	20
	29	12	16
	25	15	17
	23		23
			19
Mean	25	15	19

12.3 Proceed as in Exercise 12.1 for the following data:

	SAMPLE		
	1	2	3
	31	30	39
	38	26	45
	31	35	39
	32	29	37
		30	
Mean	33	30	40

12.4 The accompanying table presents fictitious data for three samples.

SAMPLE		
1	2	3
12	8	6
10	5	2
	3	4
	4	

a. Compute SS(between) and SS(within) using the definitional formulas.
b. Compute SS(between) and SS(within) using the computational formulas, and verify that the answers agree with those obtained in part **a.**

12.5 Proceed as in Exercise 12.4 for the following data:

SAMPLE		
1	2	3
2	7	5
5	9	4
2		7
		4

12.6 Invent examples of data with
    a. SS(between) = 0 and SS(within) > 0
    b. SS(between) > 0 and SS(within) = 0
For each example, use three samples, each of size 5. (Use definitional formulas.)

12.7 The table gives summary statistics for three fictitious samples.

	SAMPLE		
	1	2	3
$\bar{y}$	110	120	170
s	10	14	15
n	5	10	5

    a. Calculate SS(between). [*Hint:* First calculate Σy for each sample.]
    b. Calculate SS(within). [*Hint:* First calculate $\Sigma(y - \bar{y})^2$ for each sample.]

12.8 Proceed as in Exercise 12.7 for the following data:

	SAMPLE		
	1	2	3
$\bar{y}$	90	75	78
s	8	7	9
n	3	4	5

S E C T I O N 12.3

THE GLOBAL *F* TEST

The global null hypothesis is

$$H_0: \quad \mu_1 = \mu_2 = \cdots = \mu_k$$

We consider testing H_0 against the nondirectional alternative hypothesis

$$H_A: \quad \text{The } \mu\text{'s are not all equal}$$

Note that H_0 is compound (unless $k = 2$), and so rejection of H_0 does not specify *which* μ's are different. Detailed comparisons among the μ's require further analysis. Testing the global null hypothesis may be likened to looking at a microscope slide through a low-power lens to see if there is anything on it; if we find something, we switch to a greater magnification to examine its fine structure.

THE *F* DISTRIBUTIONS

The **F distributions**, named after the statistician and geneticist R. A. Fisher, are probability distributions that are used in many kinds of statistical analysis. The form of an *F* distribution depends on two parameters, called the **numerator degrees of freedom** and the **denominator degrees of freedom**. Figure 12.2 shows an *F* distribution with numerator df = 4 and denominator df = 20. Critical values for the *F* distribution are given in Table 9 at the end of this book. Note that Table 9 occupies ten pages, each page having a different value of the numerator df. As a specific example, for numerator df = 4 and denominator df = 20, we find in Table 9 that $F_{.05} = 2.87$; this value is shown in Figure 12.2.

FIGURE 12.2

The *F* Distribution with Numerator df = 4 and Denominator df = 20

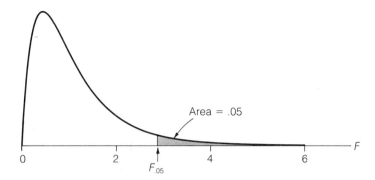

THE *F* TEST

The **F test** is a classical test of the global null hypothesis. The test statistic, the **F statistic**, is calculated as follows:

$$F_s = \frac{MS(\text{between})}{MS(\text{within})}$$

From the definitions of the mean squares (Section 12.2), it is clear that F_s will be large if the discrepancies among the group means ($\bar{y}$'s) are large relative to the variability within the groups. Thus, large values of F_s tend to provide evidence against H_0.

To carry out the *F* test of the global null hypothesis, critical values are obtained from an *F* distribution (Table 9) with

Numerator df = df(between)

and

Denominator df = df(within)

It can be shown that (when suitable conditions for validity are met) the null distribution of F_s is an *F* distribution with df as given above.

The following example illustrates the global *F* test.

EXAMPLE 12.9

WEIGHT GAIN OF LAMBS

For the lamb feeding experiment of Example 12.2, the global null hypothesis and alternative can be stated verbally as

H_0: Mean weight gain is the same on all three diets

H_A: Mean weight gain is not the same on all three diets

or symbolically as

H_0: $\mu_1 = \mu_2 = \mu_3$

H_A: The μ's are not all equal

Let us carry out the *F* test at $\alpha = .05$. From the ANOVA table (Table 12.7) we find

$$F_s = \frac{18}{23.333} = .77$$

The degrees of freedom can also be read from the ANOVA table as

Numerator df $= 2$

Denominator df $= 9$

From Table 9 we find $F_{.20} = 1.93$, so that $P > .20$. Thus, H_0 is not rejected; there is insufficient evidence to conclude that there is any difference among the diets with respect to population mean weight gain. The observed differences in the mean gains can readily be attributed to chance variation.   ■

RELATIONSHIP BETWEEN F TEST AND t TEST

Suppose only two groups are to be compared ($k = 2$). Then one could test H_0: $\mu_1 = \mu_2$ against H_A: $\mu_1 \neq \mu_2$ using either the F test or the t test. It can be shown that the F test and the pooled t test are actually equivalent procedures. The relationship between the test statistics is $t_s^2 = F_s$; that is, the value of the F statistic for any set of data is necessarily equal to the square of the value of the (pooled) t statistic. The corresponding relationship between the critical values is $t_{.05}^2 = F_{.05}$, $t_{.01}^2 = F_{.01}$, etc. For example, suppose $n_1 = 10$ and $n_2 = 7$. Then the appropriate t distribution has df $= n_1 + n_2 - 2 = 15$, so that $t_{.05} = 2.131$, whereas the F distribution has numerator df $= k - 1 = 1$ and denominator df $= n^* - k = 15$, so that $F_{.05} = 4.54$; note that $(2.131)^2 = 4.54$. Because of the equivalence of the tests, the application of the F test to compare the means of two samples will always give exactly the same P-value as the pooled t test applied to the same data.

EXERCISES 12.9–12.13

12.9 Monoamine oxidase (MAO) is an enzyme that is thought to play a role in the regulation of behavior. To see whether different categories of schizophrenic patients have different levels of MAO activity, researchers collected blood specimens from 42 patients and measured the MAO activity in the platelets. The results are summarized in the accompanying table. (Values are expressed as nmol benzylaldehyde product/ 10^8 platelets/hour.)[3] Calculations based on the raw data yielded SS(between) $= 136.12$ and SS(within) $= 418.25$.

DIAGNOSIS	MAO ACTIVITY		NO. OF PATIENTS
	Mean	SD	
Chronic undifferentiated schizophrenic	9.81	3.62	18
Undifferentiated with paranoid features	6.28	2.88	16
Paranoid schizophrenic	5.97	3.19	8

a. Construct the ANOVA table and test the global null hypothesis at $\alpha = .05$.

b. Calculate the pooled standard deviation, s_c.

12.10 It is thought that stress may increase susceptibility to illness through suppression of the immune system. In an experiment to investigate this theory, 48 rats were randomly allocated to four treatment groups: no stress, mild stress, moderate stress, and high stress. The stress conditions involved various amounts of restraint and electric shock. The concentration of lymphocytes (cells/ml $\times$ 10^{-6}) in the peripheral blood was measured for each rat, with the results given in the accompanying table.[4] Calculations based on the raw data yielded SS(between) = 89.036 and SS(within) = 340.24.

	NO STRESS	MILD STRESS	MODERATE STRESS	HIGH STRESS
$\bar{y}$	6.64	4.84	3.98	2.92
s	2.77	2.42	3.91	1.45
n	12	12	12	12

a. Construct the ANOVA table and test the global null hypothesis at $\alpha = .05$.

b. Calculate the pooled standard deviation s_c.

12.11 A plant physiologist investigated the effect of flooding on root metabolism in two tree species: flood-tolerant river birch and the intolerant European birch. Four seedlings of each species were flooded for 1 day and four were used as controls. The concentration of adenosine triphosphate (ATP) in the roots of each plant was measured. The data (nmol ATP per mg tissue) are shown in the table.[5] For these data, SS(between) = 4.5530 and SS(within) = .47438.

RIVER BIRCH		EUROPEAN BIRCH	
Flooded	Control	Flooded	Control
1.45	1.70	.21	1.34
1.19	2.04	.58	.99
1.05	1.49	.11	1.17
1.07	1.91	.27	1.30

a. Complete the ANOVA table and carry out the F test at $\alpha = .05$.

b. Assuming that each of the four populations has the same standard deviation, use the data to calculate an estimate of that standard deviation.

12.12 Refer to Exercise 12.11. Compute the sums of squares from the raw data, using the computational formulas.

12.13 Human beta-endorphin (HBE) is a hormone secreted by the pituitary gland under conditions of stress. An exercise physiologist measured the resting (unstressed) blood concentration of HBE in three groups of men: 15 who had just entered a physical fitness program, 11 who had been jogging regularly for some time, and 10 sedentary people. The HBE levels (pg/ml) are shown in the table at the top of page 408.[6] Calculations based on the raw data yielded SS(between) = 240.69 and SS(within) = 6,887.6.

	FITNESS PROGRAM ENTRANTS	JOGGERS	SEDENTARY
Mean	38.7	35.7	42.5
SD	16.1	13.4	12.8
n	15	11	10

a. Complete the ANOVA table and test the global null hypothesis. Let $\alpha = .05$.
b. Calculate the pooled standard deviation s_c.

SECTION 12.4

**APPLICABILITY
OF METHODS**

Like all methods of statistical inference, the calculations and interpretations of ANOVA are based on certain assumptions.

STANDARD ASSUMPTIONS

The ANOVA techniques described in this chapter, including the global F test, are valid if the following conditions are assumed to hold.

1. *Design conditions*
   a. It must be reasonable to regard the groups of observations as random samples from their respective populations. The observations within each sample must be independent of each other.
   b. The k samples must be independent of each other.
2. *Population conditions*   The k population distributions must be (approximately) normal with equal standard deviations:

$$\sigma_1 = \sigma_2 = \cdots = \sigma_k$$

Note that these conditions are direct extensions of the conditions given in Chapter 7 for the independent-samples t test. As with the t test, the assumption of normal populations with equal standard deviations is less crucial if the sample sizes (n) are large and approximately equal.

VERIFICATION OF CONDITIONS

The design conditions may be verified as for the independent-samples t test. To check condition 1(a), one looks for biases or hierarchical structure in the collection of the data. A completely randomized design assures independence of the samples [condition 1(b)]. If units have been allocated to treatment groups by a randomized blocks design, or if observations on the same experimental unit appear in different samples, then the samples are not independent. (In Chapter 9 we discussed dependence between samples for $k = 2$.)

As with the independent-samples t test, the population conditions can be roughly checked from the data. Equality of the population SDs is checked by comparing the sample SDs; one useful trick is to plot the SDs against the means

($\bar{y}$'s) to check for a trend. To check normality, a separate histogram or stem-and-leaf display can be made for each sample.*

FURTHER ANALYSIS

In addition to their relevance to the F test, the standard conditions underlie many classical methods for further analysis of the data.

If the k populations have the same SD, then a pooled estimate of that SD from the data is

$$s_c = \sqrt{MS(within)}$$

from the ANOVA. This pooled standard deviation s_c is a better estimate than any individual sample SD because s_c is based on more observations.

A simple way to see the advantage of s_c is to consider the standard error of an individual sample mean, which can be calculated as

$$SE_{\bar{y}} = \frac{s_c}{\sqrt{n}}$$

where n is the size of the individual sample. The df associated with this standard error is df(within), which is the sum of the degrees of freedom of all the samples. By contrast, if the individual SD were used in calculating $SE_{\bar{y}}$, it would have only $(n - 1)$ df. When the SE is used for inference, larger df yield smaller critical values (see Table 4), which in turn lead to improved power and narrower confidence intervals.

In optional sections 12.5 and 12.6 we will consider methods for detailed analysis of the group means $\bar{y}_1, \bar{y}_2, \ldots, \bar{y}_k$. Like the F test, these methods were designed for independent samples from normal populations with equal standard deviations. The methods use standard errors based on the pooled standard deviation estimate s_c.

| | | | | | | | | | | | | |

EXERCISE 12.14

12.14 Refer to the lymphocyte data of Exercise 12.10. The global F test is based on certain assumptions concerning the population distributions.
a. State the assumptions.
b. Which features of the data suggest that the assumptions may be doubtful in this case?

| | | | | | | | | | | | | |

S E C T I O N   12.5

LINEAR COMBINATIONS OF MEANS (OPTIONAL)

In many studies, interesting questions can be addressed by considering linear combinations of the group means. A **linear combination** L is a quantity of the form

$$L = m_1\bar{y}_1 + m_2\bar{y}_2 + \cdots + m_k\bar{y}_k$$

where the m's are multipliers of the $\bar{y}$'s.

*Another option is to make a single histogram of the quantities $(y - \bar{y})$ from all the samples.

LINEAR COMBINATIONS FOR ADJUSTMENT

One use of linear combinations is to "adjust" for an extraneous variable, as illustrated by the following example.

EXAMPLE 12.10
FORCED VITAL CAPACITY

One measure of lung function is forced vital capacity (FVC), which is the maximal amount of air a person can expire in one breath. In a public health survey, researchers measured FVC in a large sample of people. The results for male ex-smokers, stratified by age, are shown in Table 12.9.[7]

TABLE 12.9
FVC in Male Ex-Smokers

AGE (years)	n	FVC (liters) Mean	SD
25–34	83	5.29	.76
35–44	102	5.05	.77
45–54	126	4.51	.74
55–64	97	4.24	.80
65–74	73	3.58	.82
25–74	481	4.56	

Suppose it is desired to calculate a summary value for FVC in male ex-smokers. One possibility would be simply to calculate the grand mean of the 481 observed values, which is 4.56 liters. But the grand mean has a serious drawback: it cannot be meaningfully compared with other populations that may have different age distributions. For instance, suppose we were to compare ex-smokers with non-smokers; the observed difference in FVC would be distorted because ex-smokers as a group are (not surprisingly) older than nonsmokers. A summary measure that does not have this disadvantage is the "age-adjusted" mean, which is an estimate of the mean FVC value in a reference population with a specified age distribution. To illustrate, we will use the reference distribution in Table 12.10, which is (approximately) the distribution for the entire U.S. population.[8]

TABLE 12.10
Age Distribution in Reference Population

AGE	RELATIVE FREQUENCY
25–34	.32
35–44	.22
45–54	.18
55–64	.17
65–74	.11

The "age-adjusted" mean FVC value is the following linear combination:

$$L = .32\bar{y}_1 + .22\bar{y}_2 + .18\bar{y}_3 + .17\bar{y}_4 + .11\bar{y}_5$$

Note that the multipliers (m's) are the relative frequencies in the reference population. From Table 12.9, the value of L is

$$L = (.32)(5.29) + (.22)(5.05) + (.18)(4.51) + (.17)(4.24) + (.11)(3.58)$$
$$= 4.73 \text{ liters}$$

This value is an estimate of the mean FVC in an idealized population of people who are biologically like male ex-smokers but whose age distribution is that of the reference population.   ∎

CONTRASTS

A linear combination whose multipliers (m's) add to zero is called a **contrast**. The following example shows how contrasts can be used to describe the results of an experiment.

EXAMPLE 12.11
GROWTH OF SOYBEANS

Table 12.11 shows the treatment means and sample sizes for the soybean growth experiment of Example 12.1. We can use contrasts to describe the effects of stress in the two temperature conditions.

TABLE 12.11
Soybean Growth Data

TREATMENT	MEAN LEAF AREA (cm²)	n
1. Low light, control	245.3	13
2. Low light, stress	212.9	13
3. Moderate light, control	304.1	13
4. Moderate light, stress	268.8	13

a. First, note that an ordinary pairwise difference is a contrast. For instance, to measure the effect of stress in low light we can consider the contrast

$$L = \bar{y}_1 - \bar{y}_2 = 245.3 - 212.9 = 32.4$$

For this contrast, the multipliers are $m_1 = 1$, $m_2 = -1$, $m_3 = 0$, $m_4 = 0$; note that they add to zero.

b. To measure the effect of stress in moderate light we can consider the contrast

$$L = \bar{y}_3 - \bar{y}_4 = 304.1 - 268.8 = 35.3$$

For this contrast, the multipliers are $m_1 = 0$, $m_2 = 0$, $m_3 = 1$, $m_4 = -1$.

c. To measure the overall effect of stress, we can average the contrasts in parts **a** and **b** to obtain the contrast

$$L = \tfrac{1}{2}(\bar{y}_1 - \bar{y}_2) + \tfrac{1}{2}(\bar{y}_3 - \bar{y}_4)$$

$$= \tfrac{1}{2}(32.4) + \tfrac{1}{2}(35.3) = 33.85$$

For this contrast, the multipliers are $m_1 = \tfrac{1}{2}$, $m_2 = -\tfrac{1}{2}$, $m_3 = \tfrac{1}{2}$, $m_4 = -\tfrac{1}{2}$.   ∎

STANDARD ERROR OF A LINEAR COMBINATION

Each linear combination L is an estimate, based on the $\bar{y}$'s, of the corresponding linear combination of the population means (μ's). As a basis for statistical inference, we need to consider the standard error of a linear combination, which is calculated as follows.

STANDARD ERROR OF *L*

The standard error of the linear combination

$$L = m_1\bar{y}_1 + m_2\bar{y}_2 + \cdots + m_k\bar{y}_k$$

is

$$SE_L = \sqrt{s_c^2 \overset{*}{\sum}\left(\frac{m^2}{n}\right)}$$

where $s_c^2 = MS(within)$ from the ANOVA.

Recall that $\overset{*}{\sum}$ denotes summation across groups. The SE can be written explicitly as

$$SE_L = \sqrt{s_c^2\left(\frac{m_1^2}{n_1} + \frac{m_2^2}{n_2} + \cdots + \frac{m_k^2}{n_k}\right)}$$

If all the sample sizes (n) are equal, the SE can be written as

$$SE_L = \sqrt{\frac{s_c^2}{n}(m_1^2 + m_2^2 + \cdots + m_k^2)} = \sqrt{\frac{s_c^2}{n}\overset{*}{\sum}m^2}$$

The following two examples illustrate the application of the standard error formula.

EXAMPLE 12.12
FORCED VITAL CAPACITY

For the linear combination L defined in Example 12.10, we find that

$$\overset{*}{\sum}\left(\frac{m^2}{n}\right) = \frac{(.32)^2}{83} + \frac{(.22)^2}{102} + \frac{(.18)^2}{126} + \frac{(.17)^2}{97} + \frac{(.11)^2}{73}$$
$$= .0024291$$

The ANOVA for these data yields $s_c^2 = .59989$. Thus, the standard error of L is

$$SE_L = \sqrt{(.59989)(.0024291)} = .0382$$ ∎

EXAMPLE 12.13
GROWTH OF SOYBEANS

For the linear combination L defined in Example 12.11(a), we find that

$$\overset{*}{\sum}m^2 = (1)^2 + (-1)^2 + (0)^2 + (0)^2 = 2$$

so that

$$SE_L = \sqrt{s_c^2\left(\frac{2}{13}\right)}$$

Note that this SE is the same as would be obtained by the pooled t method of Chapter 7, except that here s_c is pooled from all four groups rather than from only two. ∎

STATISTICAL INFERENCE

Linear combinations of means can be used for testing hypotheses and for constructing confidence intervals. Critical values are obtained from Student's t distribution with

$$df = df(within)$$

from the ANOVA.* Confidence intervals are constructed using the familiar Student's t format. For instance, a 95% confidence interval is

$$L \pm t_{.05}\, SE_L$$

To test the null hypothesis that the population value of a contrast is zero, the test statistic is calculated as

$$t_s = \frac{L}{SE_L}$$

and the t test is carried out in the usual way.

The following example illustrates the construction of the confidence interval. (The t test will be illustrated in Example 12.15.)

EXAMPLE 12.14
GROWTH OF SOYBEANS

Consider the contrast defined in Example 12.11(c):

$$L = \tfrac{1}{2}(\bar{y}_1 - \bar{y}_2) + \tfrac{1}{2}(\bar{y}_3 - \bar{y}_4)$$

This contrast is an estimate of the quantity

$$\tfrac{1}{2}(\mu_1 - \mu_2) + \tfrac{1}{2}(\mu_3 - \mu_4)$$

which can be described as the true (population) effect of stress, averaged over the light conditions.† Let us construct a 95% confidence interval for this true difference.

*This method of determining critical values does not take account of multiple comparisons. See Section 14.2.

†This is sometimes called the *main effect* of stress.

We found in Example 12.11 that the value of L is

$$L = 33.85$$

To calculate SE_L, we first calculate

$$\overset{*}{\sum} m^2 = \left(\tfrac{1}{2}\right)^2 + \left(-\tfrac{1}{2}\right)^2 + \left(\tfrac{1}{2}\right)^2 + \left(-\tfrac{1}{2}\right)^2 = 1$$

From the ANOVA, which is shown in Table 12.12, we find that $s_c^2 = 895.34$; thus,

$$SE_L = \sqrt{\frac{s_c^2(1)}{n}} = \sqrt{\frac{895.34}{13}} = 8.299$$

TABLE 12.12
ANOVA for Soybean
Growth Data

SOURCE	SS	df	MS
Between treatments	57,636.4	3	19,212.13
Within treatments	42,976.3	48	895.34

From Table 4 with df $= 40 \approx 48$, we find $t_{.05} = 2.021$. The confidence interval is

$$33.85 \pm (2.021)(8.299)$$
$$33.85 \pm 16.77$$

The confidence interval is written in full as

$$17.1 \text{ cm}^2 < \tfrac{1}{2}(\mu_1 - \mu_2) + \tfrac{1}{2}(\mu_3 - \mu_4) < 50.6 \text{ cm}^2$$

We are 95% confident that the effect of stress, averaged over the light conditions, is to reduce the leaf area by an amount whose mean value is between 17.1 cm² and 50.6 cm². ∎

CONTRASTS TO ASSESS INTERACTION

Often an investigator wishes to study the separate and joint effects of two or more factors on a response variable Y. Linear contrasts can be used to study the way in which two factors interact. The following is an example.

EXAMPLE 12.15
GROWTH OF SOYBEANS

In the soybean experiment (Example 12.11), the two factors of interest are stress condition and light level. Table 12.13 shows the treatment means, arranged in a new format which permits us easily to consider the factors separately and together.

TABLE 12.13

Mean Leaf Areas for
Soybean Experiment

	STRESS CONDITION		
	Control	Stress	Difference
Low Light	245.3 (1)	212.9 (2)	32.4
Moderate Light	304.1 (3)	268.8 (4)	35.3

At each light level, the mean effect of stress can be measured by a contrast:

Effect of stress in low light: $\qquad$ $\bar{y}_1 - \bar{y}_2 = 32.4$

Effect of stress in moderate light: $\quad \bar{y}_3 - \bar{y}_4 = 35.3$

Now consider the question: Is the effect of stress the same in both light conditions? One way to address this question is to compare $(\bar{y}_1 - \bar{y}_2)$ versus $(\bar{y}_3 - \bar{y}_4)$; the difference between these two values is a contrast:

$$L = (\bar{y}_1 - \bar{y}_2) - (\bar{y}_3 - \bar{y}_4)$$
$$= 32.4 - 35.3 = -2.9$$

This contrast L can be used as the basis for a confidence interval or a test of hypothesis. We illustrate the test. The null hypothesis is

$$H_0: \quad (\mu_1 - \mu_2) = (\mu_3 - \mu_4)$$

or, in words,

H_0: The effect of stress is the same in the two light conditions

For the above L, $\overset{*}{\Sigma}m^2 = 4$, and the standard error is

$$SE_L = \sqrt{\frac{s_c^2(4)}{13}} = \sqrt{\frac{(895.34)(4)}{13}} = 16.6$$

The test statistic is

$$t_s = \frac{-2.9}{16.6} = -.2$$

From Table 4 with df $= 40$ we find $t_{.20} = 1.303$. The data provide virtually no evidence that the effect of stress is different in the two light conditions. ∎

If the joint influence of two factors is equal to the sum of their separate influences, the two factors are said to be **additive** in their effects. For instance, consider the soybean experiment of Example 12.15. If stress reduces mean leaf area by the same amount in either light condition, then the effect of stress (a negative effect, in this case) is *added* to the effect of light. To visualize this additivity of effects, consider Figure 12.3, which shows the treatment means. The two dashed lines are almost parallel because the data display a pattern of nearly perfect additivity.

FIGURE 12.3
Treatment Means for
Soybean Experiment

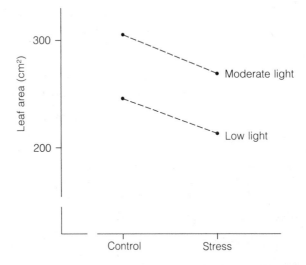

If effects are *not* additive—that is, if the joint influence of two factors is *different* from the sum of the separate influences—the factors are said to exhibit **interaction**. Thus, the null hypothesis tested in Example 12.15 can be stated as:*

H_0: The effects of stress and light level are additive (with respect to leaf area of soybeans)

or

H_0: There is no interaction between stress and light level (with respect to leaf area of soybeans)

If the hypothesis of additivity is true, then the effect of each factor can be estimated by combining levels of the other factor. For instance, in Example 12.14 we estimated the effect of stress by averaging over the light levels; this averaging process is very natural if we assume that the true effect of stress is the same at both light levels.

The concept of interaction occurs throughout biology. The terms "synergism" and "antagonism" describe interactions between biological agents. The term "epistasis" describes interaction between genes at two loci.

The *statistical* definition of interaction is rather specialized. It is defined in terms of the observed variable rather than in terms of a biological mechanism. Further, interaction as measured by a contrast is defined by *differences* between means. In some applications the biologist might feel that *ratios* of means are

*It is understood that this statement refers to only the two light levels under discussion.

more meaningful or relevant than differences. The following example shows that the two points of view can lead to different answers.

EXAMPLE 12.16

CHROMOSOMAL
ABERRATIONS

A research team investigated the separate and joint effects in mice of exposure to high temperature (35°C) and injection with the cancer drug cyclophosphamide (CTX). A completely randomized design was used, with eight mice in each treatment group. For each animal, the researchers measured the incidence of a certain chromosomal aberration in the bone marrow; the result is expressed as the number of abnormal cells per 1,000 cells. The treatment means are shown in Table 12.14.[9]

TABLE 12.14

Mean Incidence of
Chromosomal Aberrations
Following Various
Treatments

| | | INJECTION | |
		CTX	None
TEMPERATURE	Room	23.5	2.7
	High	75.4	20.9

Is the observed effect of CTX greater at room temperature or at high temperature? The answer depends on whether "effect" is measured absolutely or relatively.

Measured as a difference, the effect of CTX is

Room temperature: $23.5 - 2.7 = 20.8$

High temperature: $75.4 - 20.9 = 54.5$

Thus, the absolute effect of CTX is greater at the high temperature. However, this relationship is reversed if we express the effect of CTX as a ratio rather than as a difference:

Room temperature: $\dfrac{23.5}{2.7} = 8.70$

High temperature: $\dfrac{75.4}{20.9} = 3.61$

At room temperature CTX produces almost a nine-fold increase in chromosomal aberrations, whereas at high temperature the increase is less than four-fold; thus, in relative terms, the effect of CTX is much greater at room temperature. ■

If the phenomenon under study is thought to be multiplicative rather than additive, so that relative rather than absolute change is of primary interest, then ordinary contrasts should not be used. One simple approach in this situation is to use a logarithmic transformation—that is, to compute $Y' = \log(Y)$, and then analyze Y' using contrasts. The motivation for this approach is that relations of constant *relative* magnitude in the Y scale become relations of constant *absolute* magnitude in the Y' scale.

We have considered interaction only in a very simple setting—when there are two factors, each at two levels (for instance, stress/control and low/moderate light). For more than two factors or more than two levels, the study of interaction requires techniques beyond the scope of this book.

12.15 Refer to the FVC data of Example 12.10.

a. Verify that the grand mean of all 481 FVC values is 4.56.

b. Taking into account the age distribution among the 481 subjects and the age distribution in the U.S. population, explain intuitively why the grand mean (4.56) is smaller than the age-adjusted mean (4.73).

12.16 To see if there is any relationship between blood pressure and childbearing, researchers examined data from a large health survey. The following table shows the data on systolic blood pressure (mm Hg) for women who had borne no children and women who had borne five or more children. The pooled standard deviation from all eight groups was $s_c = 18$ mm Hg.[10]

AGE	NO CHILDREN		FIVE OR MORE CHILDREN	
	Mean Blood Pressure	No. of Women	Mean Blood Pressure	No. of Women
18–24	113	230	114	7
25–34	118	110	116	82
35–44	125	105	124	127
45–54	134	123	138	124
18–54	121	568	127	340

Carry out age-adjustment, as directed below, using the following reference distribution, which is the approximate distribution for U.S. women:[11]

AGE	RELATIVE FREQUENCY
18–24	.25
25–34	.33
35–44	.22
45–54	.20

a. Calculate the age-adjusted mean blood pressure for women with no children.

b. Calculate the age-adjusted mean blood pressure for women with five or more children.

c. Calculate the difference between the values obtained in parts **a** and **b**. Explain intuitively why the result is smaller than the unadjusted difference of $127 - 121 = 6$ mg Hg.

d. Calculate the standard error of the value calculated in part **a**.

e. Calculate the standard error of the value calculated in part **c**.

12.17 Refer to the ATP data of Exercise 12.11. The sample means and standard deviations are as follows:

	RIVER BIRCH		EUROPEAN BIRCH	
	Flooded	Control	Flooded	Control
$\bar{y}$	1.19	1.78	.29	1.20
s	.18	.24	.20	.16

Define linear combinations (that is, specify the multipliers) to measure each of the following:

a. The effect of flooding in river birch
b. The effect of flooding in European birch
c. The difference between river birch and European birch with respect to the effect of flooding (that is, the interaction between flooding and species)

12.18 (*Continuation of Exercise 12.17*)

a. Use a t test to investigate whether flooding has the same effect in river birch and in European birch. Use a nondirectional alternative and let $\alpha = .05$. (The pooled standard deviation is $s_c = .199$.)
b. If the sample sizes were $n = 10$ rather than $n = 4$ for each group, but the means, standard deviations, and s_c remained the same, how would the result of part **a** change?

12.19 (*Continuation of Exercise 12.17*)   Consider the null hypothesis that flooding has no effect on ATP level in river birch. This hypothesis could be tested in two ways: as a contrast (using the method of Section 12.5), or with a two-sample t test (as in Exercise 7.28). Answer the following questions; do not actually carry out the tests.

a. In what way or ways do the two test procedures differ?
b. In what way or ways do the conditions for validity of the two procedures differ?
c. One of the two procedures requires more assumptions for its validity, but if the assumptions are met then this procedure has certain advantages over the other one. What are these advantages?

12.20 A completely randomized double-blind clinical trial was conducted to compare two drugs, ticrynafen (T) and hydrochlorothiazide (H), for effectiveness in treatment of high blood pressure. Each drug was given at either a low or a high dosage level for 6 weeks. The accompanying table shows the results for the drop (baseline minus final value) in systolic blood pressure (mm Hg).[12] The pooled standard deviation was $s_c = 11.83$ mm Hg.

	TICRYNAFEN (T)		HYDROCHLOROTHIAZIDE (H)	
	Low Dose	High Dose	Low Dose	High Dose
Mean	13.9	17.1	15.8	17.5
No. of patients	53	57	55	58

a. The difference in response between T and H appears to be larger for the low dose than for the high dose. Use a t test to assess whether this pattern can be ascribed to chance variation. Let $\alpha = .10$.

b. If the two drugs have equal effects on blood pressure, then T might be preferable because it has fewer side effects. Construct a 95% confidence interval for the difference between the drugs (with respect to mean blood pressure reduction), averaged over the two dosage levels.

12.21 In a study of lettuce growth, 36 seedlings were randomly allocated to receive either high or low light and to be grown in either a standard nutrient solution or one containing extra nitrogen. After 16 days of growth, the lettuce plants were harvested and the dry weight of the leaves was determined for each plant. The accompanying table shows the mean leaf dry weight (gm) of the nine plants in each treatment group.[13] The MS(within) from the ANOVA was .3481.

	NUTRIENT SOLUTION	
	Standard	Extra Nitrogen
Low Light	2.16	3.09
High Light	3.26	4.48

a. Use a *t* test to compare the effect of extra nitrogen in the two light conditions (that is, to test for interaction between nitrogen level and light level). Use a nondirectional alternative and let $\alpha = .05$.
b. Construct a 95% confidence interval for the effect of extra nitrogen, averaged over the two light conditions.

12.22 Refer to the MAO data of Exercise 12.9.
a. Define a contrast to compare the MAO activity for schizophrenics without paranoid features versus the average of the two types with paranoid features.
b. Calculate the value of the contrast in part a and its standard error.
c. Apply a *t* test to the contrast in part a. Let H_A be nondirectional and $\alpha = .05$.

12.23 Are the brains of left-handed people anatomically different? To investigate this question, a neuroscientist conducted postmortem brain examinations in 42 people. Each person had been evaluated before death for hand preference and categorized as consistently right-handed (CRH) or mixed-handed (MH). The table shows the results on the area of the anterior half of the corpus callosum (the structure that links the left and right hemispheres of the brain).[14] The MS(within) from the ANOVA was 2,498.

GROUP	AREA (mm²)		
	Mean	SD	*n*
1. Males: MH	423	48	5
2. Males: CRH	367	49	7
3. Females: MH	377	63	10
4. Females: CRH	345	43	20

a. The difference between MH and CRH is 56 mm² for males and 32 mm² for females. Is this sufficient evidence to conclude that the corresponding population difference is greater for males than for females? Test an appropriate hypothesis. (Use a nondirectional alternative and let $\alpha = .10$.)

b. As an overall measure of the difference between MH and CRH, one can consider the quantity $.5(\mu_1 - \mu_2) + .5(\mu_3 - \mu_4)$. Construct a 95% confidence interval for this quantity. (This is a sex-adjusted comparison of MH and CRH, where the reference population is 50% male and 50% female.)

SECTION 12.6

THE NEWMAN–KEULS PROCEDURE (OPTIONAL)

One approach to detailed analysis of the means $\bar{y}_1, \bar{y}_2, \ldots, \bar{y}_k$ is to make every pairwise comparison among them. Suppose it is desired to test all possible pairwise hypotheses:

$$H_0: \quad \mu_1 = \mu_2$$
$$H_0: \quad \mu_1 = \mu_3$$
$$H_0: \quad \mu_2 = \mu_3$$

and so on

We saw in Section 12.1 that using repeated t tests leads to an increased overall risk of Type I error. There are several methods that can be used to control the overall risk of Type I error. One of these is the **Newman–Keuls procedure**, which we now describe. The procedure is designed for use when the k sample sizes (n) are equal.

DESCRIPTION OF THE PROCEDURE

The Newman–Keuls procedure is a decision-oriented procedure, conducted at a prespecified overall significance level α. For each pair of means (for instance, $\bar{y}_1$ vs $\bar{y}_2$), the procedure leads to a decision as to whether the corresponding null hypothesis (for instance $H_0: \mu_1 = \mu_2$) is rejected.

We give a step-by-step description of the Newman–Keuls procedure, and then illustrate with an example. (Although the description is lengthy, the procedure itself is not complicated.)

STEP 1. *Array of means* Construct an array of the sample means arranged in increasing order.

STEP 2. *Critical values* Table 10 (at the end of this book) provides constants,* denoted as q_j, for the Newman–Keuls procedure at $\alpha = .05$ or $\alpha = .01$. To use Table 10, first determine MS(within) and df(within) from the ANOVA. From the row of Table 10 corresponding to df $=$ df(within), read the values of q_j for $j = 2, 3, \ldots, k$. Next, calculate the following scale factor:

$$\sqrt{\frac{s_c^2}{n}}$$

*Technically, the constants q_j are called "percentage points of the Studentized range distribution."

where $s_c^2 = $ MS(within). (Note that this is $SE_{\bar{y}}$ as given in Section 12.4.) Finally, calculate critical values, denoted R_j, as follows:

$$R_j = q_j \sqrt{\frac{s_c^2}{n}}$$

The work for step 2 can be conveniently arranged in a table as follows:

j	2	3	$\cdots$	*k*
q_j	q_2	q_3	$\cdots$	q_k
R_j	R_2	R_3	$\cdots$	R_k

STEP 3.   *The pairwise comparisons*   The R_j's are the critical values with which the differences between sample means will be compared; larger R_j's will be used for the means that are farther apart in the array of means constructed in step 1. As a convenient way of keeping track of the results, nonrejection of a null hypothesis will be indicated by underlining the corresponding pair of means. The procedure is carried out sequentially, as follows:

a. Compare the difference between the largest and smallest of the *k* sample means with the critical value R_k. If the difference is smaller than R_k, the corresponding null hypothesis is not rejected; in this case a line is drawn under the entire array of means and the procedure is ended. If the difference is larger than R_k, proceed to step (b).

b. Ignore the smallest $\bar{y}$ and consider the remaining subarray of $(k-1)$ means. Compare the difference between the largest and smallest mean in the subarray with R_{k-1}; if the difference is less than R_{k-1}, underline the entire subarray. Now consider the other subarray of $(k-1)$ means—the means that remain if the largest $\bar{y}$ is ignored. Again, underline this subarray if the difference between its largest and smallest mean is less than R_{k-1}. (When underlining, use a *separate* line each time; never join a line to one that has already been drawn.)

c. Continue by looking at all subarrays of $(k-2)$ means and comparing with R_{k-2}, then subarrays of $(k-3)$ means and comparing with R_{k-3}, and so on until, finally, each subarray of two means is compared with R_2. During this procedure, however, *never test within any subarray that has already been underlined; all hypotheses in such a subarray are automatically not rejected.*

d. When the procedure is complete, those pairs of means not connected by an underline correspond to null hypotheses that have been rejected. All other pairwise null hypotheses are not rejected.

ILLUSTRATION OF THE PROCEDURE

The following example shows how to carry out the Newman–Keuls procedure.

**BLOOD CHEMISTRY
IN RATS**

For an evaluation of diets used for routine maintenance of laboratory rats, researchers used a completely randomized design to allocate weanling male rats to five different diets. After 4 weeks, specimens of blood were collected and various biochemical variables were measured. We consider the results for blood urea concentration (mg/dl). The group means were as follows:[15]

Diet	A	B	C	D	E
$\bar{y}$	40.0	40.7	32.9	29.6	48.8

The ANOVA is shown in Table 12.15.

TABLE 12.15
ANOVA for Blood
Urea Data

SOURCE	SS	df	MS
Between diets	894.80	4	223.70
Within diets	319.35	15	21.29
Total	1,214.15	19	

We now apply the Newman–Keuls procedure to compare every pair of diets at $\alpha = .05$.

STEP 1. The ordered array is as follows:

DIET	D	C	A	B	E
MEAN	29.6	32.9	40.0	40.7	48.8

STEP 2. The number of rats in each group was $n = 4$; obtaining MS(within) from Table 12.15, we calculate the scale factor

$$\sqrt{\frac{s_c^2}{n}} = \sqrt{\frac{21.29}{4}} = 2.307$$

We read the values q_j from Table 10 with df = 15; we then multiply each q_j by 2.307 to obtain R_j. The results are shown in Table 12.16.

TABLE 12.16
Critical Values for
Example 12.17

j	2	3	4	5
q_j	3.01	3.67	4.08	4.37
R_j	6.9	8.5	9.4	10.1

STEP 3. We first compare the largest mean against the smallest, using the critical value R_5. We find

$$\bar{y}_E - \bar{y}_D = 48.8 - 29.6 = 19.2$$
$$R_5 = 10.1$$

Because $19.2 > 10.1$, we reject the null hypothesis $H_0: \mu_D = \mu_E$. We then proceed to compare $\bar{y}_E$ against $\bar{y}_C$, using the critical value R_4. The entire sequence of comparisons is shown in Table 12.17.

TABLE 12.17
Newman–Keuls Analysis for Example 12.17

VALUE OF *j*	COMPARISON	CONCLUSION
5	$48.8 - 29.6 = 19.2 > 10.1$	Reject
4	$48.8 - 32.9 = 15.9 > 9.4$	Reject
4	$40.7 - 29.6 = 11.1 > 9.4$	Reject
3	$48.8 - 40.0 = 8.8 > 8.5$	Reject
3	$40.7 - 32.9 = 7.8 < 8.5$	Do not reject (line from C to B)
3	$40.0 - 29.6 = 10.4 > 8.5$	Reject
2	$48.8 - 40.7 = 8.1 > 6.9$	Reject
2	$40.7 - 40.0$: Already underlined; do not reject.	
2	$40.0 - 32.9$: Already underlined; do not reject.	
2	$32.9 - 29.6 = 3.3 < 6.9$	Do not reject (line from D to C)

Note that the difference $40.7 - 40.0$ is *not* compared to 6.9 because the subarray containing that pair is already underlined, and similarly for the difference $40.0 - 32.9$. This is an essential feature of the Newman–Keuls procedure.
The final array is as follows:

DIET	D	C	A	B	E
MEAN	29.6	32.9	40.0	40.7	48.8

Note that D and B are *not* joined by an underline, even though D and C are joined by an underline and C and B are joined by an underline. The overlapping of the underlines reflects the fact that we rejected the null hypothesis $H_0: \mu_D = \mu_B$ but we did not reject $H_0: \mu_D = \mu_C$ or $H_0: \mu_C = \mu_B$. This may seem contradictory, but it is not if you remember that nonrejection of a null hypothesis is absence of evidence, rather than evidence of absence, of a difference. The data provide enough information to conclude that $\mu_A > \mu_D$, but not enough to conclude that $\mu_A > \mu_C$ nor enough to conclude that $\mu_C > \mu_D$.

We can describe our conclusions verbally as follows. At $\alpha = .05$, there is sufficient evidence to conclude that diet E gives the highest mean blood urea; either diet D or C gives the lowest; μ_A and μ_B are greater than μ_D but not necessarily greater than μ_C; we cannot say which of μ_A or μ_B is greater. Note that the Newman–Keuls procedure takes a strictly decision-oriented approach; there are no *P*-values associated with the various differences. ∎

EXAMPLES OF POSSIBLE PATTERNS

In Example 12.17, the Newman–Keuls procedure yielded a moderately complicated pattern of results. Here are some other possible patterns that might emerge for five treatments A, B, C, D, and E.

Pattern 1:

D⎯⎯C   A⎯⎯B   E

In pattern 1 there is no overlapping of underlines, so the treatments fall into distinct groups.

Pattern 2:

D   C   A   B   E

In pattern 2 there are no underlines; all pairwise null hypotheses have been rejected.

Pattern 3:

D   C   A   B   E

Pattern 3 is more subtle to interpret. The overlapping indicates that adjacent treatments are not distinguishable even though more distant ones are (just as black can shade into white through imperceptible stages of grey).

RELATION TO THE *t* TEST

The critical value R_2 deserves special comment. The comparison of two means using R_2 is very similar to a pooled two-sample t test; in fact, it differs from a t test only in that the quantity s_c (which enters the SE calculation) is derived from all k samples rather than just two samples. (This relationship is explained in more detail in Appendix 12.1.)

In spite of the close link between R_2 and the t test, the Newman–Keuls method does not suffer from Type I error risks as high as those given in Table 12.2 for repeated t tests. The Newman–Keuls procedure has less chance of Type I error because many of its comparisons use the larger critical values (R_3, R_4, and so on) rather than R_2, and also because comparisons within an underline are automatically nonsignificant. [In Example 12.17, for instance, notice that the difference ($\bar{y}_B - \bar{y}_C$) and the difference ($\bar{y}_A - \bar{y}_C$) are both greater than R_2 and yet the corresponding null hypotheses are not rejected by the Newman–Keuls procedure.]

RELATION TO THE *F* TEST

Although it is customary to precede the Newman–Keuls procedure with an F test of the global null hypothesis, it is not necessary to do so. It is possible for the global F test and the first Newman–Keuls comparison (using R_k) to disagree, but this is rare.

CONDITIONS FOR VALIDITY

The conditions for validity of the Newman–Keuls procedure consist of the standard conditions given in Section 12.4, together with the requirement that the sample sizes all be equal. In practice, it often happens that the sample sizes are slightly unequal (for instance, an experimental animal may die for reasons unrelated to the experiment); in this case the procedure is approximately valid.

OTHER MULTIPLE COMPARISON PROCEDURES

In addition to the Newman–Keuls method, statisticians have developed several other methods for multiple comparison of means.* The methods differ from each other in the *degree* of protection they provide against Type I error. All of the methods have the property that the chance of Type I error is controlled (say, at .05) when the global null hypothesis is true. But Type I errors can occur even when the global null hypothesis is false. For instance, suppose $\mu_1 = 30$, $\mu_2 = 30$, and $\mu_3 = 40$; then rejection of the hypothesis $H_0: \mu_1 = \mu_2$ would be a Type I error. A conservative procedure, which guards stringently against all possible kinds of Type I error, is not as powerful as a less conservative procedure. Compared to the other procedures, the Newman–Keuls procedure is moderately conservative. (In Appendix 12.1 we give more detail on the different degrees of control of Type I error.)

Certain of the other multiple comparison methods can also be used to construct confidence intervals. The Newman–Keuls procedure does not share this advantage.

EXERCISES
12.24–12.28

12.24 A botanist used a completely randomized design to allocate 45 individually potted eggplant plants to five different soil treatments. The observed variable was the total plant dry weight without roots (gm) after 31 days of growth. The treatment means were as shown in the table.[16] The MS(within) was .2246. Use the Newman–Keuls method to compare all pairs of means at $\alpha = .05$.

TREATMENT	A	B	C	D	E
Mean	4.37	4.76	3.70	5.41	5.38
n	9	9	9	9	9

12.25 Proceed as in Exercise 12.24, but let $\alpha = .01$.

---

*Two popular methods are the LSD (least significant difference) method and the HSD (honestly significant difference), or Tukey method. The HSD procedure resembles the Newman–Keuls procedure but it uses the largest critical value, R_k, for all comparisons. The LSD procedure uses the smallest critical value, R_2, for all comparisons. For additional protection against Type I error, the LSD procedure begins with the global F test, and proceeds to pairwise comparisons only if the global null hypothesis is rejected.

12.26 In a study of the dietary treatment of anemia in cattle, researchers randomly divided 144 cows into four treatment groups. Group A was a control group, and groups B, C, and D received different regimens of dietary supplementation with selenium. After a year of treatment, blood samples were drawn and assayed for selenium. The accompanying table shows the mean selenium concentrations (μg/dl).[17] The MS(within) from the ANOVA was 2.071. Use the Newman–Keuls method to compare all pairs of means at $\alpha = .05$.

GROUP	MEAN	n
A	.8	36
B	5.4	36
C	6.2	36
D	5.0	36

12.27 Proceed as in Exercise 12.26, but let $\alpha = .01$.

12.28 Ten treatments were compared for their effect on the liver in mice. There were 13 animals in each treatment group. The ANOVA gave MS(within) = .5842. The mean liver weights are given in the table.[18]

TREATMENT	MEAN LIVER WEIGHT (gm)
1	2.59
2	2.28
3	2.34
4	2.07
5	2.40
6	2.84
7	2.29
8	2.45
9	2.76
10	2.37

a. Use the Newman–Keuls method to compare all pairs of means at $\alpha = .05$.
b. Suppose the critical value R_2 were used for all comparisons (this would be a form of repeated t tests). Which pairs of means would be declared significantly different?

S E C T I O N 12.7

PERSPECTIVE

In Chapter 12 we have introduced some statistical issues that arise when analyzing data from more than two samples, and we have considered some classical methods of analysis. In this section we review these issues and briefly mention some alternative methods of analysis.

ADVANTAGES OF GLOBAL APPROACH

Let us recapitulate the advantages of analyzing k independent samples by a global approach rather than by viewing each pairwise comparison separately.

1. *Multiple comparisons*   In Section 12.1 we saw that the use of repeated t tests can greatly inflate the overall risk of Type I error. Some control of Type I error can be gained by the simple device of beginning the data analysis with a global F test. For more stringent control of Type I error, special multiple comparison methods are available. One of these was described in optional Section 12.6.

   The problem of multiple comparisons is not confined to an ANOVA setting. In Chapter 14 we will discuss multiple comparisons and related issues from a broader perspective.

2. *Use of a pooled SD*   We have seen that pooling all of the within-sample variability into a single pooled SD leads to a better estimate of the common population SD and thus to a more precise analysis. This is particularly advantageous if the individual sample sizes (n's) are small, in which case the individual SD estimates are quite imprecise.

   It sometimes happens that one cannot take advantage of pooling the SDs because the assumption of equal population SDs is not tenable. One approach that can be helpful in this case is to analyze a transformed variable, such as $Y' = \log(Y)$; the SDs may be more nearly equal in the transformed scale.

3. *Use of structure in the treatments or groups*   Analysis of suitable combinations of group means can be very useful in interpreting data. Many of the relevant techniques are beyond the scope of this book. The discussion in optional Section 12.5 gave a hint of the possibilities. In Chapter 13 we will discuss some ideas that are applicable when the treatments themselves are quantitative (for instance, doses).

OTHER EXPERIMENTAL DESIGNS

The techniques of this chapter are valid only for independent samples. But the basic idea—partitioning variability within and between treatments into interpretable components—can be applied in many experimental designs. For instance, all of the techniques discussed in this chapter can be adapted (by suitable modification of the SE calculation) to analysis of data from a randomized blocks design. These and related techniques belong to the large subject called *analysis of variance*, of which we have discussed only a small part.

NONPARAMETRIC APPROACHES

There are k-sample analogs of the Mann–Whitney test and other nonparametric tests. These tests have the advantage of not assuming underlying normal distributions. However, many of the advantages of the parametric techniques—such as the use of linear combinations—do not easily carry over to the nonparametric setting.

RANKING AND SELECTION

In some investigations the primary aim of the investigator is not to answer research questions about the populations but simply to *select* one or several "best" populations. For instance, suppose ten populations (stocks) of laying hens are available and it is desired to select the one population with the highest egg-laying potential. The investigator will select a random sample of n chickens from each stock and will observe for each chicken Y = total number of eggs laid in 500 days.[19] One relevant question is: How large should n be so that the stock that is *actually* best (has the highest μ) is likely to also *appear* best (have the highest $\bar{y}$)? This and similar questions are addressed by a relatively new branch of statistics called *ranking and selection theory.*

SUPPLEMENTARY EXERCISES 12.29–12.37

[*Note:* Exercises preceded by an asterisk refer to optional sections.]

12.29 The table presents fictitious data for three samples.

	SAMPLE		
	1	2	3
	29	26	30
	31	23	28
	35	20	29
	29	23	26
	31		32
Mean	31	23	29

a. Use the definitional formulas to calculate the ANOVA table.
b. Carry out the global F test at $\alpha = .05$.

12.30 In a study of the mutual effects of the air pollutants ozone and sulfur dioxide, Blue Lake snap beans were grown in open-top field chambers. Some chambers were fumigated repeatedly with sulfur dioxide. The air in some chambers was carbon-filtered to remove ambient ozone. There were three chambers per treatment combination, allocated at random. After 1 month of treatment, total yield (kg) of bean pods was recorded for each chamber, with results shown in the accompanying table.[20] For these data, SS(between) = 1.3538 and SS(within) = .27513. Complete the ANOVA table and carry out the F test at $\alpha = .05$.

	OZONE ABSENT		OZONE PRESENT	
	Sulfur Dioxide		Sulfur Dioxide	
	Absent	Present	Absent	Present
	1.52	1.49	1.15	.65
	1.85	1.55	1.30	.76
	1.39	1.21	1.57	.69
Mean	1.587	1.417	1.340	.700
SD	.237	.181	.213	.056

12.31 Refer to Exercise 12.30. The F test is based on certain assumptions concerning the population distributions.
  a. State the assumptions.
  b. Which feature of the data suggests that one of the assumptions may be doubtful in this case?

*12.32 Refer to Exercise 12.30. Define contrasts to measure each effect specified below, and calculate the value of each contrast.
  a. The effect of sulfur dioxide in the absence of ozone
  b. The effect of sulfur dioxide in the presence of ozone
  c. The interaction between sulfur dioxide and ozone

*12.33 (*Continuation of Exercise* 12.32)   For the snap-bean data, MS(within) = .03439. Test the null hypothesis of no interaction against the alternative that sulfur dioxide is more harmful in the presence of ozone than in its absence. Let α = .05.

12.34 (*Computer exercise*)   Refer to the snap-bean data of Exercise 12.30. For each yield value Y, calculate $Y' = \log(Y)$. (Use base-10 logarithms.)
  a. Calculate the ANOVA table for Y' and carry out the F test.
  b. It often happens with biological data that the SDs are more nearly equal for log-transformed data than for the original data. Is this true for the snap-bean data?

*12.35 (*Computer exercise*)   (*Continuation of Exercises* 12.33 *and* 12.34)   Repeat the test in Exercise 12.33 using Y' instead of Y, and compare with the results of Exercise 12.33.

12.36 In a study of the eye disease retinitis pigmentosa (RP), 211 patients were classified into four groups according to the pattern of inheritance of their disease. Visual acuity (spherical refractive error, in diopters) was measured for each eye, and the two values were then averaged to give one observation per person. The accompanying table shows the number of people in each group and the group mean refractive error.[21] The ANOVA of the 211 observations yields SS(between) = 129.49 and SS(within) = 2,506.8. Construct the ANOVA table and carry out the F test at α = .05.

GROUP	NUMBER OF PEOPLE	MEAN REFRACTIVE ERROR
Autosomal dominant RP	27	+.07
Autosomal recessive RP	20	−.83
Sex-linked RP	18	−3.30
Isolate RP	146	−.84
Total	211	

12.37 (*Continuation of Exercise* 12.36)   Another approach to the data analysis is to use the eye, rather than the person, as the observational unit. For the 211 persons there were 422 measurements of refractive error; the accompanying table summarizes these measurements. The ANOVA of the 422 observations yields SS(between) = 258.97 and SS(within) = 5,143.9.

GROUP	NUMBER OF EYES	MEAN REFRACTIVE ERROR
Autosomal dominant RP	54	$+.07$
Autosomal recessive RP	40	$-.83$
Sex-linked RP	36	-3.30
Isolate RP	292	$-.84$
Total	422	

a. Construct the ANOVA table and bracket the P-value for the F test. Compare with the P-value obtained in Exercise 12.36. Which of the two P-values is of doubtful validity, and why?

b. The mean refractive error for the sex-linked RP patients was -3.30. Calculate the standard error of this mean two ways: (i) Regarding the person as the observational unit and using s_c from the ANOVA of Exercise 12.36; (ii) Regarding the eye as the observational unit and using s_c from the ANOVA of this exercise. Which of these standard errors is of doubtful validity, and why?

CONTENTS

LINEAR REGRESSION AND CORRELATION

SECTION 13.1

INTRODUCTION

In this chapter we discuss some methods for analyzing the relationship between two quantitative variables, X and Y. **Linear regression** and **correlation analysis** are techniques based on fitting a straight line to the data.

EXAMPLES

Data for regression and correlation analysis consist of *pairs* of observations (X, Y). Here are two examples.

EXAMPLE 13.1

AMPHETAMINE AND
FOOD CONSUMPTION

Amphetamine is a drug that suppresses appetite. In a study of this effect, a pharmacologist randomly allocated 24 rats to three treatment groups to receive an injection of amphetamine at one of two dosage levels, or an injection of saline solution. She measured the amount of food consumed by each animal in the 3-hour period following injection. The results (gm of food consumed per kg body weight) are shown in Table 13.1.[1]

TABLE 13.1

Food Consumption (Y)
of Rats (gm/kg)

	X = Dose of amphetamine (mg/kg)		
	0	2.5	5.0
	112.6	73.3	38.5
	102.1	84.8	81.3
	90.2	67.3	57.1
	81.5	55.3	62.3
	105.6	80.7	51.5
	93.0	90.0	48.3
	106.6	75.5	42.7
	108.3	77.1	57.9
Mean	100.0	75.5	55.0
SD	10.7	10.7	13.3
No. of animals	8	8	8

Figure 13.1 shows a scatterplot of

   Y = Food consumption

against

   X = Dose of amphetamine

The scatterplot suggests a definite dose–response relationship, with larger values of X tending to be associated with smaller values of Y.*

---

*In many dose–response relationships, the response depends linearly on log(dose) rather than on dose itself. We have chosen a linear portion of the dose–response curve to simplify the exposition.

FIGURE 13.1

Scatterplot of Food
Consumption Against
Dose of Amphetamine

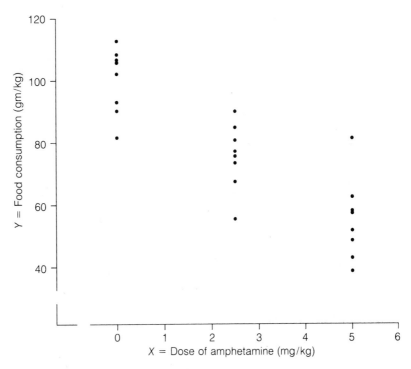

EXAMPLE 13.2

FECUNDITY OF
CRICKETS

In a study of reproductive behavior in the Mormon cricket (*Anabrus simplex*), a biologist collected a field sample of 39 females involved in active courtship. For each female, he observed the number of mature eggs (an indicator of fecundity) and the body weight.[2] Figure 13.2 (page 436) shows a scatterplot of

Y = Number of mature eggs

against

X = Body weight

The scatterplot suggests that larger values of X tend to be associated with larger values of Y; in other words, heavier females tend to be more fecund. ∎

In Examples 13.1 and 13.2 the data do not fall on a straight line, even approximately. Nevertheless, the data in these examples are suitable for linear regression analysis because a straight line is a reasonable summary of the *average* trend of Y as related to X.

TWO CONTEXTS FOR REGRESSION AND CORRELATION

Observations of pairs (X, Y) can arise in two different contexts, namely:

1. Y is an observed variable, and the values of X are specified by the experimenter.
2. Both X and Y are observed variables.

FIGURE 13.2
Scatterplot of Number of
Eggs Against Body
Weight

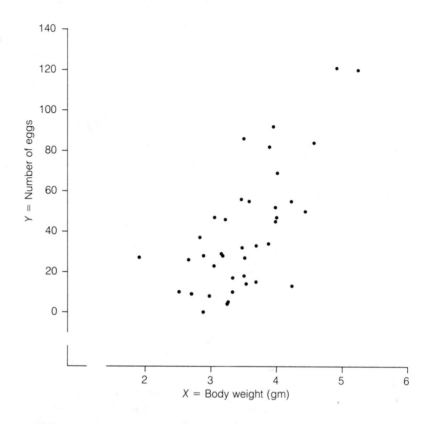

The first context is illustrated by Example 13.1, in which amphetamine dose (X) is manipulated by the experimenter, and food consumption (Y) is observed. The second context is illustrated by Example 13.2, in which both body weight (X) and number of eggs (Y) are observed variables.

The distinction between the two contexts is parallel to the distinction for contingency tables which we discussed in Sections 11.3 and 11.4. For regression as for contingency tables, the distinction between the two contexts is not always sharp. The statistical calculations are the same for the two contexts, but we will see that the emphasis and some of the interpretations can differ.

A LOOK AHEAD

In the following sections we will consider some classical methods for linear analysis of (X, Y) data. Our topics will include:

How to fit a straight line to the data

How to describe the closeness of the data points to the fitted line

How to make statistical inferences concerning the fitted line

We begin our study with the basic calculations that are used in all analyses.

In this section we explain the calculation of the basic statistics used in linear regression and correlation analysis. The data consist of n pairs (x, y). Thus,

n = Total number of observed pairs (x, y)

We will use the following example to illustrate the calculations.

EXAMPLE 13.3

LENGTH AND WEIGHT
OF SNAKES

In a study of a free-living population of the snake *Vipera berus*, researchers caught and measured nine adult females. Their body lengths (X) and weights (Y) are shown in Table 13.2, and displayed as a scatterplot in Figure 13.3.[3] The number of observations is $n = 9$.

TABLE 13.2

Body Length and
Weight of Snakes

	LENGTH X (cm)	WEIGHT Y (gm)
	60	136
	69	198
	66	194
	64	140
	54	93
	67	172
	59	116
	65	174
	63	145
Mean	63	152
SD	4.6	35.3

FIGURE 13.3

Body Length and Weight
of Nine Snakes

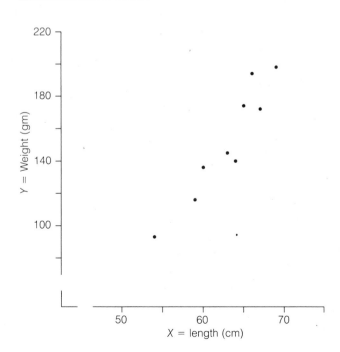

THE MARGINAL QUANTITIES

We first consider the statistics describing the distributions of X and Y separately—that is, the **marginal distributions** of X and Y. These statistics are already familiar to you from Chapter 2. The statistics $\bar{y}$, SS_Y, and s_Y are the mean, sum of squares, and standard deviation of Y; analogous statistics for X are $\bar{x}$, SS_X, and s_X. The notation and the definitional formulas are given in the box.

DEFINITIONAL FORMULAS FOR MARGINAL STATISTICS		
	X	Y
Mean	$\bar{x} = \dfrac{\sum x}{n}$	$\bar{y} = \dfrac{\sum y}{n}$
Sum of squares	$SS_X = \sum (x - \bar{x})^2$	$SS_Y = \sum (y - \bar{y})^2$
Standard deviation	$s_X = \sqrt{\dfrac{SS_X}{n-1}}$	$s_Y = \sqrt{\dfrac{SS_Y}{n-1}}$

The following example illustrates these formulas.

EXAMPLE 13.4

LENGTH AND WEIGHT
OF SNAKES

For the data of Example 13.3, the marginal statistics are as follows:

$$\bar{x} = 63 \text{ cm} \qquad \bar{y} = 152 \text{ gm}$$
$$SS_X = (60 - 63)^2 + (69 - 63)^2 + \cdots + (63 - 63)^2 = 172$$
$$SS_Y = (136 - 152)^2 + (198 - 152)^2 + \cdots + (145 - 152)^2 = 9{,}990$$
$$s_X = \sqrt{\frac{172}{8}} = 4.6 \text{ cm} \qquad s_Y = \sqrt{\frac{9{,}990}{8}} = 35.3 \text{ gm}$$ ■

When both X and Y are observed variables, it is natural to summarize each marginal distribution by its mean and standard deviation (as in Table 13.2).

Next we consider information about the **joint distribution** of X and Y—that is, how variation in Y is related to variation in X.

THE SUM OF PRODUCTS

A basic statistic that can be calculated from joint information about X and Y is the **sum of products**, denoted SP_{XY}, defined as follows:

DEFINITION OF SP_{XY}
$$SP_{XY} = \sum (x - \bar{x})(y - \bar{y})$$

The following example illustrates the computation of SP_{XY}.

EXAMPLE 13.5

LENGTH AND WEIGHT OF SNAKES

For the data of Table 13.2, $\bar{x} = 63$ and $\bar{y} = 152$. Table 13.3 shows the application of the definitional formula for SP_{XY}; the result is $SP_{XY} = 1,237$.

TABLE 13.3

Computation of SP_{XY} for the Snake Data

x	y	$(x - \bar{x})$	$(y - \bar{y})$	$(x - \bar{x})(y - \bar{y})$
60	136	−3	−16	48
69	198	6	46	276
66	194	3	42	126
64	140	1	−12	−12
54	93	−9	−59	531
67	172	4	20	80
59	116	−4	−36	144
65	174	2	22	44
63	145	0	−7	0
			$SP_{XY} =$	1,237

■

The quantity SP_{XY} plays a central role in regression and correlation analysis. Let us consider in more detail the structure of SP_{XY}. The quantity SP_{XY} can be positive or negative (or zero). In fact, the sign (+ or −) of SP_{XY} indicates the *direction* of trend in the data. Broadly speaking, scatterplots that slope upward have positive values of SP_{XY}, whereas those that slope downward have negative values. To see why this is so, let us look at a scatterplot from a special viewpoint. The **joint mean** of the data is the point whose coordinates are $(\bar{x}, \bar{y})$. Figure 13.4 shows the snake data with the X–Y plane divided into four quadrants centered at the joint mean. Because the data show an upward trend, most of the points fall in quadrants B and C.

FIGURE 13.4

Scatterplot of Snake Data Showing Quadrants Centered at the Joint Mean

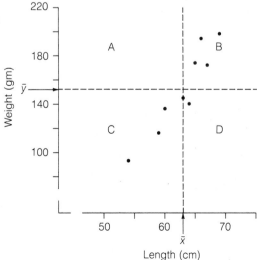

Now consider the formula for SP_{XY}. Each data point contributes a term $(x - \bar{x})(y - \bar{y})$ to the sum SP_{XY}. For a data point in quadrant B or quadrant C, the term $(x - \bar{x})(y - \bar{y})$ is necessarily *positive*, because its factors $(x - \bar{x})$ and $(y - \bar{y})$ are either both positive (in quadrant B) or both negative (in quadrant C). By contrast, for a data point in quadrant A or quadrant D the term $(x - \bar{x})(y - \bar{y})$ is necessarily *negative*, because its factors $(x - \bar{x})$ and $(y - \bar{y})$ must have opposite signs. (See Figure 13.5.) If the data show an upward trend, then the data points tend to lie mostly in quadrants B and C, the positive terms $(x - \bar{x})(y - \bar{y})$ outweigh the negative ones, and so SP_{XY} is positive. [For instance, in Table 13.3 seven of the terms $(x - \bar{x})(y - \bar{y})$ are positive, one is negative, and one is zero.] Contrariwise, if the data points show a downward trend, then they tend to lie mostly in quadrants A and D, and so SP_{XY} is negative.

FIGURE 13.5

Schematic Indication of Contributions to SP_{XY}

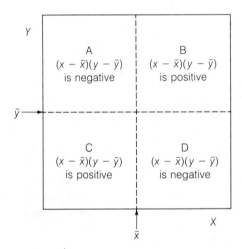

SEVERAL Y-VALUES FOR EACH X

When X is a treatment variable, there may be several observations of Y for each distinct value of X, so that the data can also be viewed as several groups of observations Y. In this case, special care is required in applying the regression notation of this chapter. In regression notation, n denotes the total number of (x, y) pairs (that is, the total number of points in the scatterplot), and $\bar{y}$ denotes the mean of *all* the Y-values. This is *not* consistent with the ANOVA notation of Chapter 12.* The discrepancies are shown in Table 13.4.

We illustrate with the data of Example 13.1.

EXAMPLE 13.6

AMPHETAMINE AND FOOD CONSUMPTION

For the data of Example 13.1, the total number of observations is $n = 24$. Each value of X occurs 8 times, each time with a new observation of Y. Thus,

---

*The only way to be completely consistent would be to give two parallel sets of regression formulas.

TABLE 13.4

Discrepancies Between Regression Notation and ANOVA Notation

	REGRESSION NOTATION (Ch. 13)	ANOVA NOTATION (Ch. 12)
Total no. of observations	n	n^*
Mean of all Y-values	$\bar{y}$	$\bar{\bar{y}}$
Size of each group	—	n
Number of groups	—	k
Sum over all data	Σ	$\overset{*}{\Sigma}\Sigma$
Sum within a group	—	Σ

$$\Sigma x = 8(0) + 8(2.5) + 8(5.0) = 60 \qquad \bar{x} = \frac{60}{24} = 2.5$$

$$\Sigma(x - \bar{x})^2 = 8(0 - 2.5)^2 + 8(2.5 - 2.5)^2 + 8(5.0 - 2.5)^2 = 100$$

$$\bar{y} = \frac{112.6 + 102.1 + \cdots + 57.9}{24} = 76.81$$

∎

COMPUTATIONAL FORMULAS (OPTIONAL)

We have given the *definitional* formulas for the basic regression quantities. If real data are to be analyzed without the aid of a computer then the following *computational* formulas should be used.

COMPUTATIONAL FORMULAS

$$SS_X = \Sigma x^2 - \frac{\left(\Sigma x\right)^2}{n}$$

$$SS_Y = \Sigma y^2 - \frac{\left(\Sigma y\right)^2}{n}$$

$$SP_{XY} = \Sigma xy - \frac{\left(\Sigma x\right)\left(\Sigma y\right)}{n}$$

The computational formulas for the sums of squares are already familiar from Chapter 2. Notice that the computational formula for SP_{XY} is analogous to those for the sums of squares. The following example illustrates the computational formulas.

EXAMPLE 13.7

LENGTH AND WEIGHT OF SNAKES

From Table 13.2 we calculate

$$\Sigma x = 567 \qquad \Sigma x^2 = 35{,}893$$
$$\Sigma y = 1{,}368 \qquad \Sigma y^2 = 217{,}926$$
$$\Sigma xy = (60)(136) + (69)(198) + \cdots + (63)(145) = 87{,}421$$
$$n = 9$$

Applying the computational formulas, we find

$$SS_X = 35{,}893 - \frac{(567)^2}{9} = 172$$

$$SS_Y = 217{,}926 - \frac{(1{,}368)^2}{9} = 9{,}990$$

$$SP_{XY} = 87{,}421 - \frac{(567)(1{,}368)}{9} = 1{,}237$$

These are exactly the same answers that we obtained from the definitional formulas in Examples 13.4 and 13.5. ∎

If there are several Y-values for each X, then the notation in the computational formulas must be interpreted with care, as indicated in Table 13.4. The sums Σx and Σx^2 each have n terms, where n is the total number of observations.

EXERCISES 13.1–13.4

13.1 The following is a fictitious set of data:

X	Y
3	5
6	5
8	9
2	4
6	7

Use the definitional formulas to compute the basic quantities $\bar{x}$, $\bar{y}$, SS_X, SS_Y, and SP_{XY}.

13.2 Proceed as in Exercise 13.1 for the following data:

X	Y
5	12
7	11
2	12
7	9
9	6

13.3 Proceed as in Exercise 13.1 for the following data:

X	Y
10	3
10	5
14	5
14	8
15	6
15	9

13.4 For each exercise indicated below, compute the quantities SS_X, SS_Y, and SP_{XY} using the *computational* formulas. (The answers should agree with the results from the definitional formulas.)

a. Exercise 13.1   **b.** Exercise 13.2   **c.** Exercise 13.3

SECTION 13.3

THE FITTED REGRESSION LINE

If a set of data points (x, y) do not fall exactly on a straight line, there are many ways in which a straight line might be drawn through the data. We will describe the classical method of "fitting" a line to the data so that (in a certain sense) it is as close as possible to the data points. The method of calculation is derived from a criterion called the **least-squares criterion**, and the fitted line is called the **least-squares line** or the **regression line** of Y on X. We will first describe how to determine the regression line, and we will then explain the least-squares criterion.

EQUATION OF THE REGRESSION LINE

The equation of a straight line can be written as

$$Y = b_0 + b_1 X$$

where b_0 is the intercept and b_1 is the slope of the line. The slope b_1 is the rate of change of Y with respect to X.

The fitted regression line of Y on X is the line whose slope and intercept are calculated from the data as follows:

REGRESSION LINE OF Y ON X

$$\text{Slope:} \quad b_1 = \frac{SP_{XY}}{SS_X}$$

$$\text{Intercept:} \quad b_0 = \bar{y} - b_1\bar{x}$$

We illustrate with the snake data from Example 13.3.

EXAMPLE 13.8

LENGTH AND WEIGHT OF SNAKES

For the data of Example 13.3, we found $\bar{x} = 63$, $\bar{y} = 152$, $SP_{XY} = 1{,}237$, and $SS_X = 172$. The slope of the fitted regression line is

$$b_1 = \frac{1{,}237}{172} = 7.19186 \approx 7.19$$

and the intercept is

$$b_0 = 152 - (7.19186)(63) = -301$$

(Note that the *unrounded* value of b_1 should be used in calculating b_0.) The equation of the fitted regression line is

$$Y = -301 + 7.19X$$

Figure 13.6 shows the data and the fitted line.

FIGURE 13.6

Length (*X*) and Weight (*Y*)
of Snakes, and the Fitted
Regression Line of *Y* on *X*

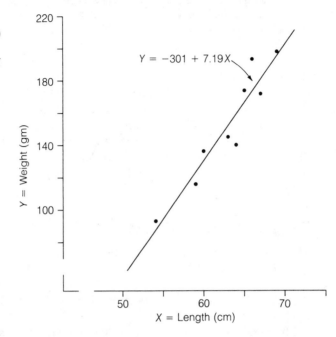

From the formula for b_1 it is clear that the sign of b_1 is determined by the sign of SP_{XY}. If SP_{XY} is positive, then b_1 is positive and the regression line slopes upward; similarly, a negative SP_{XY} leads to a negative b_1 and a downward-sloping line. The magnitude of b_1 expresses, in an average sense, the rate of change of Y with respect to X. For instance, for the snake data, $b_1 \approx 7.2$ gm/cm; on the average, each centimeter of additional length is associated with an additional 7.2 gm of weight.

The formula for the intercept b_0 has a simple interpretation. The formula can be written as

$$\bar{y} = b_0 + b_1\bar{x}$$

This means that *the regression line passes through the joint mean* $(\bar{x}, \bar{y})$ *of the data.*

PLOTTING TIP In preparing a graph like Figure 13.6, a convenient way to draw the regression line is to choose two values of X that lie near the extremes of the data, and calculate the corresponding Y's from the regression equation $Y = b_0 + b_1X$. For instance, for the snake data you could choose $X = 54$ and $X = 70$; substituting these in the regression equation yields $Y = 87$ and $Y = 202$, respectively. You would then plot the points (54, 87), and (70, 202), and use a ruler to draw a line between them. As a check, you can verify that the line passes through the joint mean $(\bar{x}, \bar{y})$.

THE RESIDUAL SUM OF SQUARES

We now consider a statistic that describes the scatter of the points about the fitted regression line. The equation of the fitted line is $Y = b_0 + b_1 X$. For points on the line whose X-values are actual observations x, we use the special notation $\hat{y}$ (read "y-hat"). Thus, for each observed x there is a value

$$\hat{y} = b_0 + b_1 x$$

Also associated with each observed pair (x, y) is a quantity called a **residual**, defined as

$$\text{Residual} = y - \hat{y}$$

Figure 13.7 shows $\hat{y}$ and the residual for a typical data point (x, y). It can be shown that the sum of the residuals, taking into account their signs, is always zero, because of "balancing" of data points above and below the fitted regression line. The *magnitude* (disregarding sign) of each residual is the vertical distance of the data point from the fitted line.

FIGURE 13.7

$\hat{y}$ and the Residual for a Typical Data Point (x, y)

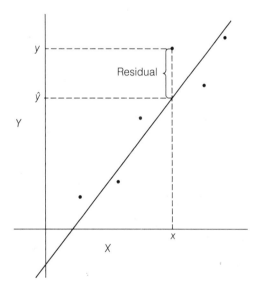

A summary measure of the distances of the data points from the regression line is the **residual sum of squares**, or **SS(resid)**, which is defined as follows:

DEFINITION OF RESIDUAL SUM OF SQUARES
$$SS(\text{resid}) = \sum(y - \hat{y})^2$$

It is clear from the definition that the residual sum of squares will be small if the data points all lie very close to the line.

It is laborious to calculate the residual sum of squares from its definition. The following computational formula should always be used instead.

COMPUTATIONAL FORMULA FOR RESIDUAL SUM OF SQUARES

$$SS(\text{resid}) = SS_Y - \frac{SP_{XY}^2}{SS_X}$$

The following example illustrates these formulas. (A proof of the equivalence of the two is given in Appendix 13.2.)

EXAMPLE 13.9

LENGTH AND WEIGHT OF SNAKES

For the snake data, Table 13.5 indicates how SS(resid) would be calculated from its definition. (The values are abbreviated to improve readability.)

TABLE 13.5

Calculation of SS(resid) from the Definitional Formula

x	y	$\hat{y}$	$y - \hat{y}$	$(y - \hat{y})^2$
60	136	130.42...	5.57...	31.08...
69	198	195.15...	2.84...	8.11...
66	194	173.57...	20.42...	417.15...
64	140	159.19...	−19.19...	368.32...
54	93	87.27...	5.72...	32.79...
67	172	180.76...	−8.76...	76.86...
59	116	123.23...	−7.23...	52.30...
65	174	166.38...	7.61...	58.00...
63	145	152.00...	−7.00...	49.00...
Sum			0	1,093.66... = SS(resid)

Next we illustrate the computational formula. In Examples 13.4 and 13.5 we found $SS_X = 172$, $SS_Y = 9,990$, and $SP_{XY} = 1,237$. We calculate the residual SS as

$$SS(\text{resid}) = 9,990 - \frac{(1,237)^2}{172} = 1,093.669$$ ∎

THE LEAST-SQUARES CRITERION

Many different criteria can be proposed to define the straight line that "best" fits a set of data points (x, y). The classical criterion is the **least-squares criterion**:

LEAST-SQUARES CRITERION

The "best" straight line is the one that minimizes the residual sum of squares.

The formulas given for b_0 and b_1 were derived from the least-squares criterion by applying calculus to solve the minimization problem. (The derivation is given in Appendix 13.1.) The fitted regression line is also called the "least-squares line."

The least-squares criterion may seem arbitrary and even unnecessary. Why not fit a straight line by eye with a ruler? Actually, unless the data lie nearly on a straight line, it can be surprisingly difficult to fit a line by eye. The least-squares criterion provides an answer which does not rely on individual judgment, and which (as we shall see in Sections 13.4 and 13.5) can be usefully interpreted in terms of estimating the distribution of Y-values for each fixed X. Furthermore, we will see in Section 13.8 that the least-squares criterion is a versatile concept, with applications far beyond the simple fitting of straight lines.

THE RESIDUAL STANDARD DEVIATION

A summary of the results of the linear regression analysis should include a measure of the closeness of the data points to the fitted line. A measure derived from SS(resid), and easier to interpret, is the **residual standard deviation**, denoted $s_{Y|X}$, which is defined as follows:

DEFINITION OF RESIDUAL STANDARD DEVIATION

$$s_{Y|X} = \sqrt{\frac{SS(resid)}{n-2}}$$

Notice the factor $(n-2)$ in the denominator, rather than the usual $(n-1)$. The following example illustrates the calculation of $s_{Y|X}$.

EXAMPLE 13.10

LENGTH AND WEIGHT OF SNAKES

For the snake data, we use SS(resid) from Example 13.9 to calculate

$$s_{Y|X} = \sqrt{\frac{1,093.669}{7}} = \sqrt{156.238} = 12.5 \text{ gm}$$ ∎

To *calculate* the residual SD, it is most convenient to use the computational formula for SS(resid). But to *interpret* the residual SD, it is better to think of it as

$$s_{Y|X} = \sqrt{\frac{\sum(y - \hat{y})^2}{n-2}}$$

This formula is closely analogous to the formula for s_Y:

$$s_Y = \sqrt{\frac{\sum(y - \bar{y})^2}{n-1}}$$

Both of these SDs measure variability in Y, but the residual SD measures variability around the regression line and the ordinary SD measures variability around the mean, $\bar{y}$. Roughly speaking, $s_{Y|X}$ is a measure of the typical vertical distance of the data points from the regression line. (Notice that the unit of measurement of $s_{Y|X}$ is the same as that of Y—for instance, grams in the case of the snake data.) Figure 13.8 shows the snake data with the residuals represented as vertical lines and the residual SD indicated as a vertical ruler line. Note that the residual SD roughly indicates the magnitude of a typical residual.

FIGURE 13.8

Length and Weight of Snakes, Showing the Residuals and the Residual SD

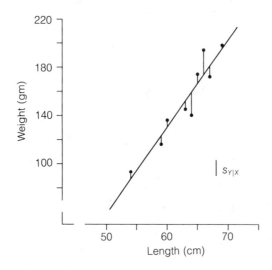

In many cases, $s_{Y|X}$ can be given a more definite quantitative interpretation. Recall from Section 2.6 that for a "nice" data set, we expect roughly 68% of the observations to be within ± 1 SD of the mean (and similarly for 95% and ± 2 SDs). Recall also that these rules work best if the data follow approximately a normal distribution. Similar interpretations hold for the residual SD: For "nice" data sets that are not too small, we expect roughly 68% of the observed y's to be within ± 1 $s_{Y|X}$ of the regression line. In other words, we expect roughly 68% of the data points to be within a vertical distance of $s_{Y|X}$ above and below the regression line (and similarly for 95% and ± 2 $s_{Y|X}$). These rules work best if the residuals follow approximately a normal distribution. The following example illustrates the 68% rule.

EXAMPLE 13.11

FECUNDITY OF CRICKETS

For the cricket fecundity data provided in Example 13.2, the fitted regression line is $Y = -72 + 31.7X$, and the residual standard deviation is $s_{Y|X} = 22.6$. Figure 13.9 shows the data and the regression line. The dotted lines are a vertical distance $s_{Y|X}$ from the regression line. Of the 39 data points, 27 are within the dotted lines; thus, $\frac{27}{39}$ or 69% of the observed y's are within ± 1 $s_{Y|X}$ of the regression line.

FIGURE 13.9
Body Weight and Number
of Eggs in 39 Female
Crickets. The dotted lines
are a vertical distance $s_{Y|X}$
from the regression line.

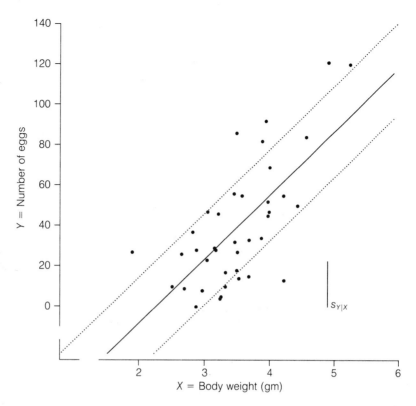

EXERCISES 13.5–13.13

13.5 The table presents a fictitious set of data.

	X	Y
	3	13
	4	15
	1	4
	2	11
	5	22
Mean	3	13
SS	10	170
	$SP_{XY} = 40$	

a. Compute the linear regression of Y on X and compute $\hat{y}$ for each data point.
b. Plot the data and also the values of $\hat{y}$.
c. Compute the residual SS two ways—from the definitional formula and from the computational formula—and verify that the results agree.

13.6 Proceed as in Exercise 13.5 for the following data:

	X	Y
	3	10
	7	2
	6	9
	7	4
	2	15
Mean	5	8
SS	22	106
	$SP_{XY} = -44$	

13.7 In a study of protein synthesis in the oocyte (developing egg cell) of the frog *Xenopus laevis*, a biologist injected individual oocytes with radioactively labelled leucine. At various times after injection, he made radioactivity measurements and calculated how much of the leucine had been incorporated into protein. The results are given in the accompanying table; each leucine value is the content of labelled leucine in two oocytes. All oocytes were from the same female.[4]

	TIME X (min)	LEUCINE Y (ng)
	0	.02
	10	.25
	20	.54
	30	.69
	40	1.07
	50	1.50
	60	1.74
Mean	30	.83
SS	2,800	2.4308
	$SP_{XY} = 81.900$	

a. Use linear regression to estimate the rate of incorporation of the labelled leucine.
b. Plot the data and draw the regression line on your graph.
c. Calculate the residual standard deviation.

13.8 In an investigation of the physiological effects of alcohol (ethanol), 15 mice were randomly allocated to three treatment groups, each to receive a different oral dose of alcohol. The dosage levels were 1.5, 3.0, and 6.0 gm alcohol per kg body weight. The body temperature of each mouse was measured immediately before the alcohol was given, and again 20 minutes afterward. The accompanying table shows the drop (before minus after) in body temperature for each mouse. (The negative value −.1 refers to a mouse whose temperature rose rather than fell.)[5]

ALCOHOL		DROP IN BODY TEMPERATURE (°C)					
Dose (gm/kg)	log(dose) X	Individual Values (Y)					Mean
1.5	.176	.2	1.9	−.1	.5	.8	.66
3.0	.477	4.0	3.2	2.3	2.9	3.8	3.24
6.0	.778	3.3	5.1	5.3	6.7	5.9	5.26

a. Plot the mean drop in body temperature versus dose. Plot the mean drop in body temperature versus log(dose). Which plot appears more nearly linear?

b. For the regression of Y on X = log(dose), preliminary calculations yield the following:

$$\bar{x} = .4771 \qquad \bar{y} = 3.053 \qquad SP_{XY} = 6.92369$$

$$SS_X = .906191 \qquad SS_Y = 63.7773$$

Calculate the fitted regression line and the residual standard deviation.

c. Plot the individual (x, y) data points and draw the regression line on your graph.

d. Draw a ruler line on your graph to show the magnitude of $s_{Y|X}$. (See Figure 13.8.)

13.9 Twenty plots, each 10 × 4 meters, were randomly chosen in a large field of corn. For each plot, the plant density (number of plants in the plot) and the mean cob weight (gm of grain per cob) were observed. The results are given in the table.[6]

PLANT DENSITY X	COB WEIGHT Y	PLANT DENSITY X	COB WEIGHT Y
137	212	173	194
107	241	124	241
132	215	157	196
135	225	184	193
115	250	112	224
103	241	80	257
102	237	165	200
65	282	160	190
149	206	157	208
85	246	119	224

Preliminary calculations yield the following results:

$$\bar{x} = 128.05 \qquad \bar{y} = 224.1 \qquad SP_{XY} = -14,563.1$$

$$SS_X = 20,209.0 \qquad SS_Y = 11,831.8$$

a. Calculate the linear regression of Y on X.

b. Plot the data and draw the regression line on your graph.

c. Calculate s_Y and $s_{Y|X}$ and specify the units of each.

13.10 Laetisaric acid is a newly discovered compound that holds promise for control of fungus diseases in crop plants. The accompanying data show the results of growing the fungus *Pythium ultimum* in various concentrations of laetisaric acid. Each growth value is the average of four radial measurements of a *P. ultimum* colony grown in a petri dish for 24 hours; there were two petri dishes at each concentration.[7]

LAETISARIC ACID CONCENTRATION X (μg/ml)	FUNGUS GROWTH Y (mm)
0	33.3
0	31.0
3	29.8
3	27.8
6	28.0
6	29.0
10	25.5
10	23.8
20	18.3
20	15.5
30	11.7
30	10.0
Mean 11.5	23.64
SS 1,303	677.349
$SP_{XY} = -927.75$	

a. Calculate the linear regression of Y on X.
b. Plot the data and draw the regression line on your graph.
c. Calculate $s_{Y|X}$. What are the units of $s_{Y|X}$?
d. Draw a ruler line on your graph to show the magnitude of $s_{Y|X}$. (See Figure 13.8.)

13.11 To investigate the dependence of energy expenditure on body build, researchers used underwater weighing techniques to determine the fat-free body mass for each of seven men. They also measured the total 24-hour energy expenditure for each man during conditions of quiet sedentary activity. The results are shown in the table.[8] (See also Exercise 13.33.)

SUBJECT	FAT-FREE MASS X (kg)	ENERGY EXPENDITURE Y (Kcal)
1	49.3	1,894
2	59.3	2,050
3	68.3	2,353
4	48.1	1,838
5	57.6	1,948
6	78.1	2,528
7	76.1	2,568
Mean 62.40		2,168
SS 877.740		570,124
$SP_{XY} = 21,953.7$		

a. Calculate the linear regression of Y on X.

b. Plot the data and draw the regression line on your graph.

c. Calculate s_Y and $s_{Y|X}$ and specify the units of each.

13.12 The rowan (*Sorbus aucuparia*) is a tree that grows in a wide range of altitudes. To study how the tree adapts to its varying habitats, researchers collected twigs with attached buds from 12 trees growing at various altitudes in North Angus, Scotland. The buds were brought back to the laboratory and measurements were made of the dark respiration rate. The accompanying table shows the altitude of origin (in meters) of each batch of buds and the dark respiration rate (expressed as μl of oxygen per hour per mg dry weight of tissue).[9]

ALTITUDE OF ORIGIN X (m)	RESPIRATION RATE Y (μl/hr·mg)
90	.11
230	.20
240	.13
260	.15
330	.18
400	.16
410	.23
550	.18
590	.23
610	.26
700	.32
790	.37

Mean	433.3	.2100
SS	506,667	.0654000
	$SP_{XY} = 161.400$	

a. Calculate the linear regression of Y on X.

b. Plot the data and draw the regression line on your graph.

c. Calculate the residual standard deviation.

13.13 For each data set indicated below, prepare a plot like Figure 13.9, showing the data, the fitted regression line, and two lines whose vertical distance above and below the regression line is $s_{Y|X}$. What percentage of the data points are within $\pm s_{Y|X}$ of the regression line?

a. The body temperature data of Exercise 13.8

b. The corn yield data of Exercise 13.9

SECTION 13.4 **PARAMETRIC INTERPRETATION OF REGRESSION: THE LINEAR MODEL**	One use of regression analysis is simply to provide a concise description of the data. The quantities b_0 and b_1 locate the regression line and $s_{Y|X}$ describes the scatter of the points about the line. For many purposes, however, data description is not enough. In this section we consider inference from the data to a larger population. In previous chapters we have spoken of one or several populations of Y values. Now, to encompass the X variable as well, we need to expand the notion of a population.

CONDITIONAL POPULATIONS AND CONDITIONAL DISTRIBUTIONS

A **conditional population** of Y values is a population of Y values associated with a fixed, or given, value of X. Within a conditional population we may speak of the **conditional distribution** of Y. The mean and standard deviation of a conditional population distribution are denoted as:

$\mu_{Y|X}$ = Population mean Y value for a given X

$\sigma_{Y|X}$ = Population SD of Y values for a given X

(Note that the "given" symbol "|" is the same one used for conditional probability in Chapter 11.) The following example illustrates this notation.

EXAMPLE 13.12

AMPHETAMINE AND FOOD CONSUMPTION

In the rat experiment of Example 13.1, the response variable Y was food consumption, and the three values of X (dose) were $X = 0$, $X = 2.5$, and $X = 5$. If we were to view the food consumption data as three independent samples (as for an ANOVA), then we would denote the three population means as μ_1, μ_2, and μ_3. In regression notation these means would be denoted as

$$\mu_{Y|X=0} \qquad \mu_{Y|X=2.5} \qquad \mu_{Y|X=5}$$

respectively. Similarly, the three population standard deviations, which would be denoted as σ_1, σ_2, and σ_3 in an ANOVA context, would be denoted as

$$\sigma_{Y|X=0} \qquad \sigma_{Y|X=2.5} \qquad \sigma_{Y|X=5}$$

respectively. In other words, the symbols

$$\mu_{Y|X} \quad \text{and} \quad \sigma_{Y|X}$$

represent the mean and standard deviation of food consumption values for rats given dose X of amphetamine. ∎

Sometimes conditional distributions pertain to actual subpopulations, as in the following example.

EXAMPLE 13.13

HEIGHT AND WEIGHT OF YOUNG MEN

Consider the variables

$$X = \text{Height}$$

and

$$Y = \text{Weight}$$

for a population of young men. The conditional means and standard deviations are

$\mu_{Y|X}$ = Mean weight of men who are X inches tall

$\sigma_{Y|X}$ = SD of weights of men who are X inches tall

Thus, $\mu_{Y|X}$ and $\sigma_{Y|X}$ are the mean and standard deviation of weight in the *subpopulation* of men whose height is X. Of course, there is a different subpopulation for each value of X. ∎

THE LINEAR MODEL

Linear regression analysis can be given a parametric interpretation if two assumptions are made. These assumptions, which constitute the **linear model**, are given in the box.

THE LINEAR MODEL

1. *Linearity.* $\mu_{Y|X}$ is a linear function of X; that is,

$$\mu_{Y|X} = \beta_0 + \beta_1 X$$

2. *Constancy of standard deviation.*

$\sigma_{Y|X}$ does not depend on X.

The following two examples show the meaning of the linear model.

EXAMPLE 13.14

AMPHETAMINE AND FOOD CONSUMPTION

For the rat food consumption experiment, the linear model asserts that (1) the population mean food consumption is a linear function of dose, and that (2) the population standard deviation of food consumption values is the same for all doses. Notice that the second assumption is closely analogous to the assumption made in ANOVA that the population SDs are equal: $\sigma_1 = \sigma_2 = \sigma_3$. ∎

EXAMPLE 13.15

HEIGHT AND WEIGHT OF YOUNG MEN

We consider an idealized fictitious population of young men whose joint height and weight distribution fits the linear model exactly. For our fictitious population we will assume that the conditional means and SDs of weight given height are as follows:

$$\mu_{Y|X} = -145 + 4.25X$$
$$\sigma_{Y|X} = 20$$

Thus, the regression parameters of the population are $\beta_0 = -145$ and $\beta_1 = 4.25$. (This fictitious population resembles that of U.S. 17-year-olds.[10])

Table 13.6 shows the conditional means and SDs of Y = weight for a few selected values of X = height. Figure 13.10 shows the conditional distributions of Y given X for these selected subpopulations.

TABLE 13.6

Conditional Means and SDs of Weight Given Height in a Population of Young Men

HEIGHT (in) X	MEAN WEIGHT (lb) $\mu_{Y|X}$	STANDARD DEVIATION OF WEIGHTS (lb) $\sigma_{Y|X}$
64	127	20
68	144	20
72	161	20
76	178	20

FIGURE 13.10
Conditional Distributions
of Weight Given Height for
a Population of Young
Men

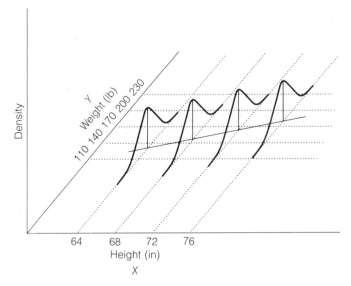

Remark   Actually, the term *regression* is not confined to linear regression. In general, the relation between $\mu_{Y|X}$ and X is called the *regression of Y on X*. The linearity assumption asserts that the regression of Y on X is linear rather than, for instance, a curvilinear function.

ESTIMATION IN THE LINEAR MODEL

Consider now the analysis of a set of (X, Y) data. Suppose we assume that the linear model is an adequate description of the true relationship of Y and X. Suppose further that we are willing to adopt the following **random subsampling model**:

> *Random subsampling model:*   For each observed pair (x, y), we regard the value y as having been sampled at random from the conditional population of Y values associated with the X value x.

(We will discuss the definition of the random subsampling model more fully in Section 13.7.)

Within the framework of the linear model and the random subsampling model, the quantities b_0, b_1, and $s_{Y|X}$ calculated from a regression analysis can be interpreted as estimates of population parameters:

b_0 is the estimate of β_0,

b_1 is the estimate of β_1,

$s_{Y|X}$ is the estimate of $\sigma_{Y|X}$.

We illustrate with the snake data.

EXAMPLE 13.16

LENGTH AND WEIGHT
OF SNAKES

For the snake data of Example 13.3, we found that $b_0 = -301$, $b_1 = 7.19$, and $s_{Y|X} = 12.5$. Thus,

-301 is our estimate of β_0,

7.19 is our estimate of β_1,

12.5 is our estimate of $\sigma_{Y|X}$.   ∎

The application of the linear model to the snake data has yielded two benefits. First, the slope of the regression line, 7.19 gm/cm, is an estimate of a morphological parameter ("weight per unit length"), which is of potential biological interest in characterizing the population of snakes. Second, we have obtained an estimate (12.5 gm) of the variability of weight among snakes of fixed length, even though no *direct* estimate of this variability was possible because no two of the observed snakes were the same length.

INTERPOLATION IN THE LINEAR MODEL

When we draw a curve or a line through a set of (X, Y) data, we are expressing a belief that the underlying dependence of Y on X is *smooth*, even though the data may show the relationship only roughly. Linear regression is one formal way of providing a smooth description of the data.

Taking advantage of this assumption of smoothness, a fitted regression is sometimes used to estimate the distribution of Y for an X for which there are no data. The following is an example.

EXAMPLE 13.17

AMPHETAMINE AND
FOOD CONSUMPTION

If we apply the formulas of Section 13.3 to the food consumption data in Table 13.1, we obtain $b_0 = 99.3$, $b_1 = -9.01$, and $s_{Y|X} = 11.4$. Figure 13.11 (page 458) shows the data and the fitted regression line. Let us predict the response of rats given amphetamine at a dose of $X = 3.5$ mg/kg. The fitted regression equation is $Y = 99.3 - 9.01X$; substituting $X = 3.5$ yields $Y = 67.8$. Thus, we estimate that rats given 3.5 mg/kg of amphetamine would show a mean food consumption of 67.8 gm/kg and an SD of 11.4 gm/kg.   ∎

Note that estimation of the mean uses the linearity assumption of the linear model, while estimation of the standard deviation uses the assumption of constant standard deviation. In some situations only the linearity assumption may be plausible, and then only the mean would be estimated.

Example 13.17 is an example of **interpolation**, because the X we chose ($X = 3.5$) was *within* the range of observed values of X. By contrast, **extrapolation** is the use of a regression line (or other curve) to predict Y for values of X that are *outside* the range of the data. Extrapolation should be avoided whenever possible, because there is usually no assurance that the relationship between $\mu_{Y|X}$ and X remains linear for X values outside the range of those observed. Many biological relationships are linear for only part of the possible range of X values. The following is an example.

FIGURE 13.11

Rat Food Consumption Data and Fitted Regression Line

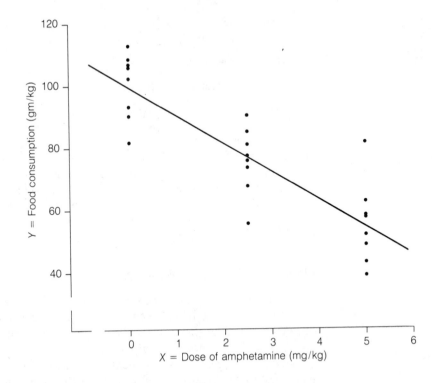

EXAMPLE 13.18

AMPHETAMINE AND FOOD CONSUMPTION

The dose–response relationship for the rat food consumption experiment looks approximately like Figure 13.12.[11] The data of Example 13.1 cover only the linear portion of the relationship. Clearly it would be unwise to extrapolate the fitted line out to $X = 10$ or $X = 15$.

FIGURE 13.12

Dose–Response Curve (Mean Response vs Dose) for Rat Food Consumption Experiment

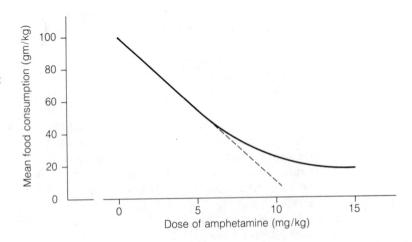

**EXERCISES
13.14–13.17**

13.14 Refer to the body temperature data of Exercise 13.8. Assuming that the linear model is applicable, estimate the mean and the standard deviation of the drop in body temperature that would be observed in mice given alcohol at a dose of 2 gm/kg.

13.15 Refer to the cob weight data of Exercise 13.9. Assume that the linear model holds.
  a. Estimate the mean cob weight to be expected in a plot containing (i) 100 plants; (ii) 120 plants.
  b. Assume that each plant produces one cob. How much grain would we expect to get from a plot containing (i) 100 plants? (ii) 120 plants?

13.16 Refer to the fungus growth data of Exercise 13.10. Assuming that the linear model is applicable, find estimates of the mean and standard deviation of fungus growth at a laetisaric acid concentration of 15 μg/ml.

13.17 Refer to the energy expenditure data of Exercise 13.11. Assuming that the linear model is applicable, estimate the mean 24-hour energy expenditure of a man whose fat-free mass is 55 kg.

| | | | | | | | | | | | |
S E C T I O N   13.5

**STATISTICAL
INFERENCE
CONCERNING β_1**

The linear model provides interpretations of b_0, b_1, and $s_{Y|X}$ that take them beyond data description into the domain of statistical inference. In this section we consider inference about the true slope β_1 of the regression line. The methods are based on the linear model and the random subsampling model. In addition, the methods assume that the conditional population distribution of Y for each value of X is a normal distribution.

THE STANDARD ERROR OF b_1

Within the context of the linear model, b_1 is an estimate of β_1. Like all estimates calculated from data, b_1 is subject to sampling error. The standard error of b_1 is calculated as follows:

STANDARD ERROR OF b_1

$$SE_{b_1} = \sqrt{\frac{s_{Y|X}^2}{SS_X}}$$

The following example illustrates the calculation of SE_{b_1}.

EXAMPLE 13.19

LENGTH AND WEIGHT
OF SNAKES

For the snake data, we found in Example 13.4 that $SS_X = 172$, and in Example 13.10 that $s_{Y|X}^2 = 156.238$. The standard error of b_1 is

$$SE_{b_1} = \sqrt{\frac{156.238}{172}} = .95308$$

To summarize, the slope of the fitted regression line (from Example 13.8) is

$$b_1 = 7.19 \text{ gm/cm}$$

and the standard error of this slope is

$$SE_{b_1} = .95 \text{ gm/cm}$$
■

STRUCTURE OF THE SE   Let us see how the standard error of b_1 depends on various aspects of the data. First, note that SE_{b_1} depends, through $s_{Y|X}$, on the scatter of the data points about the fitted regression line; naturally, smaller scatter gives more precise information about β_1. Second, note that SE_{b_1} depends on SS_X; larger SS_X gives a smaller SE. Recall that

$$SS_X = \sum (x - \bar{x})^2$$

Thus, SS_X can be made larger in two ways: (a) by increasing n (so that there are more terms in the sum), and (b) by increasing the dispersion or spread in the X values. The dependence on the spread in the X values is illustrated in Figure 13.13, which shows two data sets with the same value of $s_{Y|X}$ and the same value of n, but different values of SS_X. Imagine using a ruler to fit a straight line by eye; it is intuitively clear that the data set in case (b)—with the larger SS_X— would determine the slope of the line more precisely.

FIGURE 13.13
Two Data Sets with the Same Value of n and of $s_{Y|X}$ but Different SS_X.
(a) Smaller SS_X;
(b) Larger SS_X.

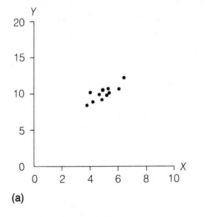

(a)

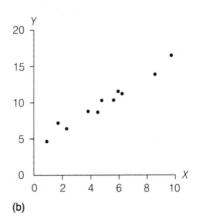

(b)

IMPLICATIONS FOR DESIGN   The above discussion implies that, for the purpose of gaining precise information about β_1, it is best to have the values of X as widely dispersed as possible. This fact can guide the experimenter when the design of the experiment includes choosing values of X. Other factors also play a role, however. For instance, if X is the dose of a drug, the criterion of widely dispersed X's would lead to using only two dosages, one very low and one very high. But in practice an experimenter would want to have at least a few observations at intermediate doses, to verify that the relation is actually linear within the range of the data.

CONFIDENCE INTERVAL FOR β_1

In many studies the quantity β_1 is a biologically meaningful parameter, and a primary aim of the data analysis is to estimate β_1. A confidence interval for β_1 can be constructed by the familiar method based on the SE and Student's t distribution. For instance, a 95% confidence interval is constructed as

$$b_1 \pm t_{.05} \, SE_{b_1}$$

where the critical value $t_{.05}$ is determined from Student's t distribution with

$$df = n - 2$$

Intervals with other confidence coefficients are constructed analogously; for instance, for a 90% confidence interval one would use $t_{.10}$.

EXAMPLE 13.20

LENGTH AND WEIGHT OF SNAKES

Let us use the snake data to construct a 95% confidence interval for β_1. We found that $b_1 = 7.19186$ and $SE_{b_1} = .95308$. There are $n = 9$ observations; we refer to Table 4 with df $= 9 - 2 = 7$, and obtain

$$t_{.05} = 2.365$$

The confidence interval is

$$7.19186 \pm (2.365)(.95308)$$

or

$$4.9 \text{ gm/cm} < \beta_1 < 9.4 \text{ gm/cm}$$

We are 95% confident that the true slope of the regression of weight on length for this snake population is between 4.9 gm/cm and 9.4 gm/cm; this is a rather wide interval because the sample size is not very large. ∎

TESTING THE HYPOTHESIS $H_0: \beta_1 = 0$

In some investigations it is not a foregone conclusion that there is *any* relationship between X and Y. It then may be relevant to consider the possibility that any apparent trend in the data is illusory and reflects only sampling variability. In this situation it is natural to formulate the null hypothesis

$$H_0: \quad \mu_{Y|X} \text{ does not depend on } X$$

Within the linear model, this hypothesis can be translated as

$$H_0: \quad \beta_1 = 0$$

A t test of H_0 is based on the test statistic

$$t_s = \frac{b_1}{SE_{b_1}}$$

Critical values are obtained from Student's t distribution with

$$\text{df} = n - 2$$

The following example illustrates the application of this t test.

EXAMPLE 13.21

**BLOOD PRESSURE AND
PLATELET CALCIUM**

It is suspected that calcium in the cells may be related to blood pressure. As part of a study of this relationship, researchers recruited 38 subjects whose blood pressure was normal (that is, not abnormally elevated). For each subject, two measurements were made: X = blood pressure (average of systolic and diastolic measurements) and Y = free calcium concentration in the blood platelets. The data are shown in Figure 13.14.[12] Calculations from the data yield $\bar{x} = 94.5$, $\bar{y} = 107.9$, $SS_X = 2{,}397.50$, $SS_Y = 9{,}564.34$, and $SP_{XY} = 2{,}792.50$, from which we can calculate

$$b_1 = 1.1648 \qquad SE_{b_1} = .27042$$

FIGURE 13.14

Blood Pressure and
Platelet Calcium for 38
People with Normal Blood
Pressure

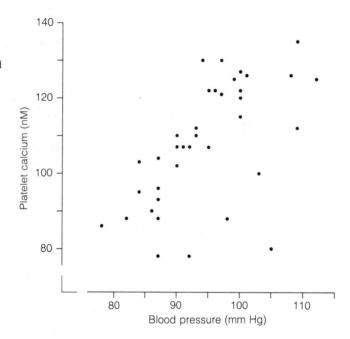

We will test the null hypothesis

$$H_0: \quad \beta_1 = 0$$

against the nondirectional alternative

$$H_A: \quad \beta_1 \neq 0$$

These hypotheses are translations, within the linear model, of the verbal hypotheses

H_0: Mean platelet calcium does not depend on blood pressure

H_A: Mean platelet calcium does depend on blood pressure

(Note, however, that "depend" does not necessarily refer to *causal* dependence. We will return to this point in Section 13.8.)

Let us choose $\alpha = .05$. The test statistic is

$$t_s = \frac{1.1648}{.27042} = 4.307$$

From Table 4 with df $= n - 2 = 36 \approx 40$, we find $t_{.001} = 3.551$, $t_{.0001} = 4.321$. Thus, we find $.0001 < P < .001$ and we reject H_0. The data provide sufficient evidence to conclude that the true slope of the regression of platelet calcium on blood pressure in this population is positive (that is, $\beta_1 > 0$). ∎

Note that the test on β_1 does not ask *whether* the relationship between $\mu_{Y|X}$ and X is linear. Rather, the test asks whether, *assuming* that the linear model holds, we can conclude that the slope is nonzero. It is therefore necessary to be careful in phrasing the conclusion from this test. For instance, the statement "There is a significant linear trend" could be easily misunderstood.*

WHY ($n - 2$)? The confidence interval and test based on b_1 have associated df $= n - 2$. Also, $(n - 2)$ is the denominator of $s_{Y|X}^2$. The origin of the $(n - 2)$ is easy to explain. It takes two points to determine a straight line, and so (under the linear model) the data provide $(n - 2)$ independent pieces of information concerning $\sigma_{Y|X}$. (Note that if $n = 2$, the regression line will fit the data exactly, but $s_{Y|X}$ cannot be calculated.) Thus, as in earlier contexts related to t distributions and F distributions (Chapters 6, 7, 9, and 12), the number of df is the number of pieces of information provided by the data about the "noise" from which the investigator wants to extract the "signal."

**EXERCISES
13.18–13.23**

13.18 Refer to the leucine data given in Exercise 13.7. Construct a 95% confidence interval for β_1.

13.19 Refer to the body temperature data of Exercise 13.8. Construct a 95% confidence interval for β_1.

13.20 Refer to the cob weight data of Exercise 13.9. Construct a 95% confidence interval for β_1.

*There are tests that can (in some circumstances) test whether the true relationship is linear. Furthermore, there are tests that can test for a linear component of trend without assuming that the relationship is linear. These tests are beyond the scope of this book.

13.21 Refer to the fungus growth data of Exercise 13.10.
    a. Calculate the standard error of the slope, b_1.
    b. Consider the null hypothesis that laetisaric acid has no effect on growth of the fungus. Assuming that the linear model is applicable, formulate this as a hypothesis about the true regression line, and test the hypothesis against the alternative that laetisaric acid inhibits growth of the fungus. Let $\alpha = .05$.

13.22 Refer to the energy expenditure data of Exercise 13.11. Construct a 95% confidence interval for β_1.

13.23 Refer to the respiration data of Exercise 13.12. Assuming that the linear model is applicable, test the null hypothesis of no relationship against the alternative that trees from higher altitudes tend to have higher respiration rates. Let $\alpha = .05$.

SECTION 13.6

THE CORRELATION COEFFICIENT

Sometimes a researcher wishes to describe how closely a set of data points (x, y) cluster about a straight line. In other words, he wants to describe the *tightness* of the linear relationship between X and Y. The residual standard deviation provides one such description. In this section we consider another descriptive statistic, the **correlation coefficient**.

DEFINITION OF r

The sample correlation coefficient is denoted r and is defined as follows:*

DEFINITION OF CORRELATION COEFFICIENT

$$r = \frac{SP_{XY}}{\sqrt{SS_X SS_Y}}$$

The following example illustrates the calculation of r.

EXAMPLE 13.22

LENGTH AND WEIGHT OF SNAKES

We found in Examples 13.4 and 13.5 that for the snake data $SS_X = 172$, $SS_Y = 9{,}990$, and $SP_{XY} = 1{,}237$. Thus, the correlation coefficient is

$$r = \frac{1{,}237}{\sqrt{(172)(9{,}990)}} = .9436 \approx .94$$

    ■

BASIC PROPERTIES OF r

From the definition of r it is clear that

    The sign of r is the same as the sign of b_1.

---

*A more complete name for this statistic is *Pearson's product-moment correlation coefficient*.

Thus, r is positive if the fitted regression line slopes upward, negative if the line slopes downward, and zero if the line is horizontal. Unlike b_1, however, r is dimensionless—that is, it is not measured in units such as gm or gm/cm.

The following fact will help to interpret the magnitude of r. (A proof of this fact is given in Appendix 13.2.)

FACT 13.1: RELATIONSHIP OF r TO SS(RESID) AND SS_Y

The correlation coefficient obeys the following relationship:

$$r^2 = 1 - \frac{SS(\text{resid})}{SS_Y}$$

The following properties of r can be seen directly from Fact 13.1:

r must be between -1 and $+1$.

r is close to $+1$ or -1 if SS(resid) is small compared to SS_Y.

$r = +1$ or -1 only if the data points lie exactly on a straight line [so that SS(resid) $= 0$].

The magnitude of r is a rough indication of the *shape* of the scatterplot. A value of r close to $+1$ or -1 suggests that the cloud of data points is long and narrow, with the points clustered close to a line. A small magnitude of r suggests that the data cloud is diffuse, with the points loosely scattered. (In Section 13.7 we will discuss exceptions to these interpretations.) The following example illustrates the general idea.

EXAMPLE 13.23
EXAMPLES OF
CORRELATIONS

Figure 13.15 (page 466) shows fictitious data sets with various values of r. Figure 13.15(a) shows 20 observations with $r = .99$; the visual impression is a long narrow cloud of points, indicating a fairly tight linear association between X and Y. Figure 13.15(b) shows 20 observations with $r = .61$; the visual impression is a loosely scattered cloud of points which nevertheless seem to show an overall upward trend. The data in Figures 13.15(c) and (d) have correlations of $r = -.99$ and $r = -.61$, respectively. ∎

Note that the value of r does *not* reflect the steepness or shallowness of the slope relating Y and X. In fact, the fitted regression lines for the data sets in Figures 13.15(a) and (b) are identical. [To see this intuitively, imagine superimposing (a) on (b).] Similarly, the regression lines in (c) and (d) are identical.

HOW r DESCRIBES THE REGRESSION

We have said that the magnitude of r describes the tightness of the linear relationship between X and Y. This interpretation is made more specific by the following fact. (This fact is proved in Appendix 13.2.)

FIGURE 13.15   Data Sets to Illustrate the Correlation Coefficient

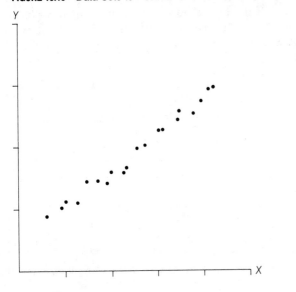

(a) *r* = .99

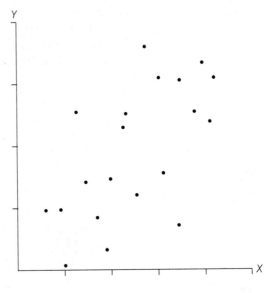

(b) *r* = .61

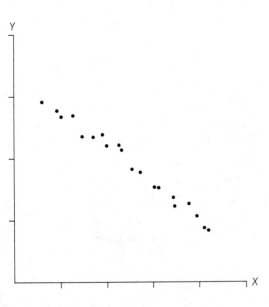

(c) *r* = −.99

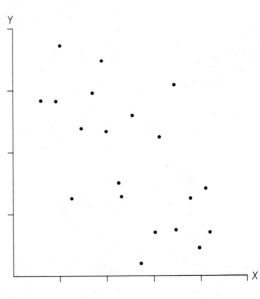

(d) *r* = −.61

> **FACT 13.2:   APPROXIMATE RELATIONSHIP OF r TO $s_{Y|X}$ AND s_Y**
>
> The correlation coefficient r obeys the following approximate relationship:
>
> $$\frac{s_{Y|X}}{s_Y} \approx \sqrt{1 - r^2}$$

(The approximation in Fact 13.2 is best for large n, but it holds reasonably well even for n as small as 5.)

If we know r, then by using Fact 13.2 we can deduce the approximate ratio of the residual SD to the ordinary (marginal) SD of Y. In turn, this ratio roughly describes the appearance of the scatterplot. The following two examples illustrate this idea.

EXAMPLE 13.24

LENGTH AND WEIGHT OF SNAKES

For the snake data, we found in Example 13.22 that the correlation coefficient is $r = .9436$. From Fact 13.2 we conclude that

$$\frac{s_{Y|X}}{s_Y} \approx \sqrt{1 - (.9436)^2} = .33$$

That is, from the value of r we can deduce that the residual SD of weight, after regression on length, is only about 33% of the overall SD of weight; this in turn means that the linear relationship is fairly tight. The two SDs are shown as ruler lines in Figure 13.16.

FIGURE 13.16

Relationship Between s_Y and $s_{Y|X}$ for Snake Data

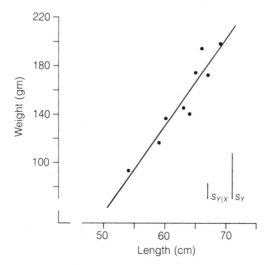

EXAMPLE 13.25

FECUNDITY OF CRICKETS

For the cricket data of Example 13.2, $r = .6873$ and $\sqrt{1 - r^2} = .73$. Thus, the value of r tells us that $s_{Y|X}$ is relatively large; it is about 73% of s_Y. This indicates that points are rather widely scattered about the regression line. The two SDs are shown as ruler lines in Figure 13.17 (page 468).

FIGURE 13.17
Relationship Between s_Y
and $s_{Y|X}$ for Cricket Data

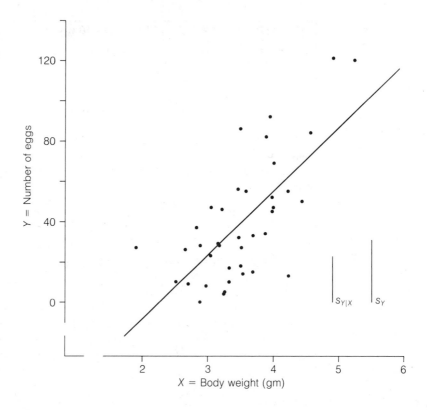

THE SYMMETRY OF r

Recall that the definition of r is

$$r = \frac{\mathrm{SP}_{XY}}{\sqrt{\mathrm{SS}_X \mathrm{SS}_Y}}$$

From this formula it is clear that *X and Y enter r symmetrically*. Therefore, if we were to interchange the labels X and Y of our variables, r would remain unchanged. In fact, this is one of the advantages of the correlation coefficient as a summary statistic: In interpreting r, it is not necessary to know (or to decide) which variable is labelled X and which is labelled Y.

A PARADOX AND ITS RESOLUTION   The symmetry of r may appear puzzling. In the preceding discussion we have interpreted r in terms of the variation in one of the variables—namely, Y—as it relates to regression on the other variable—namely, X. This asymmetric interpretation of r appears to contradict the symmetric definition or r.

We can resolve this apparent paradox by considering reverse regression. Suppose we keep our variable labels X and Y fixed, but we regress in the reverse direction—that is, we regress X on Y.* The reverse regression minimizes the sum of squares of the *horizontal* distances

$$x - \hat{x}$$

and the residual sum of squares is

$$\sum (x - \hat{x})^2$$

Because the least-squares criterion is applied to vertical distances in one case and horizontal distances in the other, there are actually *two* regression lines associated with any set of data—the regression of Y on X and the regression of X on Y. Remarkably, however, *the closeness of the data points to the lines, as measured by r, is the same for both regression lines*. Specifically, we can say that, for the regression of Y on X,

$$r^2 = 1 - \frac{\sum (y - \hat{y})^2}{\sum (y - \bar{y})^2}$$

and for the regression of X on Y,

$$r^2 = 1 - \frac{\sum (x - \hat{x})^2}{\sum (x - \bar{x})^2}$$

Similarly, either regression yields an approximate relation between r and the standard deviations. For the regression of Y on X,

$$\frac{s_{Y|X}}{s_Y} \approx \sqrt{1 - r^2}$$

and for the regression of X on Y,

$$\frac{s_{X|Y}}{s_X} \approx \sqrt{1 - r^2}$$

The following example illustrates these relationships.

EXAMPLE 13.26

FECUNDITY OF CRICKETS

Consider the cricket data of Example 13.2 on X = body weight and Y = number of eggs. The marginal means are $\bar{x} = 3.5$ eggs and $\bar{y} = 40$ gm. For the regression of Y on X, the fitted line is $Y = -71.7 + 31.7X$, the residual SD is $s_{Y|X} = 22.6$ eggs, and the marginal SD of Y is $s_Y = 30.7$ eggs. For the regression of X on Y, the fitted line is $X = 2.93 + .0149Y$, the residual SD is $s_{X|Y} = .490$ gm, and the marginal SD of X is $s_X = .665$ gm. Figure 13.18 (page 470) shows the data and the two regression lines. The marginal and residual SDs are shown as ruler lines. Note that both regression lines pass through the joint mean $(\bar{x}, \bar{y})$. (The calculations for the regressions are shown in Appendix 13.3.)

---

*The equations for the reverse regression are not given here because they are of no practical importance in data analysis. They are given in Appendix 13.3.

FIGURE 13.18
Body Weight (*X*) and
Number of Eggs (*Y*) for 39
Female Crickets, Showing
the Regression Lines of *Y*
on *X* and of *X* on *Y*

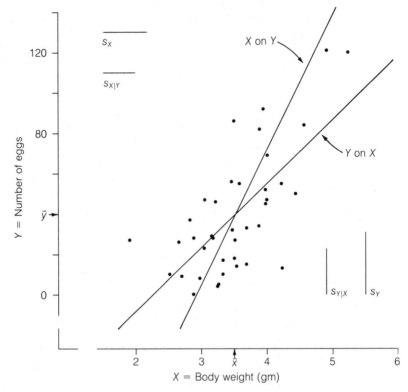

For these data, $r = .6873$, and $\sqrt{1 - r^2} = .73$. We verify that

$$\frac{s_{Y|X}}{s_Y} = \frac{22.6}{30.7} = .74 \approx .73 = \sqrt{1 - r^2}$$

$$\frac{s_{X|Y}}{s_X} = \frac{.490}{.665} = .74 \approx .73 = \sqrt{1 - r^2}$$

∎

STATISTICAL INFERENCE CONCERNING CORRELATION

We have described various ways in which the correlation coefficient describes a data set. Now we consider statistical inference based on *r*.

We saw in Section 13.4 that statistical inference about the regression line is based on the random subsampling model—that is, the view that the *Y* values were selected by random sampling from the conditional populations defined by the *X* values. But the correlation approach, unlike the regression approach, treats *X* and *Y* symmetrically. In order to regard the sample correlation coefficient *r* as an estimate of a population parameter, it must be reasonable to assume that both the *X and* the *Y* values were selected at random, as in the following **bivariate random sampling model**:

> *Bivariate random sampling model:* We regard each pair (*x, y*) as having been sampled at random from a population of (*x, y*) pairs.

In the bivariate random sampling model, the sample correlation coefficient *r* is an estimate of the population correlation coefficient, which is denoted ρ (the Greek letter rho). Also, in the bivariate random sampling model, the observed *x*'s are regarded as a random sample and the observed *y*'s are also regarded as a random sample, so that the marginal statistics $\bar{x}$, $\bar{y}$, s_X, and s_Y are estimates of corresponding population values μ_X, μ_Y, σ_X, and σ_Y.

The following example illustrates a case where the bivariate random sampling model is reasonable.

EXAMPLE 13.27

BLOOD PRESSURE AND PLATELET CALCIUM

For the data of Example 13.21, the correlation coefficient between blood pressure and platelet calcium is *r* = .58. The data were obtained from 38 adult volunteers who did not suffer from high blood pressure. One might regard the observed pairs (*x, y*) as a random sample from potential measurements on a corresponding population (adults not suffering from high blood pressure). In this setting, the observed value of *r*—.58—is an estimate of the population correlation coefficient ρ. Similarly, the observed mean and SD of blood pressure, and the observed mean and SD of platelet calcium, would be estimates of corresponding quantities in the population. ∎

For many investigations the random subsampling model is reasonable, but the additional assumption of a bivariate random sampling model is not. This is generally the case when the values of *X* are specified by the experimenter. The following is an example.

EXAMPLE 13.28

AMPHETAMINE AND FOOD CONSUMPTION

In the food consumption experiment of Example 13.1, *X* represents the dose of amphetamine. The observed *x*'s are not a random sample, but were specified by the experimenter. We can calculate $\bar{x}$ and s_X from the data, but these statistics are not estimates of any population parameters. Similarly, $\bar{y}$, s_Y, and *r* are not estimates of population parameters. ∎

We now consider statistical inference concerning the population correlation coefficient ρ. Suppose we wish to test the hypothesis

H_0: $\rho = 0$

which asserts that *X* and *Y* are uncorrelated in the population.* It turns out that no new technique is required; we can simply reinterpret the *t* test described in

*Methods are available to construct a confidence interval for ρ, but these are beyond the scope of this book.

Section 13.5. Specifically, it can be shown that, if the linear model holds, then ρ is related to the population regression slope β_1 as follows:

$$\rho = \beta_1 \frac{\sigma_X}{\sigma_Y}$$

Because of this relationship, the two hypotheses

$$H_0: \quad \beta_1 = 0 \quad \text{and} \quad H_0: \quad \rho = 0$$

are equivalent. Recall from Section 13.5 that the t test for the hypothesis

$$H_0: \quad \beta_1 = 0$$

is based on the test statistic

$$t_s = \frac{b_1}{SE_{b_1}}$$

and uses critical values from Student's t distribution with df $= n - 2$. Under the bivariate random sampling model, this test can be reinterpreted as a test of the hypothesis $H_0: \rho = 0$. It can also be shown that the t statistic can be rewritten in terms of r, as follows:

THE t STATISTIC IN TERMS OF r

$$t_s = \frac{b_1}{SE_{b_1}} = r\sqrt{\frac{n-2}{1-r^2}}$$

The following example illustrates the equivalence of the two tests.

EXAMPLE 13.29

BLOOD PRESSURE AND
PLATELET CALCIUM

For the platelet calcium data of Example 13.21, the observed correlation is $r = .58316$. Let us test the hypothesis

H_0:   Platelet calcium and blood pressure are uncorrelated in the population

that is,

$$H_0: \quad \rho = 0$$

The test statistic can be calculated from r as

$$t_s = r\sqrt{\frac{n-2}{1-r^2}}$$

$$= .58316\sqrt{\frac{36}{1-(.58316)^2}} = 4.307$$

which is the same as the value obtained by regression of platelet calcium on blood pressure in Example 13.21. The reverse regression—of blood pressure on platelet calcium—would also yield the same value of t_s. At $\alpha = .05$ there is sufficient evidence to conclude that blood pressure and platelet calcium are positively correlated in the population. (This is equivalent to asserting that β_1 is positive.) ∎

CAUTIONARY NOTE To describe the results of testing a correlation coefficient, investigators often use the term "significant," which can be misleading. For instance, a statement such as "A highly significant correlation was noted . . ." is easily misunderstood. It is important to remember that statistical significance simply indicates rejection of a null hypothesis; it does not necessarily indicate a large or important effect. A "significant" correlation may in fact be quite a weak one; its "significance" means only that it cannot easily be explained away as a chance pattern.

THE COEFFICIENT OF DETERMINATION (OPTIONAL)

In some settings it is natural to say that variation in X "accounts for" or "explains" variation in Y. For instance, consider X = height and Y = weight of people. Some people are heavier than others, and part of this variation is explained by the fact that some people are taller than others.

The square of the correlation coefficient, r^2, is called the **coefficient of determination**. The coefficient of determination can be interpreted as the proportion of the variation in Y that is "accounted for" by linear regression of Y on X. The basis for this interpretation is the following: The residual SS, $\Sigma(y - \hat{y})^2$, represents variation in Y that is *not* accounted for by the regression, and SS_Y represents the total variation in Y. Consequently, the fraction of SS_Y that is *not* accounted for by regression on X is

$$\frac{SS(\text{resid})}{SS_Y}$$

and therefore the fraction of SS_Y that *is* accounted for by regression on X is

$$1 - \frac{SS(\text{resid})}{SS_Y} = r^2 \quad \text{(because of Fact 13.1)}$$

The coefficient of determination is often expressed as a percentage, as in the following example.

EXAMPLE 13.30

LENGTH AND WEIGHT OF SNAKES

We found in Example 13.22 that for the snake data $r = .9436$. The coefficient of determination is

$$r^2 = (.9436)^2 = .89$$

Thus, one might say that 89% of the variation in weight among these snakes is accounted for by the variation in length. A complementary interpretation is that 11% of the variation in weight is "residual," or not accounted for by variation in length. (In interpreting these phrases, however, it should be remembered that "accounted for" means "accounted for by linear regression.") ∎

13.24 A plant physiologist grew 13 individually potted soybean seedlings in a greenhouse. The table gives measurements of the total leaf area (cm^2) and total plant dry weight (gm) for each plant after 16 days of growth.[13]

PLANT	LEAF AREA X	DRY WEIGHT Y
1	411	2.00
2	550	2.46
3	471	2.11
4	393	1.89
5	427	2.05
6	431	2.30
7	492	2.46
8	371	2.06
9	470	2.25
10	419	2.07
11	407	2.17
12	489	2.32
13	439	2.12
Mean	443.8	2.174
SS	28,465.7	.363708
	$SP_{XY} = 82.8977$	

a. Calculate the correlation coefficient.
b. Calculate s_Y and $s_{Y|X}$; specify the units for each. Verify the approximate relationship between s_Y, $s_{Y|X}$, and r.
c. Calculate the regression line of Y on X.
d. Construct a scatterplot of the data and draw the regression line on your graph. Place ruler lines on the scatterplot to show the magnitudes of s_Y and $s_{Y|X}$. (See Figure 13.16.)

13.25 Proceed as in Exercise 13.24, but use the cob weight data of Exercise 13.9.

13.26 Proceed as in Exercise 13.24, but use the energy expenditure data of Exercise 13.11.

13.27 In a study of 2,669 adult men, the correlation between age and systolic blood pressure was found to be $r = .43$. The SD of systolic blood pressures among all 2,669 men was 19.5 mm Hg. Assuming that the linear model is applicable, estimate the SD of systolic blood pressures among men 50 years old.[14]

13.28 A veterinary anatomist measured the density of nerve cells at specified sites in the intestine of nine horses. Each density value is the average of counts of nerve cells in five equal sections of tissue. The results are given in the accompanying table for site I (midregion of jejunum) and site II (mesenteric region of jejunum).[15] (These data were also given in Exercise 9.28.)

ANIMAL	SITE I	SITE II
1	50.6	38.0
2	39.2	18.6
3	35.2	23.2
4	17.0	19.0
5	11.2	6.6
6	14.2	16.4
7	24.2	14.4
8	37.4	37.6
9	35.2	24.4
Mean	29.36	22.02
SS	1,419.82	853.396
	$SP_{XY} = 893.689$	

a. Calculate the correlation coefficient between the densities at the two sites.

b. Construct a scatterplot of the data.

c. Four potential sources of variation in these data are: (i) errors in counting the nerve cells, (ii) sampling error due to choosing certain slices for counting, (iii) variation from one horse to another, and (iv) variation from site to site within a horse. Which of these sources of variation would tend to produce positive correlation between the sites? [*Hint:* Ask yourself how each source contributes to the appearance of the scatterplot.]

13.29 Refer to Exercise 13.28. Test the hypothesis that the true (population) correlation coefficient is zero against the alternative that it is positive. Let $\alpha = .05$.

13.30 In a study of natural variation in blood chemistry, blood specimens were obtained from 284 healthy people. The concentrations of urea and of uric acid were measured for each specimen, and the correlation between these two concentrations was found to be $r = .2291$. Test the hypothesis that the population correlation coefficient is zero against the alternative that it is positive.[16] Let $\alpha = .05$.

SECTION 13.7

GUIDELINES FOR INTERPRETING REGRESSION AND CORRELATION

Any set of (x, y) data can be submitted to a regression analysis and values of b_0, b_1, $s_{Y|X}$, and r can be calculated. But these quantities require care in interpretation. In this section we discuss guidelines and cautions for interpretation of linear regression and correlation. We first consider the use of regression and correlation for purely *descriptive* purposes, and then turn to *inferential* uses.

WHEN IS LINEAR REGRESSION DESCRIPTIVELY INADEQUATE?

Linear regression and correlation may provide inadequate description of a data set if either of the following features is present:

curvilinearity

outliers

We briefly discuss each of these.

If the dependence of Y on X is actually curvilinear rather than linear, the application of linear regression and correlation can be very misleading. The following example shows how this can happen.

EXAMPLE 13.31

Figure 13.19 shows a set of fictitious data that obey an *exact* relationship: $Y = 6X - X^2$. Nevertheless, X and Y are uncorrelated: $r = 0$ and $b_1 = 0$. The best straight line through the data would be a horizontal one, but of course the line would be a poor summary of the curvilinear relationship between X and Y. The residual SD is $s_{Y|X} = 2.43$; this value does not measure random variation but rather measures deviation from linearity.

FIGURE 13.19

Data for Which X and Y Are Uncorrelated but Have a Strong Curvilinear Relationship

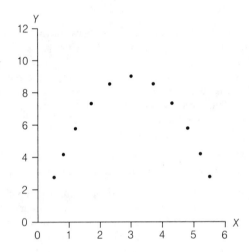

FIGURE 13.20

Data Displaying Mild Curvilinearity

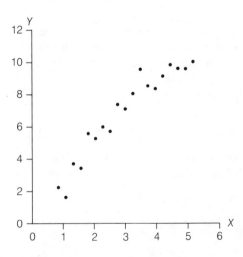

Generally, the consequences of curvilinearity are that (1) the fitted line does not adequately represent the data; (2) the correlation is misleadingly small; (3) $s_{Y|X}$ is inflated. Of course, Example 13.31 is an extreme case of this distortion. A data set with mild, but still noticeable, curvilinearity is shown in Figure 13.20.

Outliers in a regression setting are data points that are unusually far from the linear trend formed by the other data points. Outliers can distort regression analysis in two ways: (1) by inflating $s_{Y|X}$ and reducing correlation; (2) by unduly influencing the regression line. The following example illustrates both of these.

EXAMPLE 13.32

Figure 13.21(a) shows a data set with a single outlier. A straight line would fit quite well through all the data points except the outlier. Figure 13.21(b) shows the regression line fitted to all the data. Notice how poorly the line fits the points.

FIGURE 13.21
The Effect of an Outlier on the Regression Line.
(a) A data set with an outlier; (b) the regression line fitted to all the data.

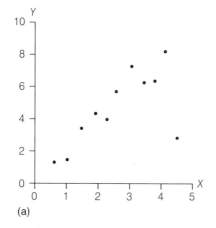 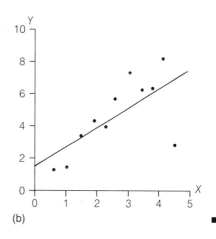

ASSUMPTIONS FOR INFERENCE

The quantities b_0, b_1, $s_{Y|X}$, and r can always be used to describe a scatterplot. However, statistical inference based on these quantities depends on certain assumptions concerning the design of the study, the parameters, and the conditional population distributions. We summarize these assumptions, and then discuss guidelines and cautions concerning them.

1. *Design assumptions* We have discussed two sampling models for regression and correlation:
   a. *Random subsampling model:* For each observed x, the corresponding observed y is viewed as randomly chosen from the conditional population distribution of Y values for that x.*

---

*If the X variable includes measurement error, then X in the linear model must be interpreted as the measured value of X rather than some underlying "true" value of X. A linear model involving the "true" value of X leads to a different kind of regression analysis.

b. *Bivariate random sampling model:* Each observed pair (x, y) is viewed as randomly chosen from the joint population distribution of bivariate pairs (X, Y).

In either sampling model, each observed pair (x, y) must be independent of the others. This means that the experimental design must not include any pairing, blocking, or hierarchical structure.

2. *Assumptions concerning parameters* The linear model states that
  a. $\mu_{Y|X} = \beta_0 + \beta_1 X$
  b. $\sigma_{Y|X}$ does not depend on X.
3. *Assumption concerning population distributions* The confidence interval and t test are based on the assumption that the conditional population distribution of Y for each fixed X is a normal distribution.

The random subsampling model is required if b_0, b_1, and $s_{Y|X}$ are to be viewed as estimates of the parameters β_0, β_1, and $\sigma_{Y|X}$ mentioned in the linear model. The bivariate random sampling model is required if r is to be viewed as an estimate of a population parameter ρ. It can be shown that, if the bivariate random sampling model is applicable, then the random subsampling model is also applicable. Thus, regression parameters can always be estimated if correlation can be estimated, but not vice versa.

GUIDELINES CONCERNING THE SAMPLING ASSUMPTIONS

Departures from the sampling assumptions not only affect the validity of formal techniques such as the confidence interval for β_1, but also can lead to faulty interpretation of the data even if no formal statistical analysis is performed. Two errors of interpretation that sometimes occur in practice are (1) failure to take into account dependency in the observations, and (2) insufficient caution in interpreting r when the x's do not represent a random sample.

The following two examples illustrate studies with dependent observations.

EXAMPLE 13.33
SERUM CHOLESTEROL AND SERUM GLUCOSE

A data set consists of 20 pairs of measurements on serum cholesterol (X) and serum glucose (Y) in humans. However, the experiment included only two subjects; each subject was measured on ten different occasions. Because of the dependency in the data, it is not correct to naively treat all 20 data points alike. Figure 13.22 illustrates the difficulty; the figure shows that there is no evidence of any correlation between X and Y, except for the modest fact that the subject who has larger X values happens also to have larger Y values. Clearly it would be impossible to properly interpret the scatterplot if all 20 points were plotted with the same symbol. By the same token, application of regression or correlation formulas to the 20 observations would be seriously misleading.[17] ∎

EXAMPLE 13.34
GROWTH OF BEEF STEERS

Figure 13.23 shows 20 pairs of measurements on the weight (Y) of beef steers at various times (X) during a feeding trial. The data represent four animals, each weighed at five different times; observations on the same animal are joined by lines in the figure. An ordinary regression analysis on the 20 data points would

FIGURE 13.22

Twenty Observations of
X = Serum Cholesterol
and Y = Serum Glucose
in Humans.

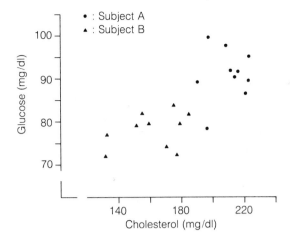

ignore the information carried in the lines and would yield inflated SEs and weak tests. Similarly, an ordinary scatterplot (without the lines) would be an inadequate representation of the data.[18]

FIGURE 13.23

Twenty Observations of
X = Days and Y =
Weight in Steers. Data for
individual animals are
joined by lines.

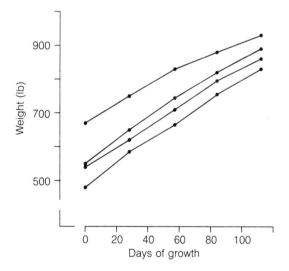

In Example 13.33, ignoring the dependency in the observations would lead to *overinterpretation* of the data—that is, concluding that a relationship exists when there is actually very little evidence for it. By contrast, ignoring the dependency in Example 13.34 would lead to *underinterpretation* of the data—that is, insufficiently extracting the "signal" from the "noise."

In interpreting the correlation coefficient r, one should recognize that r is influenced by the degree of spread in the values of X. If the regression quantities b_0, b_1, and $s_{Y|X}$ are unchanged, *more spread in the X values leads to a stronger correlation (larger magnitude of r)*. The following example shows how this happens.

EXAMPLE 13.35
Figure 13.24 shows fictitious data that illustrate how r can be affected by the distribution of X. The data points in parts (a) and (b) have been plotted together in part (c). The regression line is the same in all three scatterplots, but notice that X and Y appear more highly correlated in (c) than in either (a) or (b). The residuals associated with the data points in (a) and (b) appear relatively smaller when viewed from the perspective of (c), with its expanded range of X. The contrasting appearance of the scatterplots is reflected in the correlation coefficients; in fact, $r = .61$ for (a), $r = .61$ for (b), but $r = .84$ for (c).

FIGURE 13.24 Dependence of r on the Distribution of X. The data of (a) and (b) are plotted together in (c).

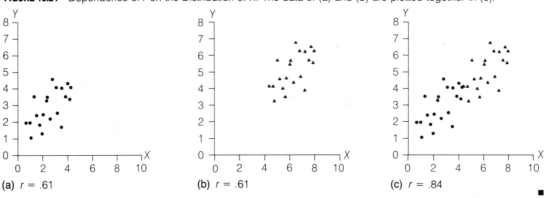

(a) $r = .61$ (b) $r = .61$ (c) $r = .84$ ∎

The fact that r depends on the distribution of X does not mean that r is invalid as a descriptive statistic. But it does mean that, when the values of X cannot be viewed as a random sample, r must be interpreted cautiously. For instance, suppose two experimenters conduct separate studies of response (Y) to various doses (X) of a drug. Each of them could calculate r as a description of his own data, but they should *not* expect to obtain *similar* values of r unless they both use the same choice of doses (X values). By contrast, they might reasonably expect to obtain similar regression lines and similar residual standard deviations, regardless of their choice of X values, as long as the dose–response relationship remains the same throughout the range of doses used.

LABELLING *X* AND *Y* If the bivariate random sampling model is applicable, then the investigator is free to decide which variable to label X and which to label Y. Of course, for calculation of r the labelling does not matter. For regression calculations, the decision depends on the purpose of the analysis. The regression of Y on X yields (within the linear model) estimates of $\mu_{Y|X}$—that is, the population mean Y value for fixed X. Similarly, the regression of X on Y is aimed at estimating $\mu_{X|Y}$—that is, the mean X value for fixed Y. These approaches do not lead to the same regression line because they are directed at answering different questions. The following is an intuitive example.

EXAMPLE 13.36

HEIGHT AND WEIGHT
OF YOUNG MEN

For the population of young men described in Example 13.15, the mean weight of young men 76″ (6′4″) tall is 178 lb. Now consider this question: What would be the mean height of young men who weigh 178 lb? There is no reason that the answer should be 76″. Intuition suggests that the answer should be *less* than 76″—and in fact it is about 71″. ∎

GUIDELINES CONCERNING THE LINEAR MODEL AND NORMALITY ASSUMPTION

The test and confidence interval for β_1 are based on the linear model and the assumption of normality. The interpretation of these inferences can be seriously degraded if the linearity assumption is not met; after all, we have seen earlier in this section that even the *descriptive* usefulness of regression is reduced if curvilinearity or outliers are present.

In addition to linearity, the linear model specifies that $\sigma_{Y|X}$ is the same for all the observations. A common pattern of departure from this assumption is a trend for larger means to be associated with larger SDs. Mild nonconstancy of the SDs does not seriously affect the interpretation of b_0, b_1, SE_{b_1}, and r (although it does invalidate the interpretation of $s_{Y|X}$ as a pooled estimate of a common SD).

Because of the Central Limit Theorem, the assumption of normality is less important if n is large. The most troublesome form of nonnormality is the presence of long tails or, especially, outliers.

DETECTING DEPARTURES FROM THE LINEAR MODEL AND NORMALITY

Formal statistical tests for curvilinearity, unequal standard deviations, nonnormality, and outliers are beyond the scope of this book. However, the single most useful instrument for detecting these features is the human eye, aided by scatterplots. For instance, notice how easily the eye detects the mild curvilinearity in Figure 13.20 and the outlier in Figure 13.21. Notice also in Figure 13.21 that examination of the *marginal* distributions of X and Y separately would *not* have revealed the outlier.

In addition to scatterplots of X vs Y, it is often useful to look at various displays of the residuals. For instance, a scatterplot of each residual $(y - \hat{y})$ against $\hat{y}$ can reveal trends in the conditional standard deviation. Also, if the assumption of normality is met, then the distribution of the residuals should look roughly like a normal distribution.* Histograms and scatterplots of the residuals are easily produced by modern statistical computing packages.

THE USE OF TRANSFORMATIONS

If the assumptions of linearity, constancy of standard deviation, and normality are not met, a remedy that is sometimes useful is to transform the scale of

*This is the basis for the 68% and 95% interpretations of $s_{Y|X}$ given in Section 13.3.

measurement of either Y, or X, or both. The following example illustrates the use of a logarithmic transformation.

EXAMPLE 13.37
GROWTH OF SOYBEANS

A botanist placed 60 one-week-old soybean seedlings in individual pots. After 12 days of growth, she harvested, dried, and weighed 12 of the young soybean plants. She weighed another 12 plants after 23 days of growth, and groups of 12 plants each after 27 days, 31 days, and 34 days. Figure 13.25 shows the 60 plant weights plotted against days of growth; the group means are connected by solid lines. It is easy to see from Figure 13.25 that the relationship between mean plant weight and time is curvilinear rather than linear and that the conditional standard deviation is not constant but is strongly increasing.[19]

FIGURE 13.25   Weight of Soybean Plants Plotted Against Days of Growth

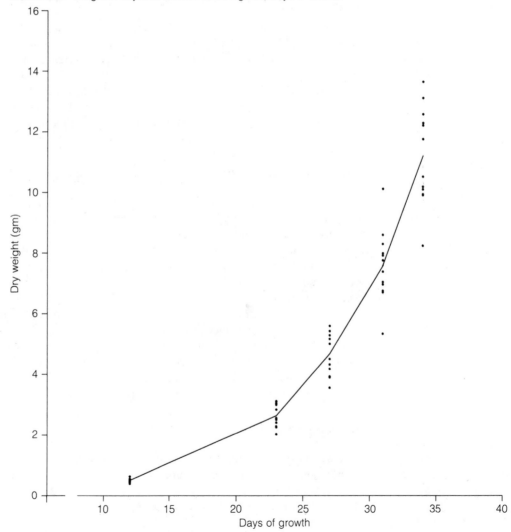

Figure 13.26 shows the logarithms (base 10) of the plant weights, plotted against days of growth; the means of the logarithms are connected by solid lines. Notice that the logarithmic transformation has simultaneously straightened the curve and more nearly equalized the standard deviations. It would not be unreasonable to assume that the linear model is valid for the variables $Y = \log(\text{dry weight})$ and $X = $ days of growth.

Table 13.7 (page 484) shows the means and standard deviations before and after the logarithmic transformation. Note especially the effect of the transformation on the equality of the SDs.

FIGURE 13.26 Log(weight) of Soybean Plants Plotted Against Days of Growth

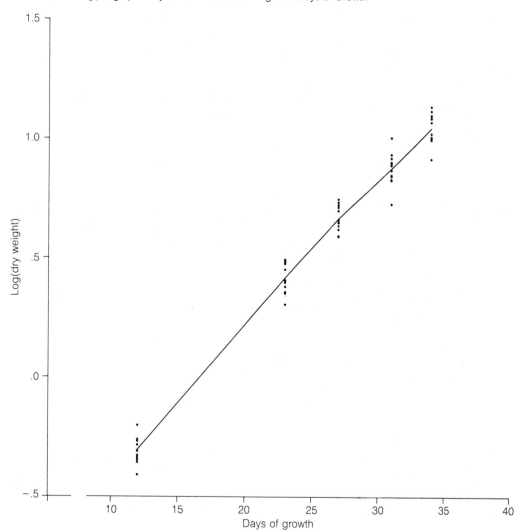

DAYS OF GROWTH	NUMBER OF PLANTS	DRY WEIGHT (gm)		LOG(DRY WEIGHT)	
		Mean	SD	Mean	SD
12	12	.50	.06	−.31	.055
23	12	2.63	.37	.42	.062
27	12	4.67	.70	.67	.066
31	12	7.57	1.19	.87	.069
34	12	11.20	1.62	1.04	.064

∎

CORRELATION AND CAUSATION

We noted in Chapter 8 that an observed association between two variables does not necessarily indicate any *causal* connection between them. It is important to remember this caution when interpreting correlation.

The following example shows that even *strongly* correlated variables may be causally *unrelated*.

EXAMPLE 13.38

REPRODUCTION OF
AN ALGA

Akinetes are sporelike reproductive structures produced by the green alga *Pithophora oedogonia*. In a study of the life cycle of the alga, researchers counted akinetes in specimens of alga obtained from an Indiana lake on 26 occasions over a 17-month period. Low counts indicated germination of the akinetes. The researchers also recorded the water temperature and the photoperiod (hours of daylight) on each of the 26 occasions.

The data showed a rather strong negative correlation between akinete counts and photoperiod; the correlation coefficient was $r = -.72$. But the researchers recognized that this observed correlation might not reflect a causal relationship. Longer days (increasing photoperiod) also tend to bring higher temperatures, and the akinetes might actually be responding to temperature rather than photoperiod. To resolve the question, the researchers conducted laboratory experiments in which temperature and photoperiod were varied independently; these experiments showed that temperature, not photoperiod, was the causal agent.[20]

∎

As Example 13.38 shows, one way to establish causality is to conduct a controlled experiment in which the putative causal factor is varied and all other factors are either held constant or controlled by randomization. When such an experiment is not possible, indirect approaches using statistical analysis can shed some light on causal relationships. (One such approach will be illustrated in Example 13.41.)

**EXERCISES
13.31–13.34**

13.31 In a metabolic study, four male swine were tested three times: when they weighed 30 kg, again when they weighed 60 kg, and again when they weighed 90 kg. During each test, the experimenter analyzed feed intake and fecal and urinary output for

15 days, and from these data calculated the nitrogen balance, which is defined as the amount of nitrogen incorporated into body tissue per day. The results are shown in the accompanying table.[21]

ANIMAL NUMBER	NITROGEN BALANCE (gm/day)		
	BODY WEIGHT		
	30 kg	60 kg	90 kg
1	15.8	21.3	16.5
2	16.4	20.8	18.2
3	17.3	23.8	17.8
4	16.4	22.1	17.5
Mean	16.47	22.00	17.50

Suppose these data are analyzed by linear regression. With X = body weight and Y = nitrogen balance, preliminary calculations yield $\bar{x} = 60$, $\bar{y} = 18.7$, $SS_X = 7,200$, $SS_Y = 77$, $SP_{XY} = 123$. The slope is $b_1 = .017$, with standard error $SE_{b_1} = .032$. The t statistic is $t_s = .53$, which is not significant at any reasonable significance level. According to this analysis, there is insufficient evidence to conclude that nitrogen balance depends on body weight under the conditions of this study.

The above analysis is flawed in two ways. What are they? [*Hint:* Look for ways in which the assumptions for inference are not met. There may be several minor departures from the assumptions, but you are asked to find two major ones. No calculation is required.]

13.32 For measuring the digestibility of forage plants, two methods can be used: The plant material can be fermented with digestive fluids in a glass container, or it can be fed to an animal. In either case, digestibility is expressed as the percentage of total dry matter that is digested. Two investigators conducted separate studies to compare the methods by submitting various types of forage to both methods and comparing the results. Investigator A reported a correlation of $r = .8$ between the digestibility values obtained by the two methods, and investigator B reported $r = .3$. The apparent discrepancy between these results was resolved when it was noted that one of the investigators had tested only varieties of canary grass (whose digestibilities ranged from 56% to 65%), whereas the other investigator had used a much wider spectrum of plants, with digestibilities ranging from 35% for corn stalks to 72% for timothy hay.[22]

Which investigator (A or B) used only canary grass? How does the different choice of test material explain the discrepancy between the correlation coefficients?

13.33 Refer to the energy expenditure data of Exercise 13.11. Each expenditure value (Y) is the average of two measurements made on different occasions. (See Example 1.7.) It might be proposed that it would be better to use the two measurements as separate data points, thus yielding 14 observations rather than 7. If this proposed approach were used, one of the assumptions for inference would be highly doubtful. Which one, and why?

13.34 Refer to the fungus growth data of Exercise 13.10.
  a. Calculate the correlation coefficient, r.

b. Suppose a second investigator were to replicate the experiment, using concentrations of 0, 2, 4, 6, 8, and 10 mg, with two petri dishes at each concentration. Would you predict that the value of r calculated by this second investigator would be about the same as that calculated in part **a**, smaller in magnitude, or larger in magnitude? Explain.

S E C T I O N   13.8

PERSPECTIVE

To put the methods of Chapter 13 in perspective, we will discuss their relationship to methods described in earlier chapters, and to methods that might be included in a second statistics course. We begin by relating regression to the methods of Chapters 7 and 12.

REGRESSION AND ANALYSIS OF VARIANCE

When there are several Y values for each value of X, one could analyze the data with an analysis of variance or with a regression analysis. Each approach uses the data to estimate the conditional mean of Y for each fixed X; these parameters are estimated by the fitted line $b_0 + b_1X$ in the regression approach and by the individual sample means $\bar{y}$ in the ANOVA approach. Also, each approach uses the data to estimate the standard deviation of Y for fixed X; this parameter is estimated by $s_{Y|X}$ in the regression approach and by s_c in the ANOVA approach. Finally, to test the null hypothesis of no dependence of Y on X, each approach translates the null hypothesis into its own terms. The following example illustrates the two approaches.

EXAMPLE 13.39

AMPHETAMINE AND
FOOD CONSUMPTION

In the food consumption experiment of Example 13.1, the treatments were doses $X = 0$, 2.5, and 5 mg/kg of amphetamine; there were eight observations of Y (food consumption) for each value of X.

Both the regression approach and the ANOVA approach use the data to estimate the population mean food consumption for each fixed dose. The regression approach estimates these means by the fitted line, which is $Y = 99.3 - 9.01X$, and the ANOVA approach estimates them by the individual sample means $\bar{y}_1 = 100.0$, $\bar{y}_2 = 75.5$, and $\bar{y}_3 = 55.0$ gm/kg.

Both approaches assume that the population standard deviation of food consumption for fixed dose is a constant. The regression approach estimates this constant by

$$s_{Y|X} = 11.4 \text{ gm/kg}$$

and the ANOVA approach estimates it by

$$s_c = 11.6 \text{ gm/kg}$$

Figure 13.27 shows the data and the fitted line. The sample means are shown as open circles. The standard deviations $s_{Y|X}$ and s_c are shown as ruler lines. Because these data agree very well with the linear model, the estimates of means and standard deviations produced by the regression approach agree very closely with those produced by the ANOVA approach.

FIGURE 13.27

Scatterplot of Food Consumption vs Dose of Amphetamine. The fitted regression line and the individual means (open circles) are shown.

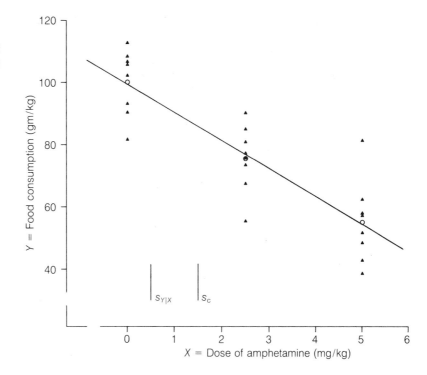

Finally, consider the null hypothesis

H_0:   Mean food consumption does not depend on dose

In a regression approach, this hypothesis would be translated as

H_0: $\beta_1 = 0$

In an ANOVA approach, the hypothesis would be translated as

H_0: $\mu_1 = \mu_2 = \mu_3$

The regression approach is more precisely focused; the change in Y as X changes is described in terms of a single parameter (β_1) rather than in terms of three parameters (μ_1, μ_2, and μ_3).   ∎

The advantage of the regression approach is that it gains power and focus by using the information about X, while ANOVA does not. The disadvantage of regression is that, if the relationship between Y and X is curvilinear rather than linear, the residual SD is inflated and the analysis loses sensitivity.

In practice, many researchers prefer to use a combination of regression and ANOVA, which takes advantage of the focused approach of regression but uses s_c from the ANOVA as a basis for calculating standard errors. This combined approach, sometimes called *response curve analysis*, permits analysis of the dependence of Y on X in a way that does not require the restrictive assumption of linearity. In fact, response curve analysis can also be used to test whether the relationship between $\mu_{Y|X}$ and X is actually linear.

The following example compares the regression approach and the two-sample approach to a data set for which (unlike Example 13.39) X varies within as well as between the samples.

EXAMPLE 13.40

BLOOD PRESSURE AND PLATELET CALCIUM

In Example 13.21 we described blood pressure (X) and platelet calcium (Y) measurements on 38 subjects. Actually, the study included *two* groups of subjects: 38 volunteers with normal blood pressure, selected from hospital lab personnel and other nonpatients, and 45 patients with a diagnosis of high blood pressure. Table 13.8 summarizes the platelet calcium measurements in the two groups and Figure 13.28 shows the blood pressure and calcium measurements for all 83 subjects.[12]

TABLE 13.8

Platelet Calcium (nM) in Two Groups of Subjects

	NORMAL BLOOD PRESSURE	HIGH BLOOD PRESSURE
$\bar{y}$	107.9	168.2
s	16.1	31.7
n	38	45

Two ways to analyze the data are (1) as two independent samples; (2) by regression analysis. To test for a relationship between blood pressure and platelet calcium, (1) a two-sample t test of H_0: $\mu_1 = \mu_2$ can be applied to Table 13.8; (2) a regression t test of H_0: $\beta_1 = 0$ can be applied to the data in Figure 13.28. The two-sample t statistic (unpooled) is $t_s = 11.2$ and the regression t statistic is $t_s = 20.8$. Both of these are highly significant, but the latter is more so because the regression analysis extracts more information from the data.

For these data, the regression approach is more enlightening and convincing than the two-sample approach. Figure 13.28 suggests that platelet calcium is correlated with blood pressure, not only between but also within the two groups. Relevant regression analyses would include (1) testing for a correlation within each group separately (as in Examples 13.21 and 13.27); (2) testing for an overall correlation (as in the previous paragraph); (3) testing whether the regression lines in the two groups are identical (using methods not described in this book).

Formal testing aside, notice the advantage of the scatterplot as a tool for understanding the data and for communicating the results. Figure 13.28 provides eloquent testimony to the reality of the relationship between blood pressure and platelet calcium. (We emphasize once again, however, that a "real" relationship is not necessarily a *causal* relationship. Further, even if the relationship is causal, the data do not indicate the direction of causality—that is, whether high calcium causes high blood pressure or vice versa.* ∎

*In fact, the authors of the study remark that "It remains possible . . . that an increased intracellular calcium concentration is a consequence rather than a cause of elevated blood pressure."

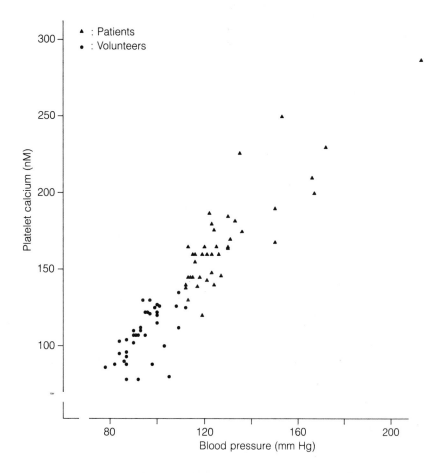

FIGURE 13.28
Blood Pressure and
Platelet Calcium for 83
Subjects

Examples 13.39 and 13.40 illustrate a general principle: If quantitative information on a variable X is available, it is usually better to use that information than to ignore it.

EXTENSIONS OF LEAST SQUARES

We have seen that the classical method of fitting a straight line to data is based on the least-squares criterion. This versatile criterion can be applied to many other statistical problems. For instance, in **curvilinear regression**, the least-squares criterion is used to fit curvilinear relationships such as

$$Y = b_0 + b_1X + b_2X^2$$

Another application is **multiple regression and correlation**, in which the least-squares criterion is used to fit an equation relating Y to *several X* variables—X_1, X_2, and so on; for instance,

$$Y = b_0 + b_1X_1 + b_2X_2$$

The following example illustrates both curvilinear and multiple regression.

EXAMPLE 13.41

SERUM CHOLESTEROL
AND BLOOD PRESSURE

As part of a large health study, various measurements of blood pressure, blood chemistry, and physique were made on 2,599 men.[23] The researchers found a positive correlation between blood pressure and serum cholesterol ($r = .23$ for systolic blood pressure). But blood pressure and serum cholesterol also are related to age and physique. To untangle the relationships, the researchers used the method of least squares to fit the following equation:

$$Y = b_0 + b_1X_1 + c_1X_1^2 + b_2X_2 + b_3X_3 + b_4X_4$$

where

Y = Systolic blood pressure

X_1 = Age

X_2 = Serum cholesterol

X_3 = Blood glucose

X_4 = Ponderal index (height divided by the cube root of weight)

Note that the regression is curvilinear with respect to age (X_1) and linear in the other X variables.

By applying multiple regression and correlation analysis, the investigators determined that there is little or no correlation between blood pressure and serum cholesterol, if age and ponderal index are held constant. They concluded that the observed correlation between serum cholesterol and blood pressure was an indirect consequence of the correlation of each of these with age and physique.

■

NONPARAMETRIC AND ROBUST REGRESSION AND CORRELATION

We have discussed the classical least-squares methods for regression and correlation analysis. There are also many excellent modern methods that are *not* based on the least-squares criterion. Some of these methods are *robust*—that is, they work well even if the conditional distributions of Y given X have long straggly tails or outliers. The nonparametric methods assume little or nothing about the form of dependence—linear or curvilinear—of Y on X, or about the form of the conditional distributions.

ANALYSIS OF COVARIANCE

Sometimes regression ideas can add greatly to the power of a data analysis, even if the relationship between X and Y is not of primary interest. The following is an example.

EXAMPLE 13.42
CATERPILLAR HEAD SIZE

Can diet affect the size of a caterpillar's head? Such an effect is plausible, because a caterpillar's chewing muscles occupy a large part of the head. To study the effect of diet, a biologist raised caterpillars (*Pseudaletia unipuncta*) on three different diets: diet 1, an artificial soft diet; diet 2, soft grasses; and diet 3, hard grasses. He measured the weight of the head and of the entire body in the final stage of larval development. The results are shown in Figure 13.29, where $Y = $ log(head weight) is plotted against $X = $ log(body weight), with different symbols for the three diets.[24] Note that the effect of diet is striking; there is very little overlap between the three groups of points. But if we were to ignore X and consider Y only, then the effect of diet would be much less clear; to see this, imagine projecting all the data points onto the Y-axis.

FIGURE 13.29
Head Weight Versus Body Weight (on Logarithmic Scales) for Caterpillars on Three Different Diets

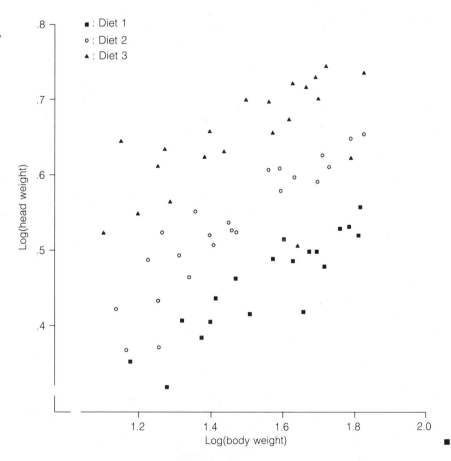

Example 13.42 shows how comparison of several groups with respect to a variable Y can be strengthened by using information on an auxiliary variable X that is correlated with Y. A classical method of statistical analysis for such data

is **analysis of covariance**, which proceeds by fitting regression lines to the (x, y) data. But even without this formal technique, an investigator can often clarify the interpretation of data simply by constructing a scatterplot like Figure 13.28. Plotting the data against X has the effect of removing that part of the variability in Y which is accounted for by X, causing the treatment effect to stand out more clearly against the residual background variation.

For convenient reference, we summarize the formulas presented in Chapter 13.

SECTION 13.9

SUMMARY OF FORMULAS

BASIC COMPUTATIONS

n = number of observed pairs (x, y)

$$\bar{x} = \frac{\sum x}{n} \qquad SS_X = \sum(x - \bar{x})^2$$

$$\bar{y} = \frac{\sum y}{n} \qquad SS_Y = \sum(y - \bar{y})^2$$

$$SP_{XY} = \sum(x - \bar{x})(y - \bar{y})$$

FITTED REGRESSION LINE

$$Y = b_0 + b_1 X$$

where

$$b_1 = \frac{SP_{XY}}{SS_X} \qquad b_0 = \bar{y} - b_1\bar{x}$$

Residuals: $y - \hat{y}$, where $\hat{y} = b_0 + b_1 x$

Residual Sum of Squares:

$$SS(\text{resid}) = \sum(y - \hat{y})^2 \quad \text{(definitional)}$$

$$= SS_Y - \frac{SP_{XY}^2}{SS_X} \quad \text{(computational)}$$

Residual Standard Deviation:

$$s_{Y|X} = \sqrt{\frac{SS(\text{resid})}{n - 2}}$$

CORRELATION COEFFICIENT

$$r = \frac{SP_{XY}}{\sqrt{SS_X SS_Y}}$$

Facts:

$Fact$ 13.1: $r^2 = 1 - \dfrac{SS(\text{resid})}{SS_Y}$

$Fact$ 13.2: $\dfrac{s_{Y|X}}{s_Y} \approx \sqrt{1 - r^2}$

INFERENCE

Standard Error of b_1:

$$SE_{b_1} = \sqrt{\frac{s_{Y|X}^2}{SS_X}}$$

95% Confidence Interval for β_1:

$$b_1 \pm t_{.05} SE_{b_1}$$

Test of H_0: $\beta_1 = 0$ or H_0: $\rho = 0$:

$$t_s = \frac{b_1}{SE_{b_1}} = r \sqrt{\frac{n - 2}{1 - r^2}}$$

Critical values for test and confidence interval are determined from Student's t distribution with df $= n - 2$.

| | | | | | | | | | | | |

SUPPLEMENTARY EXERCISES 13.35–13.47

13.35 The table presents a fictitious set of data.

	X	Y
	3	6
	4	13
	1	6
	8	19
Mean	4	11
SS	26	118
	$SP_{XY} = 52$	

a. Use the definitional formulas to verify the given values of $\bar{x}$, $\bar{y}$, SS_X, SS_Y, and SP_{XY}.

b. Compute the linear regression of Y on X and compute $\hat{y}$ for each data point.
c. Use the definitional formula to compute the residual sum of squares.
d. Use the computational formula to compute SS(resid). Verify that the answer agrees with that obtained in part c.

13.36 In a study of the Mormon cricket (*Anabrus simplex*), the correlation between female body weight and ovary weight was found to be $r = .836$. The standard deviation of the ovary weights of the crickets was .429 gm. Assuming that the linear model is applicable, estimate the standard deviation of ovary weights of crickets whose body weight is 4 gm.[25]

13.37 In a study of crop losses due to air pollution, plots of Blue Lake snap beans were grown in open-top field chambers, which were fumigated with various concentrations of sulfur dioxide. After a month of fumigation, the plants were harvested and the total yield of bean pods was recorded for each chamber. The results are shown in the table.[26]

	X = Sulfur dioxide concentration (ppm)			
	0	.06	.12	.30
Y = yield (kg)	1.15	1.19	1.21	.65
	1.30	1.64	1.00	.76
	1.57	1.13	1.11	.69
Mean	1.34	1.32	1.11	.70

Preliminary calculations yield the following results:

$\bar{x} = .12$ $\qquad$ $\bar{y} = 1.117$

$SS_X = .1512$ $\qquad$ $SS_Y = 1.069067$

$\qquad SP_{XY} = -.342$

a. Calculate the linear regression of Y on X.
b. Plot the data and draw the regression line on your graph.
c. Calculate $s_{Y|X}$. What are the units of $s_{Y|X}$?

13.38 Refer to Exercise 13.37.
a. Assuming that the linear model is applicable, find estimates of the mean and the standard deviation of yields of beans exposed to .24 ppm of sulfur dioxide.
b. Which assumption of the linear model appears doubtful for the snap bean data?

13.39 Refer to Exercise 13.37. Consider the null hypothesis that sulfur dioxide concentration has no effect on yield. Assuming that the linear model holds, formulate this as a hypothesis about the true regression line. Use the data to test the hypothesis against a directional alternative. Let $\alpha = .05$.

13.40 (*Continuation of Exercise* 13.39) Use an analysis of variance to test the hypothesis that sulfur dioxide concentration has no effect on yield. (Let $\alpha = .05$.) Compare with the results of Exercise 13.39. [*Hint:* You may verify that SS(between) = .7948. Note that SS(total) has been given to you (by another name) in Exercise 13.37. Determine SS(within) by subtraction.]

13.41 Another way to analyze the data of Exercise 13.37 is to take each treatment mean as the observation Y; then the data would be summarized as in the accompanying table.

	SULFUR DIOXIDE X (ppm)	MEAN YIELD Y (kg)
	0	1.34
	.06	1.32
	.12	1.11
	.30	.70
Mean	.12	1.117
SS	.0504	.264875
	$SP_{XY} = -.114$	

a. For the regression of mean yield on X, calculate the regression line and the residual standard deviation, and compare with the results of Exercise 13.37. Explain why the discrepancy is not surprising.

b. Calculate the correlation coefficient between mean yield and X. Also, calculate the correlation coefficient between individual chamber yield and X. Explain why the discrepancy is not surprising.

13.42 In a study of the tufted titmouse (*Parus bicolor*), an ecologist captured seven male birds, measured their wing lengths and other characteristics, and then marked and released them. During the ensuing winter, he repeatedly observed the marked birds as they foraged for insects and seeds on tree branches. He noted the branch diameter on each occasion, and calculated (from 50 observations) the average branch diameter for each bird. The results are shown in the table.[27]

BIRD	WING LENGTH X (mm)	BRANCH DIAMETER Y (cm)
1	79.0	1.02
2	80.0	1.04
3	81.5	1.20
4	84.0	1.51
5	79.5	1.21
6	82.5	1.56
7	83.5	1.29
Mean	81.4	1.26
SS	23.7143	.265486
	$SP_{XY} = 2.01571$	

a. Calculate the correlation coefficient between wing length and branch diameter.

b. Calculate s_Y and $s_{Y|X}$. Specify the units for each. Verify the approximate relationship between s_Y, $s_{Y|X}$, and r.

c. Construct a scatterplot of the data.

13.43 Refer to Exercise 13.42.

    a. Do the data provide sufficient evidence to conclude that the diameter of the forage branches chosen by male titmice is correlated with their wing length? Test an appropriate hypothesis against a nondirectional alternative. Let $\alpha = .05$.

    b. The test in part **a** was based on 7 observations, but each branch diameter value was the mean of 50 observations. If we were to test the hypothesis of part **a** using the raw numbers, we would have 350 observations rather than only 7. Why would this approach not be valid?

13.44 An exercise physiologist used skinfold measurements to estimate the total body fat, expressed as a percentage of body weight, for 19 participants in a physical fitness program. The body fat percentages and the body weights are shown in the table.[28]

PARTICIPANT	WEIGHT X (kg)	FAT Y (%)	PARTICIPANT	WEIGHT X (kg)	FAT Y (%)
1	89	28	11	57	29
2	88	27	12	68	32
3	66	24	13	69	35
4	59	23	14	59	31
5	93	29	15	62	29
6	73	25	16	59	26
7	82	29	17	56	28
8	77	25	18	66	33
9	100	30	19	72	33
10	67	23			

Actually, participants 1–10 are men, and participants 11–19 are women. Preliminary calculations yield the following results:

	MEN X	MEN Y	WOMEN X	WOMEN Y	BOTH SEXES X	BOTH SEXES Y
Mean	79.4	26.3	63.1	30.7	71.7	28.4
SS	1,578.40	62.100	268.89	66.000	3,104.1	218.42
SP_{XY}	292.80		108.33		64.211	

    a. Calculate the correlation coefficient between X and Y (i) for men; (ii) for women; and (iii) for all participants. The answers may surprise you.

    b. Draw a scatterplot with the points for men and women denoted by different symbols. After studying the scatterplot, try to sketch by eye the regression line of Y on X (pooling the sexes); it will be helpful to visually estimate the mean Y given X for a small X, an intermediate X, and a large X.

    c. Compute the regression line that you estimated in part **b** and draw it on the scatterplot.

    d. Using the insight gained from parts **b** and **c**, can you explain the discrepancy between the correlation coefficients computed in part **a**? Discuss.

13.45 (*Computer exercise*) The accompanying table gives the data plotted in Figures 13.15(a) and (b). The values of X are the same for both data sets.

X	(a) Y	(b) Y	X	(a) Y	(b) Y
.61	.88	.96	2.56	1.97	1.20
.93	1.02	.97	2.74	2.02	3.59
1.02	1.12	.07	3.04	2.26	3.09
1.27	1.10	2.54	3.13	2.27	1.55
1.47	1.44	1.41	3.45	2.43	.71
1.71	1.45	.84	3.48	2.57	3.05
1.91	1.41	.32	3.79	2.53	2.54
2.00	1.59	1.46	3.96	2.73	3.33
2.27	1.58	2.29	4.12	2.92	2.38
2.33	1.66	2.51	4.21	2.96	3.08

a. Calculate r for each data set.
b. For data set (a), multiply the values of X by 10, and multiply the values of Y by 3 and add 5. Recalculate r and compare with the value before the transformation. How is r affected by the linear transformation?
c. Verify that the regression lines for the two data sets are virtually identical (even though the correlation coefficients are very different).
d. Generate scatterplots like Figures 13.15(a) and (b).
e. Draw the regression line on each scatterplot.
f. Construct a scatterplot in which the two data sets are superimposed, using different plotting symbols for each data set.

13.46 (*Computer exercise*) This exercise shows the power of scatterplots to reveal features of the data that may not be apparent from the ordinary linear regression calculations. The accompanying table gives three fictitious data sets, A, B, and C. The values of X are the same for each data set, but the values of Y are different.[29]

	Data Set		
	A	B	C
X	Y	Y	Y
10	8.04	9.14	7.46
8	6.95	8.14	6.77
13	7.58	8.74	12.74
9	8.81	8.77	7.11
11	8.33	9.26	7.81
14	9.96	8.10	8.84
6	7.24	6.13	6.08
4	4.26	3.10	5.39
12	10.84	9.13	8.15
7	4.82	7.26	6.42
5	5.68	4.74	5.73

a. Verify that the fitted regression line is almost exactly the same for all three data sets. Are the residual standard deviations the same? Are the values of r the same?

b. Construct a scatterplot for each of the data sets. What does each plot tell you about the appropriateness of linear regression for the data set?

c. Plot the fitted regression line on each of the scatterplots.

13.47 (*Computer exercise*)   In a pharmacological study, 12 rats were randomly allocated to receive an injection of amphetamine at one of two dosage levels or an injection of saline. Shown in the table is the water consumption of each animal (ml water per kg body weight) during the 24 hours following injection.[30]

DOSE OF AMPHETAMINE (mg/kg)		
0	1.25	2.5
122.9	118.4	134.5
162.1	124.4	65.1
184.1	169.4	99.6
154.9	105.3	89.0

a. Calculate the regression line of water consumption on dose of amphetamine, and calculate the residual standard deviation.

b. Construct a scatterplot of water consumption against dose.

c. Draw the regression line on the scatterplot.

d. Use linear regression to test the hypothesis that amphetamine has no effect on water consumption against the alternative that amphetamine tends to reduce water consumption. (Use $\alpha = .05$.)

e. Use analysis of variance to test the hypothesis that amphetamine has no effect on water consumption. (Use $\alpha = .05$.) Compare with the result of part **d**.

f. What assumptions are necessary for the validity of the test in part **d** but not for the test in part **e**?

g. Calculate the pooled standard deviation from the ANOVA, and compare it with the residual standard deviation calculated in part **a**.

CHAPTER 14

CONTENTS

MULTIPLICITY IN STATISTICAL INFERENCE

INTRODUCTION

In Section 12.1 we introduced the problem of multiple comparisons in analysis of variance. Similar problems emerge whenever one is faced with **multiplicity**— that is, with a situation requiring multiple simultaneous statistical inferences because many questions are asked of the same set of data.

One consequence of multiplicity is loss of control of the risk of Type I error. For instance, if we test 20 null hypotheses, each at $\alpha = .05$, then (because $.05 = \frac{1}{20}$) it would not be surprising if one of them were rejected just by chance, even if all 20 are true. Multiplicity also compromises the interpretation of confidence intervals. If we construct 20 95% confidence intervals, it would not be surprising if one of them fails to contain its parameter.

Multiplicity arises whenever a set of data is subjected to many different manipulations in an attempt to find meaningful patterns. In any but the simplest experiment, there are many ways to search for patterns in the data. Using a computer, one can calculate differences, correlations, test statistics, confidence intervals, and P-values by the dozen. The investigator who naively regards each of these as if the others did not exist is likely to attach false importance to some patterns that just by chance happen to be "statistically significant."

In Section 14.2 we will illustrate some of the sources of multiplicity in biological investigations. In Section 14.3 we will briefly consider some approaches to making scientific inferences in the face of multiplicity.

SOURCES OF MULTIPLICITY

In analyzing data, in interpreting the results, and in reading reports of research in the life sciences, it is important to be aware of the various ways in which multiplicity can arise.

MULTIPLE GROUPS, OBSERVED VARIABLES, AND/OR TIME POINTS

Multiplicity most commonly arises because an investigation includes one or more of the following:

Observations on several treatments or groups

Observations on several response variables

Observations at several time points

In Chapter 12 we considered examples involving multiple treatments. The following example illustrates multiple response variables.

EXAMPLE 14.1
INSULIN AND LACTATION

To study the effect of insulin on lactation, investigators randomly allocated lactating rats to two treatment groups. One group (14 animals) received daily injections of insulin, while the other group (20 animals) served as controls and received injections of saline. Table 14.1 shows the treatment means for 12 observed variables, measured either on the rats themselves or on their pups.[1]

	VARIABLE	INSULIN	SALINE

TABLE 14.1 Means of Twelve Variables for Rats Given Insulin or Saline

VARIABLE	INSULIN	SALINE
Weight gain (gm)	4.5	11.6
Food intake (gm)	495	473
Milk yield (gm)	15.3	11.8
Percent fat in milk	13.7	16.4
Percent lactose in milk	4.6	3.8
Percent protein in milk	16.4	16.5
Liver weight (gm)	12.2	11.8
Mammary tissue weight (gm)	8.3	8.6
Mammary DNA (mg/gland)	22.3	18.4
Mammary protein (mg/gland)	1,108	869
Pup weight gain: Day 14 (gm)	162	151
Pup weight gain: Day 16 (gm)	182	162

For these data one could make 12 two-group comparisons (insulin versus control). In addition, one could calculate (from the raw data) various correlations between the variables; this would yield 66 correlations for the insulin group and 66 others for the control group. ∎

HIDDEN MULTIPLICITY

While it is obvious that multiplicity arises when many comparisons are made, one should also be alert for more subtle, or "hidden," sources of multiplicity.

Often a researcher inspects the data and chooses, from the multitude of possible analyses, a few that look particularly promising. Thus, the analyses are not preplanned but are "inspired" by the data. Such data-inspired analysis can give rise to serious multiplicity problems, because the investigator actually performs many "hidden" analyses while inspecting the data. The following example shows how the overall risk of Type I error, when testing a data-inspired hypothesis, is inflated just as if many hypotheses had been tested.

EXAMPLE 14.2 LIVER WEIGHT OF MICE

Ten treatments were compared for their effect on the liver in mice. There were 13 animals in each treatment group. The mean liver weights are shown in Table 14.2.[2]

TABLE 14.2 Liver Weights of Mice

TREATMENT	MEAN LIVER WEIGHT (gm)	TREATMENT	MEAN LIVER WEIGHT (gm)
1	2.59	6	2.84
2	2.28	7	2.29
3	2.34	8	2.45
4	2.07	9	2.76
5	2.40	10	2.37

Consider two investigators, A and B. Investigator A performs separate t tests (each at $\alpha = .05$) on all possible pairs of the ten treatments. She finds the following differences significant:

H_0: $\mu_4 = \mu_6$ Rejected

H_0: $\mu_4 = \mu_9$ Rejected

Investigator B begins his analysis by inspecting the treatment means. He notices that $\bar{y}_4$ is especially small, and that $\bar{y}_6$ and $\bar{y}_9$ are especially large. He then uses t tests to confirm these impressions. His conclusions are:

H_0: $\mu_4 = \mu_6$ Rejected

H_0: $\mu_4 = \mu_9$ Rejected

Among ten means there are 45 possible pairwise t tests. Investigator A conducted all 45 tests and investigator B conducted only two of them. But because investigator B's choice of those two hypotheses was "inspired" by inspection of the data, their conclusions are identical. Investigator B's procedure contains "hidden" multiplicity.   ∎

As Example 14.2 illustrates, multiplicity can be "hidden" if the investigator examines the data before deciding which manipulations to perform. Indeed, hidden multiplicity may be the most dangerous kind. There is a natural human tendency to perceive patterns even in phenomena that are purely random, and an equally natural reluctance to admit that a perceived pattern is not "real." Even in everyday life, people face this kind of hidden multiplicity. For instance, suppose you dream about your cousin Joe and then the next day he calls you long-distance. Many people would interpret this event as evidence of some paranormal power, such as telepathy or precognition. On the other hand, you might very sensibly interpret the event as a mere coincidence—especially if you force yourself to remember all the times you dreamed about people and they did *not* telephone you.

FURTHER EXAMPLES

In addition to multiple treatments, multiple variables, and multiple time points, there are other less obvious sources of multiplicity (which may or may not be "hidden"). We illustrate two of these.

ANALYSIS OF SUBSETS   Sometimes one class of experimental subject may respond differently than another. Multiplicity arises if the investigator tries to identify such classes by subdividing the data in various ways. For instance, medical researchers may search for classes of patients who respond best to a certain treatment. Here is an example.

EXAMPLE 14.3

SIMULATED "TREATMENT"
FOR CORONARY ARTERY
DISEASE

To illustrate the pitfalls of subset analysis, medical researchers conducted a simulated experiment on patients suffering from coronary artery disease. The patients were real but the "treatment" was not. Baseline and follow-up data on 1,073 patients were available in a large data bank. The investigators divided the 1,073 patients into two "treatment" groups at random, and then analyzed their survival (from time of diagnosis) using the follow-up data from the data bank. They examined the survival among all patients and also in subsets defined by various prognostic factors, such as age, the number of diseased vessels (one, two, or three), and the left ventricular contraction pattern (normal or abnormal) at baseline. This search among subsets bore fruit. Figure 14.1 shows the survival pattern for patients in the subset of patients with three-vessel disease and abnormal left ventricular contraction. The difference between "treatment" groups is both large and statistically significant ($P < .025$).* But of course the difference is not due to "treatment," because the two groups were not treated differently. The apparent treatment difference in Figure 14.1 is spurious, a result of multiplicity.[3]

FIGURE 14.1

Survival of Patients with
Three-Vessel Disease and
Abnormal Left Ventricular
Contraction

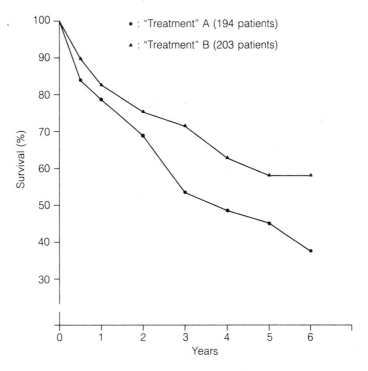

CHOOSING A CUTPOINT   Another way in which multiplicity can creep, almost undetected, into a data analysis, is in the process of defining combinations of observed

*The method for calculating this P-value, based on analysis of survival data with variable follow-up time, is beyond the scope of this book.

variables that seem best to summarize the way the data appear. Consider, for instance, the seemingly innocent act of defining a cutpoint for dividing a continuum of responses into two discrete categories. In many instances, such as the division of blood pressure readings into "high" and "normal," the cutpoint would be chosen before the experiment, perhaps at a conventional value. Suppose, however, that the dividing line is chosen by inspecting the data. For instance, suppose two treatments are being compared for their effect on the response time of rabbits to a certain stimulus. Upon examining the data, the experimenter notices that the treatments seem especially to differ in the proportion of animals that respond in 6 seconds or less; consequently, he chooses 6 seconds as a cutpoint between "fast" and "slow" responders, and he then analyzes the data as a 2 × 2 contingency table. For the chi-square test at $\alpha = .05$, it can be shown that the actual risk of Type I error using this kind of data-inspired cutpoint is not 5%, but can be as high as 40% or even 50%![4]

EXERCISE 14.1

14.1 In a sequence of digits, let π represent the probability that a digit will be even (0, 2, 4, 6, or 8). If the sequence is random then $\pi = .5$. Look at Table 1, and search through it (starting anywhere) until you find a string of ten consecutive digits that are overwhelmingly even or overwhelmingly odd (at least a 9:1 imbalance). When you find such a string, treat it as $n = 10$ independent observations and test the hypothesis $H_0: \pi = .5$ using a chi-square goodness-of-fit test at $\alpha = .05$. (This exercise illustrates how efficiently the human observer can sift through myriad patterns to find the "significant" result. How long did it take you to find a "significantly" unbalanced string of ten digits?)

SECTION 14.3

GUIDELINES FOR CONTROLLING MULTIPLICITY

We have seen that multiplicity can seriously compromise the interpretation of data. The problems can be reduced by the following approaches:

Using statistical techniques designed to take account of multiplicity

Attempting to ask focused questions whenever possible

Making a clear distinction between analyses that were inspired by the data and those that were preplanned.

We discuss each of these briefly.

For certain types of analysis, specialized techniques have been devised that take account of multiplicity. For instance, for comparing all pairs of means in an ANOVA setting, there are several techniques available which give better control of Type I error than the naive use of repeated t tests. One of these methods—the Newman–Keuls method—was described in optional Section 12.6.

A useful nonspecialized technique for controlling multiplicity is the Bonferroni method, which we now describe.

THE BONFERRONI METHOD

The **Bonferroni method** is based on a very simple and general relationship: The probability that at least one of several events will occur cannot exceed the sum of the individual probabilities. For instance, suppose we conduct six tests of hypotheses, each at $\alpha = .01$. Then the overall risk of Type I error—that is, the chance of rejecting at least one of the six hypotheses when in fact all of them are true—cannot exceed

$$.01 + .01 + .01 + .01 + .01 + .01 = (6)(.01) = .06$$

Turning this logic around, suppose an investigator plans to conduct six tests of hypotheses and wants the overall risk of Type I error not to exceed .05. A conservative approach is to conduct each of the separate tests at the significance level $\alpha = \frac{.05}{6} = .0083$; this is called a **Bonferroni adjustment**.

Note that the Bonferroni technique is very broadly applicable. The separate tests may relate to different response variables, different subsets, and so on; some may be t tests, some chi-square tests, and so on.

The Bonferroni approach can be used by a person reading a research report, if the author has included explicit P-values. For instance, if the report contains six P-values and the reader desires overall 5%-level protection against Type I error, then the reader will not regard a P-value as sufficient evidence of an effect unless it is smaller than .0083.

A Bonferroni adjustment can also be made for confidence intervals. For instance, suppose we wish to construct six confidence intervals, and desire an overall probability of 95% that *all* the intervals contain their respective parameters. Then this can be accomplished by constructing each interval at confidence level 99.17% (because $\frac{.05}{6} = .0083$ and $1 - .0083 = .9917$). Note that application of this idea requires unusual critical values, so that standard tables are not sufficient.

A disadvantage of the Bonferroni method is that it is overly conservative. If there are very many comparisons, the Bonferroni criterion becomes so strict that it is very difficult to reject any null hypothesis. In addition, the Bonferroni method does not help if the comparisons chosen were inspired by the data.

THE ADVANTAGE OF FOCUSED QUESTIONS

Often the pitfalls of multiplicity can be avoided by limiting a data analysis to address specific well-defined questions.

Biological experiments often involve many measurements—for instance, at various time points or at various doses of a chemical. A thoughtless way to analyze such data is to make all possible comparisons. Such a scattershot approach is usually undesirable for the following reasons:

1. A scattershot approach leads to multiplicity and the attendant loss of control of risk of Type I error.

2. Modifying the analysis to compensate for (1) may result in a drastic drop in power.
3. With or without modification to compensate for (1), a scattershot approach often fails to reveal the message in the data because its power is diffuse rather than focused on specific alternative hypotheses.

The following example illustrates these difficulties.

EXAMPLE 14.4

BICARBONATE SECRETION

Secretion of bicarbonate by the pancreas was studied in seven dogs. Each dog was infused with secretin and caerulein to stimulate bicarbonate secretion; after 1 hour, pancreatic polypeptide (PP) was also infused for 1 hour. Bicarbonate secretion was measured every 10 minutes. On another day, the procedure was repeated with the same animal but without the PP infusion. The question of interest was whether PP would inhibit bicarbonate secretion.*

The mean bicarbonate secretion at each time is shown in Table 14.3.[5]

TABLE 14.3

Bicarbonate Secretion (μmole/10 min) at 10-minute Intervals

TIME		MEAN SECRETION		PAIRED t	ONE-TAILED P-VALUE
		PP Day	Control Day		
0		70	53	.54	.30
10		653	69	3.28	.01
20		933	437	2.52	.02
30		967	877	.35	.37
40		1,019	1,134	−.59	.29
50		1,191	1,094	.41	.35
60		992	1,283	−1.51	.09
70		840	1,197	−1.06	.16
80		890	1,410	−1.25	.13
90	PP	819	1,134	−.76	.24
100	infusion	800	1,509	−2.47	.02
110		876	1,366	−1.45	.10
120		760	1,412	−2.89	.01

Table 14.3 contains the results of 13 paired t tests, one for each time of measurement. One way to compensate for the multiplicity of t tests would be to use a Bonferroni correction; to maintain an overall 5% significance level, the significance threshold for each P-value would be $\frac{.05}{13} = .0038$. According to this criterion, none of the null hypotheses would be rejected, and the evidence would be regarded as insufficient to conclude that PP inhibits bicarbonate secretion.

Even if we ignore the difficulties due to multiplicity and apply a separate 5% threshold to each t test, the results in Table 14.3 do not provide a coherent view of the biological situation. For instance, the P-values are less than .05 at times 100 and 120, but greater than .05 at time 110. But is it biologically plausible to

*Examples 14.4 and 14.5 are closely modeled on the discussion by J. D. Elashoff cited in Note 5.

suppose that PP inhibits bicarbonate secretion at 100 and 120 minutes but not at 110 minutes? Figure 14.2 shows the means plotted against time. Intuitively, the figure suggests that the time course of response to PP is smooth, with the zigzags in the curve representing chance fluctuations. This in turn suggests that, if PP inhibits bicarbonate secretion at 100 and 120 minutes, it probably also does so at 110 minutes.

FIGURE 14.2
Bicarbonate Secretion in Dogs

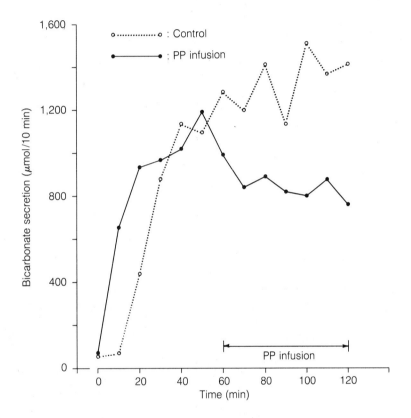

A focused data analysis is preferable to a tangle of separate comparisons. One way to focus an analysis is to define measures of response that reflect biological questions. For instance, we saw in Chapter 12 that linear combinations of means can be used to ask focused questions about structured treatments or groups. As another example, we saw in Chapter 13 how a regression line can serve as a concise summary of a relationship; for instance, comparisons of different drugs can sometimes be simply expressed in terms of the slopes and intercepts of linear dose–response relationships. Numerous other statistical methods, beyond the scope of this book, can help to streamline and clarify a data analysis.

In many situations no special statistical expertise is required in order to focus a data analysis. Rather, a suitable summary measure of response can be devised by the experimenter on the basis of his or her biological knowledge. The following is an example.

EXAMPLE 14.5

BICARBONATE
SECRETION

To better interpret the bicarbonate data of Example 14.4, the experimenter might reason (based on prior biological knowledge) as follows: The response to the background infusion of secretin and caerulein will have stabilized by 40 minutes, and PP infusion begins at 60 minutes; thus, a "before-PP" measure can be defined as the total of the bicarbonate values at 40, 50, and 60 minutes. Further, the response to PP will be well established by 100 minutes; thus, a "during-PP" measure can be defined as the total of the bicarbonate values at 100, 110, and 120 minutes. Table 14.4 shows these summary measures for each animal.

TABLE 14.4

Bicarbonate Secretion
(μmole/30 min) Before
and During PP Infusion

DOG	SECRETION ON PP DAY			SECRETION ON CONTROL DAY		
	Before	During	Change	Before	During	Change
1	2,808	1,173	−1,635	3,349	3,654	305
2	926	245	−681	2,411	1,706	−705
3	1,601	993	−608	3,276	6,818	3,542
4	4,316	2,779	−1,537	2,841	2,835	−6
5	1,809	1,143	−666	3,648	4,470	822
6	6,544	7,192	648	6,752	6,908	156
7	4,412	3,527	−885	2,292	3,621	1,329
Mean			−766			778

The effect of PP can now be assessed by a single comparison based on the "before" and "during" measures. Applying a paired t test to the "change" columns of Table 14.4 yields $t_s = 2.67$ with df $= 6$; from Table 4 we find that the one-tailed P-value is $.01 < P < .025$. At $\alpha = .05$ there is sufficient evidence to conclude that PP inhibits bicarbonate secretion under these conditions. ∎

EXPLORATORY VERSUS CONFIRMATORY ANALYSIS

We have indicated that problems of multiplicity arise from two sources:

Too many analyses

Analyses "inspired" by the data (hidden multiplicity)

We have seen that the number of analyses can be reduced by focusing attention on a few biologically meaningful measures of response. Of course, if a response measure is specifically designed to exploit a pattern noticed in the data, then any comparisons involving it contain "hidden" multiplicity and must be interpreted cautiously.

Researchers sometimes overreact to the problem of hidden multiplicity, and conclude that it is always wrong to consider hypotheses that have been "inspired" by the data. This view is far too restrictive. If scientists never explored their data, searching for inspiration, many important discoveries would be lost.

To put the matter in perspective, it is helpful to distinguish between two modes of data analysis—confirmatory analysis and exploratory analysis. **Confirmatory analysis** attempts to confirm effects that were anticipated in the planning of the investigation. By contrast, in **exploratory analysis** the investigator "snoops" through the data looking for interesting patterns.

In confirmatory analysis, the investigator works with hypotheses and summary measures formulated *before* seeing the data. The number of hypotheses tested and the number of confidence intervals constructed should be kept small, so that significance levels and confidence coefficients are not seriously disturbed by multiplicity.

By contrast, in exploratory analysis free rein is given to the imagination in devising various summary measures, scatterplots, and other devices to tease messages out of the data. In an exploratory setting, confidence intervals and tests of hypotheses cannot be taken at face value because of the large degree of multiplicity. Often it is best in exploratory analysis to bypass formal statistical inference altogether and to regard the data as "suggestive of," rather than "evidence for," a result. Confirmation of the result would then be hoped for in a future confirmatory study.

CHAPTER APPENDICES

CONTENTS

The computational formula for the standard deviation is a consequence of the following equation:

$$\sum(y - \bar{y})^2 = \sum y^2 - \frac{\left(\sum y\right)^2}{n}$$

We now prove that this equation is always valid. The proof will be easier to follow if we write the individual observations with subscripts as $y_1, y_2, \ldots, y_n$. Then

$$\sum(y - \bar{y})^2 = (y_1 - \bar{y})^2 + (y_2 - \bar{y})^2 + \cdots + (y_n - \bar{y})^2$$

Setting this up in tabular form, we can expand each squared term and add as follows:

$$
\begin{aligned}
(y_1 - \bar{y})^2 &= \quad y_1^2 - 2\bar{y}y_1 \quad + \bar{y}^2 \\
(y_2 - \bar{y})^2 &= \quad y_2^2 - 2\bar{y}y_2 \quad + \bar{y}^2 \\
&\;\;\vdots \qquad\qquad \vdots \\
\underline{(y_n - \bar{y})^2} &= \underline{\quad y_n^2 - 2\bar{y}y_n \quad + \bar{y}^2} \\
\sum(y - \bar{y})^2 &= \sum y^2 - 2\bar{y}\sum y + n\bar{y}^2
\end{aligned}
$$

Using the fact that $\bar{y} = \dfrac{\sum y}{n}$, we can simplify the above equation to

$$
\begin{aligned}
\sum(y - \bar{y})^2 &= \sum y^2 - 2\left(\frac{\sum y}{n}\right)\sum y + n\left(\frac{\sum y}{n}\right)^2 \\
&= \sum y^2 - 2\frac{\left(\sum y\right)^2}{n} + \frac{\left(\sum y\right)^2}{n} \\
&= \sum y^2 - \frac{\left(\sum y\right)^2}{n}
\end{aligned}
$$

and the computational formula is proved.

A MORE GENERAL FORM OF CHEBYSHEV'S RULE

Chebyshev's rule is actually more general than the version given in Section 2.6. The general form is given below.

CHEBYSHEV'S RULE Let C be any positive number. Then for any data set, the proportion of the observations within $\pm C$ standard deviations of the mean is at least

$$1 - \frac{1}{C^2}$$

In Chapter 2 we stated this rule for $C = 2$ and for $C = 3$:

If $C = 2$, then $1 - \dfrac{1}{C^2} = \dfrac{3}{4} = 75\%$.

If $C = 3$, then $1 - \dfrac{1}{C^2} = \dfrac{8}{9} = 89\%$.

HOW CHEBYSHEV'S RULE IS PROVED

We will prove Chebyshev's rule for $C = 2$. First, imagine separating the observations into two groups, depending on how many standard deviations they are from the mean:

Group 1: Those y's for which $|y - \bar{y}| \leq 2s$

Group 2: Those y's for which $|y - \bar{y}| > 2s$

Let m be the number of y's in group 2. We wish to prove that

$$\frac{m}{n} < \frac{1}{4}$$

This inequality is obviously true if $m = 0$ or $s = 0$; consequently, we assume in the following that both m and s are positive.

The standard deviation depends on $\Sigma(y - \bar{y})^2$. Let us break this sum into two parts, $\underset{1}{\Sigma}$ and $\underset{2}{\Sigma}$, where

$\underset{1}{\Sigma}$ is the sum over those y's in group 1;

$\underset{2}{\Sigma}$ is the sum over those y's in group 2.

This gives

$$\Sigma(y - \bar{y})^2 = \underset{1}{\Sigma}(y - \bar{y})^2 + \underset{2}{\Sigma}(y - \bar{y})^2$$

$$\geq \underset{2}{\Sigma}(y - \bar{y})^2 \qquad \text{because } \underset{1}{\Sigma} \text{ is } \geq 0.$$

$$> m(2s)^2 \qquad \begin{array}{l}\text{because each } (y - \bar{y})^2 \text{ in group 2} \\ \text{is greater than } (2s)^2 \text{ by virtue of} \\ \text{the definition of group 2.}\end{array}$$

We can rewrite the above inequality as

$$4ms^2 < \Sigma(y - \bar{y})^2$$

so it follows that

$$\frac{4ms^2}{n} < \frac{\Sigma(y - \bar{y})^2}{n} < \frac{\Sigma(y - \bar{y})^2}{n - 1} = s^2$$

or

$$\frac{4ms^2}{n} < s^2$$

or

$$\frac{m}{n} < \frac{1}{4} \quad \text{as was to be proved}$$

This proves Chebyshev's rule for $C = 2$. The proof for any value of C is exactly parallel.

| | | | | | | | | | | | | |

APPENDIX 3.1

GENERATING PSEUDO-RANDOM NUMBERS

The following is a simple method for calculating a sequence of pseudo-random numbers.

a. Arbitrarily choose any number between 0 and 1; this number is called the *seed* of the sequence. Let the seed be denoted u_0.

b. Calculate u_1 according to the formula

$$u_1 = \text{Fractional part of } [(\pi + u_0)^5]$$

where $\pi = 3.1415927\ldots$ (The fractional part of a number is the part to the right of the decimal point; thus, the fractional part of 27.403911 is .403911.)

c. Continuing in the same way, calculate $u_2, u_3, \ldots$, as follows:

$$u_2 = \text{Fractional part of } [(\pi + u_1)^5]$$
$$u_3 = \text{Fractional part of } [(\pi + u_2)^5]$$

and so on.

d. Each u (except u_0) is a pseudo-random number between 0 and 1. If you plotted the values of a long sequence of u's you would find that they would be more or less uniformly distributed between 0 and 1.

e. The digits of $u_1, u_2, u_3, \ldots$, are pseudo-random digits. They can be read singly or in pairs or in triplets, etc., as explained in Chapter 3.

Remark We do not give an example of a sequence of u's generated by this method, because the sequence may vary from one type of calculator to the next, even using the same seed. This occurs because the u's are highly sensitive to how many significant figures are used in the computations. To see this great sensitivity, try using $\pi = 3.141593$ instead of 3.1415927 and notice how much the sequence of u's changes. This instability of the u's is not a drawback of the method; on the contrary, the instability enhances the pseudo-random properties of the sequence.

MORE ON THE BINOMIAL DISTRIBUTION FORMULA

In this appendix we explain more about the reasoning behind the binomial distribution formula, and also some interesting facts about the binomial coefficients.

THE BINOMIAL DISTRIBUTION FORMULA

We begin by deriving the binomial distribution formula for $n = 3$. Suppose that we conduct 3 independent trials and that each trial results in success (S) or failure (F). On each trial the probabilities of success and failure are

$$\Pr\{S\} = \pi$$
$$\Pr\{F\} = 1 - \pi$$

There are eight possible outcomes of the three trials. Reasoning as in Example 3.19 shows that the probabilities of these outcomes are as follows:

OUTCOME	PROBABILITY
FFF	$(1 - \pi)^3$
FFS	$\pi(1 - \pi)^2$
FSF	$\pi(1 - \pi)^2$
SFF	$\pi(1 - \pi)^2$
FSS	$\pi^2(1 - \pi)$
SFS	$\pi^2(1 - \pi)$
SSF	$\pi^2(1 - \pi)$
SSS	π^3

Again by reasoning parallel to Example 3.19, these probabilities can be collapsed to obtain the binomial distribution for $n = 3$ as shown in the table:

NUMBER OF SUCCESSES j	FAILURES $n - j$	PROBABILITY
0	3	$1\pi^0(1 - \pi)^3$
1	2	$3\pi^1(1 - \pi)^2$
2	1	$3\pi^2(1 - \pi)^1$
3	0	$1\pi^3(1 - \pi)^0$

This distribution illustrates the origin of the binomial coefficients. The coefficient C_1 ($= 3$) is the number of ways in which 1 S and 2 F's can be rearranged; the coefficient C_2 ($= 3$) is the number of ways in which 2 S's and 1 F can be rearranged.

An argument similar to the above shows that the general formula (for any n) is

$$\Pr\{j \text{ successes and } n - j \text{ failures}\} = C_j\pi^j(1 - \pi)^{n-j}$$

where

C_j = Number of ways in which j S's and $(n - j)$ F's can be rearranged

THE BINOMIAL COEFFICIENTS: CONNECTIONS

The binomial coefficients are related to other ideas which may be familiar.

THE BINOMIAL EXPANSION   The binomial coefficients appear in the algebraic identity known as the **binomial expansion**. If a and b are any numbers, then the binomial expansion for the quantity $(a + b)^n$ is

$$(a + b)^n = C_0 b^n + C_1 a b^{n-1} + C_2 a^2 b^{n-2} + \cdots + C_n a^n$$

The most familiar special case is the binomial expansion for $n = 2$:

$$(a + b)^2 = a^2 + 2ab + b^2$$

COMBINATIONS   The binomial coefficient C_j is also known as the number of combinations of n items taken j at a time; it is equal to the number of different subsets of size j that can be formed from a set of n items.

PASCAL'S TRIANGLE   Pascal's triangle is a triangular array of numbers in which the borders are 1's and each interior entry is the sum of the two entries above it. The first seven rows of Pascal's triangle are shown here:

```
                   1     1
                1     2     1
             1     3     3     1
          1     4     6     4     1
       1     5    10    10     5     1
    1     6    15    20    15     6     1
 1     7    21    35    35    21     7     1
```

If you compare this array with Table 2, you will see that the numbers in Pascal's triangle are the binomial coefficients.

THE BINOMIAL COEFFICIENTS: A FORMULA

Binomial coefficients can be calculated from the formula

$$C_j = \frac{n!}{j!(n - j)!}$$

where $x!$ ("x-factorial") is defined for any positive integer x by

$$x! = x(x - 1)(x - 2) \cdots (2)(1)$$

and $0! = 1$.

For example, for $n = 7$ and $j = 4$, the formula gives

$$C_4 = \frac{7!}{4!3!} = \frac{7 \cdot 6 \cdot 5 \cdot 4 \cdot 3 \cdot 2 \cdot 1}{(4 \cdot 3 \cdot 2 \cdot 1)(3 \cdot 2 \cdot 1)}$$

$$= 35$$

To see why this is correct, let us consider in detail why the number of ways of rearranging 4 S's and 3 F's should be equal to

$$\frac{7!}{4!3!}$$

Suppose 4 S's and 3 F's were written on cards, like this:

| S_1 | | S_2 | | S_3 | | S_4 | | F_1 | | F_2 | | F_3 |

Temporarily we put subscripts on the S's and F's to distinguish them. First let us see how many ways there are to arrange the 7 cards in a row:

There are 7 choices for which card goes first;

for each of these, there are 6 choices for which card goes second;

for each of these, there are 5 choices for which card goes third;

for each of these, there are 4 choices for which card goes fourth;

for each of these, there are 3 choices for which card goes fifth;

for each of these, there are 2 choices for which card goes sixth;

for each of these, there is 1 choice for which card goes last.

It follows that there are 7! ways of arranging the 7 cards. If we were to ignore the subscripts, some of these arrangements would be indistinguishable. How many *distinguishable* arrangements would there be? The answer is

$$\frac{7!}{4!3!}$$

We have divided by 4! because there are 4! ways of rearranging the 4 S's among themselves, and so the 7! is an overcount by a factor of 4!. Similarly, we have divided by 3! because there are 3! ways of rearranging the 3 F's among themselves.

| | | | | | | | | | | | |

APPENDIX 4.1

AREAS OF INDEFINITELY EXTENDED REGIONS

Consider the region bounded between a normal curve and the horizontal axis. Because the curve never touches the axis, the region extends indefinitely far to the left and to the right. Yet the area of the region is exactly equal to 1.0. How is it possible for an indefinitely extended region to have a finite area?

To gain insight into this paradoxical situation, consider Figure A.1 (page 518), which shows a region which is simpler than that bounded by a normal curve. In this region, the width of each bar is 1.0; the height of the first bar is $\frac{1}{2}$, the

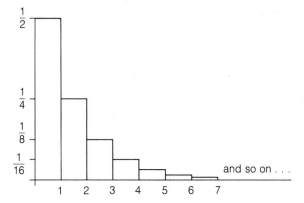

second bar is half as high as the first, the third bar is half as high as the second, and so on indefinitely. The bars form a region which is indefinitely extended. Nevertheless, we shall see that it makes sense to say that the area of the region is equal to 1.0.

Let us first consider the areas of the individual bars. The area of the first bar is $\frac{1}{2}$, the area of the second bar is $\frac{1}{4}$, the third $\frac{1}{8}$, and so on. Now imagine that we choose a number, say k, and add up the areas of the first k bars, as follows:

BAR	HEIGHT OF BAR	CUMULATIVE TOTAL AREA
1	$\frac{1}{2}$	$\frac{1}{2}$
2	$\frac{1}{4}$	$\frac{3}{4}$
3	$\frac{1}{8}$	$\frac{7}{8}$
4	$\frac{1}{16}$	$\frac{15}{16}$
$\vdots$	$\vdots$	$\vdots$
k	$\frac{1}{2^k}$	$\frac{2^k - 1}{2^k}$

The total area of the first two bars is $\frac{3}{4}$, the total area of the first three bars is $\frac{7}{8}$, and so on; in fact, the total area of the first k bars is equal to

$$\frac{2^k - 1}{2^k} = 1 - \frac{1}{2^k}$$

If k is very large, this area is very close to 1.0; in fact, we can make the area as close to 1.0 as we might wish, simply by choosing k large enough. In these circumstances it is reasonable to say that the total area of the entire, indefinitely extended region is equal to exactly 1.0.

The preceding example shows that an indefinitely extended region can have (in a certain sense) a finite area. In a similar sense, the total area under the normal curve is 1.0 (but the proof of this fact requires fairly advanced calculus).

| | | | | | | | | | | | |

APPENDIX 5.1

RELATIONSHIP BETWEEN CENTRAL LIMIT THEOREM AND NORMAL APPROXIMATION TO BINOMIAL DISTRIBUTION

Consider sampling from a dichotomous population. Theorem 5.2 (Section 5.5) states that the sampling distribution of $\hat{\pi}$, and the equivalent binomial distribution, can be approximated by normal distributions. In this appendix we show how these approximations are related to Theorem 5.1 and the Central Limit Theorem (Section 5.3).

Imagine that we replace the dichotomous population by an equivalent population of quantitative observations Y in which every value of Y is either 0 or 1— that is, every "success" has $Y = 1$ and every "failure" has $Y = 0$. Then the population mean value of Y is the same as the proportion of 1's in the population (that is, π) and the sample mean $\bar{Y}$ is the same as the proportion of 1's in the sample (that is, $\hat{\pi}$). Thus, the sample proportion $\hat{\pi}$ can be regarded as a sample mean, and so its sampling distribution is described by Theorem 5.1.

From part 3 of Theorem 5.1 (the Central Limit Theorem), the sampling distribution of $\hat{\pi}$ is approximately normal if n is large. From part 1 of Theorem 5.1, the mean of the sampling distribution of $\hat{\pi}$ is equal to the population mean— that is, π; this is the value given in Theorem 5.2(b). From part 2 of Theorem 5.1, the standard deviation of the sampling distribution of $\hat{\pi}$ is equal to

$$\frac{\sigma}{\sqrt{n}}$$

where σ represents the standard deviation of our imaginary population of 0's and 1's. In order to relate this to Theorem 5.2, we need to express σ in terms of π. Recall from Section 2.9 that the definition of σ is

$$\sigma = \sqrt{\text{Population mean value of } (Y - \mu)^2}$$

In the population of 0's and 1's, the quantity $(Y - \mu)^2$ is equal to $(Y - \pi)^2$ and can have only the following two possible values:

$$(Y - \pi)^2 = \begin{cases} \pi^2 & \text{if } Y = 0 \\ (1 - \pi)^2 & \text{if } Y = 1 \end{cases}$$

Furthermore, these values occur in the proportions $(1 - \pi)$ and π, so that the population mean value of $(Y - \pi)^2$ is equal to

$$(1 - \pi)\pi^2 + \pi(1 - \pi)^2$$

which can be simplified to

$$\pi(1 - \pi)$$

so that

$$\sigma = \sqrt{\pi(1 - \pi)}$$

The standard deviation of the sampling distribution of $\hat{\pi}$ is then equal to

$$\sqrt{\frac{\pi(1 - \pi)}{n}}$$

which is the value given in Theorem 5.2(b).

Since the binomial distribution is just a rescaled version of the sampling distribution of $\hat{\pi}$, it follows that the binomial distribution also can be approximated by a normal curve with suitably rescaled mean and standard deviation. Because the binomial variable (number of successes in n trials) is equal to $n\hat{\pi}$, the rescaling requires multiplying the mean and standard deviation of the sampling distribution of $\hat{\pi}$ by n; this gives the mean $n\pi$ and standard deviation

$$n\sqrt{\frac{\pi(1 - \pi)}{n}} = \sqrt{n\pi(1 - \pi)}$$

which are as given in Theorem 5.2(a).

APPENDIX 6.1

SIGNIFICANT DIGITS

In this appendix we review the concept of significant digits. Let us begin with an example.

Suppose a university president reports that there are now 23,000 students at the university. How many significant digits are in the number

23,000?

When the number is expressed this way, in ordinary notation rather than scientific notation, it is not possible to tell for sure how many significant digits it has. Does the president *really* mean

23,000　rather than　23,001　or　22,999?

If he does, then all 5 of the digits are significant. If (as is probable) he really means

23,000　rather than　22,000　or　24,000,

then only the 2 and the 3 are significant digits. Scientific notation removes the ambiguity:

2.3×10^4 　　has 2 significant digits;

2.3000×10^4 　has 5 significant digits.

As the above example illustrates, you can clarify how many significant digits are in a number by expressing the number in scientific notation. Here are some examples:

ORDINARY NOTATION	SCIENTIFIC NOTATION	NUMBER OF SIGNIFICANT DIGITS
60,700	6.07×10^4	3
60,700	6.0700×10^4	5
60.7	6.07×10^1	3
60.70	6.070×10^1	4
.0607	6.07×10^{-2}	3
.06070	6.070×10^{-2}	4

In the above numbers, note that the interior zero (between 6 and 7) is always a significant digit; the leading zeros (before the 6) are not significant; the terminal zeros (after the 7) are significant in scientific notation and ambiguous in ordinary notation. Digits other than zero are always significant.

Here are some examples of rounding a number to two significant digits:

NUMBER	ROUNDED TO TWO SIGNIFICANT DIGITS	
60,700	61,000	(that is, 6.1×10^4)
60.7	61	
.0607	.061	
.0592	.059	
.0596	.060	

APPENDIX 7.1

HOW POWER IS CALCULATED

The required sample sizes given in Table 5 were determined by calculating the power of the t test. For large samples, an approximate power calculation can be based on the normal curve (Table 3). In this appendix we indicate how such an approximate calculation is done.

Recall that the power is the probability of rejecting H_0 when H_A is true. In order to calculate power, therefore, we need to know the sampling distribution of t_s when H_A is true. For large samples, the sampling distribution can be approximated by a normal curve, as shown by the following theorem.

THEOREM A.1   Suppose we choose independent random samples, each of size n, from normal populations with means μ_1 and μ_2 and a common standard deviation σ. If n is large, the sampling distribution of t_s can be approximated by a normal distribution with

$$\text{Mean} = \sqrt{\frac{n}{2}} \left(\frac{\mu_1 - \mu_2}{\sigma} \right)$$

and

$$\text{Standard deviation} = 1$$

To illustrate the use of Theorem A.1 for power calculations, suppose we are considering a one-tailed t test at $\alpha = .025$. The hypotheses are

$$H_0: \quad \mu_1 = \mu_2$$
$$H_A: \quad \mu_1 > \mu_2$$

If we want a power of .80 for an effect size of .4, then Table 5 recommends sample size $n = 100$. Let us confirm this recommendation using Theorem A.1.

If H_0 is true, so that $\mu_1 = \mu_2$, then the sampling distribution of t_s is approximately a normal distribution with mean equal to 0 and SD equal to 1. This is the null distribution of t_s; it is shown as the dashed curve in Figure A.2.

FIGURE A.2

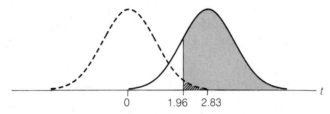

Suppose that in fact H_A is true, that the effect size is

$$\frac{\mu_1 - \mu_2}{\sigma} = .4$$

and that we are using samples of size $n = 100$. Then, according to Theorem A.1, the sampling distribution of t_s will be approximately a normal distribution with SD equal to 1 and mean equal to

$$\sqrt{\frac{n}{2}} \left(\frac{\mu_1 - \mu_2}{\sigma} \right) = \sqrt{\frac{100}{2}} (.4) = 2.83$$

This distribution is the solid curve in the figure.

For $n_1 = n_2 = 100$, we have df $\approx \infty$, so from Table 4 the critical value is equal to 1.96. Thus, we would reject H_0 if

$$t_s > 1.96$$

Using the dotted curve, the probability of this event is equal to .025; this is shown in the figure as a hatched area. Using the solid curve, the probability that $t_s > 1.96$ is the shaded area in the figure. The shaded area can be determined from Table 3 using

$$Z = 1.96 - 2.83 = -.87$$

from Table 3, the area is

$$.3078 + .5000 = .8078 \approx .81$$

Thus, we have shown that, for $n_1 = n_2 = 100$,

$$\text{If } \frac{\mu_1 - \mu_2}{\sigma} = .4, \text{ then } Pr\{\text{reject } H_0\} = .81.$$

We have found that the power against the specified alternative is approximately equal to .81; this agrees well with Table 5, which claims that the power is equal to .80.

If we were concerned with a two-tailed test at $\alpha = .05$, the critical value would again be 1.96, and so the power would again be approximately equal to .81, because the area under the solid curve corresponding to the left-hand tail of the dashed curve is negligible.

Of course, in constructing Table 5, one begins with the specified power (.80) and determines n, rather than the other way around. This "inverse" problem can be solved using an approach similar to the above. In the figure, the shaded area (.80) would be given; this would determine the Z value and in turn determine n, once the effect size is specified.

| | | | | | | | | | | | | |

APPENDIX 7.2

MORE ON THE MANN–WHITNEY TEST

In Section 7.11 we saw how critical values for the Mann–Whitney test are related to the null distributions of K_1, K_2, and U_s. In this appendix we indicate how these null distributions can be determined by simple counting methods.

Let us consider the sample sizes $n = 5$, $n' = 4$. In Figure 7.14 (page 253), the Y_1's and Y_2's are plotted as dots. To save space, let us now represent the data in a more compact way: We will represent each Y_1 by a "1" and each Y_2 by a "2". Thus, the arrangement in Figure 7.14(a) (where the Y_1's are entirely to the left of the Y_2's) would be represented as

1   1   1   1   1   2   2   2   2

For sample sizes $n = 5$, $n' = 4$, there are 126 possible arrangements of the Y_1's and the Y_2's. Here is a partial list of those arrangements and the associated values of K_1 and K_2:

NUMBER	ARRANGEMENT									K_1	K_2
1	1	1	1	1	1	2	2	2	2	0	20
2	1	1	1	1	2	1	2	2	2	1	19
3	1	1	1	1	2	2	1	2	2	2	18
4	1	1	1	2	1	1	2	2	2	2	18
5	1	1	2	1	1	1	2	2	2	3	17
6	1	1	1	2	1	2	1	2	2	3	17
7	1	1	1	1	2	2	2	1	2	3	17
8	1	2	1	1	1	1	2	2	2	4	16
9	1	1	2	1	1	2	1	2	2	4	16
10	1	1	1	2	1	2	2	1	2	4	16
11	1	1	1	2	2	1	1	2	2	4	16
12	1	1	1	1	2	2	2	2	1	4	16
	. . . and so on . . .										
126	2	2	2	2	1	1	1	1	1	20	0

To determine the null distributions from this list, we need to know the likelihood of the various arrangements, assuming that H_0 is true. According to H_0, all 9 observations (Y's) were drawn at random from the same population. Under this assumption, it can be shown that the 126 arrangements are all *equally likely*. (This is not self-evident, but we will ask you to believe it.) Because of this simple and elegant fact, the null distribution of K_1 and K_2 (and therefore U_s) can be determined by straightforward counting. Working from the above list, we find the following probabilities:

K_1	K_2	PROBABILITY
0	20	$\dfrac{1}{126}$
1	19	$\dfrac{1}{126}$
2	18	$\dfrac{2}{126}$
3	17	$\dfrac{3}{126}$
4	16	$\dfrac{5}{126}$
. . . and so on . . .		
20	0	$\dfrac{1}{126}$
	Total	1

These probabilities constitute the null distribution of K_1 and K_2—plotted in Figure 7.15(a). For instance, the first probability in the null distribution is

$$\Pr\{K_1 = 0, K_2 = 20\} = \frac{1}{126} = .008$$

as stated in Section 7.11.

Why is the Mann–Whitney test distribution-free? The reason should be clear from the preceding discussion. If the two population distributions are the same, then all possible arrangements of the Y's are equally likely and the specific shape of the population distributions does not matter (except, of course, that we have assumed that there would be no ties; the null distributions are altered if ties are possible).

The Mann–Whitney null distribution can always be determined by straightforward counting such as illustrated above (although for larger sample sizes the counting is very tedious, and approximate methods are used instead). The number of possible arrangements for samples of size n and n' is equal to

$$\frac{(n + n')!}{n! n'!}$$

For example, for sample sizes 5 and 4 (as above), we find

$$\frac{9!}{5!4!} = 126$$

(To see why this formula works, refer to the discussion of the formula for binomial coefficients at the end of Appendix 3.2; the reasoning is exactly parallel.)

| | | | | | | | | | | |

A P P E N D I X 11.1

CONDITIONAL PROBABILITY AND INDEPENDENCE

The topics of conditional probability and independence were treated informally in Chapter 11. In this appendix we give formal definitions, which may be familiar to you from an earlier study of probability. Also, we link these definitions to the notions of "independent" observations and "independent" samples.

DEFINITION OF CONDITIONAL PROBABILITY

For any two events E_1 and E_2, the **conditional probability** of E_1 given E_2 is defined as*

$$\Pr\{E_1 \mid E_2\} = \frac{\Pr\{E_1 \text{ and } E_2 \text{ both occur}\}}{\Pr\{E_2\}}$$

For instance, suppose a person is chosen at random from a population, and consider $\Pr\{CB \mid M\}$—the conditional probability that the person is color-blind, given that the person is male. According to the above definition,

$$\Pr\{CB \mid M\} = \frac{\Pr\{\text{the person is color-blind } and \text{ male}\}}{\Pr\{\text{the person is male}\}}$$

In terms of numbers of people in this population, this would be

$$\Pr\{CB \mid M\} = \frac{\text{Number of color-blind males}}{\text{Number of males}}$$

DEFINITION OF INDEPENDENCE

Two events E_1 and E_2 are said to be **independent** of each other if the probability that they both occur is the product of their separate probabilities:

$$\Pr\{E_1 \text{ and } E_2 \text{ both occur}\} = \Pr\{E_1\} \cdot \Pr\{E_2\}$$

Using this definition and the above definition of conditional probability, it is straightforward to show that each of the following statements is equivalent to asserting that E_1 and E_2 are independent:

(1) $\Pr\{E_1 \mid E_2\} = \Pr\{E_1\}$
(2) $\Pr\{E_2 \mid E_1\} = \Pr\{E_2\}$

*Of course, the definition will not work if $\Pr\{E_2\} = 0$. To simplify the discussion in this appendix, we assume that all events under consideration have positive probabilities.

(3) $\Pr\{E_1 \mid E_2\} = \Pr\{E_1 \mid \text{not-}E_2\}$

(4) $\Pr\{E_2 \mid E_1\} = \Pr\{E_2 \mid \text{not-}E_1\}$

Here, "not-E_1" means "E_1 does not occur," and similarly for "not-E_2." Statements (3) and (4) are the ones we used to describe independence in Section 11.3.

INDEPENDENCE OF RANDOM VARIABLES

Independence can be defined for random variables as well as for events.

DEFINITION　Let Y_1 and Y_2 be random variables. For any two numbers c_1 and c_2, consider the following events:

E_1:　$Y_1 > c_1$

E_2:　$Y_2 > c_2$

The random variables Y_1 and Y_2 are said to be **independent** of each other if E_1 and E_2 are independent events for all possible choices of the numbers c_1 and c_2. Less formally, Y_1 and Y_2 are independent if the events

$\{Y_1 \text{ is large}\}$　and　$\{Y_2 \text{ is large}\}$

are independent events, no matter how we define "large."

Throughout the text, when we have spoken of observations being "independent," it has been in the sense we have just defined. Also, in speaking of "independent" samples, we mean that any observation Y_1 in the first sample must be independent of any observation Y_2 in the second sample. This would not be true if the samples were matched; typically, matching produces a situation in which the members (Y_1, Y_2) of a pair tend to be similar, so that

$$\Pr\{Y_2 \text{ is large} \mid Y_1 \text{ is large}\} > \Pr\{Y_2 \text{ is large} \mid Y_1 \text{ is small}\}$$

APPENDIX 12.1　THE NEWMAN–KEULS PROCEDURE AND THE t TEST

MORE ON THE NEWMAN–KEULS PROCEDURE

In Section 12.6 we mentioned the link between the Newman–Keuls procedure and the t test. We will now show the link explicitly.

Suppose $\bar{y}_1$ and $\bar{y}_2$ are to be compared using a pooled two-sample t test at $\alpha = .05$. The null hypothesis is H_0: $\mu_1 = \mu_2$. H_0 would be rejected if

$$\frac{|\bar{y}_1 - \bar{y}_2|}{\text{SE}_{(\bar{y}_1 - \bar{y}_2)}} > t_{.05}$$

If $n_1 = n_2 = n$, then

$$\text{SE}_{(\bar{y}_1 - \bar{y}_2)} = \sqrt{s_c^2\left(\frac{1}{n} + \frac{1}{n}\right)} = \sqrt{2}\,\sqrt{\frac{s_c^2}{n}}$$

and consequently the t test rejects H_0 if

$$|\bar{y}_1 - \bar{y}_2| > \left(\sqrt{2}\, t_{.05}\right)\left(\sqrt{\frac{s_c^2}{n}}\right) \tag{1}$$

On the other hand, suppose $\bar{y}_1$ and $\bar{y}_2$ are to be compared using the critical value R_2 from the Newman–Keuls procedure. H_0 would be rejected if

$$|\bar{y}_1 - \bar{y}_2| > R_2$$

that is, if

$$|y_1 - y_2| > q_2\, \sqrt{\frac{s_c^2}{n}} \tag{2}$$

Let us compare conditions (1) and (2). The second factor on the right-hand side is the same in both expressions—although of course s_c is based on two samples for the t test and on all k samples for the Newman–Keuls procedure. To complete the correspondence, it can be shown that the first factors on the right-hand sides of (1) and (2) are equal; that is,

$$\sqrt{2}\, t_{.05} = q_2$$

where q_2 is determined from Table 10 with $\alpha = .05$ and with the same df as $t_{.05}$. For instance, suppose df $= 15$. Then Table 10 gives $q_2 = 3.01$ and Table 4 gives $t_{.05} = 2.131$; and indeed it is true that

$$(\sqrt{2})\,(2.131) = 3.01$$

Thus, except for the difference in computing s_c, a comparison of two means using R_2 is equivalent to a pooled two-sample t test against a nondirectional alternative. Of course, the Newman–Keuls procedure does not use R_2 for all its comparisons.

VARIETIES OF TYPE I ERROR

In Section 12.6 we mentioned that different multiple comparison procedures differ with respect to the degree of control of Type I error. To be more specific about this, let us consider various ways in which the risk of a Type I error might be computed. Suppose the number of groups to be compared is $k = 4$. Let H_{ij} represent the null hypothesis H_0: $\mu_i = \mu_j$. The following are a few of the probability calculations that express various aspects of the risk of Type I error:

1. Assume $\mu_1 = \mu_2 = \mu_3 = \mu_4$ and calculate

Pr{at least one of the H_{ij} is rejected}

2. Assume $\mu_1 = \mu_2 = \mu_3$ and calculate

Pr{H_{12} or H_{13} or H_{23} is rejected}

3. Assume $\mu_1 = \mu_2$ and calculate

Pr{H_{12} is rejected}

4. Assume $\mu_1 = \mu_2$ and $\mu_3 = \mu_4$, and calculate

$$\Pr\{H_{12} \text{ or } H_{34} \text{ is rejected}\}$$

If the Newman–Keuls procedure is performed at $\alpha = .05$, then probabilities such as those defined in (1), (2), and (3) are held at .05, but probabilities such as that in (4) are not controlled. Some multiple comparison procedures control only the probability in (1), whereas some procedures control all the probabilities in (1), (2), (3), and (4).

Researchers do not all agree on how stringently a multiple comparison procedure should control the risk of Type I error. Consequently, there is no multiple comparison procedure that is unambiguously the "best" one.

| | | | | | | | | | | | | |

APPENDIX 13.1

DERIVATION OF FORMULAS FOR b_0 AND b_1

In this appendix we use calculus to show that the least-squares criterion leads to the formulas

$$b_1 = \frac{SP_{XY}^2}{SS_X} \quad \text{and} \quad b_0 = \bar{y} - b_1 \bar{x}$$

COMPUTATIONAL FORMULAS

In finding formulas for b_0 and b_1 we will use the following formulas:

$$SS_X = \sum x^2 - \frac{\left(\sum x\right)^2}{n} \tag{1}$$

$$SP_{XY} = \sum xy - \frac{\left(\sum x\right)\left(\sum y\right)}{n} \tag{2}$$

Equation (1) is the same (using x instead of y) as the computational formula for the standard deviation, which was derived in Appendix 2.1. Equation (2) can be derived in an exactly analogous way.

DERIVATION OF FORMULAS FOR b_0 AND b_1

Suppose we have n data points (x, y). We consider a straight line whose equation is $Y = v_0 + v_1 X$. To satisfy the least-squares criterion, we want to find the values of v_0 and v_1 that *minimize* the sum of squares of the vertical distances of the data points from the line.

The quantity we want to minimize is

$$Q = \sum [y - (v_0 + v_1 x)]^2$$

The quantity Q is a function of v_0, v_1, and the data. To find values of v_0 and v_1 that minimize Q, we will compute the partial derivatives of Q with respect to v_0

and v_1, set these derivatives equal to zero, and solve the resulting equations for v_0 and v_1. We will find that the solutions are b_0 and b_1.

First we differentiate Q with respect to v_0 and set the derivative equal to zero:

$$\frac{\partial Q}{\partial v_0} = \sum 2[y - (v_0 + v_1 x)](-1) = 0$$

which leads to

$$\sum [y - (v_0 + v_1 x)] = 0$$

or

$$\sum y = \sum (v_0 + v_1 x) = n v_0 + v_1 \sum x$$

so that

$$v_0 = \frac{\sum y}{n} - v_1 \frac{\sum x}{n} \qquad (3)$$

or

$$v_0 = \bar{y} - v_1 \bar{x} \qquad (4)$$

We now differentiate Q with respect to v_1 and set the derivative equal to zero:

$$\frac{\partial Q}{\partial v_1} = \sum 2[y - (v_0 + v_1 x)](-x) = 0$$

which gives

$$\sum x[y - (v_0 + v_1 x)] = 0$$

or

$$\sum xy = \sum x(v_0 + v_1 x)$$

or

$$\sum xy = v_0 \sum x + v_1 \sum x^2 \qquad (5)$$

Substituting v_0 from (3) into (5) gives

$$\sum xy = \left(\frac{\sum y}{n} - v_1 \frac{\sum x}{n} \right) \sum x + v_1 \sum x^2$$

$$= \frac{\sum x \sum y}{n} - v_1 \frac{\left(\sum x \right)^2}{n} + v_1 \sum x^2$$

or

$$\sum xy - \frac{\sum x \sum y}{n} = v_1 \left[\sum x^2 - \frac{\left(\sum x \right)^2}{n} \right] \qquad (6)$$

Using (1) and (2), (6) becomes

$$SP_{XY} = v_1 SS_X$$

or

$$v_1 = \frac{SP_{XY}}{SS_X} \tag{7}$$

Now (4) and (7) give the formulas for b_0 and b_1.

| | | | | | | | | | | | | |

APPENDIX 13.2

ALGEBRAIC RELATIONS AMONG REGRESSION QUANTITIES

In this appendix we derive the various relations among the regression quantities. No calculus is used.

COMPUTATIONAL FORMULA FOR SS(resid)

The definition of the residual sum of squares is

$$SS(resid) = \sum (y - \hat{y})^2 \tag{1}$$

We will derive the computational formula.

First, recall that $\hat{y} = b_0 + b_1 x$ and that $b_0 = \bar{y} - b_1 \bar{x}$. Consequently,

$$
\begin{aligned}
y - \hat{y} &= y - (b_0 + b_1 x) \\
&= y - (\bar{y} - b_1 \bar{x}) - b_1 x \\
&= (y - \bar{y}) - b_1 (x - \bar{x})
\end{aligned} \tag{2}
$$

From (1) and (2),

$$SS(resid) = \sum [(y - \bar{y}) - b_1(x - \bar{x})]^2$$

Expanding the square yields

$$
\begin{aligned}
SS(resid) &= \sum [(y - \bar{y})^2 - 2b_1(x - \bar{x})(y - \bar{y}) + b_1^2(x - \bar{x})^2] \\
&= \sum (y - \bar{y})^2 - 2b_1 \sum (x - \bar{x})(y - \bar{y}) + b_1^2 \sum (x - \bar{x})^2 \\
&= SS_Y - 2b_1 SP_{XY} + b_1^2 SS_X
\end{aligned} \tag{3}
$$

But recall that

$$b_1 = \frac{SP_{XY}}{SS_X}$$

Substituting this in (3) gives

$$
\begin{aligned}
SS(resid) &= SS_Y - 2 \frac{SP_{XY}}{SS_X} SP_{XY} + \left(\frac{SP_{XY}}{SS_X}\right)^2 SS_X \\
&= SS_Y - \frac{SP_{XY}^2}{SS_X}
\end{aligned} \tag{4}
$$

which is the computational formula.

DERIVATION OF FACT 13.1

The definition of r is

$$r = \frac{SP_{XY}}{\sqrt{SS_X SS_Y}}$$

Consequently,

$$1 - r^2 = 1 - \frac{SP_{XY}^2}{SS_X SS_Y} = \frac{SS_Y - \dfrac{SP_{XY}^2}{SS_X}}{SS_Y}$$

Substituting the computational formula for SS(resid) from (4), we find

$$1 - r^2 = \frac{SS(\text{resid})}{SS_Y} \tag{5}$$

which is Fact 13.1.

DERIVATION OF FACT 13.2

From the definitions of s_Y and $s_{Y|X}$, it follows that

$$SS_Y = (n - 1)s_Y^2$$

and

$$SS(\text{resid}) = (n - 2)s_{Y|X}^2$$

Substituting these in (5) gives

$$1 - r^2 = \frac{n - 2}{n - 1}\left(\frac{s_{Y|X}^2}{s_Y^2}\right)$$

so that

$$\sqrt{1 - r^2} = \frac{s_{Y|X}}{s_Y} f$$

where

$$f = \sqrt{\frac{n - 2}{n - 1}}$$

The factor f is close to 1 unless n is quite small. Here are some values of f:

n	f
5	.87
10	.94
15	.96

Thus, we have shown that

$$\sqrt{1 - r^2} \approx \frac{s_{Y|X}}{s_Y}$$

The approximation is reasonably good if $n \geq 5$.

APPENDIX 13.3

CALCULATIONS FOR EXAMPLE 13.26

In this appendix we describe the calculations for the regression analysis presented in Example 13.26, including the regression of X on Y.

We first recall the formulas for regression of Y on X. The regression line is

$$Y = b_0 + b_1 X$$

where

$$b_1 = \frac{SP_{XY}}{SS_X} \quad \text{and} \quad b_0 = \bar{y} - b_1 \bar{x}$$

The residual sum of squares is

$$\sum (y - \hat{y})^2 = SS_Y - \frac{SP_{XY}^2}{SS_X}$$

and the residual standard deviation is

$$s_{Y|X} = \sqrt{\frac{\sum (y - \hat{y})^2}{n - 2}}$$

The formulas for the reverse regression—of X on Y—are exactly analogous. The regression line can be written as

$$X = b_0' + b_1' Y$$

where

$$b_1' = \frac{SP_{XY}}{SS_Y} \quad \text{and} \quad b_0' = \bar{x} - b_1' \bar{y}$$

The residual sum of squares for the regression of X on Y is

$$\sum (x - \hat{x})^2 = SS_X - \frac{SP_{XY}^2}{SS_Y}$$

and the residual standard deviation is

$$s_{X|Y} = \sqrt{\frac{\sum (x - \hat{x})^2}{n - 2}}$$

The regression analysis displayed in Figure 13.18 is based on the above formulas. The basic statistics for the cricket data are

$$n = 39 \qquad \bar{x} = 3.5186 \qquad \bar{y} = 39.846$$
$$SS_X = 16.8040 \qquad SS_Y = 35{,}769.1 \qquad SP_{XY} = 532.858$$

Application of the above formulas to these basic statistics gives the regression lines and residual SDs as described in Example 13.26.

C H A P T E R   N O T E S

CHAPTER 1

1. Nicolle, J. (1961). *Louis Pasteur: The Story of His Major Discoveries.* New York: Basic Books. p. 170. © 1961 by Jacques Nicolle. © 1961 English translation Hutchinson & Co. (Publishers) Ltd. Reprinted by permission of Basic Books, Inc.

2. Mizutani, T. and Mitsuoka, T. (1979). Effect of intestinal bacteria on incidence of liver tumors in gnotobiotic C3H/He male mice. *Journal of the National Cancer Institute 63*, 1365–1370.

3. Tripepi, R. R. and Mitchell, C. A. (1984). Metabolic response of river birch and European birch roots to hypoxia. *Plant Physiology 76*, 31–35. Raw data courtesy of the authors.

4. Adapted from Potkin, S. G., Cannon, H. E., Murphy, D. L., and Wyatt, R. J. (1978). Are paranoid schizophrenics biologically different from other schizophrenics? *New England Journal of Medicine 298*, 61–66. The data given are approximate, having been reconstructed from the histograms and summary information given by Potkin et al. Reprinted by permission of the *New England Journal of Medicine.*

5. Wolfson, J. L. (1987). Impact of *Rhizobium* nodules on *Sitona hispidulus,* the clover root curculio. *Entomologia Experimentalis et Applicata 43*, 237–243. Data courtesy of the author. The experiment actually included 11 dishes.

6. Heggestad, H. E. and Bennett, J. H. (1981). Photochemical oxidants potentiate yield losses in snap beans attributable to sulfur dioxide. *Science 213*, 1008–1010. Copyright 1981 by the AAAS. Raw data courtesy of H. E. Heggestad.

7. Webb, P. (1981). Energy expenditure and fat-free mass in men and women. *American Journal of Clinical Nutrition 34*, 1816–1826.

CHAPTER 2

1. Stewart, R. N. and Arisumi, T. (1966). Genetic and histogenic determination of pink bract color in poinsettia. *Journal of Heredity 57*, 217–220.

2. Data of Wiener, A. S., Moor-Jankowski, J., and Gordon, E. B. (1966). Reproduced in Erskine, A. G. and Socha, W. W. (1978), *The Principles and Practices of Blood Grouping,* 2nd edition. St. Louis: Mosby, p. 64.

3. Unpublished data courtesy of C. M. Cox and K. J. Drewry.

4. Unpublished data courtesy of W. F. Jacobson.

5. Unpublished data courtesy of J. F. Nash, Jr. and J. E. Zabik. Similar data are reported in Maickel, R. P. and Zabik, J. E. (1980). A simple animal test system to predict the likelihood of a drug causing human physical dependence. *Substance and Alcohol Actions/Misuse 1*, 259–267.

6. Knoll, A. E. and Barghoorn, E. S. (1977). Archean microfossils showing cell division from the Swaziland system of South Africa. *Science 198*, 396–398.

7. Nurse, C. A. (1981). Interactions between dissociated rat sympathetic neurons and skeletal muscle cells developing in cell culture. II. Synaptic mechanisms. *Developmental Biology 88*, 71–79.

8. Abraham, S., Johnson, C. L., and Najjar, M. F. (1979). Weight and height of adults 18–74 years of age, United States 1971–1974. *U.S. National Center for Health Statistics, Vital and Health Statistics Series 11, No. 211.* Washington, D.C.: U.S. Department of Health, Education and Welfare. The data have been slightly adjusted to account for certain biases in the sampling scheme.

9. Johannsen, W. (1903). *Ueber Erblichkeit in Populationen und in reinen Linien.* Jena: G. Fischer. Data reproduced in Strickberger, M. W. (1976). *Genetics,* New York: Macmillan, p. 277; and Peters, J. A. (ed.) (1959). *Classic Papers in Genetics,* Englewood Cliffs, New Jersey: Prentice-Hall, p. 23.

10. Unpublished data courtesy of W. F. Jacobson.

11. Simpson, G. G., Roe, A., and Lewontin, R. C. (1960). *Quantitative Zoology*. New York: Harcourt, Brace. p. 51.

12. Adapted from Potkin, S. G., Cannon, H. E., Murphy, D. L., and Wyatt, R. J. (1978). Are paranoid schizophrenics biologically different from other schizophrenics? *New England Journal of Medicine* 298, 61–66. The data given are approximate, having been reconstructed from the histogram and summary information given by Potkin et al. Reprinted by permission of the *New England Journal of Medicine*.

13. Peters, H. G. and Bademan, H. (1963). The form and growth of stellate cells in the cortex of the guinea-pig. *Journal of Anatomy (London)* 97, 111–117.

14. Data courtesy of R. E. Jones, Indiana State Dairy Association, Inc.

15. Unpublished data courtesy of D. J. Honor and W. A. Vestre.

16. Connolly, K. (1968). The social facilitation of preening behaviour in *Drosophila melanogaster*. *Animal Behaviour 16*, 385–391.

17. Shields, D. R. (1981). The influence of niacin supplementation on growing ruminants and *in vivo* and *in vitro* rumen parameters. Ph.D. thesis, Purdue University. Raw data courtesy of the author and D. K. Colby.

18. Gwynne, D. T. (1981). Sexual difference theory: Mormon crickets show role reversal in mate choice. *Science 213*, 779–780. Copyright 1981 by the AAAS. Raw data courtesy of the author.

19. Unpublished data courtesy of M. A. Morse and G. P. Carlson.

20. Adapted from Anderson, J. W., Story, L., Sieling, B., Chen, W. L., Petro, M. S., and Story, J. (1984). Hypocholesterolemic effects of oat-bran or bean intake for hypercholesterolemic men. *American Journal of Clinical Nutrition 40*, 1146–1155. There were actually 20 men in the study.

21. Unpublished data courtesy of C. H. Noller.

22. Luria, S. E. and Delbruck, M. (1943). Mutations of bacteria from virus sensitivity to virus resistance. *Genetics 28*, 491–511.

23. Fictitious but realistic data. See Roberts, J. (1975). Blood pressure of persons 18–74 years, United States, 1971–72. *U.S. National Center for Health Statistics, Vital and Health Statistics Series 11, No. 150*. Washington, D.C.: U.S. Department of Health, Education and Welfare.

24. Fictitious but realistic data. Based on Beyl, C. A. and Mitchell, C. A. (1977). Characterization of mechanical stress dwarfing in chrysanthemum. *Journal of the American Society for Horticultural Science 102*, 591–594.

25. Nelson, L. A. (1980). *Report of the Indiana Beef Evaluation Program, Inc.* Purdue University, West Lafayette, Indiana.

26. Reem, G. H., Cook, L. A., and Vilcek, J. (1983). Gamma interferon synthesis by human thymocytes and T lymphocytes inhibited by Cyclosporin A. *Science 221*, 63–64. Copyright 1983 by the AAAS.

27. Unpublished data courtesy of F. Delgado.

28. Unpublished data courtesy of J. Y. Latimer and C. A. Mitchell.

29. Day, K. M., Patterson, F. L., Luetkemeier, O. W., Ohm, H. W., Polizotto, K., Roberts, J. J., Shaner, G. E., Huber, D. M., Finney, R. E., Foster, J. E., and Gallun, R. L. (1980). Performance and adaptation of small grains in Indiana. Station Bulletin No. 290. West Lafayette, Indiana: Agricultural Experiment Station of Purdue University. Raw data provided courtesy of W. E. Nyquist.

30. Tripepi, R. R. and Mitchell, C. A. (1984). Metabolic response of river birch and European birch roots to hypoxia. *Plant Physiology 76*, 31–35. Raw data courtesy of the authors.

31. Ogilvie, R. I., Macleod, S., Fernandez, P., and McCullough, W. (1974). Timolol in essential hypertension. In *Beta-Adrenergic Blocking Agents in the Management of Hypertension and Angina Pectoris*, B. Magnani (ed.). New York: Raven Press. pp. 31–43.

32. Unpublished data courtesy of J. F. Nash and J. E. Zabik.

33. Schall, J. J., Bennett, A. F., and Putnam, R. W. (1982). Lizards infected with malaria: Physiological and behavioral consequences. *Science 217*, 1057–1059. Copyright 1982 by the AAAS. Raw data courtesy of J. J. Schall.

34. Schaeffer, J., Andrysiak, T., and Ungerleider, J. T. (1981). Cognition and long-term use of ganja (cannabis). *Science 213*, 465–466.

35. Fictitious but realistic data. Each observation is the average of several measurements made on the same woman at different times. See Royston, J. P. and Abrams, R. M. (1980). An objective method for detecting the shift in basal body temperature in women. *Biometrics 36*, 217–224.

36. Adapted from data in Cicirelli, M. F., Robinson, K. R., and Smith, L. D. (1983). Internal pH of *Xenopus* oocytes: A study of the mechanism and role of pH changes during meiotic maturation. *Developmental Biology 100*, 133–146.

37. Adapted from Royston, J. P. and Abrams, R. M. (1980). An objective method for detecting the shift in basal body temperature in women. *Biometrics 36*, 217–224.

38. Adapted from data provided courtesy of L. A. Nelson.

39. Ikin, E. W., Prior, A. M., Race, R. R., and Taylor, G. L. (1939). The distribution of the A_1A_2BO blood groups in England. *Annals of Eugenics, London 9*, 409–411. Reprinted with permission of Cambridge University Press.

40. Borg, S., Kvande, H., and Sedvall, G. (1981). Central norepinephrine metabolism during alcohol intoxication in addicts and healthy volunteers. *Science 213*, 1135–1137. Copyright 1981 by the AAAS. Raw data courtesy of S. Borg.

41. Long, T. F. and Murdock, L. L. (1983). Stimulation of blowfly feeding behavior by octopaminergic drugs. *Proceedings of the National Academy of Sciences 80*, 4159–4163. Raw data courtesy of the authors and L. C. Sudlow.

42. Fictitious but realistic population. Adapted from LeClerg, E. L., Leonard, W. H., and Clark, A. G. (1962). *Field Plot Technique*. Minneapolis: Burgess.

43. Selawry, O. S. (1974). The role of chemotherapy in the treatment of lung cancer. *Seminars in Oncology 1*, No. 3, 259–272.

44. Hayes, H. K., East, E. M., and Bernhart, E. G. (1913). *Connecticut Agricultural Experiment Station Bulletin 176*. Data reproduced in Strickberger, M. W. (1976). *Genetics*, New York: Macmillan, p. 288.

45. The population is fictitious, but resembles the population of American women aged 18–24, excluding known or suspected diabetics, as reported in Gordon, T. (1964). Glucose tolerance of adults, United States 1960–62. *U.S. National Center for Health Statistics, Vital and Health Statistics Series 11, No. 2.* Washington, D.C.: U.S. Department of Health, Education and Welfare.

46. Meyer, W. H. (1930). Diameter distribution series in even-aged forest stands. *Yale University School of Forestry Bulletin 28*. The curve is fitted in Bliss, C. I., and Reinker, K. A. (1964). A lognormal approach to diameter distributions in even-aged stands. *Forest Science 10*, 350–360.

47. The results of similar assays are reported in Pascholati, S. F. and Nicholson, R. L. (1983). *Helminthosporum maydis* suppresses expression of resistance to *Helminthosporum carbonum* in corn. *Phytopathologische Zeitschrift 107*, 97–105. Unpublished data courtesy of the investigators.

48. Richens, A. and Ahmad, S. (1975). Controlled trial of valproate in severe epilepsy. *British Medical Journal 4*, 255–256.

49. Fleming, W. E. and Baker, F. E. (1936). A method for estimating populations of larvae of the Japanese beetle in the field. *Journal of Agricultural Research 53*, 319–331. Data reproduced in *Statistical Ecology, Volume 1* (1971). University Park: The Pennsylvania State University Press, p. 327.

50. Chiarotti, R. M. (1972). An investigation of the energy expenditure of women squash players. Master's thesis, The Pennsylvania State University. Raw data courtesy of R. M. Lyle (nee Chiarotti).

51. Masty, J. (1983). Innervation of the equine small intestine. Master's thesis, Purdue University. Raw data courtesy of the author.

52. Bruce, D., Harvey, D., Hamerton, A. E., and Bruce, L. (1913). Morphology of various strains of the trypanosome causing disease in man in Nyasaland. I. The human strain. *Proceedings of the Royal Society of London, Series B 86*, 285–302. See also Pearson, K. (1914). On the probability that two independent distributions of frequency are really samples of the same population, with reference to recent work on the identity of trypanosome strains. *Biometrika 10*, 85–143.

53. Fictitious but realistic data. Adapted from data presented in Falconer, D. S. (1981). *Introduction to Quantitative Genetics,* 2nd edition. New York: Longman, Inc. p. 97.

54. Dow, T. G. B., Rooney, P. J., and Spence, M. (1975). Does anaemia increase the risks to the fetus caused by smoking in pregnancy? *British Medical Journal 4*, 253–254.

55. Christophers, S. R. (1924). The mechanism of immunity against malaria in communities living under hyper-endemic conditions. *Indian Journal of Medical Research 12*, 273–294. Data reproduced in Williams, C. B. (1964). *Patterns in the Balance of Nature*. London: Academic Press. p. 243.

CHAPTER 3

1. Parks, N. J., Krohn, K. A., Mathis, C. A., Chasko, J. H., Geiger, K. R., Gregor, M. E., and Peek, N. F. (1981). Nitrogen-13-labelled nitrite and nitrate: Distribution and metabolism after intratracheal administration. *Science 212*, 58–61.

2. Fictitious but realistic population. Adapted from Hubbs, C. L. and Schultz, L. P. (1932). *Cottus tubulatus*, a new sculpin from Idaho. *Occasional Papers of the Museum of Zoology, University of Michigan,* 242, 1–9. Data reproduced in Simpson, G. G., Roe, A., and Lewontin, R. C. (1960). *Quantitative Zoology.* New York: Harcourt, Brace. p. 81.

3. Fictitious but realistic situation. Based on data given by Lack, D. (1948). Natural selection and family size in the starling. *Evolution 2*, 95–110. Data reproduced by Riclefs, R. E. (1973). *Ecology.* Newton, Massachusetts: Chiron Press. p. 37.

4. Pearson, K. (1914). On the probability that two independent distributions of frequency are really samples of the same population, with reference to recent work on the identity of trypanosome strains. *Biometrika 10*, 85–143. Reprinted by permission of the Biometrika Trustees.

5. This is one of the crosses performed by Gregor Mendel in his classic studies of heredity; heterozygous plants (which are yellow-seeded because yellow is dominant) are crossed with each other.

6. Fictitious but realistic value. See Hutchison, J. G. P., Johnston, N. M., Plevey, M. V. P., Thangkhiew, I., and Aidney, C. (1975). Clinical trial of Mebendazole, a broad-spectrum anthelminthic. *British Medical Journal 2*, 309–310.

7. Fictitious but realistic population. Adapted from Owen, D. F. (1963). Polymorphism and population density in the African land snail, *Limicolaria martensiana. Science 140*, 666–667.

8. *The World Almanac and Book of Facts*, 1982. New York: Newspaper Enterprise Association.

9. Adapted from discussion in Galen, R. S. and Gambino, S. R. (1980). *Beyond Normality: The Predictive Value and Efficiency of Medical Diagnoses.* New York: Wiley. pp. 71–74.

10. This would be true for some central-city populations. See Annest, J. L., Mahaffey, K. R., Cox, D. H., and Roberts, J. (1982). Blood lead levels for persons 6 months–74 years of age: United States, 1976–80. *U.S. National Center for Health Statistics, Advance Data from Vital and Health Statistics, No. 79.* Hyattsville, Maryland: U.S. Department of Health and Human Services.

11. Geissler, A. (1889). Beitrage zur Frage des Geschlechtsverhaltnisses der Geborenen. *Zeitschrift des K. Sachsischen Statistischen Bureaus 35*, 1–24. Data reproduced by Edwards, A. W. F. (1958). An analysis of Geissler's data on the human sex ratio. *Annals of Human Genetics 23*, 6–15. The data are also discussed by Stern, C. (1960). *Human Genetics.* San Francisco: Freeman.

12. Haseman, J. K. and Soares, E. R. (1976). The distribution of fetal death in control mice and its implications on statistical tests for dominant lethal effects. *Mutation Research 41*, 277–288.

13. Data courtesy of S. N. Postlethwaite.

14. This is typical for U.S. populations. See, for example, Maccready, R. A. and Mannin, M. C. (1951). A typing study of one hundred and fifty thousand bloods. *Journal of Laboratory and Clinical Medicine 37*, 634–636.

15. Fictitious but realistic situation. See Krebs, C. J. (1972). *Ecology: The Experimental Analysis of Distribution and Abundance.* New York: Harper & Row. p. 142.

16. See Mather, K. (1943). *Statistical Analysis in Biology.* London: Methuen. p. 38.

17. The technique is described in Waid, W. M., Orne, E. C., Cook, M. R., and Orne, M. T. (1981). Meprobamate reduces accuracy of physiological detection of deception. *Science 212*, 71–73.

18. Fictitious but realistic population, closely resembling the population of males aged 45–59 years as described in Roberts, J. (1975). Blood pressure of persons 18–74 years, United States, 1971–72. *U.S. National Center for Health Statistics, Vital and Health Statistics Series 11, No. 150.* Washington, D.C.: U.S. Department of Health, Education and Welfare.

CHAPTER 4

1. Levy, P. S., Hamill, P. V. V., Heald, F., and Rowland, M. (1976). Total serum cholesterol values of youths 12–17 years, United States. *U.S. National Center for Health Statistics, Vital and Health Statistics Series 11, No. 156.* Washington, D.C.: U.S. Department of Health, Education and Welfare.

2. Ikeme, A. I., Roberts, C., Adams, R. L., Hester, P. Y., and Stadelman, W. J. (1983). Effects of supplementary water-administered vitamin D_3 on egg shell thickness. *Poultry Science 62*, 1120–1122. The normal curve was fitted to raw data provided courtesy of W. J. Stadelman and A. I. Ikeme.

3. Hengstenberg, R. (1971). Das Augenmuskelsystem der Stubenfliege *Musca domestica*. I. Analyse der "clock-spikes" und ihrer Quellen. *Kybernetik 2*, 56–57.

4. Adapted from Magath, T. B. and Berkson, J. (1960). Electronic blood-cell counting. *American Journal of Clinical Pathology 34*, 203–213. Actually, the percentage error is somewhat less for high counts and somewhat more for low counts. Described in Coulter Electronics (1982). *Performance Characteristics and Specifications for Coulter Counter Model S-560*. Hialeah, Florida: Coulter Electronics.

5. Fictitious but realistic population. Adapted from data given by Hildebrand, S. F. and Schroeder, W. C. (1927). Fishes of Chesapeake Bay. *Bulletin of the United States Bureau of Fisheries 43*, Part 1, p. 88. The fish are young of the year, observed in October; they are quite small. (The distribution of lengths in older populations is not approximately normal.)

6. Adapted from Pearl, R. (1905). Biometrical studies on man. I. Variation and correlation in brain weight. *Biometrika 4*, 13–104.

7. Adapted from Swearingen, M. L. and Holt, D. A. (1976). Using a "blank" trial as a teaching tool. *Journal of Agronomic Education 5*, 3–8. The standard deviation given in this problem is realistic for an idealized "uniform" field, in which yield differences between plots are due to local random variation rather than large-scale and perhaps systematic variation.

8. Adapted from Coulter Electronics (1982). *Performance Characteristics and Specifications for the Coulter Counter Model S-560*. Hialeah, Florida: Coulter Electronics.

9. Fictitious but realistic population. Adapted from data given by Falconer, D. S. (1981). *Introduction to Quantitative Genetics*, 2nd edition. New York: Longman. p. 97.

10. Fictitious but realistic population. Adapted from data given in Falconer, D. S. (1981). *Introduction to Quantitative Genetics*, 2nd edition. New York: Longman. p. 97.

11. Fictitious but realistic population. Based on unpublished data provided by W. F. Jacobson.

12. Long, E. C. (1976). *Liquid Scintillation Counting Theory and Techniques*. Irvine, California: Beckman Instruments. The distribution is actually a discrete distribution called a Poisson distribution; however, a Poisson distribution with large mean is approximately normal.

13. Fictitious but realistic population, based on data of Emerson, R. A. and East, E. M. (1913). Inheritance of quantitative characters in maize. *Nebraska Experimental Station Research Bulletin 2*. Data reproduced by Mather, K. (1943). *Statistical Analysis in Biology*. London: Methuen. pp. 29, 34. Modern hybrid corn is taller and less variable than this population.

14. Abraham, S., Johnson, C. L., and Najjar, M. F. (1979). Weight and height of adults 18–74 years of age, United States, 1971–74. *U.S. National Center for Health Statistics Series 11, No. 211*. Washington, D.C.: U.S. Department of Health, Education and Welfare.

15. This is the standard reference distribution for Stanford–Binet scores. See Sattler, J. M. (1982). *Assessment of Children's Intelligence and Special Abilities*, 2nd edition. Boston: Allyn and Bacon. p. 19 and back cover.

CHAPTER 5

1. This value is approximately correct for American adults. See Roberts, J. (1964). Binocular visual acuity of adults, United States 1960–62. *U.S. National Center for Health Statistics, Vital and Health Statistics Series 11, No. 3*. Washington, D.C.: U.S. Department of Health, Education and Welfare.

2. Fictitious but realistic population. See Example 2.10.

3. The mean and standard deviation are realistic for American women aged 25–34. See O'Brien, R. J. and Drizd, T. A. (1981). Basic data on spirometry in adults 25–74 years of age: United States, 1971–75. *U.S. National Center for Health Statistics, Vital and Health Statistics Series 11, No. 222*. Washington, D.C.: U.S. Department of Health and Human Services. The normality assumption may or may not be realistic.

4. Adapted from data given in Sebens, K. P. (1981). Recruitment in a sea anemone population; juvenile substrate becomes adult prey. *Science 213*, 785–787.

5. Fictitious but realistic data. Adapted from distribution given for men aged 45–59 in Roberts, J. (1975). Blood pressure of persons 18–74 years, United States, 1971–72. *U.S. National Center for Health Statistics, Vital and Health Statistics Series 11, No. 150*. Washington, D.C.: U.S. Department of Health, Education and Welfare.

6. The distribution in Figure 5.12 is based on data given in Zeleny, C. (1922). The effect of selection for eye facet number in the white bar-eye race of *Drosophila melanogaster*. *Genetics* 7, 1–115. The data are displayed in Falconer, D. S. (1981). *Introduction to Quantitative Genetics*, 2nd edition. New York: Longman. p. 97.

7. The distribution in Figure 5.14 is adapted from data described by Bradley, J. V. (1980). Nonrobustness in one-sample Z and t tests: A large-scale sampling study. *Bulletin of the Psychonomic Society* 15 (1), 29–32, used by permission of the Psychonomic Society, Inc.; and Bradley, J. V. (1977). A common situation conducive to bizarre distribution shapes. *American Statistician* 31, 147–150. Bradley's distribution included additional peaks, because sometimes the subject fumbled the button more than once on a single trial.

8. Fictitious but realistic situation, adapted from data given in Bradley, D. D., Krauss, R. M., Petitte, D. B., Ramcharin, S., and Wingird, I. (1978). Serum high-density lipoprotein cholesterol in women using oral contraceptives, estrogens, and progestins. *New England Journal of Medicine* 299, 17–20.

9. Kahneman, D. and Tversky, A. (1972). Subjective probability: A judgment of representativeness. *Cognitive Psychology 3*, 430–454.

10. Strickberger, M. W. (1976). *Genetics*, 2nd edition. New York: Macmillan. p. 206.

11. Fictitious but realistic situation. See Waugh, G. D. (1954). The occurrence of *Mytilicola intestinalis* (Steuer) on the east coast of England. *Journal of Animal Ecology 23*, 364–367.

12. Mosteller, F. and Tukey, J. W. (1977). *Data Analysis and Regression*. Reading, Massachusetts: Addison-Wesley. p. 25.

13. This is typical for U.S. populations. See, for example, Maccready, R. A. and Mannin, M. C. (1951). A typing study of one hundred and fifty thousand bloods. *Journal of Laboratory and Clinical Medicine 37*, 634–636.

14. Fictitious but realistic population, resembling the population of young American men aged 18–24, as described in Abraham, S., Johnson, C. L., and Najjar, M. F. (1979). Weight and height of adults 18–74 years of age: United States 1971–1974. *U.S. National Center for Health Statistics Series 11, No. 211*. Washington, D.C.: U.S. Department of Health, Education and Welfare.

15. The mean and standard deviation are realistic, based on unpublished data provided courtesy of J. Y. Latimer and C. A. Mitchell. The normality assumption may or may not be realistic.

16. Fictitious but realistic situation. See Krebs, C. J. (1972). *Ecology: The Experimental Analysis of Distribution and Abundance*. New York: Harper and Row.

17. The mean and standard deviation are realistic, based on unpublished data provided courtesy of S. Newman and D. L. Harris. The normality assumption may or may not be realistic.

CHAPTER 6

1. Pappas, T. and Mitchell, C. A. (1984). Effects of seismic stress on the vegetative growth of *Glycine max* (L.) Merr. cv. Wells II. *Plant, Cell and Environment 8*, 143–148. Reprinted with permission of Blackwell Scientific Publications Limited. Raw data courtesy of the authors. The actual experiment included several groups of plants grown under different environmental conditions.

2. Student (W. S. Gosset) (1908). The probable error of a mean. *Biometrika 6*, 1–25.

3. Based on data reported in Rea, T. M., Nash, J. F., Zabik, J. E., Born, G. S., and Kessler, W. V. (1984). Effects of toluene inhalation on brain biogenic amines in the rat. *Toxicology 31*, 143–150.

4. Based on an experiment by M. Morales.

5. Adapted from Cherney, J. H., Volenec, J. J., and Nyquist, W. E. (1985). Sequential fiber analysis of forage as influenced by sample weight. *Crop Science 25*, 6, Nov./Dec. 1985, 1113–1115 (Table 1). By permission of the Crop Science Society of America, Inc. Raw data courtesy of W. E. Nyquist.

6. Bockman, D. E. and Kirby, M. L. (1984). Dependence of thymus development on derivatives of the neural crest. *Science 223*, 498–500. Copyright 1984 by the AAAS.

7. Brown, S. A., Riviere, J. E., Coppoc, G. L., Hinsman, E. J., Carlton, W. W., and Steckel, R. R. (1985). Single intravenous and multiple intramuscular dose pharmacokinetics and tissue residue profile of gentamicin in sheep. *American Journal of Veterinary Research 46*, 69–74. Raw data courtesy of S. A. Brown and G. L. Coppoc.

8. Lobstein, D. D. (1983). A multivariate study of exercise training effects on beta-endorphin and emotionality in psychologically normal, medically healthy men. Ph.D. thesis, Purdue University. Raw data courtesy of the author.

9. Nicholson, R. L. and Moraes, W. B. C. (1980). Survival of *Colletotrichum graminicola*: Importance of the spore matrix. *Phytopathology 70*, 255–261.

10. Adapted from Morris, J. G., Cripe, W. S., Chapman, H. L., Jr., Walker, D. F., Armstrong, J. B., Alexander, J. D., Jr., Miranda, R., Sanchez, A., Jr., Sanchez, B., Blair-West, J. R., and Denton, D. A. (1984). Selenium deficiency in cattle associated with Heinz bodies and anemia. *Science 223*, 491–492. Copyright 1984 by the AAAS.

11. Shaffer, P. L. and Rock, G. C. (1983). Tufted apple budmoth (Lepidoptera: Tortricidae): Effects of constant daylengths and temperatures on larval growth rate and determination of larval-pupal ecdysis. *Environmental Entomology 12*, 76–80.

12. Newman, S., Everson, D. O., Gunsett, F. C., and Christian, R. E. (1984). Analysis of two- and three-way crosses among Rambouillet, Targhee, Columbia and Suffolk sheep for three preweaning traits. Unpublished manuscript. Raw data courtesy of S. Newman.

13. Adapted from the following two papers. Potkin, S. G., Cannon, H. E., Murphy, D. L., and Wyatt, R. J. (1978). Are paranoid schizophrenics biologically different from other schizophrenics? *New England Journal of Medicine 298*, 61–66. Murphy, D. L., Wright, C., Buchsbaum, M., Nichols, A., Costa, J. L., and Wyatt, R. J. (1976). Platelet and plasma amine oxidase activity in 680 normals: Sex and age differences and stability over time. *Biochemical Medicine 16*, 254–265. The data displayed are fictitious but realistic, having been reconstructed from the histograms and summary information given by Potkin et al. and Murphy et al.

14. Dice, L. R. (1932). Variation in the geographic race of the deermouse, *Peromyscus maniculatus bairdii*. *Occasional Papers of the Museum of Zoology, University of Michigan*, No. 239. Data reproduced in Simpson, G. G., Roe, A., and Lewontin, R. C. (1960). *Quantitative Zoology*. New York: Harcourt, Brace. p. 79.

15. Bodor, N. and Simpkins, J. W. (1983). Redox delivery system for brain-specific, sustained release of dopamine. *Science 221*, 65–67.

16. Based on data provided by C. H. Noller.

17. This is roughly the SD for the U.S. population of middle-aged men. See Moore, F. E. and Gordon, T. (1973). Serum cholesterol levels in adults, United States 1960–62. *U.S. National Center for Health Statistics, Vital and Health Statistics Series 11, No. 22*. Washington, D.C.: U.S. Department of Health, Education and Welfare.

18. Schaeffer, J., Andrysiak, T., and Ungerleider, J. T. (1981). Cognition and long-term use of ganja (cannabis). *Science 213*, 465–466.

19. Desai, R. (1982). An anatomical study of the canine male and female pelvic diaphragm and the effect of testosterone on the status of the levator ani of male dogs. *Journal of the American Animal Hospital Association 18*, 195–202.

20. The probabilities in Table 6.4 were estimated by computer simulation carried out by the author and R. P. Becker. The standard error of each probability estimate is less than .0015. The sources of the parent distributions are given in Notes 6 and 7 to Chapter 5.

21. Hessell, E. A., Johnson, D. D., Ivey, T. D., and Miller, D. W. (1980). Membrane vs bubble oxygenator for cardiac operations. *Journal of Thoracic and Cardiovascular Surgery 80*, 111–122.

22. Peters, H. G. and Bademan, H. (1963). The form and growth of stellate cells in the cortex of the guinea-pig. *Journal of Anatomy (London) 97*, 111–117.

23. Kaneto, A., Kosaka, K., and Nakao, K. (1967). Effects of stimulation of the vagus nerve on insulin secretion. *Endocrinology 80*, 530–536. Copyright © 1967 by The Endocrine Society.

24. Simmons, F. J. (1943). Occurrence of superparasitism in *Nemeritis canescens*. *Revue Canadienne de Biologie 2*, 15–40. Data reproduced in Williams, C. B. (1964). *Patterns in the Balance of Nature*. London: Academic Press. p. 223.

25. An appropriate standard error for the trimmed mean (assuming that the decision to trim is made before seeing the data) is based on a modified standard deviation called the *Winsorized* standard deviation. The method is described in the following textbooks. Koopmans, L. H. (1981). *An Introduction to Contemporary Statistics*. Boston: Duxbury. Byrkit, D. (1987). *Statistics Today: A Comprehensive Introduction*. Menlo Park, California: Benjamin Cummings.

26. Parks, N. J., Krohn, K. A., Mathis, C. A., Chasko, J. H., Geiger, K. R., Gregor, M. E., and Peek, N. F. (1981). Nitrogen-13-labelled nitrite and nitrate: Distribution and metabolism after intratracheal administration. *Science 212*, 58–61. Copyright 1981 by the AAAS. Raw data courtesy of N. J. Parks.

27. Krick, J. A. (1982). Effects of seeding rate on culm diameter and the inheritance of culm diameter in soft red winter wheat (*Triticum aestivum* L. em Thell). M. S. thesis, Department of Agronomy, Purdue University. Raw data courtesy of J. A. Krick and H. W. Ohm. Each diameter is the mean of measurements taken at six prescribed locations on the stem.

28. Bailey, J. and Marshall, J. (1970). The relationship of the post-ovulatory phase of the menstrual cycle to total cycle length. *Journal of Biosocial Science 2*, 123–132.

29. Unpublished data courtesy of W. F. Jacobson.

30. Dale, E. M. and Housley, T. L. (1986). Sucrose synthase activity in developing wheat endosperms differing in maximum weight. *Plant Physiology 82*, 7–10. Raw data courtesy of the authors.

CHAPTER 7

1. Heald, F. (1974). Hematocrit values of youths 12–17 years, United States. *U.S. National Center for Health Statistics, Vital and Health Statistics Series 11, No. 146*. Washington, D.C.: U.S. Department of Health, Education and Welfare. Actually, the data were obtained by a sampling scheme more complicated than simple random sampling.

2. Long, T. F. and Murdock, L. L. (1983). Stimulation of blowfly feeding behavior by octopaminergic drugs. *Proceedings of the National Academy of Sciences 80*, 4159–4163. Raw data courtesy of the authors and L. C. Sudlow.

3. Knight, S. L. and Mitchell, C. A. (1983). Enhancement of lettuce yield by manipulation of light and nitrogen nutrition. *Journal of the American Society for Horticultural Science 108*, 750–754. Raw data courtesy of the authors. The actual sample sizes were equal; in this example some observations have been omitted to improve the exposition.

4. Rea, T. M., Nash, J. F., Zabik, J. E., Born, G. S., and Kessler, W. V. (1984). Effects of toluene inhalation on brain biogenic amines in the rat. *Toxicology 31*, 143–150. Raw data courtesy of J. F. Nash and J. E. Zabik.

5. Unpublished data courtesy of G. P. Carlson and M. A. Morse.

6. Hagerman, A. E. and Nicholson, R. L. (1982). High-performance liquid chromatographic determination of hydroxycinnamic acids in the maize mesocotyl. *Journal of Agricultural and Food Chemistry 30*, 1098–1102. Reprinted with permission. Copyright 1982 American Chemical Society.

7. Patel, C., Marmot, M. M., and Terry, D. J. (1981). Controlled trial of biofeedback-aided behavioral methods in reducing mild hypertension. *British Medical Journal 282*, 2005–2008.

8. Lipsky, J. J., Lewis, J. C., and Novick, W. J., Jr. (1984). Production of hypoprothrombinemia by Moxalactam and 1-methyl-5-thiotetrazole in rats. *Antimicrobial Agents and Chemotherapy 25*, 380–381.

9. Gwynne, D. T. (1981). Sexual difference theory: Mormon crickets show role reversal in mate choice. *Science 213*, 779–780. Copyright 1981 by the AAAS. Data provided courtesy of the author.

10. Lemenager, R. P., Nelson, L. A., and Hendrix, K. S. (1980). Influence of cow size and breed type on energy requirements. *Journal of Animal Science 51*, 566–576. Some of the animals *lost* weight during the 78 days, so that the mean weight gains are based on both positive and negative values.

11. Sagan, C. (1977). *The Dragons of Eden*. New York: Ballantine. p. 7.

12. Adapted from Miyada, V. S. (1978). Uso da levedura seca de distilarias de alcool de cana de acucar na alimentacao de suinos em crescimento e acabamento. Master's thesis, University of São Paulo, Brazil.

13. Kalsner, S. and Richards, R. (1984). Coronary arteries of cardiac patients are hyperreactive and contain stores of amines: A mechanism for coronary spasm. *Science 223*, 1435–1437. Copyright 1984 by the AAAS.

14. Adapted from Dybas, H. S. and Lloyd, M. (1962). Isolation by habitat in two synchronized species of periodical cicadas (Homoptera, Cicadidae, *Magicicada*). *Ecology 43*, 444–459.

15. Bockman, D. E. and Kirby, M. L. (1984). Dependence of thymus development on derivatives of the neural crest. *Science 223*, 498–500. Copyright 1984 by the AAAS.

16. Tripepi, R. R. and Mitchell, C. A. (1984). Metabolic response of river birch and European birch roots to hypoxia. *Plant Physiology 76*, 31–35. Raw data courtesy of the authors.

17. Lamke, L. O. and Liljedahl, S.-O. (1976). Plasma volume changes after infusion of various plasma expanders. *Resuscitation 5*, 93–102.

18. Anderson, J. W., Story, L., Sieling, B., Chen, W. J. L., Petro, M. S., and Story, J. (1984). Hypo-cholesterolemic effects of oat-bran or bean intake for hypercholesterolemic men. *The American Journal of Clinical Nutrition 40*, 1146–1155.

19. Adapted from data provided courtesy of D. R. Shields and D. K. Colby. See Shields, D. R. (1981). The influence of niacin supplementation on growing ruminants and *in vivo* and *in vitro* rumen parameters. Ph.D. thesis, Purdue University.

20. Schall, J. J., Bennett, A. F., and Putnam, R. W. (1982). Lizards infected with malaria: Physiological and behavioral consequences. *Science 217*, 1057–1059. Copyright 1982 by the AAAS.

21. Agosti, E. and Camerota, G. (1965). Some effects of hypnotic suggestion on respiratory function. *International Journal of Clinical and Experimental Hypnosis 13*, 149–156.

22. Adapted from Knight, S. L. and Mitchell, C. A. (1983). Enhancement of lettuce yield by manipulation of light and nitrogen nutrition. *Journal of the American Society for Horticultural Science 108*, 750–754.

23. Unpublished data courtesy of J. L. Wolfson.

24. Unpublished data courtesy of M. B. Nichols and R. P. Maickel.

25. Fictitious but realistic data.

26. Williams, G. Z., Widdowson, G. M., and Penton, J. (1978). Individual character of variation in time-series studies of healthy people. II. Differences in values for clinical chemical analytes in serum among demographic groups, by age and sex. *Clinical Chemistry 24*, 313–320. Copyright American Association for Clinical Chemistry, Inc. Reprinted with permission from AACC.

27. Fictitious but realistic data. See Abraham, S., Johnson, C. L., and Najjar, M. F. (1979). Weight and height of adults 18–74 years of age, United States 1971–74. *U.S. National Center for Health Statistics, Vital and Health Statistics Series 11, No. 211*. Washington, D.C.: U.S. Department of Health, Education and Welfare.

28. Example communicated by D. A. Holt.

29. Petrie, B. and Segalowitz, S. J. (1980). Use of fetal heart rate, other perinatal and maternal factors as predictors of sex. *Perceptual and Motor Skills 50*, 871–874. Reprinted by permission of the authors and publisher.

30. Hagerman, A. E. and Nicholson, R. L. (1982). High-performance liquid chromatographic determination of hydroxycinnamic acids in the maize mesocotyl. *Journal of Agricultural and Food Chemistry 30*, 1098–1102. Copyright 1982 American Chemical Society. Reprinted with permission.

31. Adapted from Williams, G. Z., Widdowson, G. M., and Penton, J. (1978). Individual character of variation in time-series studies of healthy people. II. Difference in values for clinical chemical analytes in serum among demographic groups, by age and sex. *Clinical Chemistry 24*, 313–320. Copyright American Association for Clinical Chemistry, Inc. Reprinted with permission from AAAC.

32. Hamill, P. V. V., Johnston, F. E., and Lemeshow, S. (1973). Height and weight of youths 12–17 years, United States. *U.S. National Center for Health Statistics, Vital and Health Statistics Series 11, No. 124*. Washington, D.C.: U.S. Department of Health, Education and Welfare.

33. Roberts, J. (1975). Blood pressure of persons 18–74 years, United States, 1971–72. *U.S. National Center for Health Statistics, Vital and Health Statistics Series 11, No. 150*. Washington, D.C.: U.S. Department of Health, Education and Welfare. However, the distribution of systolic blood pressure is more skewed (see Exercise 5.21).

34. Galen, R. S. and Gambino, S. R. (1975). *Beyond Normality: The Predictive Value and Efficiency of Medical Diagnoses*. New York: Wiley.

35. Lobstein, D. D. (1983). A multivariate study of exercise training effects on beta-endorphin and emotionality in psychologically normal, medically healthy men. Ph.D. thesis, Purdue University. Raw data courtesy of the author.

36. It is sometimes stated that the validity of the Mann–Whitney test requires that the two population distributions have the same shape, and differ only by a shift. This is not correct. The computations underlying Table 6 require only that the common population distribution be continuous. A further property, technically called *consistency* of the test, requires that the two distributions be *stochastically ordered*, which is the technical way of saying that one of the variables has a consistent tendency to be larger than the other. In fact, the title of Mann and Whitney's original paper is "On a test of whether one of two random variables is stochastically larger than the other" (*Annals of Mathematical Statistics 18*, 1947). In Section 7.12 we discuss the requirement of stochastic ordering, calling it an "implicit assumption." (The confidence interval procedure mentioned at the end of Section 7.11 does require the stronger assumption that the distributions have the same shape.)

37. Rea, T. M., Nash, J. F., Zabik, J. E., Born, G. S., and Kessler, W. V. (1984). Effects of toluene inhalation on brain biogenic amines in the rat. *Toxicology 31*, 143–150. Raw data courtesy of J. F. Nash.

38. Agosti, E. and Camerota, G. (1965). Some effects of hypnotic suggestion on respiratory function. *International Journal of Clinical and Experimental Hypnosis 13*, 149–156.

39. Connolly, K. (1968). The social facilitation of preening behaviour in *Drosophila melanogaster*. *Animal Behaviour 16*, 385–391.

40. Erne, P., Bolli, P., Buergisser, E., and Buehler, F. R. (1984). Correlation of platelet calcium with blood pressure. *New England Journal of Medicine 310*, 1084–1088. Reprinted by permission of the *New England Journal of Medicine*. Summary statistics calculated from raw data provided courtesy of F. R. Buehler.

41. Adapted from unpublished data provided by F. Delgado. The extremely high somatic cell counts probably represent cases of mastitis.

42. Pappas, T. and Mitchell, C. A. (1985). Effects of seismic stress on the vegetative growth of *Glycine max* (L.) Merr. cv. Wells II. *Plant, Cell and Environment 8*, 143–148. Reprinted with permission of Blackwell Scientific Publications Limited. Raw data courtesy of the authors. The original experiment included more than two treatment groups.

43. Cicirelli, M. F., Robinson, K. R., and Smith, L. D. (1983). Internal pH of *Xenopus* oocytes: A study of the mechanism and role of pH changes during meiotic maturation. *Developmental Biology 100*, 133–146. Raw data courtesy of M. F. Cicirelli.

44. Unpublished data courtesy of J. A. Henricks and V. J. K. Liu.

45. Schall, J. J., Bennett, A. F., and Putnam, R. W. (1982). Lizards infected with malaria: Physiological and behavioral consequences. *Science 217*, 1057–1059. Copyright 1982 by the AAAS. Raw data courtesy of J. J. Schall.

CHAPTER 8

1. Yerushalmy, J. (1971). The relationship of parents' cigarette smoking to outcome of pregnancy—implications as to the problem of inferring causation from observed associations. *American Journal of Epidemiology 93*, 443–456.

2. Based on a study described in Glickman, L. T. and Appel, M. J. G. (1982). A controlled field trial of an attenuated canine origin parvovirus vaccine. *The Compendium of Continuing Education for the Practicing Veterinarian 4*, 888–892.

3. This is the design described in the following papers. Rosenzweig, M. R., Bennett, E. L., and Diamond, M. R. (1972). Brain changes in response to experience. *Scientific American 226*, 22–29. Bennett, E. L., Diamond, M. C., Krech, D., and Rosenzweig, M. R. (1964). Chemical and anatomical plasticity of brain. *Science 146*, 610–619.

4. Based on an experiment by Resh, W. and Stoughton, R. B. (1976). Topically applied antibiotics in acne vulgaris. *Archives of Dermatology 112*, 182–184.

5. Swearingen, M. L. and Holt, D. A. (1976). Using a "blank" trial as a teaching tool. *Journal of Agronomic Education 5*, 3–8. Reprinted by permission of the American Society of Agronomy, Inc. In fact, in order to demonstrate the variability of plot yields, the experimenters planted the *same* variety of barley in all 16 plots.

6. Peto, R., Pike, M. C., Armitage, P., Breslow, N. E., Cox, D. R., Howard, S. V., Mantel, N., McPherson, K., Peto, J., and Smith, P. G. (1976). Design and analysis of randomized clinical trials requiring prolonged observation of each patient. I. Introduction and design. *British Journal of Cancer 34*, 585–612.

7. Sacks, H., Chalmers, T. C., and Smith, H. (1982). Randomized versus historical controls for clinical trials. *American Journal of Medicine 72*, 233–240.

8. Chalmers, T. C., Celano, P., Sacks, H. S., and Smith, H. (1983). Bias in treatment assignment in controlled clinical trials. *New England Journal of Medicine 309*, 1358–1361.

9. Nicholson, R. L. and Moraes, W. B. C. (1980). Survival of *Colletotrichum graminicola*: Importance of the spore matrix. *Phytopathology 70*, 255–261. Raw data courtesy of R. L. Nicholson.

10. Gould, S. J. (1981). *The Mismeasure of Man*. New York: Norton. pp. 50ff. The SDs were estimated from the ranges reported by Gould.

11. Yerushalmy, J. (1972). Infants with low birth weight born before their mothers started to smoke cigarettes. *American Journal of Obstetrics and Gynecology 112*, 277–284.

12. Anderson, G. D., Blidner, I. N., McClemont, S., and Sinclair, J. C. (1984). Determinants of size at birth in a Canadian population. *American Journal of Obstetrics and Gynecology 150*, 236–244.

13. Mochizuki, M., Maruo, T., Masuko, K., and Ohtsu, T. (1984). Effects of smoking on fetoplacental–maternal system during pregnancy. *American Journal of Obstetrics and Gynecology 149*, 413–420.

14. Wainright, R. L. (1983). Change in observed birth weight associated with a change in maternal cigarette smoking. *American Journal of Epidemiology 117*, 668–675.

15. Haenszel, W., Kurihara, M., Segi, M., and Lee, R. K. C. (1972). Stomach cancer among Japanese in Hawaii. *Journal of the National Cancer Institute 49*, 969–988.

16. Yerushalmy, J. and Hilleboe, H. E. (1957). Fat in the diet and mortality from heart disease. *New York State Journal of Medicine 57*, 2343–2354. Reprinted by permission. Copyright by the Medical Society of the State of New York.

CHAPTER 9

1. Adapted from Bodian, D. (1947). Nucleic acid in nerve-cell regeneration. *Symposia of the Society for Experimental Biology, No. 1. Nucleic Acid.* 163–178. Reprinted with permission of Cambridge University Press.

2. It is true, however, that a substantial discrepancy between s_1 and s_2 can suggest that some other analysis may be more appropriate than the straightforward analysis by Student's t. For instance, if the ratio of the SDs is the same as the ratio of the means, then the treatment effect may be multiplicative rather than additive, in which case it is more appropriate (and more powerful) to analyze $\log(Y)$ rather than Y.

3. Day, K. M., Patterson, F. L., Luetkemeier, O. W., Ohm, H. W., Polizotto, K., Roberts, J. J., Shaner, G. E., Huber, D. M., Finney, R. E., Foster, J. E., and Gallun, R. L. (1980). Performance and adaptation of small grains in Indiana. *Station Bulletin No. 290*. West Lafayette, Indiana: Agricultural Experiment Station of Purdue University. Raw data provided courtesy of W. E. Nyquist. The actual trial included more than two varieties.

4. Golden, C. J., Graber, B., Blose, I., Berg, R., Coffman, J., and Block, S. (1981). Difference in brain densities between chronic alcoholic and normal control patients. *Science 211*, 508–510. Raw data courtesy of C. J. Golden. Copyright 1981 by the AAAS.

5. Cicirelli, M. F. and Smith, L. D. (1985). Cyclic AMP levels during the maturation of *Xenopus* oocytes. *Developmental Biology 108*, 254–258. Raw data courtesy of M. F. Cicirelli.

6. Judge, M. D., Aberle, E. D., Cross, H. R., and Schanbacher, B. D. (1984). Thermal shrinkage temperature of intramuscular collagen of bulls and steers. *Journal of Animal Science 59*, 706–709. Raw data courtesy of the authors and E. W. Mills.

7. Wainright, R. L. (1983). Change in observed birth weight associated with a change in maternal cigarette smoking. *American Journal of Epidemiology 117*, 668–675.

8. Schriewer, H., Guennewig, V., and Assmann, G. (1983). Effect of 10 weeks endurance training on the concentration of lipids and lipoproteins as well as on the composition of high-density lipoproteins in blood serum. *International Journal of Sports Medicine 4*, 109–115. Reprinted with permission of Georg Thieme Verlag Stuttgart.

9. Data from experiments reported in several papers, for example, Fout, G. S. and Simon, E. H. (1983). Antiviral activities directed against wild-type and interferon-sensitive mengovirus. *Journal of*

General Virology 64, 1543–1555. Raw data courtesy of E. H. Simon. The unit of measurement is proportional to the number of plaques formed by the virus on a monolayer of mouse cells. Because they are obtained by a serial dilution technique, the measurements have varying numbers of significant digits; the final zeroes of the three-digit numbers are not significant digits.

10. Sallan, S. E., Cronin, C., Zelen, M., and Zinberg, N. E. (1980). Antiemetics in patients receiving chemotherapy for cancer. *New England Journal of Medicine* 302, 135–138. Reprinted by permission.

11. Adapted from Batchelor, J. R. and Hackett, M. (1970). HL-A matching in treatment of burned patients with skin allografts. *Lancet 2*, 581–583.

12. Rosenzweig, M. R., Bennett, E. L., and Diamond, M. R. (1972). Brain changes in response to experience. *Scientific American* 226, 22–29. Also Bennett, E. L., Diamond, M. C., Krech, D., and Rosenzweig, M. R. (1964). Chemical and anatomical plasticity of brain. *Science 146*, 610–619. Copyright 1964 by the AAAS.

13. Richens, A. and Ahmad, S. (1975). Controlled trial of valproate in severe epilepsy. *British Medical Journal 4*, 255–256.

14. Wiedenmann, R. N. and Rabenold, K. N. (1987). The effects of social dominance between two subspecies of dark-eyed juncos, *Junco hyemalis*. *Animal Behavior 35*, 856–864. Raw data courtesy of the authors.

15. Patel, C., Marmot, M. G., and Terry, D. J. (1981). Controlled trial of biofeedback-aided behavioural methods in reducing mild hypertension. *British Medical Journal 282*, 2005–2008.

16. Knowlen, G. G., Kittleson, M. D., Nachreiner, R. F., and Eyster, G. E. (1983). Comparison of plasma aldosterone concentration among clinical status groups of dogs with chronic heart failure. *Journal of the American Veterinary Medical Association 183*, 991–996.

17. Førde, O. H., Knutsen, S. F., Arnesen, E., and Thelle, D. S. (1985). The Tromsø heart study: Coffee consumption and serum lipid concentrations in men with hypercholesterolaemia: A randomised intervention study. *British Medical Journal 290*, 893–895. (The sample sizes are unequal because the 25 no-coffee men actually represent three different treatment groups, which followed the same regimen for the first 5 weeks of the study and different regimens thereafter.)

18. Dalvit, S. P. (1981). The effect of the menstrual cycle on patterns of food intake. *American Journal of Clinical Nutrition 34*, 1811–1815.

19. Unpublished data courtesy of D. J. Honor and W. A. Vestre.

20. Sesin, G. P. (1984). Pharmacokinetic dosing of Tobramycin sulfate. *American Pharmacy NS24*, 778. Vakoutis, J., Stein, G. E., Miller, P. B., and Clayman, A. E. (1981). Aminoglycoside monitoring program. *American Journal of Hospital Pharmacy 38*, 1477–1480. Copyright 1981, American Society of Hospital Pharmacists, Inc. All rights reserved. Reprinted with permission.

21. Masty, J. (1983). Innervation of the equine small intestine. Master's thesis, Purdue University. Raw data courtesy of the author.

22. Dale, E. M. and Housley, T. L. (1986). Sucrose synthase activity in developing wheat endosperms differing in maximum weight. *Plant Physiology 82*, 7–10. Raw data courtesy of the authors.

23. Salib, N. M. (1985). The effect of caffeine on the respiratory exchange ratio of separate submaximal arms and legs exercise of middle distance runners. Master's thesis, Purdue University.

24. Unpublished data courtesy of C. H. Noller.

25. Robinson, L. R. (1985). The effects of electrical fields on wound healing in *Notophthalmus viridescens*. Master's thesis, Purdue University. Raw data courtesy of the author and J. W. Vanable, Jr.

26. Agosti, E. and Camerota, G. (1965). Some effects of hypnotic suggestion on respiratory function. *International Journal of Clinical and Experimental Hypnosis 13*, 149–156. The experiment actually included a third phase.

CHAPTER 10

1. Oldfield, R. C. (1971). The assessment and analysis of handedness: The Edinburgh inventory. *Neuropsychologia 9*, 97–113.

2. Adapted from McCloskey, R. V., Goren, R., Bissett, D., Bentley, J., and Tutlane, V. (1982). Cefotaxime in the treatment of infections of the skin and skin structure. *Reviews of Infectious Diseases 4, Supp.*, S444–S447.

3. Adapted from Petras, M. L. (1967). Studies of natural populations of *Mus*. III. Coat color polymorphisms. *Canadian Journal of Genetic Cytology 9*, 287–296.

4. Miller, C. L., Pollock, T. M., and Clewer, A. D. E. (1974). Whooping-cough vaccination: An assessment. *The Lancet ii*, 510–513.

5. Erskine, A. G. and Socha, W. W. (1978). *The Principles and Practices of Blood Grouping*. St. Louis: Mosby. p. 209.

6. Curtis, H. (1983). *Biology*, 4th edition. New York: Worth. p. 908.

7. Mourant, A. E., Kopec, A. C., and Domaniewska-Sobczak, K. (1976). *The Distribution of Human Blood Groups and Other Polymorphisms*, 2nd edition. London: Oxford University Press. p. 44.

8. Based on an experiment described in Oellerman, C. M., Patterson, F. L., and Gallun, R. L. (1983). Inheritance of resistance in "Luso" wheat to Hessian fly. *Crop Science 23*, 221–224.

9. Rabenold, K. R. and Rabenold, P. P. (1985). Variation in altitudinal migration, winter segregation, and site tenacity in two subspecies of dark-eyed juncos in the Southern Appalachians. *The Auk 102*, 805–819.

10. Adapted from Jacobs, G. H. (1978). Spectral sensitivity and colour vision in the ground-dwelling sciurids: Results from golden mantled ground squirrels and comparisons for five species. *Animal Behaviour 26*, 409–421. See also Jacobs, G. H. (1981). *Comparative Color Vision*. New York: Academic Press.

11. Sinnott, E. W. and Durham, G. B. (1922). Inheritance in the summer squash. *Journal of Heredity 13*, 177–186.

12. Adapted from Gould, J. L. (1985). How bees remember flower shapes. *Science 227*, 1492–1494. Figure copyright 1985 by the American Association for the Advancement of Science; used by permission.

13. Adapted from 1983 birth data for West Lafayette, Indiana.

14. Bateson, W. and Saunders, E. R. (1902). *Reports to the Evolution Committee of the Royal Society 1*, 1–160. Feather color and comb shape are controlled independently; white feather is dominant and small comb is dominant. The parents in the experiment were first-generation hybrids (F_1) and thus were necessarily double heterozygotes.

15. This is a realistic value. See Exercise 3.12.

16. Adapted from Mantel, N., Bohidar, N. R., and Ciminera, J. L. (1977). Mantel–Haenszel analyses of litter-matched time-to-response data, with modifications for recovery of inter-litter information. *Cancer Research 37*, 3863–3868. (A more powerful analysis, which uses the partially informative triplets, is described in the paper.)

17. Described in the *Lafayette Journal and Courier*, Lafayette, Indiana, January 23 and August 19, 1976.

18. Wolfson, J. L. (1987). Impact of *Rhizobium* nodules on *Sitona hispidulus*, the clover root curculio. *Entomologia Experimentalis et Applicata 43*, 237–243. Raw data courtesy of the author. The experiment actually contained 11 dishes.

19. Table 10.2 was calculated using the criterion that the approximate 95% confidence interval should differ by no more than 15% from the 95% confidence interval calculated by an exact method. See Lu, T.-F. C. and Samuels, M. L. (1988). Sample size requirements for approximation of binomial confidence intervals. *Purdue University Department of Statistics Technical Report Series*, #88–8.

20. Collins, R. L. (1970). The sound of one paw clapping: An inquiry into the origin of left-handedness. In Lindzey, G. and Thiessen, D. D. (eds.) *Contributions to Behavior-Genetic Analysis: The Mouse as Prototype*. New York: Appleton-Century-Crofts.

21. Salzman, A. G. (1985). Habitat selection in a clonal plant. *Science 228*, 603–604. Copyright 1985 by the AAAS.

22. Thornberry, O. T., Wilson, R. W., and Golden, P. (1986). Health promotion and disease prevention. Provisional data from the National Health Interview Survey: United States, January–June 1985. *Advance Data from Vital and Health Statistics No. 119*. Washington, D.C.: U.S. Department of Health and Human Services.

23. Freeland, W. J. (1981). Parasitism and behavioral dominance among male mice. *Science 213*, 461–462. Copyright 1981 by the AAAS.

24. Cheatum, E. L. and Severinghaus, C. W. (1950). Variations in fertility of white-tailed deer related to range conditions. *Transactions of the North American Wildlife Conference 15*, 170–189.

25. Adapted from Finkel, M. P., Jinkins, P. R., and Biskis, B. O. (1964). Parameters of radiation dosage that influence production of osteogenic sarcomas in mice. *National Cancer Institute Monograph* 14, 243–270.

26. Floersheim, G. L., Weber, O., Tschumi, P., and Ulbrich, M. (1983). Research cited in *Scientific American*, April 1983, Vol. 248, No. 4, p. 75.

27. Kohn, J. and Weaver, P. C. (1974). Serum alpha$_1$-fetoprotein in hepatocellular carcinoma. *Lancet* ii, 334–336.

28. Roberts, J. (1964). Binocular visual acuity of adults, United States 1960–62. *U.S. National Center for Health Statistics Series 11, No. 3.* Washington, D.C.: U.S. Department of Health, Education and Welfare.

29. Fuchs, J. A., Smith, J. D., and Bird, L. S. (1972). Genetic basis for an 11:5 dihybrid ratio observed in *Gossypium hirsutum. Journal of Heredity 63*, 300–303. The genetic basis for the 13:3 and 11:5 ratios is explained in Strickberger, M. W. (1976). *Genetics*, 2nd edition. New York: Macmillan. pp. 206–208.

30. Adapted from Goodyear, C. P. (1970). Terrestrial and aquatic orientation in the starhead top-minnow, *Fundulus noti. Science 168*, 603–605. Copyright 1970 by the AAAS.

31. See Batschelet, E. (1981). *Circular Statistics in Biology.* New York: Academic Press.

CHAPTER 11

1. Brailovsky, D. (1974). Timolol maleate (MK-950): A new beta-blocking agent for the prophylactic management of angina pectoris. A multicenter, multinational, cooperative trial. In *Beta-Adrenergic Blocking Agents in the Management of Hypertension and Angina Pectoris*, B. Magnani (ed.). New York: Raven Press.

2. Waaler, G. H. M. (1927). Uber die Erblichkeitsverhältnisse der verschiedenen Arten von ange-borener Rotgrünblindheit. *Zeitschrift für induktive Abstammungs- und Vererbungslehre 45*, 279–333.

3. Brodie, E. D., Jr. and Brodie, E. D. III. (1980). Differential avoidance of mimetic salamanders by free-ranging birds. *Science 208*, 181–182. Copyright 1980 by the AAAS.

4. Karban, R., Adamchak, R., and Schnathorst, W. C. (1987). Induced resistance and interspecific competition between spider mites and a vascular wilt fungus. *Science 235*, 678–680. Copyright 1987 by the AAAS.

5. Turnbull, D. M., Rawlins, M. D., Weightman, D., and Chadwick, D. W. (1982). A comparison of phenytoin and valproate in previously untreated adult epileptic patients. *Journal of Neurology, Neurosurgery, and Psychiatry 45*, 55–59.

6. Unpublished data courtesy of W. Singleton and K. Hendrix.

7. Mizutani, T. and Mitsuoka, T. (1979). Effect of intestinal bacteria on incidence of liver tumors in gnotobiotic C3H/He male mice. *Journal of the National Cancer Institute 63*, 1365–1370.

8. Selawry, O. S. (1974). The role of chemotherapy in the treatment of lung cancer. *Seminars in Oncology 1*, 259–272.

9. Adapted from Ammon, O. (1899). *Zur Anthropologie der Badener.* Jena: G. Fischer. Ammon's data appear in Goodman, L. A. and Kruskal, W. H. (1954). Measures of association for cross classifications. *Journal of the American Statistical Association 49*, 732–764. Light hair was blonde or red; dark hair was brown or black. Light eyes were blue, grey, or green; dark eyes were brown.

10. These data are fictitious, but the proportions of left-handed males and females are realistic, and the independence of the twins is in agreement with published data. See Porac, C. and Coren, S. (1981). *Lateral Preferences and Human Behavior.* New York: Springer-Verlag, p. 36; and Morgan, M. C. and Corballis, M. J. (1978). On the biological basis of human laterality: I. Evidence for a maturational left–right gradient. *The Behavioral and Brain Sciences 2*, p. 274.

11. Davies, D. F., Johnson, A. P., Rees, B. W. G., Elwood, P. C., and Abernethy, M. (1974). Food antibodies and myocardial infarction. *The Lancet i*, 1012–1014. See also Galen, R. S. (1974). Food antibodies and myocardial infarction. *The Lancet ii*, 832.

12. Adapted from Porac, C. and Coren, S. (1981). *Lateral Preferences and Human Behavior.* New York: Springer-Verlag. The frequencies given are approximate, having been deduced from percentages on pages 36 and 45. People with neutral preference were counted as Left.

13. Upton, G. and Fingleton, B. (1985). *Spatial Data Analysis by Example: Point Pattern and Quantitative Data, Vol. 1.* New York: Wiley, p. 230. Adapted from Diggle, P. J. (1979). Statistical methods for spatial point patterns in ecology, pp. 95–150 in *Spatial and Temporal Analysis in Ecology,* R. M. Cormack and J. K. Ord (eds.). Fairland, Maryland: International Cooperative Publishing House.

14. Yerushalmy, J. (1971). The relationship of parents' cigarette smoking to outcome of pregnancy—implications as to the problem of inferring causation from observed associations. *American Journal of Epidemiology 93,* 443–456.

15. The three data sets, collected in the 1950s by three different investigators, are reproduced in Mourant, A. E., Kopec, A. C., and Domaniewska-Sobczak, K. (1976). *The Distribution of the Human Blood Groups and Other Polymorphisms.* London: Oxford University Press. p. 204.

16. Visintainer, M. A., Volpicelli, J. R., and Seligman, M. E. P. (1982). Tumor rejection in rats after inescapable or escapable shock. *Science 216,* 437–439. Copyright 1982 by the AAAS.

17. Inglesfield, C. and Begon, M. (1981). Open-ground individuals and population structure in *Drosophila subobscura* Collin. *Biological Journal of the Linnean Society 15,* 259–278.

18. Aird, I., Bentall, H. H., Mehigan, J. A., and Roberts, J. A. F. (1954). The blood groups in relation to peptic ulceration and carcinoma of colon, rectum, breast, and bronchus: An association between the ABO blood groups and peptic ulceration. *British Medical Journal ii,* 315–321.

19. Govind, C. K. and Pearce, J. (1986). Differential reflex activity determines claw and closer muscle asymmetry in developing lobsters. *Science 233,* 354–356. Copyright 1986 by the AAAS.

20. Adapted from Paige, K. N. and Whitham, T. G. (1985). Individual and population shifts in flower color by scarlet gilia: A mechanism for pollinator tracking. *Science 227,* 315–317. The raw data given are fictitious, but have been constructed to agree with the summary statistics given by Paige and Whitham.

21. Beck, S. L. and Gavin, D. L. (1976). Susceptibility of mice to audiogenic seizures is increased by handling their dams during gestation. *Science 193,* 427–428. Copyright 1976 by the AAAS.

22. Pittet, P. G., Acheson, K. J., Wuersch, P., Maeder, E., and Jequier, E. (1981). Effects of an oral load of partially hydrolyzed wheatflour on blood parameters and substrate utilization in man. *The American Journal of Clinical Nutrition 34,* 2438–2445.

23. Saunders, M. C., Dick, J. S., Brown, I. M., McPherson, K., and Chalmers, I. (1985). The effects of hospital admission for bed rest on the duration of twin pregnancy: A randomised trial. *The Lancet ii,* 793–795.

24. Allison, A. C. and Clyde, D. F. (1961). Malaria in African children with deficient erythrocyte dehydrogenase. *British Medical Journal 1,* 1346–1349.

25. Conover, D. O. and Kynard, B. E. (1981). Environmental sex determination: Interaction of temperature and genotype in a fish. *Science 213,* 577–579. Copyright 1981 by the AAAS.

26. Carson, J. L., Collier, A. M., and Hu, S. S. (1985). Acquired ciliary defects in nasal epithelium of children with acute viral upper respiratory infections. *New England Journal of Medicine 312,* 463–468. Reprinted by permission.

27. Larson, E. B., Roach, R. C., Schoene, R. B., and Hornbein, T. F. (1982). Acute mountain sickness and acetazolamide. *Journal of the American Medical Association 248,* 328–332. Copyright 1982 American Medical Association.

28. Data from Wiener, A. S. (1962). *Blood Groups and Transfusion.* Cited by Erskine, A. G. and Socha, W. W. (1978). *The Principles and Practices of Blood Grouping.* St. Louis: Mosby.

29. Kluger, M. J., Ringler, D. H., and Anver, M. R. (1975). Fever and survival. *Science 188,* 166–168. Copyright 1975 by the AAAS. The original article contains a misprint, but Dr. Kluger has kindly provided the correct mortality at 40°C.

30. Collaborative Group for the Study of Stroke in Young Women (1973). Oral contraception and increased risk of cerebral ischemia or thrombosis. *New England Journal of Medicine 288,* 871–878. Reprinted by permission.

31. Shorrocks, B. and Nigro, L. (1981). Microdistribution and habitat selection in *Drosophila subobscura* collin. *Biological Journal of the Linnean Society 16,* 293–301.

32. Savin, R. C. and Turner, M. C. (1966). Antibiotics and the placebo reaction in acne. *Journal of the American Medical Association 196,* 365–367. Copyright 1966 American Medical Association.

33. Mochizuki, M., Maruo, T., Masuko, K., and Ohtsu, T. (1984). Effects of smoking on fetoplacental–maternal system during pregnancy. *American Journal of Obstetrics and Gynecology 149,* 413–420.

CHAPTER 12

1. Pappas, T. and Mitchell, C. A. (1985). Effects of seismic stress on the vegetative growth of *Glycine max* (L.) Merr. cv. Wells II. *Plant, Cell and Environment 8*, 143–148. Reprinted with permission of Blackwell Scientific Publications Limited. Raw data courtesy of the authors. The original experiment included more than four treatments.

2. Shields, D. R. (1981). The influence of niacin supplementation on growing ruminants and *in vivo* and *in vitro* rumen parameters. Ph.D. thesis, Purdue University. Adapted from raw data provided courtesy of the author and D. K. Colby.

3. Adapted from Potkin, S. G., Cannon, H. E., Murphy, D. L., and Wyatt, R. J. (1978). Are paranoid schizophrenics biologically different from other schizophrenics? *New England Journal of Medicine 298*, 61–66. Reprinted by permission. The calculations are based on the data in Example 1.4 in this book, which are an approximate reconstruction from the histograms and summary information given by Potkin et al.

4. Adapted from Keller, S. E., Weiss, J. M., Schleifer, S. J., Miller, N. E., and Stein, M. (1981). Suppression of immunity by stress: Effect of a graded series of stressors on lymphocyte stimulation in the rat. *Science 213*, 1397–1400. Copyright 1981 by the AAAS. The SDs and SSs were estimated from the SEs given by Keller et al.

5. Tripepi, R. R. and Mitchell, C. A. (1984). Metabolic response of river birch and European birch roots to hypoxia. *Plant Physiology 76*, 31–35. Raw data courtesy of the authors.

6. Lobstein, D. D. (1983). A multivariate study of exercise training effects on beta-endorphin and emotionality in psychologically normal, medically healthy men. Ph.D. thesis, Purdue University. Raw data courtesy of the author.

7. Fictitious but realistic data, adapted from O'Brien, R. J. and Drizd, T. A. (1981). Basic data on spirometry in adults 25–74 years of age: United States, 1971–75. *U.S. National Center for Health Statistics, Vital and Health Statistics Series 11, No. 222*. Washington, D.C.: U.S. Department of Health and Human Services.

8. U.S. Bureau of the Census (1982). *Statistical Abstract of the United States, 1982–83*, 103rd edition. Washington, D.C: U.S. Government Printing Office. The age distribution varies from year to year; we have given the 1981 distribution.

9. Chrisman, C. L. and Baumgartner, A. P. (1980). Micronuclei in bone-marrow cells of mice subjected to hyperthermia. *Mutation Research 77*, 95–97. The original experiment included six treatments.

10. Baird, J. T. and Quinlivan, L. G. (1973). Parity and hypertension. *U.S. National Center for Health Statistics, Vital and Health Statistics Series 11, No. 38*. Washington, D.C.: U.S. Department of Health, Education and Welfare.

11. U.S. Bureau of the Census (1982). *Statistical Abstract of the United States, 1982–83*, 103rd edition. Washington, D.C.: U.S. Government Printing Office.

12. Adapted from Veterans Administration Cooperative Study Group on Antihypertensive Agents (1979). Comparative effects of ticrynafen and hydrochlorothiazide in the treatment of hypertension. *New England Journal of Medicine 301*, 293–297. Reprinted by permission. The value of s_c was calculated from the SEs given in the paper.

13. Knight, S. L. and Mitchell, C. A. (1983). Enhancement of lettuce yield by manipulation of light and nitrogen nutrition. *Journal of the American Society for Horticultural Science 108*, 750–754. Calculations based on raw data provided by the authors.

14. Adapted from Witelson, S. F. (1985). The brain connection: The corpus callosum is larger in left-handers. *Science 229*, 665–668. Copyright 1985 by the AAAS. Reprinted with permission of *Science* and S. F. Witelson. The SDs and MS(within) have been calculated from the standard errors given by Witelson.

15. Adapted from Clapp, M. J. L. (1980). The effect of diet on some parameters measured in toxicological studies in the rat. *Laboratory Animals 14*, 253–261.

16. Latimer, J. (1985). Adapted from unpublished data provided by the investigator.

17. Adapted from Morris, J. G., Cripe, W. S., Chapman, H. L., Jr., Walker, D. F., Armstrong, J. B., Alexander, J. D., Jr., Miranda, R., Sanchez, A., Jr., Sanchez, B., Blair-West, J. R., and Denton, D. A. (1984). Selenium deficiency in cattle associated with Heinz bodies and anemia. *Science 223*, 491–492. Copyright 1984 by the AAAS. The MS(within) is fictitious but agrees with the standard errors given by Morris et al.

18. Fictitious but realistic data. Adapted from Mizutani, T. and Mitsuoka, T. (1979). Effect of intestinal bacteria on incidence of liver tumors in gnotobiotic C3H/He male mice. *Journal of the National Cancer Institute 63*, 1365–1370.

19. Becker, W. A. (1961). Comparing entries in random sample tests. *Poultry Science 40*, 1507–1514.

20. Heggestad, H..E. and Bennett, J. H. (1981). Photochemical oxidants potentiate yield losses in snap beans attributable to sulfur dioxide. *Science 213*, 1008–1010. Copyright 1981 by the AAAS. Raw data courtesy of H. E. Heggestad.

21. Adapted from Rosner, B. (1982). Statistical methods in ophthalmology: An adjustment for the intraclass correlation between eyes. *Biometrics 38*, 105–114. Reprinted with permission from the Biometric Society. The medical study is reported in Berson, E. L., Rosner, B., and Simonoff, E. (1980). An outpatient population of retinitis pigmentosa and their normal relatives: Risk factors for genetic typing and detection derived from their ocular examinations. *American Journal of Ophthalmology 89*, 763–775. The means and sums of squares have been estimated from data given by Rosner, after estimating missing values for two patients for whom only one eye was measured.

CHAPTER 13

1. Unpublished data courtesy of M. B. Nichols and R. P. Maickel. The original experiment contained more than three treatment groups.

2. Gwynne, D. T. (1981). Sexual difference theory: Mormon crickets show role reversal in mate choice. *Science 213*, 779–780. Copyright 1981 by the AAAS. Raw data courtesy of the author.

3. Adapted from Andren, C. and Nilson, G. (1981). Reproductive success and risk of predation in normal and melanistic colour morphs of the adder, *Vipera berus. Biological Journal of the Linnean Society 15*, 235–246. (The data are for the melanistic females; the values have been manipulated slightly to simplify the exposition.)

4. Cicirelli, M. F., Robinson, K. R., and Smith, L. D. (1983). Internal pH of *Xenopus* oocytes: A study of the mechanism and role of pH changes during meiotic maturation. *Developmental Biology 100*, 133–146. Raw data courtesy of M. F. Cicirelli.

5. Maickel, R. P. and Nash, J. F., Jr. (1985). Differing effects of short-chain alcohols on body temperature and coordinated muscular activity in mice. *Neuropharmacology 24*, 83–89. Reprinted with permission. Copyright 1985, Pergamon Journals, Ltd. Raw data courtesy of J. F. Nash, Jr.

6. Smith, R. D. (1978–1979). *Institute of Agricultural Engineering Annual Report.* Salisbury, Zimbabwe: Department of Research and Specialist Services, Ministry of Agriculture. Raw data courtesy of R. D. Smith.

7. Bowers, W. S., Hoch, H. C., Evans, P. H., and Katayama, M. (1986). Thallophytic allelopathy: Isolation and identification of laetisaric acid. *Science 232*, 105–106. Copyright 1986 by the AAAS. Raw data courtesy of the authors.

8. Webb, P. (1981). Energy expenditure and fat-free mass in men and women. *American Journal of Clinical Nutrition 34*, 1816–1826.

9. Adapted from Barclay, A. M. and Crawford, R. M. M. (1984). Seedling emergence in the rowan (*Sorbus aucuparia*) from an altitudinal gradient. *Journal of Ecology 72*, 627–636. Reprinted with permission of Blackwell Scientific Publications Limited.

10. Hamill, P. V. V., Johnston, F. E., and Lemeshow, S. (1973). Height and weight of youths 12–17 years, United States. *U.S. National Center for Health Statistics, Vital and Health Statistics Series 11, No. 124*. Washington, D.C.: U.S. Department of Health, Education and Welfare. The conditional distributions of weight given height are plotted in Figure 13.10 as normal distributions. The fictitious population agrees well with the real population (as described by Hamill et al.) in the central portion of each conditional distribution, but the real conditional distributions have shorter left tails and longer right tails than the fictitious normal conditional distributions.

11. Maickel, R. P. Personal communication.

12. Erne, P., Bolli, P., Buergisser, E., and Buehler, F. R. (1984). Correlation of platelet calcium with blood pressure. *New England Journal of Medicine 310*, 1084–1088. Reprinted by permission. Raw data courtesy of F. R. Buehler. To simplify the discussion, we have omitted nine patients with "borderline" high blood pressure.

13. Pappas, T. and Mitchell, C. A. (1985). Effects of seismic stress on the vegetative growth of *Glycine max* (L.) Merr. cv. Wells II. *Plant, Cell and Environment 8*, 143–148. Reprinted with permission of Blackwell Scientific Publications Limited. Raw data courtesy of the authors.

14. Florey, C. du V. and Acheson, R. M. (1969). Blood pressure as it relates to physique, blood glucose, and serum cholesterol. *U.S. National Center for Health Statistics, Series 11, No. 34.* Washington, D.C.: U.S. Department of Health, Education and Welfare.

15. Masty, J. (1983). Innervation of the equine small intestine. Master's thesis, Purdue University. Raw data courtesy of the author.

16. Albert, A. (1981). Atypicality indices as reference values for laboratory data. *American Journal of Clinical Pathology 76*, 421–425.

17. Fictitious but realistic data, based on inter- and intra-individual variation as described in Williams, G. Z., Widdowson, G. M., and Penton, J. (1978). Individual character of variation in time-series studies of healthy people. II. Differences in values for clinical chemical analytes in serum among demographic groups, by age and sex. *Clinical Chemistry 24*, 313–320.

18. Stewart, T. S., Nelson, L. A., Perry, T. W., and Martin, T. G. (1985). Unpublished data provided courtesy of T. S. Stewart.

19. Pappas, T. and Mitchell, C. A. (1985). Effects of seismic stress on the vegetative growth of *Glycine max* (L.) Merr. cv. Wells II. *Plant, Cell and Environment 8*, 143–149. Reprinted with permission of Blackwell Scientific Publications Limited. Raw data courtesy of the authors.

20. Adapted from Spencer, D. F., Volpp, T. R., and Lembi, C. A. (1980). Environmental control of *Pithophora oedogonia* (Chlorophyceae) akinete germination. *Journal of Phycology 16*, 424–427. The value $r = -.72$ was calculated from data displayed graphically by Spencer et al.

21. Fialho, E. T., Ferreira, A. S., Freitas, A. R., and Albino, L. F. T. (1982). Energy and nitrogen balance of ration (corn–soybean meal) for male castrated and non-castrated swine of different weights and breeds (in Portuguese). *Revista Sociedade Brasileira de Zootecnia 11*, 405–419. Raw data courtesy of E. T. Fialho.

22. Example communicated by D. A. Holt.

23. Florey, C. du V. and Acheson, R. M. (1969). Blood pressure as it relates to physique, blood glucose, and serum cholesterol. *U.S. National Center for Health Statistics, Series 11, No. 34.* Washington, D.C.: U.S. Department of Health, Education and Welfare.

24. Bernays, E. A. (1986). Diet-induced head allometry among foliage-chewing insects and its importance for graminovores. *Science 231*, 495–497. Copyright 1986 by the AAAS. Raw data courtesy of the author.

25. Gwynne, D. T. (1981). Sexual difference theory: Mormon crickets show role reversal in mate choice. *Science 213*, 779–780. Copyright 1981 by the AAAS. Calculations based on raw data provided courtesy of the author.

26. Heggestad, H. E. and Bennett, J. H. (1981). Photochemical oxidants potentiate yield losses in snap beans attributable to sulfur dioxide. *Science 213*, 1008–1010. Copyright 1981 by the AAAS. Raw data courtesy of H. E. Heggestad.

27. Thirakhupt, K. (1985). Foraging ecology of sympatric parids: Individual and population responses to winter food scarcity. Ph.D. thesis, Purdue University. Raw data courtesy of the author and K. N. Rabenold.

28. Unpublished data courtesy of A. H. Ismail and L. S. Verity.

29. These data sets were invented by F. J. Anscombe. See Anscombe, F. J. (1973). Graphs in statistical analysis. *The American Statistician 27*, 17–21.

30. Unpublished data courtesy of M. B. Nichols and R. P. Maickel. The experiment actually contained more than three groups. The data in Example 13.1 are from another part of the same study, using a different chemical form of amphetamine.

CHAPTER 14

1. Raskin, R. L., Raskin, M., and Baldwin, R. L. (1973). Effect of chronic insulin and cortisol administration on lactational performance and mammary metabolism in rats. *Journal of Dairy Science 56*, 1033–1041. The sample sizes varied slightly for some of the variables.

2. Fictitious but realistic data. Adapted from Mizutani, T. and Mitsuoka, T. (1979). Effect of intestinal bacteria on incidence of liver tumors in gnotobiotic C3H/He male mice. *Journal of the National Cancer Institute 63*, 1365–1370.

3. Lee, K. L., McNeer, J. F., Starmer, C. F., Harris, P. J., and Rosati, R. A. (1980). Clinical judgment and statistics: Lessons from a simulated randomized trial in coronary artery disease. *Circulation 61*, 508–515. By permission of the American Heart Association.

4. These values are based on results described in the following two papers: Miller, R. and Siegmund, D. (1982). Maximally selected chi-square statistics. *Biometrics 38*, 1011–1016; Halperin, J. (1982). Maximally selected chi-square statistics for small samples. *Biometrics 38*, 1017–1023.

5. The data are discussed by Elashoff, J. D. (1981). Down with multiple *t*-tests! *Gastroenterology 80*, 615–619. The experiment is reported in Taylor, I. L., Solomon, T. E., Walsh, J. H., and Grossman, M. I. (1979). Pancreatic polypeptide: Metabolism and effect on pancreatic secretion in dogs. *Gastroenterology 76*, 524–528.

STATISTICAL TABLES

CONTENTS

TABLE 1 Random Digits

	01	06	11	16	21	26	31	36	41	46
01	06048	96063	22049	86532	75170	65711	29969	06826	39208	80631
02	25636	73908	85512	78073	19089	66458	06597	93985	14193	69366
03	61378	45410	43511	54364	97334	01267	28304	35047	38789	84896
04	15919	71559	12310	00727	54473	51547	09816	83641	72973	75367
05	47328	20405	88019	82276	33679	10328	25116	59176	64675	95141
06	72548	80667	53893	64400	81955	15163	06146	58549	75530	19582
07	87154	04130	55985	44508	37515	71689	80765	46598	45539	12792
08	68379	96636	32154	94718	22845	80265	92747	66238	58474	23783
09	89391	54041	70806	36012	30833	83132	39338	54753	00722	44568
10	15816	60231	28365	61924	66934	21243	09896	92428	51611	46756
11	29618	55219	18394	11625	27673	08117	89314	42581	36897	03738
12	30723	42988	30002	95364	45473	46107	34222	00739	84847	49096
13	54028	04975	92323	53836	76128	84762	32050	59516	40831	59687
14	40376	02036	48087	05216	26684	97959	85601	86622	70750	15603
15	64439	37357	90935	57330	79738	65361	85944	23619	30504	61564
16	83037	30144	29166	20915	53462	42573	75204	50064	08847	07082
17	71071	01636	31085	71638	77357	14256	89174	15184	81701	21592
18	67891	43187	58159	24144	29683	04276	02987	04571	18334	04291
19	52487	39499	97330	40045	47304	98528	00422	82693	87547	73525
20	67550	82107	27302	79145	73213	27217	19211	59784	63929	04609
21	86472	80165	70773	90519	49710	31921	36102	45042	04203	01439
22	08699	38051	60404	06609	98435	91560	22634	98014	43316	61099
23	59596	13000	07655	74837	81211	71530	28341	83110	72289	25180
24	31810	54868	92799	09893	97499	96509	71548	06462	40498	22628
25	71753	90756	21382	84209	95900	11119	34507	61241	17641	83147

(continued)

TABLE 1 continued

	01	06	11	16	21	26	31	36	41	46
26	17155	07370	65655	04824	53417	20737	70510	92615	89967	50216
27	36211	24724	94769	16940	43138	25260	75318	69037	95982	28631
28	94777	66946	16120	56382	58416	92391	81457	28101	69766	32436
29	52994	58881	81841	51844	75566	48567	18552	66829	91230	39141
30	84643	32635	51440	96854	35739	66440	82806	82841	56302	31640
31	95690	34873	11297	60518	72717	47616	55751	37187	31413	31132
32	64093	92948	21565	51686	40368	66151	82877	99951	85069	54503
33	89484	50055	67586	16439	96385	67868	66597	51433	44764	66573
34	70184	38164	74646	90244	83169	85276	07598	69242	90088	32308
35	75601	91867	80848	94484	98532	36183	28549	17704	28653	80027
36	99044	78699	34681	31049	40790	50445	79897	68203	11486	93676
37	10272	18347	89369	02355	76671	34097	03791	93817	43142	24974
38	69738	85488	34453	80876	43018	59967	84458	71906	54019	70023
39	93441	58902	17871	45425	29066	04553	42644	54624	34498	27319
40	25814	74497	75642	58350	64118	87400	82870	26143	46624	21404
41	29757	84506	48617	48844	35139	97855	43435	74581	35678	69793
42	56666	86113	06805	09470	07992	54079	00517	19313	53741	25306
43	26401	71007	12500	27815	86490	01370	47826	36009	10447	25953
44	40747	59584	83453	30875	39509	82829	42878	13844	84131	48524
45	99434	51563	73915	03867	24785	19324	21254	11641	25940	92026
46	50734	88330	39128	14261	00584	94266	99677	19852	49673	18680
47	89728	32743	19102	83279	68308	41160	32365	25774	39699	50743
48	71395	61945	41082	93648	99874	82577	26507	07054	29381	16995
49	50945	68182	23108	95765	81136	06792	13322	41631	37118	35881
50	36525	26551	28457	75699	74537	68623	50099	91909	23508	35751

TABLE 1 continued

	51	56	61	66	71	76	81	86	91	96
01	64825	74126	86159	26710	49256	04655	06001	73192	67463	16746
02	46184	63916	89160	87844	53352	43318	70766	23625	09906	65847
03	79976	48891	69431	86571	25979	58755	08884	36704	01107	12308
04	10656	47210	48512	06805	42114	98741	51440	06070	49071	02700
05	18058	84528	56753	02623	81077	60045	06678	53748	10386	37895
06	58979	98046	88467	27762	24781	12559	98384	40926	79570	34746
07	12705	41974	14473	49872	29368	80556	95833	20766	76643	35656
08	39660	83664	18592	82388	27899	24223	36462	61582	95173	36155
09	00360	42077	84161	04464	45042	29560	37916	29889	00342	82533
10	09873	64084	34685	53542	09254	23257	14713	44295	94139	00403
11	12957	84063	79808	23633	77133	41422	26559	29131	74402	82213
12	06090	71584	48965	60201	02786	88929	19861	99361	27535	38297
13	66812	57167	28185	19708	74672	25615	61640	18955	40854	50749
14	91701	36216	66249	04256	31694	33127	67529	73254	72065	74294
15	02775	78899	36471	37098	50270	58933	91765	95157	01384	75388
16	75892	53340	92363	58300	77300	08059	63743	12159	05640	87014
17	18581	70057	82031	68349	55759	46851	33632	28855	74633	08598
18	69698	18177	52824	61742	58119	04168	57843	37870	50988	80316
19	30023	30731	00803	09336	87709	39307	09732	66031	04904	91929
20	94334	05698	97910	37850	77074	56152	67521	48973	29448	84115
21	64133	14640	28418	45405	86974	06666	07879	54026	92264	23418
22	93895	83557	17326	28030	09113	56793	79703	18804	75807	20144
23	54438	83097	52533	86245	02182	11746	58164	90520	99255	44830
24	90565	76710	42456	22612	00232	18919	24019	32254	30703	06678
25	90848	81871	24382	16218	98216	42323	75061	68261	09071	68776

(continued)

TABLE 1 continued

	51	56	61	66	71	76	81	86	91	96
26	41169	08175	69938	61958	72578	31791	74952	71055	40369	00429
27	84627	70347	41566	00019	24481	15677	54506	54545	89563	50049
28	67460	49111	54004	61428	61034	47197	90084	88113	39145	94757
29	99231	60774	52238	05102	71690	72215	61323	13326	01674	81510
30	95775	73679	04900	27666	18424	59793	14965	22220	30682	35488
31	42179	98675	69593	17901	48741	59902	98034	12976	60921	73047
32	91196	05878	92346	45886	31080	21714	19168	94070	77375	10444
33	18794	03741	17612	65467	27698	20456	91737	36008	88225	58013
34	88311	93622	34501	70402	12272	65995	66086	04938	52966	71909
35	17904	33710	42812	72105	91848	39724	26361	09634	50552	98769
36	05905	28509	69631	69177	39081	58818	01998	53949	47884	91326
37	23432	22211	65648	71866	49532	45529	00189	80025	68956	26445
38	29684	43229	54771	90604	48938	13663	24736	83199	41512	43364
39	26506	65067	64252	49765	87650	72082	48997	04845	00136	98941
40	08807	43756	01579	34508	94082	68736	67149	00209	76138	95467
41	50636	70304	73556	32872	07809	20787	85921	41748	10553	97988
42	32437	41588	46991	36667	98127	05072	63700	51803	77262	31970
43	32571	97567	78420	04633	96574	88830	01314	04811	10904	85923
44	28773	22496	11743	23294	78070	20910	86722	50551	37356	92698
45	65768	76188	07781	05314	26017	07741	22268	31374	53559	46971
46	68601	06488	73776	45361	89059	59775	59149	64095	10352	11107
47	98364	17663	85972	72263	93178	04284	79236	04567	31813	82283
48	95308	70577	96712	85697	55685	19023	98112	96915	50791	31107
49	68681	24419	15362	60771	09962	45891	03130	09937	15775	51935
50	30721	22371	65174	57363	37851	71554	19708	23880	86638	05880

TABLE 2 Binomial Coefficients C_j

n	0	1	2	3	4	5	j 6	7	8	9	10
1	1	1									
2	1	2	1								
3	1	3	3	1							
4	1	4	6	4	1						
5	1	5	10	10	5	1					
6	1	6	15	20	15	6	1				
7	1	7	21	35	35	21	7	1			
8	1	8	28	56	70	56	28	8	1		
9	1	9	36	84	126	126	84	36	9	1	
10	1	10	45	120	210	252	210	120	45	10	1
11	1	11	55	165	330	462	462	330	165	55	11
12	1	12	66	220	495	792	924	792	495	220	66
13	1	13	78	286	715	1,287	1,716	1,716	1,287	715	286
14	1	14	91	364	1,001	2,002	3,003	3,432	3,003	2,002	1,001
15	1	15	105	455	1,365	3,003	5,005	6,435	6,435	5,005	3,003
16	1	16	120	560	1,820	4,368	8,008	11,440	12,870	11,440	8,008
17	1	17	136	680	2,380	6,188	12,376	19,448	24,310	24,310	19,448
18	1	18	153	816	3,060	8,568	18,564	31,824	43,758	48,620	43,758
19	1	19	171	969	3,876	11,628	27,132	50,388	75,582	92,378	92,378
20	1	20	190	1,140	4,845	15,504	38,760	77,520	125,970	167,960	184,756

TABLE 3　Areas Under the Normal Curve

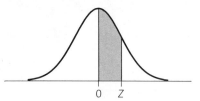

Z	Area	Z	Area	Z	Area	Z	Area	Z	Area	Z	Area
.00	.0000	.30	.1179	.60	.2257	.90	.3159	1.20	.3849	1.50	.4332
.01	.0040	.31	.1217	.61	.2291	.91	.3186	1.21	.3869	1.51	.4345
.02	.0080	.32	.1255	.62	.2324	.92	.3212	1.22	.3888	1.52	.4357
.03	.0120	.33	.1293	.63	.2357	.93	.3238	1.23	.3907	1.53	.4370
.04	.0160	.34	.1331	.64	.2389	.94	.3264	1.24	.3925	1.54	.4382
.05	.0199	.35	.1368	.65	.2422	.95	.3289	1.25	.3944	1.55	.4394
.06	.0239	.36	.1406	.66	.2454	.96	.3315	1.26	.3962	1.56	.4406
.07	.0279	.37	.1443	.67	.2486	.97	.3340	1.27	.3980	1.57	.4418
.08	.0319	.38	.1480	.68	.2517	.98	.3365	1.28	.3997	1.58	.4429
.09	.0359	.39	.1517	.69	.2549	.99	.3389	1.29	.4015	1.59	.4441
.10	.0398	.40	.1554	.70	.2580	1.00	.3413	1.30	.4032	1.60	.4452
.11	.0438	.41	.1591	.71	.2611	1.01	.3438	1.31	.4049	1.61	.4463
.12	.0478	.42	.1628	.72	.2642	1.02	.3461	1.32	.4066	1.62	.4474
.13	.0517	.43	.1664	.73	.2673	1.03	.3485	1.33	.4082	1.63	.4484
.14	.0557	.44	.1700	.74	.2704	1.04	.3508	1.34	.4099	1.64	.4495
.15	.0596	.45	.1736	.75	.2734	1.05	.3531	1.35	.4115	1.65	.4505
.16	.0636	.46	.1772	.76	.2764	1.06	.3554	1.36	.4131	1.66	.4515
.17	.0675	.47	.1808	.77	.2794	1.07	.3577	1.37	.4147	1.67	.4525
.18	.0714	.48	.1844	.78	.2823	1.08	.3599	1.38	.4162	1.68	.4535
.19	.0753	.49	.1879	.79	.2852	1.09	.3621	1.39	.4177	1.69	.4545
.20	.0793	.50	.1915	.80	.2881	1.10	.3643	1.40	.4192	1.70	.4554
.21	.0832	.51	.1950	.81	.2910	1.11	.3665	1.41	.4207	1.71	.4564
.22	.0871	.52	.1985	.82	.2939	1.12	.3686	1.42	.4222	1.72	.4573
.23	.0910	.53	.2019	.83	.2967	1.13	.3708	1.43	.4236	1.73	.4582
.24	.0948	.54	.2054	.84	.2995	1.14	.3729	1.44	.4251	1.74	.4591
.25	.0987	.55	.2088	.85	.3023	1.15	.3749	1.45	.4265	1.75	.4599
.26	.1026	.56	.2123	.86	.3051	1.16	.3770	1.46	.4279	1.76	.4608
.27	.1064	.57	.2157	.87	.3078	1.17	.3790	1.47	.4292	1.77	.4616
.28	.1103	.58	.2190	.88	.3106	1.18	.3810	1.48	.4306	1.78	.4625
.29	.1141	.59	.2224	.89	.3133	1.19	.3830	1.49	.4319	1.79	.4633

TABLE 3 continued

Z	Area	Z	Area	Z	Area	Z	Area	Z	Area
1.80	.4641	2.10	.4821	2.40	.4918	2.70	.4965	3.00	.4987
1.81	.4649	2.11	.4826	2.41	.4920	2.71	.4966	3.10	.4990
1.82	.4656	2.12	.4830	2.42	.4922	2.72	.4967	3.20	.4993
1.83	.4664	2.13	.4834	2.43	.4925	2.73	.4968	3.30	.4995
1.84	.4671	2.14	.4838	2.44	.4927	2.74	.4969	3.40	.4997
1.85	.4678	2.15	.4842	2.45	.4929	2.75	.4970	3.50	.499767
1.86	.4686	2.16	.4846	2.46	.4931	2.76	.4971	4.00	.499968
1.87	.4693	2.17	.4850	2.47	.4932	2.77	.4972	4.50	.499997
1.88	.4699	2.18	.4854	2.48	.4934	2.78	.4973		
1.89	.4706	2.19	.4857	2.49	.4936	2.79	.4974		
1.90	.4713	2.20	.4861	2.50	.4938	2.80	.4974		
1.91	.4719	2.21	.4864	2.51	.4940	2.81	.4975		
1.92	.4726	2.22	.4868	2.52	.4941	2.82	.4976		
1.93	.4732	2.23	.4871	2.53	.4943	2.83	.4977		
1.94	.4738	2.24	.4875	2.54	.4945	2.84	.4977		
1.95	.4744	2.25	.4878	2.55	.4946	2.85	.4978		
1.96	.4750	2.26	.4881	2.56	.4948	2.86	.4979		
1.97	.4756	2.27	.4884	2.57	.4949	2.87	.4979		
1.98	.4761	2.28	.4887	2.58	.4951	2.88	.4980		
1.99	.4767	2.29	.4890	2.59	.4952	2.89	.4981		
2.00	.4772	2.30	.4893	2.60	.4953	2.90	.4981		
2.01	.4778	2.31	.4896	2.61	.4955	2.91	.4982		
2.02	.4783	2.32	.4898	2.62	.4956	2.92	.4982		
2.03	.4788	2.33	.4901	2.63	.4957	2.93	.4983		
2.04	.4793	2.34	.4904	2.64	.4959	2.94	.4984		
2.05	.4798	2.35	.4906	2.65	.4960	2.95	.4984		
2.06	.4803	2.36	.4909	2.66	.4961	2.96	.4985		
2.07	.4808	2.37	.4911	2.67	.4962	2.97	.4985		
2.08	.4812	2.38	.4913	2.68	.4963	2.98	.4986		
2.09	.4817	2.39	.4916	2.69	.4964	2.99	.4986		

TABLE 4 Critical Values of Student's t-Distribution

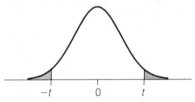

Notes: (1) Two-tailed area is shaded area in the illustration. (2) If H_A is directional, column headings should be multiplied by $\frac{1}{2}$ when bracketing the P-value.

df				TWO-TAILED AREA				df
	.20	.10	.05	.02	.01	.001	.0001	
1	3.078	6.314	12.706	31.821	63.657	636.619	6,366.198	1
2	1.886	2.920	4.303	6.695	9.925	31.598	99.992	2
3	1.638	2.353	3.182	4.541	5.841	12.924	28.000	3
4	1.533	2.132	2.776	3.747	4.604	8.610	15.544	4
5	1.476	2.015	2.571	3.365	4.032	6.869	11.178	5
6	1.440	1.943	2.447	3.143	3.707	5.959	9.082	6
7	1.415	1.895	2.365	2.998	3.499	5.408	7.885	7
8	1.397	1.860	2.306	2.896	3.355	5.041	7.120	8
9	1.383	1.833	2.262	2.821	3.250	4.781	6.594	9
10	1.372	1.812	2.228	2.764	3.169	4.587	6.211	10
11	1.363	1.796	2.201	2.718	3.106	4.437	5.921	11
12	1.356	1.782	2.179	2.681	3.055	4.318	5.694	12
13	1.350	1.771	2.160	2.650	3.012	4.221	5.513	13
14	1.345	1.761	2.145	2.624	2.977	4.140	5.363	14
15	1.341	1.753	2.131	2.602	2.947	4.073	5.239	15
16	1.337	1.746	2.120	2.583	2.921	4.015	5.134	16
17	1.333	1.740	2.110	2.567	2.898	3.965	5.044	17
18	1.330	1.734	2.101	2.552	2.878	3.922	4.966	18
19	1.328	1.729	2.093	2.539	2.861	3.883	4.897	19
20	1.325	1.725	2.086	2.528	2.845	3.850	4.837	20
21	1.323	1.721	2.080	2.518	2.831	3.819	4.784	21
22	1.321	1.717	2.074	2.508	2.819	3.792	4.736	22
23	1.319	1.714	2.069	2.500	2.807	3.767	4.693	23
24	1.318	1.711	2.064	2.492	2.797	3.745	4.654	24
25	1.316	1.708	2.060	2.485	2.787	3.725	4.619	25

TABLE 4 continued

df	.20	.10	.05	.02	.01	.001	.0001	df
				TWO-TAILED AREA				
26	1.315	1.706	2.056	2.479	2.779	3.707	4.587	26
27	1.314	1.703	2.052	2.473	2.771	3.690	4.558	27
28	1.313	1.701	2.048	2.467	2.763	3.674	4.530	28
29	1.311	1.699	2.045	2.462	2.756	3.659	4.506	29
30	1.310	1.697	2.042	2.457	2.750	3.646	4.482	30
40	1.303	1.684	2.021	2.423	2.704	3.551	4.321	40
60	1.296	1.671	2.000	2.390	2.660	3.460	4.169	60
100	1.290	1.660	1.984	2.364	2.626	3.390	4.053	100
140	1.288	1.656	1.977	2.353	2.611	3.361	4.006	140
∞	1.282	1.645	1.960	2.326	2.576	3.291	3.891	∞

TABLE 5   Number of Observations for Independent-Samples t-Test

	SIGNIFICANCE LEVEL (TWO-TAILED TEST)																			
	$\alpha = .01$					$\alpha = .02$					$\alpha = .05$					$\alpha = .10$				
POWER →	.99	.95	.90	.80	.50	.99	.95	.90	.80	.50	.99	.95	.90	.80	.50	.99	.95	.90	.80	.50
$\frac{\mu_1 - \mu_2}{\sigma}$																				
.20																				137
.25															124					88
.30										123					87					61
.35					110					90					64				102	45
.40					85					70				100	50			108	78	35
.45				118	68				101	55			105	79	39		108	86	62	28
.50				96	55			106	82	45		106	86	64	32		88	70	51	23
.55			101	79	46		106	88	68	38		87	71	53	27	112	73	58	42	19
.60		101	85	67	39		90	74	58	32	104	74	60	45	23	89	61	49	36	16
.65		87	73	57	34	104	77	64	49	27	88	63	51	39	20	76	52	42	30	14
.70	100	75	63	50	29	90	66	55	43	24	76	55	44	34	17	66	45	36	26	12
.75	88	66	55	44	26	79	58	48	38	21	67	48	39	29	15	57	40	32	23	11
.80	77	58	49	39	23	70	51	43	33	19	59	42	34	26	14	50	35	28	21	10
.85	69	51	43	35	21	62	46	38	30	17	52	37	31	23	12	45	31	25	18	9
.90	62	46	39	31	19	55	41	34	27	15	47	34	27	21	11	40	28	22	16	8
.95	55	42	35	28	17	50	37	31	24	14	42	30	25	19	10	36	25	20	15	7
1.00	50	38	32	26	15	45	33	28	22	13	38	27	23	17	9	33	23	18	14	7
	$\alpha = .005$					$\alpha = .01$					$\alpha = .025$					$\alpha = .05$				
	SIGNIFICANCE LEVEL (ONE-TAILED TEST)																			

TABLE 5 continued

SIGNIFICANCE LEVEL (TWO-TAILED TEST)

POWER →	α = .01					α = .02					α = .05					α = .10				
$\frac{\mu_1 - \mu_2}{\sigma}$	.99	.95	.90	.80	.50	.99	.95	.90	.80	.50	.99	.95	.90	.80	.50	.99	.95	.90	.80	.50
1.1	42	32	27	22	13	38	28	23	19	11	32	23	19	14	8	27	19	15	12	6
1.2	36	27	23	18	11	32	24	20	16	9	27	20	16	12	7	23	16	13	10	5
1.3	31	23	20	16	10	28	21	17	14	8	23	17	14	11	6	20	14	11	9	5
1.4	27	20	17	14	9	24	18	15	12	8	20	15	12	10	6	17	12	10	8	4
1.5	24	18	15	13	8	21	16	14	11	7	18	13	11	9	5	15	11	9	7	4
1.6	21	16	14	11	7	19	14	12	10	6	16	12	10	8	5	14	10	8	6	4
1.7	19	15	13	10	7	17	13	11	9	6	14	11	9	7	4	12	9	7	6	3
1.8	17	13	11	10	6	15	12	10	8	5	13	10	8	6	4	11	8	7	5	
1.9	16	12	11	9	6	14	11	9	8	5	12	9	7	6	4	10	7	6	5	
2.0	14	11	10	8	6	13	10	9	7	5	11	8	7	6	4	9	7	6	4	
2.1	13	10	9	8	5	12	9	8	7	5	10	8	6	5	3	8	6	5	4	
2.2	12	10	8	7	5	11	9	7	6	4	9	7	6	5		8	6	5	4	
2.3	11	9	8	7	5	10	8	7	6	4	9	7	6	5		7	5	5	4	
2.4	11	9	8	6	5	10	8	7	6	4	8	6	5	4		7	5	4	4	
2.5	10	8	7	6	4	9	7	6	5	4	8	6	5	4		6	5	4	3	
3.0	8	6	6	5	4	7	6	5	4	3	6	5	4	4		5	4	3		
3.5	6	5	5	4	3	6	5	4	4		5	4	4	3		4	3			
4.0	6	5	4	4		5	4	4	3		4	4	3			4				
	α = .005					α = .01					α = .025					α = .05				

SIGNIFICANCE LEVEL (ONE-TAILED TEST)

Source: "Number of observations for *t*-test of difference between two means." *Research*, Volume 1 (1948), pp. 520–525. Used with permission of the Longman Group UK Ltd. and Butterworth Scientific Publications.

TABLE 6 Critical Values of U, the Mann–Whitney Statistic

Notes: (1) Because the Mann–Whitney null distribution is discrete, the actual tail probability corresponding to a given critical value is typically somewhat *less* than the column heading. (2) If H_A is directional, column headings should be multiplied by $\frac{1}{2}$ when bracketing the P-value.

		NOMINAL TAIL PROBABILITY						
n	n'	.20	.10	.05	.02	.01	.002	.001
3	2	6						
	3	8	9					
4	2	8						
	3	11	12					
	4	13	15	16				
5	2	9	10					
	3	13	14	15				
	4	16	18	19	20			
	5	20	21	23	24	25		
6	2	11	12					
	3	15	16	17				
	4	19	21	22	23	24		
	5	23	25	27	28	29		
	6	27	29	31	33	34		
7	2	13	14					
	3	17	19	20	21			
	4	22	24	25	27	28		
	5	27	29	30	32	34		
	6	31	34	36	38	39	42	
	7	36	38	41	43	45	48	49
8	2	14	15	16				
	3	19	21	22	24			
	4	25	27	28	30	31		
	5	30	32	34	36	38	40	
	6	35	38	40	42	44	47	48
	7	40	43	46	49	50	54	55
	8	45	49	51	55	57	60	62
9	1	9						
	2	16	17	18				
	3	22	23	25	26	27		
	4	27	30	32	33	35		
	5	33	36	38	40	42	44	45
	6	39	42	44	47	49	52	53
	7	45	48	51	54	56	60	61
	8	50	54	57	61	63	67	68
	9	56	60	64	67	70	74	76
10	1	10						
	2	17	19	20				
	3	24	26	27	29	30		
	4	30	33	35	37	38	40	
	5	37	39	42	44	46	49	50
	6	43	46	49	52	54	57	58
	7	49	53	56	59	61	65	67
	8	56	60	63	67	69	74	75
	9	62	66	70	74	77	82	83
	10	68	73	77	81	84	90	92

TABLE 6   continued

n	n'	.20	.10	.05	.02	.01	.002	.001
				NOMINAL TAIL PROBABILITY				
11	1	11						
	2	19	21	22				
	3	26	28	30	32	33		
	4	33	36	38	40	42	44	
	5	40	43	46	48	50	53	54
	6	47	50	53	57	59	62	64
	7	54	58	61	65	67	71	73
	8	61	65	69	73	75	80	82
	9	68	72	76	81	83	89	91
	10	74	79	84	88	92	98	100
	11	81	87	91	96	100	106	109
12	1	12						
	2	20	22	23				
	3	28	31	32	34	35		
	4	36	39	41	42	45	48	
	5	43	47	49	52	54	58	59
	6	51	55	58	61	63	68	69
	7	58	63	66	70	72	77	79
	8	66	70	74	79	81	87	89
	9	73	78	82	87	90	96	98
	10	81	86	91	96	99	106	108
	11	88	94	99	104	108	115	117
	12	95	102	107	113	117	124	127
13	1	13						
	2	22	24	25	26			
	3	30	33	35	37	38		
	4	39	42	44	47	49	51	52
	5	47	50	53	56	58	62	63
	6	55	59	62	66	68	73	74
	7	63	67	71	75	78	83	85
	8	71	76	80	84	87	93	95
	9	79	84	89	94	97	103	106
	10	87	93	97	103	106	113	116
	11	95	101	106	112	116	123	126
	12	103	109	115	121	125	133	136
	13	111	118	124	130	135	143	146
14	1	14						
	2	24	25	27	28			
	3	32	35	37	40	41		
	4	41	45	47	50	52	55	56
	5	50	54	57	60	63	67	68
	6	59	63	67	71	73	78	79
	7	67	72	76	81	83	89	91
	8	76	81	86	90	94	100	102
	9	85	90	95	100	104	111	113
	10	93	99	104	110	114	121	124
	11	102	108	114	120	124	132	135
	12	110	117	123	130	134	143	146
	13	119	126	132	139	144	153	157
	14	127	135	141	149	154	164	167

(continued)

TABLE 6 continued

n	n′	.20	.10	.05	.02	.01	.002	.001
15	1	15						
	2	25	27	29	30			
	3	35	38	40	42	43		
	4	44	48	50	53	55	59	60
	5	53	57	61	64	67	71	72
	6	63	67	71	75	78	83	85
	7	72	77	81	86	89	95	97
	8	81	87	91	96	100	106	109
	9	90	96	101	107	111	118	120
	10	99	106	111	117	121	129	132
	11	108	115	121	128	132	141	144
	12	117	125	131	138	143	152	155
	13	127	134	141	148	153	163	167
	14	136	144	151	159	164	174	178
	15	145	153	161	169	174	185	189
16	1	16						
	2	27	29	31	32			
	3	37	40	42	45	46		
	4	47	50	53	57	59	62	63
	5	57	61	65	68	71	75	77
	6	67	71	75	80	83	88	90
	7	76	82	86	91	94	101	103
	8	86	92	97	102	106	113	115
	9	96	102	107	113	117	125	128
	10	106	112	118	124	129	137	140
	11	115	122	129	135	140	149	152
	12	125	132	139	146	151	161	165
	13	134	143	149	157	163	173	177
	14	144	153	160	168	174	185	189
	15	154	163	170	179	185	197	201
	16	163	173	181	190	196	208	213
17	1	17						
	2	28	31	32	34			
	3	39	42	45	47	49	51	
	4	50	53	57	60	62	66	67
	5	60	65	68	72	75	80	81
	6	71	76	80	84	87	93	95
	7	81	86	91	96	100	106	109
	8	91	97	102	108	112	119	122
	9	101	108	114	120	124	132	135
	10	112	119	125	132	136	145	148
	11	122	130	136	143	148	158	161
	12	132	140	147	155	160	170	174
	13	142	151	158	166	172	183	187
	14	153	161	169	178	184	195	199
	15	163	172	180	189	195	208	212
	16	173	183	191	201	207	220	225
	17	183	193	202	212	219	232	238

NOMINAL TAIL PROBABILITY spans columns .20–.001.

TABLE 6 continued

n	n'	NOMINAL TAIL PROBABILITY						
		.20	.10	.05	.02	.01	.002	.001
18	1	18						
	2	30	32	34	36			
	3	41	45	47	50	52	54	
	4	52	56	60	63	66	69	71
	5	63	68	72	76	79	84	86
	6	74	80	84	89	92	98	100
	7	85	91	96	102	105	112	115
	8	96	103	108	114	118	126	129
	9	107	114	120	126	131	139	142
	10	118	125	132	139	143	153	156
	11	129	137	143	151	156	166	170
	12	139	148	155	163	169	179	183
	13	150	159	167	175	181	192	197
	14	161	170	178	187	194	206	210
	15	172	182	190	200	206	219	224
	16	182	193	202	212	218	232	237
	17	193	204	213	224	231	245	250
	18	204	215	225	236	243	258	263
19	1	18	19					
	2	31	34	36	37	38		
	3	43	47	50	53	54	57	
	4	55	59	63	67	69	73	74
	5	67	72	76	80	83	88	90
	6	78	84	89	94	97	103	106
	7	90	96	101	107	111	118	120
	8	101	108	114	120	124	132	135
	9	113	120	126	133	138	146	150
	10	124	132	138	146	151	161	164
	11	136	144	151	159	164	175	178
	12	147	156	163	172	177	188	193
	13	158	167	175	184	190	202	207
	14	169	179	188	197	203	216	221
	15	181	191	200	210	216	230	235
	16	192	203	212	222	230	244	249
	17	203	214	224	235	242	257	263
	18	214	226	236	248	255	271	277
	19	226	238	248	260	268	284	291
20	1	19	20					
	2	33	36	38	39	40		
	3	45	49	52	55	57	60	
	4	58	62	66	70	72	77	78
	5	70	75	80	84	87	93	95
	6	82	88	93	98	102	108	111
	7	94	101	106	112	116	124	126
	8	106	113	119	126	130	139	142
	9	118	126	132	140	144	154	157
	10	130	138	145	153	158	168	172
	11	142	151	158	167	172	183	187
	12	154	163	171	180	186	198	202
	13	166	176	184	193	200	212	217
	14	178	188	197	207	213	226	231
	15	190	200	210	220	227	241	246
	16	201	213	222	233	241	255	261
	17	213	225	235	247	254	270	275
	18	225	237	248	260	268	284	287
	19	237	250	261	273	281	298	304
	20	249	262	273	286	295	312	319

TABLE 7 Critical Values of *B* for the Sign Test

Notes: (1) Because the sign-test null distribution is discrete, the actual tail probability corresponding to a given critical value is typically somewhat *less* than the column heading. (2) If H_A is directional, column headings should be multiplied by $\frac{1}{2}$ when bracketing the *P*-value.

n_d	NOMINAL TAIL PROBABILITY						
	.20	.10	.05	.02	.01	.002	.001
1							
2							
3							
4							
5	5	5					
6	6	6	6				
7	6	7	7	7			
8	7	7	8	8	8		
9	7	8	8	9	9		
10	8	9	9	10	10	10	
11	9	9	10	10	11	11	11
12	9	10	10	11	11	12	12
13	10	10	11	12	12	13	13
14	10	11	12	12	13	13	14
15	11	12	12	13	13	14	14
16	12	12	13	14	14	15	15
17	12	13	13	14	15	16	16
18	13	13	14	15	15	16	17
19	13	14	15	15	16	17	17
20	14	15	15	16	17	18	18
21	14	15	16	17	17	18	19
22	15	16	17	17	18	19	19
23	16	16	17	18	19	20	20
24	16	17	18	19	19	20	21
25	17	18	18	19	20	21	21

TABLE 8 Critical Values of the Chi-Square Distribution

Note: If H_A is directional (for df = 1), column headings should be multiplied by $\frac{1}{2}$ when bracketing the *P*-value.

df	TAIL PROBABILITY						
	.20	.10	.05	.02	.01	.001	.0001
1	1.64	2.71	3.84	5.41	6.63	10.83	15.14
2	3.22	4.61	5.99	7.82	9.21	13.82	18.42
3	4.64	6.25	7.81	9.84	11.34	16.27	21.11
4	5.99	7.78	9.49	11.67	13.28	18.47	23.51
5	7.29	9.24	11.07	13.39	15.09	20.51	25.74
6	8.56	10.64	12.59	15.03	16.81	22.46	27.86
7	9.80	12.02	14.07	16.62	18.48	24.32	29.88
8	11.03	13.36	15.51	18.17	20.09	26.12	31.83
9	12.24	14.68	16.92	19.68	21.67	27.88	33.72
10	13.44	15.99	18.31	21.16	23.21	29.59	35.56
11	14.63	17.28	19.68	22.62	24.72	31.26	37.37
12	15.81	18.55	21.03	24.05	26.22	32.91	39.13
13	16.98	19.81	22.36	25.47	27.69	34.53	40.87
14	18.15	21.06	23.68	26.87	29.14	36.12	42.58
15	19.31	22.31	25.00	28.26	30.58	37.70	44.26
16	20.47	23.54	26.30	29.63	32.00	39.25	45.92
17	21.61	24.77	27.59	31.00	33.41	40.79	47.57
18	22.76	25.99	28.87	32.35	34.81	42.31	49.19
19	23.90	27.20	30.14	33.69	36.19	43.82	50.80
20	25.04	28.41	31.41	35.02	37.57	45.31	52.39
21	26.17	29.62	32.67	36.34	38.93	46.80	53.96
22	27.30	30.81	33.92	37.66	40.29	48.27	55.52
23	28.43	32.01	35.17	38.97	41.64	49.73	57.08
24	29.55	33.20	36.42	40.27	42.98	51.18	58.61
25	30.68	34.38	37.65	41.57	44.31	52.62	60.14
26	31.79	35.56	38.89	42.86	45.64	54.05	61.66
27	32.91	36.74	40.11	44.14	46.96	55.48	63.16
28	34.03	37.92	41.34	45.42	48.28	56.89	64.66
29	35.14	39.09	42.56	46.69	49.59	58.30	66.15
30	36.25	40.26	43.77	47.96	50.89	59.70	67.63

TABLE 9 Critical Values of the *F* Distribution

Numerator df = 1

Denom. df	TAIL PROBABILITY						
	.20	.10	.05	.02	.01	.001	.0001
1	9.47	39.86	161	101^1	405^1	406^3	405^5
2	3.56	8.53	18.51	48.51	98.50	998	100^2
3	2.68	5.54	10.13	20.62	34.12	167	784
4	2.35	4.54	7.71	14.04	21.20	74.14	242
5	2.18	4.06	6.61	11.32	16.26	47.18	125
6	2.07	3.78	5.99	9.88	13.75	35.51	82.49
7	2.00	3.59	5.59	8.99	12.25	29.25	62.17
8	1.95	3.46	5.32	8.39	11.26	25.41	50.69
9	1.91	3.36	5.12	7.96	10.56	22.86	43.48
10	1.88	3.29	4.96	7.64	10.04	21.04	38.58
11	1.86	3.23	4.84	7.39	9.65	19.69	35.06
12	1.84	3.18	4.75	7.19	9.33	18.64	32.43
13	1.82	3.14	4.67	7.02	9.07	17.82	30.39
14	1.81	3.10	4.60	6.89	8.86	17.14	28.77
15	1.80	3.07	4.54	6.77	8.68	16.59	27.45
16	1.79	3.05	4.49	6.67	8.53	16.12	26.36
17	1.78	3.03	4.45	6.59	8.40	15.72	25.44
18	1.77	3.01	4.41	6.51	8.29	15.38	24.66
19	1.76	2.99	4.38	6.45	8.18	15.08	23.99
20	1.76	2.97	4.35	6.39	8.10	14.82	23.40
21	1.75	2.96	4.32	6.34	8.02	14.59	22.89
22	1.75	2.95	4.30	6.29	7.95	14.38	22.43
23	1.74	2.94	4.28	6.25	7.88	14.20	22.03
24	1.74	2.93	4.26	6.21	7.82	14.03	21.66
25	1.73	2.92	4.24	6.18	7.77	13.88	21.34
26	1.73	2.91	4.23	6.14	7.72	13.74	21.04
27	1.73	2.90	4.21	6.11	7.68	13.61	20.77
28	1.72	2.89	4.20	6.09	7.64	13.50	20.53
29	1.72	2.89	4.18	6.06	7.60	13.39	20.30
30	1.72	2.88	4.17	6.04	7.56	13.29	20.09
40	1.70	2.84	4.08	5.87	7.31	12.61	18.67
60	1.68	2.79	4.00	5.71	7.08	11.97	17.38
100	1.66	2.76	3.94	5.59	6.90	11.50	16.43
140	1.66	2.74	3.91	5.54	6.82	11.30	16.05
∞	1.64	2.71	3.84	5.41	6.63	10.83	15.14

Notation: 406^3 means 406×10^3

TABLE 9 continued

Numerator df = 2

Denom. df	TAIL PROBABILITY						
	.20	.10	.05	.02	.01	.001	.0001
1	12.00	49.50	200	125[1]	500[1]	500[3]	500[5]
2	4.00	9.00	19.00	49.00	99.00	999	100[2]
3	2.89	5.46	9.55	18.86	30.82	149	695
4	2.47	4.32	6.94	12.14	18.00	61.25	198
5	2.26	3.78	5.79	9.45	13.27	37.12	97.03
6	2.13	3.46	5.14	8.05	10.92	27.00	61.63
7	2.04	3.26	4.74	7.20	9.55	21.69	45.13
8	1.98	3.11	4.46	6.64	8.65	18.49	36.00
9	1.93	3.01	4.26	6.23	8.02	16.39	30.34
10	1.90	2.92	4.10	5.93	7.56	14.91	26.55
11	1.87	2.86	3.98	5.70	7.21	13.81	23.85
12	1.85	2.81	3.89	5.52	6.93	12.97	21.85
13	1.83	2.76	3.81	5.37	6.70	12.31	20.31
14	1.81	2.73	3.74	5.24	6.51	11.78	19.09
15	1.80	2.70	3.68	5.14	6.36	11.34	18.11
16	1.78	2.67	3.63	5.05	6.23	10.97	17.30
17	1.77	2.64	3.59	4.97	6.11	10.66	16.62
18	1.76	2.62	3.55	4.90	6.01	10.39	16.04
19	1.75	2.61	3.52	4.84	5.93	10.16	15.55
20	1.75	2.59	3.49	4.79	5.85	9.95	15.12
21	1.74	2.57	3.47	4.74	5.78	9.77	14.74
22	1.73	2.56	3.44	4.70	5.72	9.61	14.41
23	1.73	2.55	3.42	4.66	5.66	9.47	14.12
24	1.72	2.54	3.40	4.63	5.61	9.34	13.85
25	1.72	2.53	3.39	4.59	5.57	9.22	13.62
26	1.71	2.52	3.37	4.56	5.53	9.12	13.40
27	1.71	2.51	3.35	4.54	5.49	9.02	13.21
28	1.71	2.50	3.34	4.51	5.45	8.93	13.03
29	1.70	2.50	3.33	4.49	5.42	8.85	12.87
30	1.70	2.49	3.32	4.47	5.39	8.77	12.72
40	1.68	2.44	3.23	4.32	5.18	8.25	11.70
60	1.65	2.39	3.15	4.18	4.98	7.77	10.78
100	1.64	2.36	3.09	4.07	4.82	7.41	10.11
140	1.63	2.34	3.06	4.02	4.76	7.26	9.84
∞	1.61	2.30	3.00	3.91	4.61	6.91	9.21

(continued)

TABLE 9   continued

Numerator df = 3

Denom. df	TAIL PROBABILITY						
	.20	.10	.05	.02	.01	.001	.0001
1	13.06	53.59	216	135^1	540^1	540^3	540^5
2	4.16	9.16	19.16	49.17	99.17	999	100^2
3	2.94	5.39	9.28	18.11	29.46	141	659
4	2.48	4.19	6.59	11.34	16.69	56.18	181
5	2.25	3.62	5.41	8.67	12.06	33.20	86.29
6	2.11	3.29	4.76	7.29	9.78	23.70	53.68
7	2.02	3.07	4.35	6.45	8.45	18.77	38.68
8	1.95	2.92	4.07	5.90	7.59	15.83	30.46
9	1.90	2.81	3.86	5.51	6.99	13.90	25.40
10	1.86	2.73	3.71	5.22	6.55	12.55	22.04
11	1.83	2.66	3.59	4.99	6.22	11.56	19.66
12	1.80	2.61	3.49	4.81	5.95	10.80	17.90
13	1.78	2.56	3.41	4.67	5.74	10.21	16.55
14	1.76	2.52	3.34	4.55	5.56	9.73	15.49
15	1.75	2.49	3.29	4.45	5.42	9.34	14.64
16	1.74	2.46	3.24	4.36	5.29	9.01	13.93
17	1.72	2.44	3.20	4.29	5.18	8.73	13.34
18	1.71	2.42	3.16	4.22	5.09	8.49	12.85
19	1.70	2.40	3.13	4.16	5.01	8.28	12.42
20	1.70	2.38	3.10	4.11	4.94	8.10	12.05
21	1.69	2.36	3.07	4.07	4.87	7.94	11.73
22	1.68	2.35	3.05	4.03	4.82	7.80	11.44
23	1.68	2.34	3.03	3.99	4.76	7.67	11.19
24	1.67	2.33	3.01	3.96	4.72	7.55	10.96
25	1.66	2.32	2.99	3.93	4.68	7.45	10.76
26	1.66	2.31	2.98	3.90	4.64	7.36	10.58
27	1.65	2.30	2.96	3.87	4.60	7.27	10.41
28	1.65	2.29	2.95	3.85	4.57	7.19	10.26
29	1.65	2.28	2.93	3.83	4.54	7.12	10.12
30	1.64	2.28	2.92	3.81	4.51	7.05	9.99
40	1.62	2.23	2.84	3.67	4.31	6.59	9.13
60	1.60	2.18	2.76	3.53	4.13	6.17	8.35
100	1.58	2.14	2.70	3.43	3.98	5.86	7.79
140	1.57	2.12	2.67	3.38	3.92	5.73	7.57
∞	1.55	2.08	2.60	3.28	3.78	5.42	7.04

TABLE 9 continued

Numerator df = 4

Denom. df	TAIL PROBABILITY						
	.20	.10	.05	.02	.01	.001	.0001
1	13.64	55.83	225	141^1	562^1	562^3	562^5
2	4.24	9.24	19.25	49.25	99.25	999	100^2
3	2.96	5.34	9.12	17.69	28.71	137	640
4	2.48	4.11	6.39	10.90	15.98	53.44	172
5	2.24	3.52	5.19	8.23	11.39	31.09	80.53
6	2.09	3.18	4.53	6.86	9.15	21.92	49.42
7	1.99	2.96	4.12	6.03	7.85	17.20	35.22
8	1.92	2.81	3.84	5.49	7.01	14.39	27.49
9	1.87	2.69	3.63	5.10	6.42	12.56	22.77
10	1.83	2.61	3.48	4.82	5.99	11.28	19.63
11	1.80	2.54	3.36	4.59	5.67	10.35	17.42
12	1.77	2.48	3.26	4.42	5.41	9.63	15.79
13	1.75	2.43	3.18	4.28	5.21	9.07	14.55
14	1.73	2.39	3.11	4.16	5.04	8.62	13.57
15	1.71	2.36	3.06	4.06	4.89	8.25	12.78
16	1.70	2.33	3.01	3.97	4.77	7.94	12.14
17	1.68	2.31	2.96	3.90	4.67	7.68	11.60
18	1.67	2.29	2.93	3.84	4.58	7.46	11.14
19	1.66	2.27	2.90	3.78	4.50	7.27	10.75
20	1.65	2.25	2.87	3.73	4.43	7.10	10.41
21	1.65	2.23	2.84	3.69	4.37	6.95	10.12
22	1.64	2.22	2.82	3.65	4.31	6.81	9.86
23	1.63	2.21	2.80	3.61	4.26	6.70	9.63
24	1.63	2.19	2.78	3.58	4.22	6.59	9.42
25	1.62	2.18	2.76	3.55	4.18	6.49	9.24
26	1.62	2.17	2.74	3.52	4.14	6.41	9.07
27	1.61	2.17	2.73	3.50	4.11	6.33	8.92
28	1.61	2.16	2.71	3.47	4.07	6.25	8.79
29	1.60	2.15	2.70	3.45	4.04	6.19	8.66
30	1.60	2.14	2.69	3.43	4.02	6.12	8.54
40	1.57	2.09	2.61	3.30	3.83	5.70	7.76
60	1.55	2.04	2.53	3.16	3.65	5.31	7.06
100	1.53	2.00	2.46	3.06	3.51	5.02	6.55
140	1.52	1.99	2.44	3.02	3.46	4.90	6.35
∞	1.50	1.94	2.37	2.92	3.32	4.62	5.88

(continued)

TABLE 9 continued

Numerator df = 5

Denom. df	TAIL PROBABILITY						
	.20	.10	.05	.02	.01	.001	.0001
1	14.01	57.24	230	144[1]	576[1]	576[3]	576[5]
2	4.28	9.29	19.30	49.30	99.30	999	100[2]
3	2.97	5.31	9.01	17.43	28.24	135	628
4	2.48	4.05	6.26	10.62	15.52	51.71	166
5	2.23	3.45	5.05	7.95	10.97	29.75	76.91
6	2.08	3.11	4.39	6.58	8.75	20.80	46.75
7	1.97	2.88	3.97	5.76	7.46	16.21	33.06
8	1.90	2.73	3.69	5.22	6.63	13.48	25.63
9	1.85	2.61	3.48	4.84	6.06	11.71	21.11
10	1.80	2.52	3.33	4.55	5.64	10.48	18.12
11	1.77	2.45	3.20	4.34	5.32	9.58	16.02
12	1.74	2.39	3.11	4.16	5.06	8.89	14.47
13	1.72	2.35	3.03	4.02	4.86	8.35	13.29
14	1.70	2.31	2.96	3.90	4.69	7.92	12.37
15	1.68	2.27	2.90	3.81	4.56	7.57	11.62
16	1.67	2.24	2.85	3.72	4.44	7.27	11.01
17	1.65	2.22	2.81	3.65	4.34	7.02	10.50
18	1.64	2.20	2.77	3.59	4.25	6.81	10.07
19	1.63	2.18	2.74	3.53	4.17	6.62	9.71
20	1.62	2.16	2.71	3.48	4.10	6.46	9.39
21	1.61	2.14	2.68	3.44	4.04	6.32	9.11
22	1.61	2.13	2.66	3.40	3.99	6.19	8.87
23	1.60	2.11	2.64	3.36	3.94	6.08	8.65
24	1.59	2.10	2.62	3.33	3.90	5.98	8.46
25	1.59	2.09	2.60	3.30	3.85	5.89	8.28
26	1.58	2.08	2.59	3.28	3.82	5.80	8.13
27	1.58	2.07	2.57	3.25	3.78	5.73	7.99
28	1.57	2.06	2.56	3.23	3.75	5.66	7.86
29	1.57	2.06	2.55	3.21	3.73	5.59	7.74
30	1.57	2.05	2.53	3.19	3.70	5.53	7.63
40	1.54	2.00	2.45	3.05	3.51	5.13	6.90
60	1.51	1.95	2.37	2.92	3.34	4.76	6.25
100	1.49	1.91	2.31	2.82	3.21	4.48	5.78
140	1.48	1.89	2.28	2.78	3.15	4.37	5.59
∞	1.46	1.85	2.21	2.68	3.02	4.10	5.15

TABLE 9 continued

Numerator df = 6

Denom. df	TAIL PROBABILITY						
	.20	.10	.05	.02	.01	.001	.0001
1	14.26	58.20	234	146[1]	586[1]	586[3]	586[5]
2	4.32	9.33	19.33	49.33	99.33	999	100[2]
3	2.97	5.28	8.94	17.25	27.91	133	620
4	2.47	4.01	6.16	10.42	15.21	50.53	162
5	2.22	3.40	4.95	7.76	10.67	28.83	74.43
6	2.06	3.05	4.28	6.39	8.47	20.03	44.91
7	1.96	2.83	3.87	5.58	7.19	15.52	31.57
8	1.88	2.67	3.58	5.04	6.37	12.86	24.36
9	1.83	2.55	3.37	4.65	5.80	11.13	19.97
10	1.78	2.46	3.22	4.37	5.39	9.93	17.08
11	1.75	2.39	3.09	4.15	5.07	9.05	15.05
12	1.72	2.33	3.00	3.98	4.82	8.38	13.56
13	1.69	2.28	2.92	3.84	4.62	7.86	12.42
14	1.67	2.24	2.85	3.72	4.46	7.44	11.53
15	1.66	2.21	2.79	3.63	4.32	7.09	10.82
16	1.64	2.18	2.74	3.54	4.20	6.80	10.23
17	1.63	2.15	2.70	3.47	4.10	6.56	9.75
18	1.62	2.13	2.66	3.41	4.01	6.35	9.33
19	1.61	2.11	2.63	3.35	3.94	6.18	8.98
20	1.60	2.09	2.60	3.30	3.87	6.02	8.68
21	1.59	2.08	2.57	3.26	3.81	5.88	8.41
22	1.58	2.06	2.55	3.22	3.76	5.76	8.18
23	1.57	2.05	2.53	3.19	3.71	5.65	7.97
24	1.57	2.04	2.51	3.15	3.67	5.55	7.79
25	1.56	2.02	2.49	3.13	3.63	5.46	7.62
26	1.56	2.01	2.47	3.10	3.59	5.38	7.48
27	1.55	2.00	2.46	3.07	3.56	5.31	7.34
28	1.55	2.00	2.45	3.05	3.53	5.24	7.22
29	1.54	1.99	2.43	3.03	3.50	5.18	7.10
30	1.54	1.98	2.42	3.01	3.47	5.12	7.00
40	1.51	1.93	2.34	2.88	3.29	4.73	6.30
60	1.48	1.87	2.25	2.75	3.12	4.37	5.68
100	1.46	1.83	2.19	2.65	2.99	4.11	5.24
140	1.45	1.82	2.16	2.61	2.93	4.00	5.06
∞	1.43	1.77	2.10	2.51	2.80	3.74	4.64

(continued)

TABLE 9 continued

Numerator df = 7

Denom. df	TAIL PROBABILITY						
	.20	.10	.05	.02	.01	.001	.0001
1	14.44	58.91	237	148[1]	593[1]	593[3]	593[5]
2	4.34	9.35	19.35	49.36	99.36	999	100[2]
3	2.97	5.27	8.89	17.11	27.67	132	614
4	2.47	3.98	6.09	10.27	14.98	49.66	159
5	2.21	3.37	4.88	7.61	10.46	28.16	72.61
6	2.05	3.01	4.21	6.25	8.26	19.46	43.57
7	1.94	2.78	3.79	5.44	6.99	15.02	30.48
8	1.87	2.62	3.50	4.90	6.18	12.40	23.42
9	1.81	2.51	3.29	4.52	5.61	10.70	19.14
10	1.77	2.41	3.14	4.23	5.20	9.52	16.32
11	1.73	2.34	3.01	4.02	4.89	8.66	14.34
12	1.70	2.28	2.91	3.85	4.64	8.00	12.89
13	1.68	2.23	2.83	3.71	4.44	7.49	11.79
14	1.65	2.19	2.76	3.59	4.28	7.08	10.92
15	1.64	2.16	2.71	3.49	4.14	6.74	10.23
16	1.62	2.13	2.66	3.41	4.03	6.46	9.66
17	1.61	2.10	2.61	3.34	3.93	6.22	9.19
18	1.60	2.08	2.58	3.27	3.84	6.02	8.79
19	1.58	2.06	2.54	3.22	3.77	5.85	8.45
20	1.58	2.04	2.51	3.17	3.70	5.69	8.16
21	1.57	2.02	2.49	3.13	3.64	5.56	7.90
22	1.56	2.01	2.46	3.09	3.59	5.44	7.68
23	1.55	1.99	2.44	3.05	3.54	5.33	7.48
24	1.55	1.98	2.42	3.02	3.50	5.23	7.30
25	1.54	1.97	2.40	2.99	3.46	5.15	7.14
26	1.53	1.96	2.39	2.97	3.42	5.07	6.99
27	1.53	1.95	2.37	2.94	3.39	5.00	6.86
28	1.52	1.94	2.36	2.92	3.36	4.93	6.75
29	1.52	1.93	2.35	2.90	3.33	4.87	6.64
30	1.52	1.93	2.33	2.88	3.30	4.82	6.54
40	1.49	1.87	2.25	2.74	3.12	4.44	5.86
60	1.46	1.82	2.17	2.62	2.95	4.09	5.27
100	1.43	1.78	2.10	2.52	2.82	3.83	4.84
140	1.42	1.76	2.08	2.48	2.77	3.72	4.67
∞	1.40	1.72	2.01	2.37	2.64	3.47	4.27

TABLE 9 continued

Numerator df = 8

Denom. df	TAIL PROBABILITY						
	.20	.10	.05	.02	.01	.001	.0001
1	14.58	59.44	239	149[1]	598[1]	598[3]	598[5]
2	4.36	9.37	19.37	49.37	99.37	999	100[2]
3	2.98	5.25	8.85	17.01	27.49	131	609
4	2.47	3.95	6.04	10.16	14.80	49.00	157
5	2.20	3.34	4.82	7.50	10.29	27.65	71.23
6	2.04	2.98	4.15	6.14	8.10	19.03	42.54
7	1.93	2.75	3.73	5.33	6.84	14.63	29.64
8	1.86	2.59	3.44	4.79	6.03	12.05	22.71
9	1.80	2.47	3.23	4.41	5.47	10.37	18.50
10	1.75	2.38	3.07	4.13	5.06	9.20	15.74
11	1.72	2.30	2.95	3.91	4.74	8.35	13.80
12	1.69	2.24	2.85	3.74	4.50	7.71	12.38
13	1.66	2.20	2.77	3.60	4.30	7.21	11.30
14	1.64	2.15	2.70	3.48	4.14	6.80	10.46
15	1.62	2.12	2.64	3.39	4.00	6.47	9.78
16	1.61	2.09	2.59	3.30	3.89	6.19	9.23
17	1.59	2.06	2.55	3.23	3.79	5.96	8.76
18	1.58	2.04	2.51	3.17	3.71	5.76	8.38
19	1.57	2.02	2.48	3.12	3.63	5.59	8.04
20	1.56	2.00	2.45	3.07	3.56	5.44	7.76
21	1.55	1.98	2.42	3.02	3.51	5.31	7.51
22	1.54	1.97	2.40	2.99	3.45	5.19	7.29
23	1.53	1.95	2.37	2.95	3.41	5.09	7.09
24	1.53	1.94	2.36	2.92	3.36	4.99	6.92
25	1.52	1.93	2.34	2.89	3.32	4.91	6.76
26	1.52	1.92	2.32	2.86	3.29	4.83	6.62
27	1.51	1.91	2.31	2.84	3.26	4.76	6.50
28	1.51	1.90	2.29	2.82	3.23	4.69	6.38
29	1.50	1.89	2.28	2.80	3.20	4.64	6.28
30	1.50	1.88	2.27	2.78	3.17	4.58	6.18
40	1.47	1.83	2.18	2.64	2.99	4.21	5.53
60	1.44	1.77	2.10	2.51	2.82	3.86	4.95
100	1.41	1.73	2.03	2.41	2.69	3.61	4.53
140	1.40	1.71	2.01	2.37	2.64	3.51	4.36
∞	1.38	1.67	1.94	2.27	2.51	3.27	3.98

(continued)

TABLE 9　continued

Numerator df = 9

Denom. df	TAIL PROBABILITY						
	.20	.10	.05	.02	.01	.001	.0001
1	14.68	59.86	241	151^1	602^1	602^3	602^5
2	4.37	9.38	19.38	49.39	99.39	999	100^2
3	2.98	5.24	8.81	16.93	27.35	130	606
4	2.46	3.94	6.00	10.07	14.66	48.47	155
5	2.20	3.32	4.77	7.42	10.16	27.24	70.13
6	2.03	2.96	4.10	6.05	7.98	18.69	41.73
7	1.93	2.72	3.68	5.24	6.72	14.33	28.99
8	1.85	2.56	3.39	4.70	5.91	11.77	22.14
9	1.79	2.44	3.18	4.33	5.35	10.11	18.00
10	1.74	2.35	3.02	4.04	4.94	8.96	15.27
11	1.70	2.27	2.90	3.83	4.63	8.12	13.37
12	1.67	2.21	2.80	3.66	4.39	7.48	11.98
13	1.65	2.16	2.71	3.52	4.19	6.98	10.92
14	1.63	2.12	2.65	3.40	4.03	6.58	10.09
15	1.61	2.09	2.59	3.30	3.89	6.26	9.42
16	1.59	2.06	2.54	3.22	3.78	5.98	8.88
17	1.58	2.03	2.49	3.15	3.68	5.75	8.43
18	1.56	2.00	2.46	3.09	3.60	5.56	8.05
19	1.55	1.98	2.42	3.03	3.52	5.39	7.72
20	1.54	1.96	2.39	2.98	3.46	5.24	7.44
21	1.53	1.95	2.37	2.94	3.40	5.11	7.19
22	1.53	1.93	2.34	2.90	3.35	4.99	6.98
23	1.52	1.92	2.32	2.87	3.30	4.89	6.79
24	1.51	1.91	2.30	2.83	3.26	4.80	6.62
25	1.51	1.89	2.28	2.81	3.22	4.71	6.47
26	1.50	1.88	2.27	2.78	3.18	4.64	6.33
27	1.49	1.87	2.25	2.76	3.15	4.57	6.21
28	1.49	1.87	2.24	2.73	3.12	4.50	6.09
29	1.49	1.86	2.22	2.71	3.09	4.45	5.99
30	1.48	1.85	2.21	2.69	3.07	4.39	5.90
40	1.45	1.79	2.12	2.56	2.89	4.02	5.26
60	1.42	1.74	2.04	2.43	2.72	3.69	4.69
100	1.40	1.69	1.97	2.33	2.59	3.44	4.29
140	1.39	1.68	1.95	2.29	2.54	3.34	4.12
∞	1.36	1.63	1.88	2.19	2.41	3.10	3.75

TABLE 9 continued

Numerator df = 10

Denom. df	TAIL PROBABILITY						
	.20	.10	.05	.02	.01	.001	.0001
1	14.77	60.19	242	151[1]	606[1]	606[3]	606[5]
2	4.38	9.39	19.40	49.40	99.40	999	100[2]
3	2.98	5.23	8.79	16.86	27.23	129	603
4	2.46	3.92	5.96	10.00	14.55	48.05	154
5	2.19	3.30	4.74	7.34	10.05	26.92	69.25
6	2.03	2.94	4.06	5.98	7.87	18.41	41.08
7	1.92	2.70	3.64	5.17	6.62	14.08	28.45
8	1.84	2.54	3.35	4.63	5.81	11.54	21.68
9	1.78	2.42	3.14	4.26	5.26	9.89	17.59
10	1.73	2.32	2.98	3.97	4.85	8.75	14.90
11	1.69	2.25	2.85	3.76	4.54	7.92	13.02
12	1.66	2.19	2.75	3.59	4.30	7.29	11.65
13	1.64	2.14	2.67	3.45	4.10	6.80	10.60
14	1.62	2.10	2.60	3.33	3.94	6.40	9.79
15	1.60	2.06	2.54	3.23	3.80	6.08	9.13
16	1.58	2.03	2.49	3.15	3.69	5.81	8.60
17	1.57	2.00	2.45	3.08	3.59	5.58	8.15
18	1.55	1.98	2.41	3.02	3.51	5.39	7.78
19	1.54	1.96	2.38	2.96	3.43	5.22	7.46
20	1.53	1.94	2.35	2.91	3.37	5.08	7.18
21	1.52	1.92	2.32	2.87	3.31	4.95	6.94
22	1.51	1.90	2.30	2.83	3.26	4.83	6.73
23	1.51	1.89	2.27	2.80	3.21	4.73	6.54
24	1.50	1.88	2.25	2.77	3.17	4.64	6.37
25	1.49	1.87	2.24	2.74	3.13	4.56	6.23
26	1.49	1.86	2.22	2.71	3.09	4.48	6.09
27	1.48	1.85	2.20	2.69	3.06	4.41	5.97
28	1.48	1.84	2.19	2.66	3.03	4.35	5.86
29	1.47	1.83	2.18	2.64	3.00	4.29	5.76
30	1.47	1.82	2.16	2.62	2.98	4.24	5.66
40	1.44	1.76	2.08	2.49	2.80	3.87	5.04
60	1.41	1.71	1.99	2.36	2.63	3.54	4.48
100	1.38	1.66	1.93	2.26	2.50	3.30	4.08
140	1.37	1.64	1.90	2.22	2.45	3.20	3.93
∞	1.34	1.60	1.83	2.12	2.32	2.96	3.56

TABLE 10 Critical Constants
for the Newman–Keuls
Procedure

$\alpha = .05$

df \ j	2	3	4	5	6	7	8	9	10
1	18.0	27.0	32.8	37.1	40.4	43.1	45.4	47.4	49.1
2	6.08	8.33	9.80	10.9	11.7	12.4	13.0	13.5	14.0
3	4.50	5.91	6.82	7.50	8.04	8.48	8.85	9.18	9.46
4	3.93	5.04	5.76	6.29	6.71	7.05	7.35	7.60	7.83
5	3.64	4.60	5.22	5.67	6.03	6.33	6.58	6.80	6.99
6	3.46	4.34	4.90	5.30	5.63	5.90	6.12	6.32	6.49
7	3.34	4.16	4.68	5.06	5.36	5.61	5.82	6.00	6.16
8	3.26	4.04	4.53	4.89	5.17	5.40	5.60	5.77	5.92
9	3.20	3.95	4.41	4.76	5.02	5.24	5.43	5.59	5.74
10	3.15	3.88	4.33	4.65	4.91	5.12	5.30	5.46	5.60
11	3.11	3.82	4.26	4.57	4.82	5.03	5.20	5.35	5.49
12	3.08	3.77	4.20	4.51	4.75	4.95	5.12	5.27	5.39
13	3.06	3.73	4.15	4.45	4.69	4.88	5.05	5.19	5.32
14	3.03	3.70	3.11	4.41	4.64	4.83	4.99	5.13	5.25
15	3.01	3.67	4.08	4.37	4.59	4.78	4.94	5.08	5.20
16	3.00	3.65	4.05	4.33	4.56	4.74	4.90	5.03	5.15
17	2.98	3.63	4.02	4.30	4.52	4.70	4.86	4.99	5.11
18	2.97	3.61	4.00	4.28	4.49	4.67	4.82	4.96	5.07
19	2.96	3.59	3.98	4.25	4.47	4.65	4.79	4.92	5.04
20	2.95	3.58	3.96	4.23	4.45	4.62	4.77	4.90	5.01
24	2.92	3.53	3.90	4.17	4.37	4.54	4.68	4.81	4.92
30	2.89	3.49	3.85	4.10	4.30	4.46	4.60	4.72	4.82
40	2.86	3.44	3.79	4.04	4.23	4.39	4.52	4.63	4.73
60	2.83	3.40	3.74	3.98	4.16	4.31	4.44	4.55	4.65
120	2.80	3.36	3.68	3.92	4.10	4.24	4.36	4.47	4.56
∞	2.77	3.31	3.63	3.86	4.03	4.17	4.29	4.39	4.47

TABLE 10   continued $\alpha = .01$

df \ j	2	3	4	5	6	7	8	9	10
1	90.0	135	164	186	202	216	227	237	246
2	14.0	19.0	22.3	24.7	26.6	28.2	29.5	30.7	31.7
3	8.26	10.6	12.2	13.3	14.2	15.0	15.6	16.2	16.7
4	6.51	8.12	9.17	9.96	10.6	11.1	11.5	11.9	12.3
5	5.70	6.97	7.80	8.42	8.91	9.32	9.67	9.97	10.2
6	5.24	6.33	7.03	7.56	7.97	8.32	8.61	8.87	9.10
7	4.95	5.92	6.54	7.01	7.37	7.68	7.94	8.17	8.37
8	4.74	5.63	6.20	6.63	6.96	7.24	7.47	7.68	7.87
9	4.60	5.43	5.96	6.35	6.66	6.91	7.13	7.32	7.49
10	4.48	5.27	5.77	6.14	6.43	6.67	6.87	7.05	7.21
11	4.39	5.14	5.62	5.97	6.25	6.48	6.67	6.84	6.99
12	4.32	5.04	5.50	5.84	6.10	6.32	6.51	6.67	6.81
13	4.26	4.96	5.40	5.73	5.98	6.19	6.37	6.53	6.67
14	4.21	4.89	5.32	5.63	5.88	6.08	6.26	6.41	6.54
15	4.17	4.83	5.25	5.56	5.80	5.99	6.16	6.31	6.44
16	4.13	4.78	5.19	5.49	5.72	5.92	6.08	6.22	6.35
17	4.10	4.74	5.14	5.43	5.66	5.85	6.01	6.15	6.27
18	4.07	4.70	5.09	5.38	5.60	5.79	5.94	6.08	6.20
19	4.05	4.67	5.05	5.33	5.55	5.73	5.89	6.02	6.14
20	4.02	4.64	5.02	5.29	5.51	5.69	5.84	5.97	6.09
24	3.96	4.54	4.91	5.17	5.37	5.54	5.69	5.81	5.92
30	3.89	4.45	4.80	5.05	5.24	5.40	5.54	5.65	5.76
40	3.82	4.37	4.70	4.93	5.11	5.27	5.39	5.50	5.60
60	3.76	4.28	4.60	4.82	4.99	5.13	5.25	5.36	5.45
120	3.70	4.20	4.50	4.71	4.87	5.01	5.12	5.21	5.30
∞	3.64	4.12	4.40	4.60	4.76	4.88	4.99	5.08	5.16

Source: Harter, H. L. "Tables of range and Studentized range." *Annals of Mathematical Statistics*, Volume 31 (1960), pp. 1122–1147.

CHAPTER 2

2.1 There is no single correct answer. One possibility is:

MOLAR WIDTH (MM)	FREQUENCY (NO. OF SPECIMENS)
5.4–5.5	1
5.6–5.7	5
5.8–5.9	7
6.0–6.1	12
6.2–6.3	8
6.4–6.5	2
6.6–6.7	1
Total	36

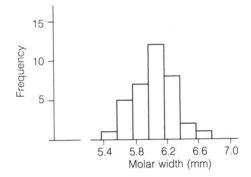

2.5 There is no single correct answer. One possibility is:

GLUCOSE (%)	FREQUENCY (NO. OF DOGS)
70–74	3
75–79	5
80–84	10
85–89	5
90–94	2
95–99	2
100–104	1
105–109	1
110–114	0
115–119	1
120–124	0
125–129	0
130–134	1
Total	31

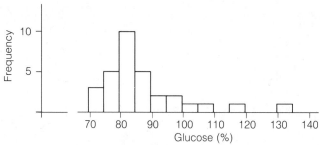

2.6 7 | 8 4 0 6 0 9 5 5
      8 | 1 8 5 4 1 4 2 6 9 9 0 2 1 4 2
      9 | 3 3 9 6
    10 | 2 6
    11 | 5
    12 |
    13 | 1

2.10 $\bar{y} = 6.40$ nmol/gm; median $= 6.3$ nmol/gm   **2.11** $\bar{y} = 293.8$ mg/dl; median $= 283$ mg/dl
2.14 Median $= 10.5$ piglets   **2.15** Range $= 65.7$ l; quartiles $= 63.7, 82.6,$ and 102.9 l; interquartile range $= 39.2$ l
2.17b. $s = 3.32$;   **c.** $s = 2.83$ **2.20a.** $\bar{y} = 33.10$ lb; $s = 3.44$ lb   **b.** Coeff. of var. $= 10.4\%$
2.22 $\bar{y} = -12.4$ mm Hg; $s = 17.6$ mm Hg   **2.31** Mean $= 7.373$; SD $= .129$

2.37a. 9 | 8 8 9
    10 | 6 7 7 7 0 0 8
    11 | 1 6 9 5 6 5 4 0 6 0
    12 | 1 0 3 4 2 3
    13 | 0
2.45a. Median $= 38$ **b.** First quartile $= 36$; third quartile $= 41$ **c.** 66.4%

CHAPTER 3

3.5a. .51   **b.** .94   **3.8a.** .62   **3.12a.** .3746   **b.** .0688   **c.** .1254
3.15a. .1181   **b.** .2699   **c.** .2891   **d.** .3229
3.16 Expected frequencies: 939.5; 5,982.5; 15,873.1; 22,461.8; 17,879.3; 7,590.2; 1,342.6.
3.21 .3369   **3.25** .0376   **3.27a.** .0209

CHAPTER 4

4.3a. 84.13%   **b.** 61.47%   **c.** 77.34%   **d.** 22.66%   **e.** 20.38%   **f.** 20.38%
4.4a. .7734   **b.** .2038   **4.7a.** 90.7 lb   **b.** 85.3 lb   **4.11c.** 1.24%   **4.14a.** .2843   **b.** .1256   **c.** .4980
4.19a. 97.98%   **b.** 12.71%   **c.** 46.39%   **d.** 10.69%   **e.** 35.31%   **f.** 5.59%   **g.** 59.10%
4.20 173.2 cm   **4.21** .122   **4.22** 1.96   **4.29** .1056

CHAPTER 5

5.2a. .2501   **b.** .0352
5.3a. (i) .3164 (ii) .4219 (iii) .2109 (iv) .0469 (v) .0039   **b.**

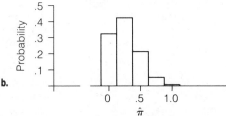

5.7 .5053   **5.12a.** 25.86%   **b.** 68.26%   **c.** .6826   **5.13a.** .6680   **5.17a.** .2710   **b.** .9587
5.20a. .1056   **b.** .0150   **5.22a.** .66   **b.** .29   **5.24a.** .1762   **b.** .1742   **5.25a.** .7198
5.28a. .6102   **5.32** .2611   **5.37a.** .2182   **b.** .5981   **5.38a.** .2206   **b.** .5990   **5.40** 9.68%

CHAPTER 6

6.1a. 51.3 ng/gm   **b.** 26.5 ng/gm   **6.6** $\bar{y} = 31.720$ mg; $s = 8.729$ mg; SE $= 3.904$ mg
6.7a. $23.4 < \mu < 40.0$ mg   **b.** $20.9 < \mu < 42.6$ mg
6.9 $4.1 < \mu < 21.9$ pg/ml   **6.12** $1.17 < \mu < 1.23$ mm   **6.14** 2.81
6.15a. $\bar{y} = 167.4$, $s = 27.6$, SE $= 4.4$ **b.** $\bar{y} = 3,062$, $s = 287$, SE $= 26$ **c.** $\bar{y} = .03762$, $s = .00624$, SE $= .00069$
d. $\bar{y} = 759,000$ or 75.9×10^4, $s = 35,000$ or 3.5×10^4, SE $= 12,000$ or 1.2×10^4
6.18 About 3 mm   **6.24a.** 6 or 7 rats   **6.27** The outlier (1,060)
6.30a. $\bar{y} = 2.275$ mm, $s = .2375$ mm, SE $= .08399$ mm   **b.** $2.08 < \mu < 2.47$ mm
c. $\mu =$ mean diameter of plants of Tetrastichon wheat 3 weeks after flowering   **6.31** Approximately 63 plants
6.35a. We assume that the data can be viewed as a random sample of independent observations from a large population which is approximately normal.   **b.** Normality of the population   **c.** Independence of the observations would be questionable, because birthweights of the members of a twin pair might be dependent.

CHAPTER 7

Remark concerning tests of hypotheses   The answer to a hypothesis testing exercise includes verbal statements of the hypotheses and a verbal statement of the conclusion from the test. In phrasing these statements, we have tried to capture the essence of the biological question being addressed; nevertheless the statements are necessarily oversimplified and they gloss over many issues which in reality might be quite important. For instance, the hypotheses and conclusion may refer to a *causal* connection between treatment and response; in reality the validity of such a causal interpretation usually depends on a number of factors related to the design of the investigation (such as unbiased allocation of animals to treatment groups) and to the specific experimental procedures (such as the accuracy of assays or measurement techniques). In short, the student should be aware that the verbal statements are intended to clarify the *statistical* concepts; their *biological* content may be open to question.

7.1a. 1.28   **b.** 1.12   **7.4a.** 2.04   **b.** 2.09   **7.6a.** 3.80   **b.** 3.80   **7.8a.** 12.8   **b.** 11.9   **7.10** 5
7.13 $.52 < \mu_1 - \mu_2 < 1.96$ nmol/gm, where 1 denotes group-housed and 2 denotes singly housed
7.17a. $-32.4 < \mu_1 - \mu_2 < -30.8$ mg, where 1 denotes control and 2 denotes Pargyline.
7.19 .1528   **7.22a.** $.02 < P < .05$ **b.** $P > .20$ **c.** $P < .0001$
7.24a. $.10 < P < .20$ **b.** $.02 < P < .05$ **c.** $.001 < P < .01$
7.25a. H_0: Mean serotonin concentration is the same in heart patients and in controls ($\mu_1 = \mu_2$); H_A: Mean serotonin concentration is not the same in heart patients as in controls ($\mu_1 \neq \mu_2$). $t_s = -1.41$. H_0 is not rejected. There is insufficient evidence ($.10 < P < .20$) to conclude that serotonin levels are different in heart patients than in controls.
7.28 H_0: Flooding has no effect on ATP ($\mu_1 = \mu_2$); H_A: Flooding has some effect on ATP ($\mu_1 \neq \mu_2$). $t_s = -3.92$. H_0 is rejected. There is sufficient evidence ($.001 < P < .01$) to conclude that flooding tends to lower ATP in birch seedlings.
7.31 .0968   **7.33a.** $.00005 < P < .0005$
7.37 H_0: Wounding the plant has no effect on larval growth ($\mu_1 = \mu_2$); H_A: Wounding the plant tends to diminish larval growth ($\mu_1 < \mu_2$), where 1 denotes wounded and 2 denotes control. $t_s = -2.65$. H_0 is rejected. There is sufficient evidence ($.005 < P < .01$) to conclude that wounding the plant tends to diminish larval growth.
7.38 H_0: Amphetamine has no effect on water consumption ($\mu_1 = \mu_2$); H_A: Amphetamine tends to suppress water consumption ($\mu_1 < \mu_2$), where 1 denotes amphetamine and 2 denotes control. $t_s = -1.41$. H_0 is not rejected. There is insufficient evidence ($P > .10$) to conclude that amphetamine tends to suppress water consumption.
7.46 No, according to the confidence interval the data do not indicate whether the true difference is "important."
7.47 Yes, according to the confidence interval the data indicate that the difference is "clinically important."
7.49a. 23   **b.** 11   **7.52a.** 71   **b.** 101   **c.** 58   **7.54** .5   **7.58a.** $P > .20$ **b.** $.02 < P < .05$ **c.** $.002 < P < .01$
7.60a. H_0: Toluene has no effect on dopamine in rat striatum; H_A: Toluene has some effect on dopamine in rat striatum. $U_s = 32$. H_0 is rejected. There is sufficient evidence ($.02 < P < .05$) to conclude that toluene increases dopamine in rat striatum.
7.65 H_0: Mean platelet calcium is the same in people with high blood pressure as in people with normal blood pressure ($\mu_1 = \mu_2$); H_A: Mean platelet calcium is different in people with high blood pressure than in people with normal blood pressure ($\mu_1 \neq \mu_2$). $t_s = 11.2$. H_0 is rejected. There is sufficient evidence ($P < .0001$) to conclude that platelet calcium is higher in people with high blood pressure.
7.66 $49.6 < \mu_1 - \mu_2 < 71.0$ nM, where 1 denotes high blood pressure and 2 denotes normal blood pressure.
7.70 H_0: Stress has no effect on growth; H_A: Stress tends to retard growth. $U_s = 148.5$. H_0 is rejected. There is sufficient evidence ($P < .0005$) to conclude that stress tends to retard growth.

CHAPTER 8

8.2 There is no single correct answer. One possibility is as follows:
Group 1: Animals 2, 5, 6
Group 2: Animals 1, 3, 7
Group 3: Animals 4, 8

8.6 There is no single correct answer. One possibility is as follows:

TREATMENT	PIGLET				
	Litter 1	Litter 2	Litter 3	Litter 4	Litter 5
1	2	5	2	4	5
2	1	4	1	1	2
3	4	2	5	2	4
4	5	3	3	3	3
5	3	1	4	5	1

CHAPTER 9

9.1a. .3379

9.3 H_0: Progesterone has no effect on cAMP ($\mu_1 = \mu_2$); H_A: Progesterone has some effect on cAMP ($\mu_1 \neq \mu_2$). $t_s = 3.34$. H_0 is rejected. There is sufficient evidence ($.02 < P < .05$) to conclude that progesterone decreases cAMP under these conditions.

9.4a. $-.50 < \mu_1 - \mu_2 < .74°C$, where 1 denotes treated and 2 denotes control.

9.12a. $P > .20$ **b.** $.10 < P < .20$ **c** $.02 < P < .05$ **d.** $.002 < P < .01$

9.14 H_0: HL-A compatibility does not affect survival of graft ($\pi = .5$); H_A: Close HL-A compatibility improves survival of graft ($\pi > .5$). $B_s = 9$. H_0 is rejected. There is sufficient evidence ($.025 < P < .05$) to conclude that close HL-A compatibility improves graft survival. **9.15** .0327

9.22a. .0156 **b.** With $n_d = 7$, the smallest possible P-value is .0156; thus, P cannot be less than .01.

9.32 $2.10 < \mu_1 - \mu_2 < 2.97$, where 1 denotes central and 2 denotes top.

9.33 H_0: Caffeine has no effect on RER ($\mu_1 = \mu_2$); H_A: Caffeine has some effect on RER ($\mu_1 \neq \mu_2$). $t_s = 3.94$. H_0 is rejected. There is sufficient evidence ($.001 < P < .01$) to conclude that caffeine tends to decrease RER under these conditions.

CHAPTER 10

10.1a. .0518 **b.** .0259 **10.4** $.031 < \pi < .066$ **10.7** 150 **10.12** 100

10.14 H_0: The population ratio is 12:3:1 (Pr{white} = .75, Pr{yellow} = .1875, Pr{green} = .0625); H_A: The ratio is not 12:3:1. $\chi_s^2 = .69$. H_0 is not rejected. There is little or no evidence ($P > .20$) that the model is incorrect; the data are consistent with the model.

10.15 H_0 and H_A as in Exercise 10.14. $\chi_s^2 = 6.9$. H_0 is rejected. There is sufficient evidence ($.02 < P < .05$) to conclude that the model is incorrect; the data are not consistent with the model.

10.20 H_0: The drug does not cause tumors (Pr{T} = $\frac{1}{3}$); H_A: The drug does cause tumors (Pr{T} > $\frac{1}{3}$), where T denotes the event that a tumor occurs first in the treated rat. $\chi_s^2 = 6.4$. H_0 is rejected. There is sufficient evidence ($.005 < P < .01$) to conclude that the drug does cause tumors.

10.25 .0038 (.38 percentage point) **10.32a.** $.77 < \pi < .91$ **10.33** 619

10.35a. H_0: Directional choice is random (Pr{toward} = .25, Pr{away} = .25, Pr{right} = .25, Pr{left} = .25); H_A: Directional choice is not random. $\chi_s^2 = 4.88$. H_0 is not rejected. There is insufficient evidence ($.10 < P < .20$) to conclude that the directional choice is not random.

CHAPTER 11

11.3a.

5	20
10	40

b. $\hat{\pi}_1 = \frac{1}{3}$, $\hat{\pi}_2 = \frac{1}{3}$; yes

11.6 H_0: Mites do not induce resistance to wilt ($\pi_1 = \pi_2$); H_A: Mites do induce resistance to wilt ($\pi_1 < \pi_2$), where π denotes the probability of wilt and 1 denotes mites and 2 denotes no mites. $\chi_s^2 = 7.21$. H_0 is rejected. There is sufficient evidence ($.0005 < P < .005$) to conclude that mites do induce resistance to wilt.

11.10 H_0: The two timings are equally effective ($\pi_1 = \pi_2$); H_A: The two timings are not equally effective ($\pi_1 \neq \pi_2$). $\chi_s^2 = 4.48$. H_0 is rejected. There is sufficient evidence ($.02 < P < .05$) to conclude that the simultaneous timing is superior to the sequential timing. **11.11a.** $\Pr\{S \mid B\} > \Pr\{S \mid G\}$

11.13a. $\widehat{\Pr}\{D \mid P\} = .266$, $\widehat{\Pr}\{D \mid N\} = .096$, $\widehat{\Pr}\{P \mid D\} = .744$, $\widehat{\Pr}\{P \mid A\} = .460$.

b. H_0: There is no relationship between antibody and survival ($\Pr\{D \mid P\} = \Pr\{D \mid N\}$); H_A: There is some relationship between antibody and survival ($\Pr\{D \mid P\} \neq \Pr\{D \mid N\}$). H_0 is rejected. There is sufficient evidence ($.001 < P < .01$) to conclude that men with antibody are less likely to survive 6 months than men without antibody ($\Pr\{D \mid P\} > \Pr\{D \mid N\}$).

11.14 $\widehat{\Pr}\{\text{correct prediction}\} = .577$

11.15a. $\widehat{\Pr}\{RF \mid RH\} = .934$ **b.** $\widehat{\Pr}\{RF \mid LH\} = .511$ **c.** $\chi_s^2 = 398$ **d.** $\chi_s^2 = 1{,}623$

11.16a. $\Pr\{DY \mid EY\} > \Pr\{DY \mid EN\}$ **d.** $\Pr\{EY \mid DY\} > .5$ **g.** $\Pr\{DY \mid EY\} > \Pr\{DN \mid EY\}$

11.25a. H_0: The blood type distributions are the same for ulcer patients and controls ($\Pr\{O \mid UP\} = \Pr\{O \mid C\}$, $\Pr\{A \mid UP\} = \Pr\{A \mid C\}$, $\Pr\{B \mid UP\} = \Pr\{B \mid C\}$, $\Pr\{AB \mid UP\} = \Pr\{AB \mid C\}$); H_A: The blood type distributions are not the same. H_0 is rejected. There is sufficient evidence ($P < .0001$) to conclude that the blood type distribution of ulcer patients is different from that of controls.

b.

		PERCENT FREQUENCY	
		Ulcer Patients	Controls
	O	55.0	45.8
BLOOD TYPE	A	35.0	42.2
	B	7.5	8.9
	AB	2.5	3.1
	TOTAL	100.0	100.0

11.32 $.003 < \pi_1 - \pi_2 < .233$. No; the confidence interval suggests that bed rest may actually be harmful.

11.34 $.067 < \pi_1 - \pi_2 < .119$

11.36a. H_0: Sex ratio is 1:1 in warm environment ($\pi_1 = .5$); H_A: Sex ratio is not 1:1 in warm environment ($\pi_1 \neq .5$), where π_1 denotes the probability of a female in the warm environment. $\chi_s^2 = .18$. H_0 is not rejected. There is insufficient evidence ($P > .20$) to conclude that the sex ratio is not 1:1 in the warm environment.   **c.** H_0: Sex ratio is the same in the two environments ($\pi_1 = \pi_2$); H_A: Sex ratio is not the same in the two environments ($\pi_1 \neq \pi_2$), where π denotes the probability of a female and 1 and 2 denote the warm and cold environments. $\chi_s^2 = 4.20$. H_0 is rejected. There is sufficient evidence to conclude that the probability of a female is higher in the cold than in the warm environment.

11.41 H_0: Fever does not have survival value ($\pi_1 = \pi_2$); H_A: Fever does have survival value ($\pi_1 > \pi_2$), where π denotes probability of death and 1 denotes 38°C and 2 denotes 40°C. $\chi_s^2 = 4.11$. H_0 is rejected. There is sufficient evidence ($.01 < P < .025$) to conclude that fever has survival value.

11.42 H_0: There is no association between oral contraceptive use and stroke ($\pi = .5$); H_A: There is an association between oral contraceptive use and stroke ($\pi \neq .5$), where π denotes the probability that a discordant pair will be Yes(case)/No(control). $\chi_s^2 = 6.72$. H_0 is rejected. There is sufficient evidence ($.001 < P < .01$) to conclude that stroke victims are more likely than controls to be oral contraceptive users ($\pi > .5$).

11.43 H_0: Site of capture and site of recapture are independent ($\Pr\{CI \mid RI\} = \Pr\{CI \mid RII\}$); H_A: Flies preferentially tend to return to their site of capture ($\Pr\{CI \mid RI\} > \Pr\{CI \mid RII\}$), where C and R denote capture and recapture and I and II denote the sites. H_0 is rejected. There is sufficient evidence ($.0005 < P < .005$) to conclude that the flies preferentially tend to return to their site of capture.

CHAPTER 12

12.1a. SS(between) $= 228$, SS(within) $= 120$ **b.** SS(total) $= 348$ **c.** MS(between) $= 114$, MS(within) $= 15$, $s_c = 3.87$

12.4 SS(between) $= 66$, SS(within) $= 24$ **12.7a.** 11,000   **b.** 3,064

12.10a. H_0: The stress conditions all produce the same mean lymphocyte concentration ($\mu_1 = \mu_2 = \mu_3 = \mu_4$); H_A: Some of the stress conditions produce different mean lymphocyte concentrations (the μ's are not all equal). $F_s = 3.84$. H_0 is rejected.

There is sufficient evidence ($.01 < P < .02$) to conclude that some of the stress conditions produce different mean lymphocyte concentrations.   **b.** $s_c = 2.78$ cells/ml $\times 10^{-6}$.

12.13a. H_0: Mean HBE is the same in all three populations ($\mu_1 = \mu_2 = \mu_3$); H_A: Mean HBE is not the same in all three populations (not all μ's are equal). $F_s = .58$. H_0 is not rejected. There is insufficient evidence ($P > .20$) to conclude that mean HBE is not the same in all three populations.   **b.** $s_c = 14.4$ pg/ml

12.16a. 121.49 mm Hg   **b.** 121.66 mm Hg   **d.** .8146 mm Hg   **e.** 2.053 mm Hg

12.20a. H_0: The difference between T and H is the same for both doses ($\mu_{T,Hi} - \mu_{H,Hi} = \mu_{T,Lo} - \mu_{H,Lo}$); H_A: The difference between T and H is not the same for both doses ($\mu_{T,Hi} - \mu_{H,Hi} \neq \mu_{T,Lo} - \mu_{H,Lo}$). $t_s = .47$. H_0 is not rejected. There is insufficient evidence ($P > .20$) to conclude that the difference between T and H is not the same at both doses. (The observed discrepancy can readily be ascribed to chance variation.)
b. $-4.26 < \mu_T - \mu_H < 1.96$ mm Hg, where $\mu_T = \frac{1}{2}(\mu_{T,Lo} + \mu_{T,Hi})$ and $\mu_H = \frac{1}{2}(\mu_{H,Lo} + \mu_{H,Hi})$.

12.22b. $L = 3.685$ nmol/10^8 platelets/hour; $SE_L = 1.048$ nmol/10^8 platelets/hour

12.24 The following hypotheses are rejected: H_0: $\mu_C = \mu_D$; H_0: $\mu_A = \mu_D$; H_0: $\mu_C = \mu_E$; H_0: $\mu_B = \mu_D$; H_0: $\mu_A = \mu_E$; H_0: $\mu_C = \mu_B$; H_0: $\mu_C = \mu_A$; H_0: $\mu_B = \mu_E$. The following hypotheses are not rejected: H_0: $\mu_A = \mu_B$; H_0: $\mu_D = \mu_E$. Summary:

C   A   B   E   D

There is sufficient evidence to conclude that treatments D and E give the largest means, treatments A and B the next largest, and treatment C the smallest. There is insufficient evidence to conclude that treatments A and B give different means; there is insufficient evidence to conclude that treatments D and E give different means.

12.29a. MS(between) = 75; MS(within) = 5.636

12.30 H_0: The four treatments produce the same mean yield ($\mu_1 = \mu_2 = \mu_3 = \mu_4$); H_A: Some of the treatments produce different mean yields (not all μ's are equal). $F_s = 13.1$. H_0 is rejected. There is sufficient evidence ($.001 < P < .01$) to conclude that some of the treatments produce different mean yields.

12.36 H_0: The mean refractive error is the same in the four populations of patients ($\mu_1 = \mu_2 = \mu_3 = \mu_4$); H_A: Some of the populations have different mean refractive errors (not all μ's are equal). $F_s = 3.56$. H_0 is rejected. There is sufficient evidence ($.01 < P < .02$) to conclude that some of the populations have different mean refractive errors.

CHAPTER 13

13.2 $\bar{x} = 6$, $\bar{y} = 10$, $SS_X = 28$, $SS_Y = 26$, $SP_{XY} = -22$ **13.3** $\bar{x} = 13$, $\bar{y} = 6$, $SS_X = 28$, $SS_Y = 24$, $SP_{XY} = 19$

13.5a. $Y = 1 + 4X$. The $\hat{y}$'s are 13, 17, 5, 9, 21.   **c.** The residuals ($y - \hat{y}$) are 0, -2, -1, 2, 1; SS(resid) = 10.

13.8b. $Y = -.592 + 7.640X$; $s_{Y|X} = .915°C$.

13.11a. $Y = 607 + 25.01X$ **b.**

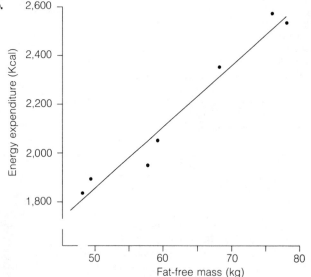

c. $s_Y = 308$ Kcal, $s_{Y|X} = 64.8$ Kcal

13.16 Estimated mean $= 21.1$ mm; estimated SD $= 1.3$ mm

13.18 $.0252 < \beta_1 < .0333$ ng/min   **13.22** $19.4 < \beta_1 < 30.6$ Kcal/kg

13.23 H_0: There is no relationship between respiration rate and altitude of origin ($\beta_1 = 0$); H_A: Trees from higher altitudes tend to have higher respiration rates ($\beta_1 > 0$). $t_s = 6.06$. H_0 is rejected. There is sufficient evidence ($.00005 < P < .0005$) to conclude that trees from higher altitudes tend to have higher respiration rates.

13.25a. $-.942$ **b.** $s_Y = 25.0$ gm per cob, $s_{Y|X} = 8.62$ gm per cob   **c.** $Y = 316 - .7206X$ **13.28a.** $.812$

13.30 H_0: There is no correlation between blood urea and uric acid concentration ($\rho = 0$); H_A: Blood urea and uric acid are positively correlated ($\rho > 0$). $t_s = 3.953$. H_0 is rejected. There is sufficient evidence ($P < .00005$) to conclude that blood urea and uric acid are positively correlated.

13.36 $.24$ gm   **13.38a.** Estimated mean $= .85$ kg; estimated SD $= .17$ kg   **13.42a.** $.803$ **b.** $s_Y = .210$ cm, $s_{Y|X} = .137$ cm

INDEX OF EXAMPLES